Marine Biochemistry

This book provides the latest comprehensive methods for isolation and other novel techniques for marine product development. Furthermore, this book offers knowledge on the biological, medical, and industrial applications of marine-derived medicinal food substances.

There has been a tremendous increase in the products derived from marine organisms for commercial application in industries every year. Functional foods of medicinal value are particularly in demand as new technology allows the stabilization of natural ingredients and their availability in pure forms to solve various human diseases. Marine flora and fauna have essential elements and trace minerals that nurture various hormones produced in the endocrine system to regulate the respective metabolisms, thereby providing a safe and healthy life to humans.

The overall presentation and clear demarcation of the contents by worldwide contributions is a novel entry point into the market of medicinal foods from the sea. The exploration of marine habitats for novel materials are discussed throughout the book.

The exploration and exploitation of the biochemistry of sea flora and fauna are limited, and this book extends the research possibilities into numerous marine habitats.

Various approaches for extracting and applying the flora and fauna are discussed. This book will be of value to researchers, marine biotechnologists, and medical practitioners, due to the vast information, as well as industrial and medical applications of marine substances all in one place.

Marine Biochemistry

Applications

Edited by

Se-Kwon Kim

CRC Press is an imprint of the
Taylor & Francis Group, an **informa** business

First edition published 2023
by CRC Press
6000 Broken Sound Parkway NW, Suite 300, Boca Raton, FL 33487–2742

and by CRC Press
4 Park Square, Milton Park, Abingdon, Oxon, OX14 4RN

CRC Press is an imprint of Taylor & Francis Group, LLC

ISBN: 978-1-032-30033-7 (hbk)
ISBN: 978-1-032-30214-0 (pbk)
ISBN: 978-1-003-30391-6 (ebk)

DOI: 10.1201/9781003303916

Typeset in Times
by Apex CoVantage, LLC

Contents

Preface

Over 70% of the earth is covered by water. The ocean is a habitat for several living organisms. The current book will provide an overview of these species, including their biochemistry, culture, and application, to better understand their use.

Furthermore, several tons of seaweed are grown for commercial purposes all over the world. Moreover, seaweed is an essential resource for a country's economic growth, and the diversity and structure of seaweeds and their applications are discussed. In addition, there is the discussion of marine-derived carbohydrates, alginate, chondroitin sulfate, astaxanthin, fucoxanthin, ulvan, and lipids.

The rising global temperature poses a severe threat to humanity. Some of the chapters indirectly aid in resolving these problems such as seaweeds, sustainable resources for food packaging, toxic issues with marine products of animals, and the transformation of marine products derived from marine-derived organisms. It is shown how marine-derived fungi can catalyze the reduction of organic compounds.

This book will be helpful to novice students, graduate students, marine biologists, those working in biotechnology, and industrialists. I am grateful to all of the people who contributed to this book.

Se-Kwon Kim

Acknowledgments

Dr. Jae-Chul Kim is the chairman, president, and founder of the Dongwon group. The company was established in 1969 to explore and utilize the oceans and marine resources. Chairman Kim is the pioneer in the deep-sea fishing boats in South Korea. He started tuna fishing in 1958 as the first mate of Korea's first deep-sea fishing vessel.

After graduating from National Fisheries University of Busan, he jumped into tuna fishing, and the university was renamed Pukyong National University. Further, he became the captain of a fleet of deep-sea fishing boats and a new tuna fishing method. He developed and succeeded in catching large amounts of tuna in the South Pacific and Indian Oceans.

He founded Dongwon Industries in 1969 at the age of 35 and started tuna fishing as the first president. After that, he built a tuna processing plant and started to produce canned food. He immersed himself in business management in earnest based on the experience he learned in the sea from his youth.

Dongwon Group has successfully expanded its business from the fishery industry as the primary industry to the manufacturing industry as the secondary industry, and financial services as the tertiary industry. Currently, Mr. Kim runs 30 affiliated companies.

To contribute to social welfare, he also established the Dongwon Educational Foundation. He devoted himself, his heart, and his soul to nurturing competent people who are the backbone of our society. He has provided scholarships to numerous college students and grants R&D funds for researchers. He recognized the great value and potential of underutilized marine resources from his youth. He dedicated himself to publishing technical books emphasizing the scientific importance of marine life and related research.

With his help, this book has been published, providing readers with how the scientific values of marine life can enhance human health and wellbeing.

I want to thank him sincerely for his support in publishing this book.

Se-Kwon Kim
Distinguished Professor
Dept. of Marine Science and Convergence Engineering
Hanyang University ERICA, South Korea

Editor

Se-Kwon Kim, PhD, is a distinguished professor at Hanyang University and Kolmar Korea Company. He worked as a distinguished professor in the Department of Marine Bio Convergence Science and Technology and director of Marine Bioprocess Research Center (MBPRC) at Pukyong National University, Busan, South Korea.

He earned an MSc and a PhD at Pukyong National University and conducted his postdoctoral studies at the Laboratory of Biochemical Engineering, University of Illinois, Urbana-Champaign, Illinois, USA. Later, he became a visiting scientist at the Memorial University of Newfoundland and University of British Colombia in Canada.

Dr. Kim served as president of the Korean Society of Chitin and Chitosan from 1986 to 1990 and the Korean Society of Marine Biotechnology from 2006 to 2007. As an acknowledgment of his research, he won the best paper award from the American Oil Chemists' Society in 2002. Dr. Kim was also the chairman for the Seventh Asia-Pacific Chitin and Chitosan Symposium, which was held in South Korea in 2006. He was the chief editor of the Korean Society's *Fisheries and Aquatic Science* from 2008 to 2009. In addition, he is a board member of the International Society of Marine Biotechnology Associations (IMBA) and International Society of Nutraceuticals and Functional Food (ISNFF).

His major research interests are investigation and development of bioactive substances from marine resources. His immense experience of marine bio-processing and mass-production technologies for the marine bioindustry is the key asset of holding majorly funded marine bio projects in Korea. Furthermore, he expanded his research field to the development of bioactive materials from marine organisms for their applications in oriental medicine, cosmeceuticals, and nutraceuticals. To date, he has authored more than 750 research papers, 70 books, and 120 patents.

Contributors

Bakrudeen Ali Ahmed Abdul
Department of Biochemistry
PRIST Deemed University
Thanjavur, India

Snezana Agatonovic-Kustrin
A.P. Arzamastsev Department of Pharmaceutical and Toxicological Chemistry
I.M. Sechenov
First Moscow State Medical University
Sechenov University
Moscow, Russian Federation

Ahmad Homaei
Department of Biology
Faculty of Marine Science and Technology
University of Hormozgan
Bandar Abbas, Iran

Hari Eko Irianto
Research and Development Center for Marine and Fisheries Product Processing and Biotechnology
and
Sahid University
Jakarta, Indonesia

Sougata Jana
Department of Pharmaceutics
Gupta College of Technological Sciences
Asansol, India
and
Department of Health and Family Welfare
Directorate of Health Services
Kolkata, India

Kannan Kamala
Department of Pharmacology
Saveetha Dental College and Hospitals
Saveetha Institute of Medical and Technical Sciences
Chennai, India

Hui-Jing Li
Weihai Marine Organism and Medical Technology Research Institute
Harbin Institute of Technology
Weihai, China

Kenji Nakajima
Laboratory of Biochemistry
Department of Nutrition
Graduate School of Koshien University
Takarazuka, Japan

Ngo Dang Nghia
Nha Trang University
Nha Trang, Vietnam

Nidhi Pareek
Department of Microbiology
School of Life Sciences
Central University of Rajasthan
Ajmer, India

Mauro Sérgio Gonçalves Pavão
Laboratory of Biochemistry and Cellular Biology of Glycoconjugates
Program of Glycobiology
Institute of Medical Biochemistry
Federal University of Rio de Janeiro
Rio de Janeiro, Brazil

Suman Samaddar
Research Institute BGS Global Institute of Medical Sciences
Bangalore, India

I. Wickramasinghe
Department of Food Science and Technology
Faculty of Applied Sciences
University of Sri Jayewardenepura
Nugegoda, Sri Lanka

Morteza Yousefzadi
Department of Marine Biology
Faculty of Marine Science and Technology
University of Hormozgan
Bandar Abbas, Iran

1 Marine-Based Carbohydrates as a Valuable Resource for Nutraceuticals and Biotechnological Application

Rajni Kumari, V. Vivekanand, and Nidhi Pareek

CONTENTS

DOI: 10.1201/9781003303916-1

1.1 INTRODUCTION

Over 80% of living beings on earth are found in the marine environment. The biggest environment on earth is the sea, which consists of more than 40,000 types of marine species (Kim and Venkatesan, 2015). Marine species address around one portion of the worldwide biodiversity, containing unique species and having a place with the principal taxa additionally involving countless classes of species such as microorganisms, algae, corals, and seagrasses. These marine species are considered as huge reservoirs for the discovery of valuable compounds, which are being used as nutraceuticals in the present era. Among all the marine compounds present in the marine sediments, marine carbohydrates are viewed as significant organic components. Carbohydrates play a major role in the biogeochemical cycle, which occurs at the interface between the seawater column and sediment water. In the marine environment carbohydrates are found as polysaccharides, disaccharides, and monosaccharides (Bhosle et al., 1998). Among carbohydrates, polysaccharides are one of the most abundant bioactive compounds in marine organisms. Indeed, many marine macro- and micro-organisms are good sources of carbohydrates with multiple applications due to their biofunctional properties. By acting on cell proliferation and the cycle and modulating various metabolic pathways, marine polysaccharides (including especially, chitosan, carrageenan, fucoidan, and laminarin) also have many pharmaceutical activities, such as antioxidant, anticoagulant, anticancer, antiviral, antibacterial, and immunostimulating (Tanna and Mishra, 2019). The composition and multiplicity of polysaccharides play an essential role in these biological activities. Furthermore, these polysaccharides have a plausible impact on human health, which is why they have been developed as potential nutraceuticals. A series of toxicological tests and clinical research reports have shown that consumption of seaweed as a functional food should be considered worldwide to enhance the immune response (Tanna and Mishra, 2019). Macroalgae are also being used as attractive feedstock in liquid biofuel production because of its unique property, low lignin content, and high structural polysaccharides. Biogas is produced by anaerobic digestion, and liquid biofuel is produced by fermentation.

1.2 CARBOHYDRATES: CLASSIFICATION AND SOURCES

Carbohydrates are the major sources of the energy on which humans and animals thrive for their survival. It is made up of three elements: carbon, hydrogen, and oxygen, which are present abundantly in the ocean (Knudsen et al., 2013). Carbohydrates are the promising compound obtained from marine environments but are very poorly explored. It is not only utilized as the main source of energy stored in the form of starch and glycogen but also contributes functionally, structurally, and biologically in several biological activities such as antioxidant, antibacterial, anticancer, antifungal, immune enhancing, etc. As represented by cellulose and chitin, carbohydrates are the most abundant and complex biomolecule in this biosphere. The structural complexity of carbohydrates is increased by many factors involving conformational fluctuations, dynamic behavior, numerous types of glycosidic bonds, and diverse monomeric units along with various enantiomers and huge post-polymerization modifications. Carbohydrates can carry neutral (starch, glycogen), negative (alginate, carrageenan), and positive charges (chitosan) (Pomin, 2014).

It is classified on the basis of molecular size and degree of polymerization into monosaccharides, disaccharides, oligosaccharides, and polysaccharides (Knudsen et al., 2013). Monosaccharides are the simplest sugar and have the chemical formula (CH2O)n, where n is the number of carbon atoms in a molecule (Vaclavik et al., 2008), that cannot be further hydrolyzed. The rest of the other saccharides are linked by glycosidic bonds and hydrolyzed into simpler units. For example, fructose, galactose, and glucose are the main source of energy preferentially utilized by the brain and red blood cells (Ferrier, 2014). Disaccharides comprise two monomer sugar units linked by glycosidic bonds. Sucrose, lactose, trehalose, and maltose belong to disaccharides. Oligosaccharides are composed of a few monosaccharide units (2 to 20 units) (Roberfroid and Slavin, 2000) which are soluble in 80% ethanol, but intestinal enzymes are unable to digest them. Fructo-oligosaccharides, galacto-oligosaccharides, and mannan-oligosaccharides are examples of oligosaccharides (Englyst et al., 2007). Polysaccharides belong to high-molecular-weight polymeric monosaccharide units, and the degree of polymerization ranges from 70,000 to 90,000, depending on the type of polysaccharide (BeMiller, 2018). They are neither sweet in taste nor utilized directly like other carbohydrates. They may be linear (starch, cellulose) or branched (amylopectin, glycogen), homopolysaccharides (cellulose, glycogen) or heteropolysaccharides (hyaluronic acid, arabinoxylans) (Slavin, 2012).

Agar, alginate, ascophyllan, carrageenan, fucoidan, laminarian, ulvan, fucan, chitin, and chitosan are bioactive polysaccharides of marine origin. Marine polysaccharides belong to animal-origin chitin and chitosan (crustaceans) and plant-origin sulfated polysaccharides (macroalgae) (Udayangani et al., 2020). Chitin and chitosan have N-acetylated groups extracted from crustaceans such as crab and krill (Thakur, 2019). Sulfated polysaccharides have a degree of sulfation isolated from brown algae, including fucoidan, alginate, and laminarian, while ulvan and carrageenan are derived from green and red algae, respectively (Goyal et al., 2019).

1.3 MARINE-BASED CARBOHYDRATES AS NUTRACEUTICALS

Nutraceutical is a mixed term derived from the words nutrition and pharmaceutical to meet the increasing demand of nutritional supplements with enhanced medicinal values and developed therapeutic agents for selective potential health benefits with minimum toxicity. Nutraceuticals consist of naturally obtained food and supplements with antimicrobial, antioxidant, anti-inflammatory, and antitumor activities and immense health benefits. Nutraceuticals comprise a variety of nutritional products which are physiologically beneficial with great effects on diseases (Ogle et al., 2013). They include dietary supplements, phytochemicals, medical foods, functional foods, and specific dietary patterns. On the basis of chemical constituents, it is classified as polyphenols (anthocyanins, flavonoids, isoflavones, coumarins, lignins, tannins), isoprenoid derivatives (carotenoids, saponins, terpenoids, tocopherols, tocotrienols), carbohydrate derivatives (ascorbic acid, non-starch polysaccharides, oligosaccharides), amino acid derivatives, structural lipids and fatty acids, microbes (prebiotics, probiotics), and micronutrients (minerals, vitamins) (Singh et al., 2020). Here we focus on carbohydrates obtained from marine animals (crustaceans like crabs and shrimp), such as chitosan, and sulfated polysaccharides from marine algae (brown algae), which play a tremendous role in various biological activities as nutraceuticals.

1.3.1 Chitosan

Chitosan is the most abundant naturally occurring marine mucopolysaccharide, composed of a linear chain having 2-amino-2 deoxy-D glucopyronase residues linked with one to four linkages (Aiba, 1989). Chitosan is obtained from the second most abundant polymer, chitin, by the conversion of N-acetyl D-glucosamine units into a chitosan polymer of D-glucosamine monomeric units. A copolymer of N-acetyl D-glucosamine and D-glucosamine units is are commercially generated by deacetylation under treatment with 40% to 50% concentrated NaOH alkaline conditions (Kurita and Goosen, 1997). The degree of deacetylation defines the percentage of all D-glucosamine units linked by the β-1–4 glycosidic bond present in a polysaccharide, whereas the degree of acetylation includes repeated units of N-acetyl D-glucosamine units (Joseph et al., 2021). Removal of less than 50% of the degree of acetylation makes the chitosan soluble in an acidic aqueous solvent (Dash et al., 2011), but it is insoluble in water and organic solvents, which control its applications. The degree of deacetylation (DDA) also contributes to bioactivity, chemical reactivity, flexibility, polymer confirmation, and viscosity (Bajaj et al., 2011). It is a cationic biopolymer and has versatile biological properties, including biodegradability, biocompatibility, and nontoxicity, and it is also stable, sterilizable, and non-allergic in nature (Chandur et al., 2011). The presence of a polycation charge on the surface of chitosan makes it unique in its properties and different from other polysaccharides that are either negative or neutral in nature in an acidic environment. The advantage of this property leads to formation of a multilayer structure or causes electrostatic interactions between chitosan and other natural or synthetic polymers having a negative charge on their surface (Venkatesan et al., 2010). It acts as a mucoadhesive because of electrostatic interactions and conformational flexibility of linear D-glucosamine units (Huang et al., 2020). It has been reported that chitosan contributes to biomedical applications in the form of aerogels, hydrogels, microspheres, membranes, nanoparticles, rods, and scaffolds (Denkbas et al., 2006). Because of the physicochemical properties and other biological properties such as antibacterial (Martins et al., 2014), antioxidant (Ngo et al., 2014), anticoagulant (Vongchan et al., 2003), and antitumor (Karagozlu et al., 2014) activities, it has promising applications in biomedical and biotechnological fields as a drug carrier (Felt et al., 1998), for wound healing (Chandy et al., 1990), and in obesity treatment (Li et al., 1999). It is used as an excellent biomaterial in wastewater treatment (Onsøyen and Skaugrud 1990) and in various industrial sectors like agriculture, cosmetic, chemical, food, medical, and pharmaceutical industries (Cheung et al., 2015). In order to increase the range of application at the industrial level, several modifications have been done to chitosan through many mechanisms such as addition/coupling, crosslinking, carboxylation, methacrylation, and benzoylation, to form substituted derivatives like thiolated, phosphorylated, and N-phthaloylated derivatives, crosslinked derivatives (chitosan-ethylenediamine tetraacetic acid, epichlorohydrin, and glutaraldehyde), carboxylic acid derivatives, and sulfated derivatives. These properties of chitosan make it a key attribute and highly selective for a wide spectrum of unique applications (Negm et al., 2020).

1.3.1.1 Sources

Chitosan is the deacetylated product of chitin. Basically, chitin is the raw material for the production of chitosan. The sources of chitin are extensively distributed among the animal and plant kingdoms. Chitin is a natural polymer made up of linear polysaccharides having β-1–4 linked with 2-acetamido-2-deoxy-β-D-glucose units (Kumar et al., 2017). First, chitin has been isolated from mushrooms about two centuries ago by Henry Braconnot in 1811 (Muzzarelli et al., 2012). Chitin has been obtained from crustaceans (for example, barnacles, crabs, crayfish, krill, shrimp, lobster), mollusks (clams, cuttlefish, fossils, geoducks, oysters, snails, squids), invertebrate animals, algae (brown algae, diatoms, green algae), and a few microorganisms (bacteria and fungi). Chitin is mostly obtained from constituents of marine invertebrates and insect exoskeletons and the cell wall components of fungus. Many researchers have reported the crustacean shell as the most prominent source for the extraction of chitin and chitosan, as it is an abundant biomaterial for the well-settled industrial-scale extraction. The Food and Agriculture Organization (FAO) already reported about the generation of 8.4 million tons of waste of crustacean shells (FAO, 2019). Mainly chitin is extracted from the shells of crustaceans, about 25% to 40% from shrimp shells and 15% to 20% from crab shells, becoming the main sources for the production of chitosan (Klongthong et al., 2020). Chitosan produced from shell waste has limited therapeutic applications because of protein contamination and high molecular weight (Manni et al., 2010). Fisheries waste is also used as a sustainable biomaterial source for chitin and chitosan (Kumari et al., 2015). Chitosan that is extracted from squid pen is colorless, very pure, and lacks a copolymer like minerals, calcium carbonate, and carotenoids—all these properties increase its demand at the commercial level for several applications (Liang et al., 2015). It is most commonly used in food and medical industries because of its properties like better solubility, purity, and high reactivity than chitosan obtained from other crustaceans (Cuong et al., 2016). However, crustacean-extracted chitosan has been widely used as an antimicrobial and anticoagulant agent for tissue formation from

TABLE 1.1
Amount of Chitosan Production from Different Species

Kingdom	Species Name	Amount of Chitosan Production	References
Fungi (mycelia)	*Mucor racemosus* *Cunninghamella elegans*	35.1 mg/g dry mycelia weight 20.5 mg/g dry mycelia weight	Amorim et al., 2001
Fungi (submerged and solid-state fermentation + urea)	*Gongronella butleri*	11.4/100 g mycelia	Nwe and Stevens, 2004
Fungi (fermentation by using molasses salt medium)	*Mucor rouxii*	6 to 7.7% of 100 g dry biomass	Chatterjee et al., 2005
Fungal (submerged fermentation supplemented with GA3 plant growth hormone)	*Rhizopus oryzae*	14.3% chitosan content/g mycelia	Chatterjee et al., 2008
Fungal strain grown on mung bean and soybean residue	*Aspergillus niger* TISTR3245, *Rhizopus oryzae* TISTR3189, *Zygosaccharomyces rouxii* TISTR5058, and *Candida albicans* TISTR5239	0.4–4.3 g chitosan per kg of soybean residue and 0.5–1.6 g of chitosan per kg of mung bean residue. *R. oryzae* produced maximum amount 4.3 g of chitosan per kg on soybean residue	Suntornsuk et al., 2002
Arthropoda	Shrimp waste	15.4% chitosan yield	Hossain et al., 2014

artificial skins and bones (Huang et al., 2020). The cuticles and wings of insects such as cockroaches, honeybees, mosquitoes, and silkworms are good sources for chitin production (Gonil and Sajomsang, 2012). Insect cuticles can be easily used in pharmaceutical applications because of the low content of minerals as compared to crustacean shells (Liu et al., 2012). However, insect larvae have become a new promising source for chitosan extraction (Hassan et al., 2016). Table 1.1 shows the species along with their kingdom that have been reported for production of chitosan.

1.3.1.2 Synthesis

Seafood such as crabs, shrimps, lobsters, and oysters are rich in protein content and thus consumed every year to gain protein. However, half of the body mass is represented by non-edible parts and external shells of crustaceans that are usually considered as waste and discarded. Coastal waste and shrimp industries generate a large amount of biowaste that can be transformed to value-added products as chitosan (Hamed et al., 2016). Chitin is used as a raw material for the production of chitosan. Crustacean shells have a rich chemical composition of 30% to 50% of calcium carbonate/phosphate, 20% to 30% of chitin, and 30% to 40% of protein that attracts people's attention towards chitin extraction (Kumari et al., 2015). The process of converting chitin to chitosan causes a good change, lowering the molecular weight and altering the distribution of charge and DDA that influence several applications (Younes and Rinaudo, 2015). Extraction of chitin involves two methods: chemical and biological methods. Both methods involve deacetylation for the formation of chitosan from chitin. But chemical deacetylation is more preferred for the industrial production of chitosan because of the low cost and utility of mass production (Cheung et al., 2015).

With the chemical method, chitosan is produced commercially by isolating chitin from its crustacean sources such as shrimps and crabs. Initially, the crustacean shells are grouped, washed, dried, and then ground into powder form (Younes and Rinaudo, 2015). Conventionally, the extraction protocol is followed in three steps: deproteinization, demineralization, and deacetylation which can be utilized by chemical (Percot et al., 2003) and biological (enzymatic and fermentation treatment) processes (Jung et al., 2006). First, deproteinization is done by alkali treatment of 2 N NaOH at 60° to 100°C for 1 to 72 hours. Second, deproteinization is followed by demineralization treated under acidic conditions of 2 M HCL, 100°C, from 0.5 to 48 hours. Third, deacetylation is carried out with alkaline conditions of 40% NaOH at 25°C to 100°C for 1 to 72 hours (Antonino et al., 2017). Chitin has β-1–4 linkages and fungus cell walls consist of a branched glucan linked with β-1–3 and β-1–6 linkages. Thus, the chitin extraction procedure from fungus is different and has a low chemical dependence with higher yields up to 22% to 44% (Mane et al., 2017). Extraction of chitin from a fungal source follows a typical procedure under alkaline conditions treated with 1 M NaOH for 0.5 to 12 hours at 60°C to 120°C; then the deacetylation process is carried out at 50°C to 95°C in presence of 2% to 10% of acetic acid and lastly precipitated with 2 N NaOH. In the case of a few fungi, decoloration is also done for the removal of pigments, which organisms exhibit naturally (Abo Elsoud and El Kady, 2019).

The biological method involves microbial species for the enzymatic action and is used as an alternative to the chemical one for the production of chitin and chitosan. It avoids the use of harsh chemicals like acid and alkali treatment at high temperature for a long time, as followed in chemical methods that alter the functionality and physicochemical properties of chitosan (Santos et al., 2019). Nowadays, microbial extraction of chitin is an interesting way for researchers to develop something that is feasible for industrial production. Demineralization of crustacean shells have been carried out by using bacteria which produce lactic acid and act on calcium carbonate, which is a component of biomass waste and then turned into calcium lactate, which can be further precipitated and washed (Arbia et al., 2013). Deproteinization, in which protein will be eliminated by enzyme proteases produced from bacteria, is another technique used. Microbial fermentation treatment is also studied for the extraction of chitin and chitosan. The fermentation treatment is carried by both lactic and non-lactic bacteria, for example, *Lactobacillus paracasei,*

Lactobacillus helveticus, Lactobacillus plantarum, Pseudomonas aeruginosa K-187, Bacillus subtilis, and *Serratia marcescens FS-3* (Jo et al., 2008). It has been reported that 99.6% demineralization and 95.3% deproteinization could result after extraction of chitin from *Allopetrolisthes punctatus* (porcelain crab) through lactic bacteria *L. plantarum* (Castro et al., 2018). Most importantly, the bioextraction method is safe for the environment and produces high-quality and homogeneous products (Kaur and Dhillon, 2015).

1.3.1.3 Physical and Chemical Properties

Chitosan is a mucopolysaccharide, biodegradable, biocompatible, nontoxic, and a polycation (Anaya et al., 2013). Structurally, it is a partially deacetylated product of chitin, having a linear copolymer chain of D-glucosamine and N-acetyl D-glucosamine with different DDAs and molecular weights ranging from 50% to 99% (Mati-Baouche et al., 2014). The physicochemical properties are influenced by many factors including molecular weight, DDA, polymer chain length and arrangement, charge densities, and crystallinity. The physical changes occurring in chitosan products can be affected by different conditions such as concentration and type of reagents, temperature, and time (Ahmed and Ikram, 2015). Contrary to chitin, chitosan has a unique structural integrity and high porosity that increase its utilization from biomarkers to quantum dots (Kumar et al., 2020). It has three functional groups which are highly reactive, at the C-2 position with a free amino group, the C-3 position with a primary hydroxyl group, and the C-6 position with a primary OH group (Younes and Rinaudo, 2015).

1.3.1.3.1 Degree of Deacetylation

DDA is defined by the present ratio of D-glucosamine repeating units to the sum of the copolymer of N-acetyl-D-glucosamine and D-glucosamine in a polymer chain. During the deacetylation process, the amino group (-NH2) replaces the acetyl group (-C2H3O) in a polymer chain and results in the formation of chitosan. The percentage of D-glucosamine units is more than 50% in a copolymer, named chitosan. Hence, 60% DDA in a polymer chain indicates that the chitosan contains 60% D-glucosamine units and 40% N-acetyl D-glucosamine units (Kumar et al., 2020). DDA is an extensive property that increases by increasing the strength and temperature of alkaline solutions (Ahmed and Ikram, 2015).

1.3.1.3.2 Solubility

Solubility basically depends on the presence of intermolecular and intramolecular hydrogen bonds that exist between the acetyl groups; thus, solubility increases by increasing its number. At physiological pH, the type of chemical modification, such as hydrophilic modification, increases the solubility of chitosan (Wang et al., 2011). It is determined by the pH of solution and DDA exhibited by chitosan. Chitosan is a basic polysaccharide having a 6.3 pKa value of the amino group; thus, it shows solubility in dilute acidic solutions such as formic acid, acetic acid, lactic acid, and succinic acid along with diluted

HCl (Tungtong et al., 2012). At pH less than 7, - groups become protonated and form polycationic (amine groups) -; thus, it rises in an interelectric repulsion between molecules and makes chitosan water-soluble polymers (Ahmadi et al., 2015). By replacing the amide group with a carboxylic group, chitosan shows solubility in a complete pH range and then is suitable for use in various biological applications (Cheung et al., 2015).

1.3.1.3.3 Molecular Weight

The molecular weight of any molecule is defined by the average of all the molecules present in a sample. Molecular weight is one of the significant key characteristics of chitosan that influences various biological and physicochemical properties such as biodegradability, immunogenicity, biocompatibility, mucoadhesion, hydrophilicity, viscosity, solubility, adsorption on solids, elasticity, and water uptake ability. The molecular weight of commercially available chitosan varies from 200 to 1000 KDa according to the use of the initial source material, and it also depends on the type of preparation method used (Ibrahim and EL-Zairy, 2015; Szymańska and Winnicka, 2015).

Chitosan can be classified on the basis of DDA and molecular weight: higher degree of deacetylation (HDD: 70% to 90%), lower degree of deacetylation (LDD: 55% to 70%), and higher molecular weight (HMW: >300 KDa) and lower molecular weight (LMW: 300 KDa) (Cheung et al., 2015). Molecular weight causes changes in physicochemical properties. For example, high molecular weight leads to a rise in conductivity, viscosity, surface tension, and adsorption on lipophilic molecules. Low molecular weight leads to lower viscosity and density, better penetration capacity into the cell, and more solubility in water (Joseph et al., 2021). On the basis of percentage of DDA and molecular weight, chitosan is utilized in different biological applications, as shown in Table 1.2.

1.3.1.4 Potential Health Benefit of Nutraceuticals

Chitosan shows a tremendous role in a wide variety of biological activities such as antioxidant, antimicrobial, antifungal, and antiviral activity and is thus

TABLE 1.2
Classification of Chitosan on the Basis of DDA and Molecular Weight

Degree of Acetylation	HMW (>300 KDa)	References	LMW (<300 KDa)	References
HDD (70%–90%)	Drug delivery system and antioxidant properties	Ivanova and Yaneva, 2020	Anti-infective properties Wound healing	Park et al., 2015
LDD (55%–70%)	Formulations of edible film Antibacterial activity	Liu et al., 2020	Thermal-resistant treatment Inhibitory activities towards phytopathogens	Hosseinnejad and Jafari, 2016 Meng et al., 2010

used for different purposes in different industrial sectors like cosmetic, food, pharmaceutical, and biomedical industries. Here, we have discussed various chitosan nutraceutical properties.

In terms of antioxidant activity, reactive oxygen species (ROS) generated by abnormal metabolic action as well as ultraviolet (UV) light, sunlight, and unwanted chemical reactions that produce free radicals can cause the deterioration of the DNA structure, membrane lipid, proteins, and carbohydrates, which may lead to severe diseases such as cancer, cardiovascular diseases, neurodegenerative diseases, diabetes, inflammation, aging, and atherosclerosis (Moskovitz et al., 2002). Activation of unanticipated enzymes can cause ROS-mediated oxidative stress that results in oxidative damage to our cellular macromolecules. Antioxidant substances such as chitosan play an important role in preventing health disorders by breaking chains of free radical oxidation (Ngo et al., 2015). With the help of the chelating ability of metal ions, chitosan acts as an antioxidant polymer where it scavenges free radicals and overcomes the ROS-mediated problem. Chitosan with HDD, LMW, and more free exposed amino groups have powerful antioxidant properties rather than the HMW chitosan (Anraku et al., 2018). Chitosan, a dietary supplement, having 90% DDA and approximately 100 KDa, leads to down the level of uremic toxins and lipid hydroperoxides in the gastrointestinal tract that protects against renal failure and oxidative damage to the human circulatory system (Anraku et al., 2011). Previous studies indicate that antioxidant properties of chitosan varied according to DDA, molecular weight, and amino groups. It has been reported that chitosan extracted from mushrooms shows higher antioxidant activity than those of shrimp against free radicals of 1,1-diphenyl-2-picrylhydrazyl (DPPH) and 2,2′-azino-bis (3-ethylbenzthiazoline-6-sulfonic acid) (ABTS). This might result in better antioxidant properties of chitosan depending on the source and, importantly, which extraction method is followed for the preparation of chitosan (Savin et al., 2020).

Many researchers have highlighted the antibacterial properties of chitosan against gram-negative organisms, yeast, and filamentous fungi. The mechanism behind the antimicrobial activities of chitosan is still not clear, but there have been a few hypotheses proposed: Due to its polycationic nature, chitosan behaves as a good antimicrobial product, as groups attract the negative charge of the phosphate group of phospholipid membranes for many microorganisms that lead to the leakage of intracellular substances and ultimately cause microbial cell death (Ma et al., 2017). The second proposed hypothesis is that the chelation function of chitosan blocks the active centers of metabolic enzymes and then stops the microbial cell growth (Kumar et al., 2020). Another hypothesis is LMW chitosan can easily penetrate into the cell and then interact with DNA and inhibit transcription and translation (Sudarshan et al., 1992), whereas HMW chitosan may bind to negatively charged components present on the microbial cell wall that changes the membrane permeability by covering a cell with an impermeable layer, thus obstructing transport in and out of the cell (Zheng et al., 2003; Cheung et al., 2015). The strength of antibacterial activities is decided on the basis of many factors such as DDA, molecular weight, degree of polymerization and pH, and the presence of a more positive charge of the amino group of chitosan (Ma et al., 2017). Gram-negative bacteria (*Escherichia coli and Klebsiella pneumonia*) show stronger inhibition against chitosan than gram-positive (*Staphylococcus aureus and Streptococcus*) because of the higher negative charge and hydrophilicity on their cell surface. Environmental pH can affect the number of positive charges of chitosan, as a lower pH increases the number of positively charged amino groups and influences the binding of bacterial cells and chitosan (Rhoades and Roller, 2000). Due to antibacterial, biodegradable, and film-forming properties, chitosan is attracted towards the formation of antibacterial packaging biomaterial in the food industry and becomes an alternative to plastics and safer for our health and environment (Kumar et al., 2020).

Researchers have reported that chitosan also has antifungal activity that inhibits the growth of many phytopathogenic fungus such as *Fusarium oxysporum, Phytophthora infestans* (Atia et al., 2005), and *Alternaria solani* (Saharan et al., 2015) in tomatoes; *Botrytis cinerea* and *Botrytis conidia* (gray mold) in cucumber plants (Ben-Shalom et al., 2003); and *Penicillium digitatum* (green mold) and *Penicillium italicum* (blue mold) in citrus fruit (Tayel et al., 2016). Earlier studies showed that chitosan reduces mycelial growth, fungal infection, sporangial production, germination of fungi, and release of zoospores. Antifungal activity is also influenced by molecular weight and degree of acetylation of homogenous chitosan, but it varies according to type of fungus; for example, *Fusarium oxysporum* is influenced by only molecular weight, *Alternaria solani* is affected by only acetylation degree and no effect of molecular weight, and degree of acetylation is observed on *Aspergillus niger* (Younes et al., 2014). The suggested mechanism is that chitosan forms a permeable layer over the crop surface, which controls the fungal growth and induces the activation of many defense actions like callus synthesis, chitinase accumulation, inhibitor of protein synthesis, and callus lignification (Bai et al., 1988). Chitosan shows potent fungicidal synergistic activity with fluconazole and is a promising therapy for *Candida albicans* and *Candida tropicalis* (Lo et al., 2020).

Cancer is still one the leading cause of morbidity around the world. The antitumor activity is gained by preventing the inhibition of apoptosis and angiogenesis from DNA fragmentation (Ngo et al., 2015) and by potent immunological activity of prepared chitosan-derived oligosaccharides (Tsvetkov et al., 2020). Chitosan could be a promising biomaterial in the treatment of targeting melanoma cells. It has limited the growth of primary melanoma cancerous cells by decreasing the adhesion property of the A375 cell line, inhibiting the proliferation of the SKMEL28 cell line, and developing pro-apoptotic effects such as upregulation of Bax molecules and anti-apoptotic effects like downregulation of BcL-2 and Bcl-XL proteins. This results in a potent pro-apoptotic condition for RPMI7951 and increases the number of CD95 receptors on its surface that rise in susceptibility towards FasL-induced apoptosis (Gibot et al., 2015). Many researchers have worked on chitosan of both HMW and LMW and found that both effectively inhibited proliferation of three cell lines, MCF-7 (breast cancer cell), HeLa (cervical cancer), and Saos-2 (osteosarcoma), and decrease their viability irrespective of chitosan concentration and molecular weight and also observed a necrosis death pattern towards MCF-7 and apoptosis by HeLa and Saos-2 (Abedian et al., 2019). It has been reported that antitumor activity shown by seleno short-chain chitosan, when exposed with the human gastric cancer BGC823 cell line, arrested the G2/M phase of the cell cycle and activated the mitochondrial signaling pathway in which the cascade reaction occurs such as the membrane potential of mitochondria is disrupted; then ROS species accumulate in large numbers; the ratio of Bax/BcL2 increases; and there is further activation of caspase 3, caspase 9, and cytochrome C that leads to apoptosis of BGC823 cells (Dong et al., 2019).

1.3.2 Sulfated Polysaccharides

Sulfated polysaccharides include a wide range of anionic polymers found in many divergent groups of organisms, from macroalgae to mammals, but are not found in terrestrial plants (Ciancia et al., 2020). Marine organisms such as marine invertebrates, seaweed and sea grasses, and sulfated polysaccharides are believed to be a physiological adaptation to the high ionic strength of the marine environment (Aquino et al., 2005; Pomin and Mourão, 2008). Sulfated polysaccharides are found in the cell wall of seaweeds or algae mainly composed of cellulose and hemicellulose having a high carbohydrate content but low in fat and calories. Seaweed biosynthesize sulfated polysaccharides are an important component of their cell walls. Sulfate groups are found in the backbone of seaweed's sugar structure to retain extremely high marine conditions such as high salinity, leading them to change their polymer structure into sulfated polysaccharides (Kraan, 2012). Sulfated polysaccharides are found most commonly in three groups of marine algae, *Chlorophyta* (green algae), *Phaeophyta* (brown algae), and *Rhodophyta* (red algae). Red algae's primary sulfated polysaccharides are galactans such as agar and carrageenan, while brown algae are fucans such as fucoidan, sargassum, ascophyllan, and glucuronoxylofucan. However, green algae generally have sulfated heteropolysaccharides that contain galactose, xylose, arabinose, mannose, glucuronic acid, or glucose as the main sulfated polysaccharides (Ngo and Kim, 2013). Sulfated polysaccharides encompass a complex group of macromolecules with a wide range of important biological activities, such as anti-inflammatory (Vo et al., 2012), antiviral, antioxidant (Souza et al., 2012), antitumor (Sithranga et al., 2010), antithrombotic, and antiallergic effects. Many research studies have reported on the composition, nutrient value, and biological properties of seaweed sulfated polysaccharides, which can be influenced by the maturation stage or environmental factors such as seasonal variability, nutritional value of seawater, geographic location, and other post-harvest factors such as drying of the seaweed and extraction procedures for the preparation of phycocolloid (Rioux et al., 2007). Sulfated polysaccharides are negatively charged because they have a crosslinking of the ions of the sulfate group with the complex polysaccharide molecules (Muthukumar et al., 2021).

1.3.2.1 Sources

Sulfated polysaccharides can be obtained mainly from different sources like algae, marine invertebrates, and marine bacteria. Most of the algal cell walls consist of more than 40% sulfated polysaccharides, which is relatively above average compared to other sulfated polysaccharide sources. The polysaccharide content varies between different species, for example, 35% to 70% in brown algae, 38% to 76% in red algae, and 15% to 65% in green algae (Saravana and Chun, 2017). Some typical algae polysaccharides, including fucoidan from brown algae, carrageenan and agar from red algae, and ulvan from green algae, have been shown to be effective.

1.3.2.1.1 Fucoidan

Fucoidan is found mainly in the matrix of the cell walls of brown algae. Although it has also been reported to be found in sea urchins and sea cucumbers (Pradhan et al., 2020), brown algae–based fucoidan has more bioactive energy and a higher yield. Brown algae represent almost 40% w/v sulfated polysaccharides relative to the dry mass of the cell wall in the form of fucoidan. The sulfate content of fucoidan depends on the raw material, ranging from 7.66% to 38.3% (Wen et al., 2021). Fucoidan also has a wide molecular weight range ranging from 7 kDa to 2379 kDa depending on the species, geographic location, and harvest seasons (Lim and Wan Aida, 2017). It is reported that in the reproductive stage fucoidan is found more abundantly (do-Amaral et al., 2020). The species of algae that produce fucoidan in India are *Sargassum ilicifolium, Turbinaria conoides, S. myriocystum, Padina boergesenii, P. gymnospora, T. graceful, Stoechospermum marginatum, Dictyota dichotoma, S. marginatum, S. wightii,* and *T. decurrens*. Out of all these species *Sargassum wightii* gives 71.5 mg fucoidan from 1 g dry weight algae. The cell wall formation of brown algae consists of alginate, cellulose, and sulfated fucans in the ratio 3:1:1 (Li et al., 2008).

1.3.2.1.2 Carrageenan

Carrageenan is found in red algae with HMW sulfated polysaccharides composed of a linear chain substitution of 1,4,3,6-anhydrogalactose and 1,3-galactose with ester sulfates (15% to 40%) and is a structural part of the cell membranes of red macroalgae (Qureshi et al., 2019). The cell wall of red algae is made up of microfibrils (cellulose and ß1, 3 xylan) and a matrix, where sulfated polysaccharides are present 38% in the form of carrageenan in the matrix (Muthukumar et al., 2021). Commercially used carrageenan's average molecular weight ranges from 100 to 1000 kDa (Nishinari and Fang, 2021). Carrageenan has three main structures i.e., kappa (κ), iota (ι), and lambda (λ) which contain one, two, and three sulfate groups respectively, in every repeating disaccharide unit (Dong et al., 2021). Usually, commercial kappa, iota, and lambda carrageenan have 22% (w/w), 32% (w/w), and 38% (w/w) sulfate content, respectively; however, this may vary due to the extraction method, algal species, and age groups (Kang et al., 2021). The properties of carrageenan are largely influenced by the number andposition of sulfate ester groups and by the content of 3.6 AG. It is reported that when carrageenan has a higher sulfate ester content, the gel strength reduces (Guan et al., 2017). Carrageenan is used in drug delivery, bone and cartilage tissue regeneration, and wound healing because of its physicochemical properties and gelation mechanism (Yegappan et al., 2018).

1.3.2.1.3 Agar

Like carrageenan, agar is another sulfated polysaccharide present in the cell wall of red algae (20% dry weight basis), which is responsible for their structure. They are a mixture of two polysaccharides, namely agarose, the gelling part, and agaropectin, the non-gelling part, but help to increase the strength of the gel. The agar is generally extracted from water at a high pressure, temperature (ranging from 100°C to 130°C), and pH between 5 and 6. On boiling, Agar hydrates; however, it is not soluble in cold water. When cooled below 40°C it turns into firm and breakable gels (Muthukumar et al., 2021). Agar is composed of a linear polysaccharide of 3,6-anhydro-L-Gal and D-Gal units, which are alternatively linked by α (1,3) and β (1,4) glycosidic bonds (Nishinari and Fang, 2017). The main characteristic that differentiates highly sulfated carrageenan from less sulfated agars is the presence of D-Gal and AnhydroDGal in the first and DGal, LGal, or Anhydro-L-Gal in the second (Chen et al., 2021).

1.3.2.1.4 Laminarin

Laminarin is a brown algae–based sulfated polysaccharides with a low molecular weight (~5 kDa). It is present in cell walls of *Saccharina* and *Laminaria* and to some extent in *Fucus* and *Ascophyllum* species, and it serves as the main carbohydrate reserve of brown algae species (Kraan, 2012; Kadam et al., 2014). Laminarin is the first discovered species of *Saccharina* and *Laminaria*—nowadays it is known as the most abundant source of carbon for marine ecosystems (Sanjeewa et al., 2017). Typically, laminarin accounts for up to 35% of the dry weight of algae. However, the percentage of laminarin in these algae depends on several factors such as algae species, extraction method, habitat, and harvest season.

1.3.2.1.5 Marine Invertebrates and Bacteria

Marine invertebrates like sea cucumber and abalone are important sources of sulfated polysaccharides. Sea cucumbers are benthic marine invertebrates that belong to the Echinodermata phylum and the Holothuroidea class. Sulfated polysaccharides from sea cucumber have a molecular weight ranging from 52 to 180 kDa (Zhao et al., 2020). Marine bacteria can also produce a few sulfated polysaccharides. Polysaccharides of the genus *Pseudoalteromonas*, a gram-negative bacteria, contain unusual sugar and non-sugar substituents with no structural similarity of repeating units (Kang et al., 2021). For example, the group configurations of sulfate, D-galacto, L-gulo, L-iduronic acid, (R)-lactic acid, and glycerol phosphorus have all been recognized as unusual acidic components of the polysaccharide *Pseudoalteromonas marinoglutinosa* KMM 232 (Kang et al., 2021).

1.3.2.2 Bioactive Properties of Fucoidan

Fucoidan is attributed to nutraceuticals that promote health and improve the quality and longevity of a person. Fucoidan moderately reduces the death of cholinergic neurons and maintains dopamine levels to support the work of the central nervous system, reducing the levels of inflammatory cytokines in the blood, interrupting the development of hormonal imbalances, and promoting the maintenance of pro-inflammatory levels. The function of the insulin cascade signaling systems at the mTOR kinase level activating the translation process leads to a reduction of free radicals (antioxidant), and their restoration

TABLE 1.3
Various Biological Activities of Fucoidan

Biological Activities	Model	Mechanism of Action	References
Antitumor activity	Human colon cancer cells (HCT116 cells) Serum from fucoidan-treated female Sprague Dawley rats for the growth of human breast cancer MCF-7 cells (Michigan Cancer Foundation -7).	LMW fucoidan induces apoptosis via p53-independent mechanism and triggers G1 arrest. Inhibit cell proliferation, migration, and induced apoptosis of MCF-7 cells via decreased epithelial mesenchymal transition process (EMT).	Park et al., 2017 He et al., 2019
Anti-inflammatory activity	Fucoidan-treated 2,4-dinitrofluorobenzene (DFNB)–induced atopic dermatitis BALB/c mice	Decreases the level of IgE and IL-4, infiltration of CD+4 cells, inducing Treg cells	Tian et al., 2019
Antioxidant activity	*Caenorhabditis elegans* Alzheimer's diseases transgenic model	Inhibits amyloid-β-induced toxicity by reducing its accumulation in cells and by decreasing the production of ROS.	Wang et al., 2018
Anticoagulant and antithrombotic activity	C57BL/6 male mice model of venous thrombosis targeted fucoidan nanoparticles loaded with recombinant tissue plasminogen activator (rt-PA)	Increases the accumulation of rt-PA on thrombus and thereby increases thrombolysis efficiency, as fucoidan has great affinity with P-selectin	Juenet et al., 2018
Angiogenesis	Human umbilical vein endothelial cells	Fucoidan induces angiogenesis in the presence of fibroblast growth factor-2 (FGF-2) through (matrix metalloproteinase -2) MMP-2/AKT (protein kinase-B) signaling by activating c-Jun-N terminal kinase (JNK) and p38.	Kim et al., 2014

(a pro-oxidant) supports the homeostasis of the coagulation system due to anticoagulant and procoagulant properties (Mukhamejanov and Kurilenko, 2020). The biological activity of fucoidan is largely dependent on the content of L-fucose and sulfate groups (Fitton et al., 2015). The content of fucose can vary from 19% to 51% and sulfates from 19% to 43%, depending on the type of algae and the isolation methods of the polysaccharide (Ale et al., 2011).

Fucoidan activates neutrophil, which helps the body to fight infections (Kim and Joo, 2015). Furthermore, fucoidan has been found to improve immunization in the elderly by increasing antibody levels (Tanino et al., 2016). It has been reported that fucoidan has a clear antiallergic effect. Edema develops after subcutaneous administration of 2,4,6-trinitrochlorobenzol to mice—the value is reduced three times after administration of fucoidan at a dose of 200 μg and six times at a dose of 400 μg (Kawashima et al., 2012).

Alkaline phosphatase plays a vital role in bone formation. Its activation results in phosphorus, mineralization of protomers, and a decrease in the concentration of extracellular pyrophosphate (inhibitor of mineralization). Acceptance of fucoidan at a dosage of 2 mg/mL increases bone mineralization by raising alkaline phosphatase activity by 35%. (Boskey et al., 1998). Osteocalcin advances bone development, speeding up the development of hydroxyapatite crystals (Min et al., 2012). Fucoidan expanded the discharge of osteocalcin in 7F2 cells and the worth of its mineralization in a portion subordinate way. Apart from these applications, fucoidan play a vital role in various biological activities that has been summarized in Table 1.3 along with mechanism of action.

1.4 BIOTECHNOLOGICAL APPLICATIONS

1.4.1 Biopolymer Nanoparticles: Chitosan DNA Nanoparticles

Nanoparticles ranging from 1 to 100 nm in size have attracted attention for pertinent advantages over chemically synthesized and biomaterials (McClean et al., 1998). But biomaterials offer suitable and viable routes in clinical application because of their versatile characteristics such as biodegradability, low immunogenicity, biocompatibility, and easy production. Many factors are responsible for the creation of pivotal scores in the preparation and selection of nanoparticles, such as size, surface charge, morphology, stability, cytotoxicity, and the rate at which loaded molecules are released. Biopolymer nanoparticles are mainly divided into two classes: polysaccharides and protein biopolymers. Polysaccharide nanoparticles include chitosan, dextran, alginate, pollunan, hyaluronic acid, and heparin (Saravanakumar et al., 2012). These biopolymer nanoparticles are typically used for drug delivery and treating diabetes, cancer, allergy, and inflammation (Jacob et al., 2018).

Chitosan is one of the most commonly and widely used carbohydrates for the fabrication of nanoparticles. Extraction of chitosan nanoparticles as compared to chemically produced chitosan has been reported from waste shrimp shells to inhibit MCF-7 proliferation and showed minimal cytotoxicity when treated with L929 fibroblast cells (Resmi et al., 2021). Liposomal or cationic lipid systems and cationic polymers are two non-viral delivery systems that have been developed as alternatives to viral vectors for safer gene carrying and gene delivery, low immunogenicity, production at large scale, to raise the loading efficiency of bioactives, and possibility to make smaller gene material (Hidai and Kitano, 2018). The formation of a cationic polymer DNA complex is more favorable than cationic lipid due to more stability—it protects DNA from nuclease degradation and also provides efficient gene delivery into the cells by condensing the DNA to a smaller size. DNA condensation depends on the electrostatic interactions between polycationic polymers and the negatively charged phosphate backbone of DNA (Borchard, 2001). Various polycationic polymers have been identified as gene carriers such as polyethyleneimine, poly-L-lysine, polyamidoamine dendrimers, and chitosan (Bozkir and Saka, 2004). But chitosan emerged as the ideal polycationic polymer and is preferred over them because DNA chitosan nanoparticles complex (DCNP) formations are more stable, cheaper, biocompatible, biodegradable, nontoxic, have a variable size, and most importantly, have high efficacy for gene transfer into cells and nuclei (Ishii et al., 2001). In recent years, DCNP has been explored as a gene carrier (Mao et al., 2001) and for delivery of drugs (Nitta and Numata, 2013), plasmid, and vaccines (Illum et al., 2001).

Malfunctioning of genes in any particular cell leads to association with approximately 4,000 diseases such as neurodegenerative diseases, periodontal diseases, cancer, and Alzheimer's disease (Cao et al., 2019; Mansouri et al., 2004). However, gene therapy emerged as a platform to treat these diseases by modifying the expression of associated disease genes at a cellular level. Unfortunately, gene therapy also faces many systemic barriers to reach the target cell due to lack of efficient and safe gene delivery vectors. To overcome these challenges of gene therapy, researchers have developed DCNP to meet the required demand of highly efficient vectors at certain levels. Chitosan transfection efficacy depends on its molecular weight, DDA, derivatives, and amount of nitrogen per gene phosphate ratio (N/P ratio) (Alameh et al., 2018). Two types of chitosan nanoparticle fabrication techniques utilized are a polyelectrolyte complex and ionic gelation. Polyelectrolyte complexes involve electrostatic interaction between a chitosan positive amino group and negatively charged DNA (Inamdar and Mourya, 2011), but ionic gelation uses an anionic crosslinker agent like tripolyphosphate (TPP) that enhances binding ability and nanoparticle stability and also encapsulates the macromolecules or protein, which improves intracellular trafficking of DNA. It has been found that the ionic gelation method using TPP followed to form encapsulated small interfering RNA (siRNA) nanoparticles reaches >96% transfection efficiency and an observed N/P ratio 100:1, which results in complete binding of chitosan and siRNA (Sharma et al., 2013). It has reported that encapsulation of plasmid DNA with chitosan nanoparticles for respiratory syncytial virus (RSV) administered intravenously in mouse tissues yielded 94.7% encapsulation efficiency and 15% to 20% loading capacity, which resulted in a higher expression level of RSV protein in mouse tissues rather than naked DNA (Boyoglu et al., 2009).

Inhalation of dry chitosan interferon-β gene complex powder is an effective way to overcome lung metastasis in CT26 cells in mice with CT26 (Okamoto et al., 2011).

Oligomeric chitosan plasmid DNA nanoparticles crosslinked with TPP are utilized for the development of gene-activated collagen-based scaffolds for tissue engineering in mesenchymal stem cells (MSCs) (Raftery et al., 2015). Tissue engineering involved an efficient chitosan DNA complex loaded onto an activated collagen scaffold, which provides a platform to carry a gene encoding angiogenic vascular endothelial growth factor (VEGF) and osteogenic bone morphogenetic promoter (BMP-2) proteins, further inducing MSC osteogenesis that accelerates the bone regeneration in large segmental bone defects (Raftery et al., 2017).

1.4.2 Bioenergy: Ethanol and Biobutanol Production from Mannitol and Laminarin

Currently, the growing world population is a big reason behind the increasing rate of liquid fuel consumption, and its demand will always be at peak in the transport sector. Global warming, increasing the price rate of fuel, and fossil fuel depletion becoming a great concern have forced us to search for alternatives which are renewable, sustainable, cost-effective, and efficient with lesser emissions of greenhouse gases (Nigam and Singh, 2011). Marine biomass can serve as alternatives to meet the increasing fuel demand of present and future generations. Marine biomass comprises abundant seaweed/macroalgae that represent a promising raw source for the production of biofuels (John et al., 2011). Regarding biofuel production, macroalgae has major merits of greater productivity over land crops; no need of arable land, irrigation water, and fertilizers; and low or no lignin in macroalgae, which means a lack of the harsh pre-treatment generally used for lignocellulosic biomasses (Van Hal et al., 2014). Carbon neutrality is another important fact to generate biofuel from seaweed. These properties of macroalgae emphasize the use of biomasses as feedstock for the production of third-generation liquid biofuels (Nigam and Singh, 2011).

Macroalgae are of three types, green, brown, and red algae, among them brown and red algae have been reported as having the highest content of carbohydrates. However, research is going on to develop a suitable process at large scale for producing biofuels by utilizing carbohydrates from these algae. Brown algae, including *Laminaria, Alaria,* and *Saccorhiza* have stored food material in the form of mannitol and laminarin and can grow meters in length (Nobe et al., 2003; Adams et al., 2009; Horn et al., 2000). Brown algae may have a high content of mannitol and laminarin—for example, *Laminaria hyperborea* has 25% mannitol and 30% laminarin as dry weight. This high content of carbohydrate in brown algae makes it a potential source for the generation of liquid fuel such as ethanol. The concentration of mannitol and laminarin in any seaweed varies throughout the year (Jensen and Haug, 1956). These storage sugars can be easily extracted from brown algae under low pH and high temperature (Percival and McDowell, 1967). Laminarin can be hydrolyzed enzymatically by laminaranase and cellulase into its glucose monomer by use of many microorganisms during fermentation. Mannitol is a sugar alcohol which has to be first oxidized to fructose by the enzyme mannitol dehydrogenase and then hydrolyzed into its monomer. The oxidation of mannitol requires oxygen; thus many microorganisms are unable to hydrolyze it anaerobically (Van Dijken and Scheffers, 1986). Various hydrolysis treatments such as dilute acid thermal, dilute alkaline thermal, and enzymatic can be used to hydrolyze complex sugars like mannitol, laminarin, cellulose, etc., to simple sugars mannose, glucose, and galactose and are further followed by fermentation to ethanol production. Two types of fermentation techniques, separate hydrolysis and fermentation (SHF) and simultaneous saccharification and fermentation (SSF) can be utilized for the production of ethanol from sugar in the presence of microorganisms (Offei et al., 2018).

Ethanol production in the fermentation process is highly dependent on the concentration of oxygen, and higher oxygen leads to a reduction of its yield. *Zymobacter palmae*, a facultative anaerobe, is able to grow on mannitol substrate extracted from *Laminaria hyperborea* and yields 0.38 g ethanol (Horn et al., 2000). Among all yeast strains, *Saccharomyces ceréviciae* (KCCM50550) has been reported to produce the highest ethanol formation: 2.59 g/L from 10.0 g/L of mannitol substrate of brown algae (Lee et al., 2012).

Mannitol can be rapidly extracted through fermentation and yields <40 mM from *Ascophyllum nodosum* and >70 mM from *Laminaria digitata*, which is further used as a carbon source to produce 75% ethanol by use of thermophilic *Clostridia* anaerobes. *Thermoanaerobacter pseudethanolicus* was found to be the best ethanol (88%) producer from pure mannitol (Chades et al., 2018). The liberation of glucose monomer from laminarin storage compounds of *Laminaria digitata* further ferments to biobutanol production by using *Clostridium species* through the acetone-butanol-ethanol (ABE) fermentation pathway (Hou et al., 2017).

1.5 CONCLUSION

The marine environment contains a series of micro-organisms and macro-organisms, which have developed certain metabolic mechanisms for the biosynthesis of secondary metabolites with specific activities that are useful for their survival. Marine environment functional ingredients include polysaccharides, vitamins, polyunsaturated fatty acids, enzymes, minerals, antioxidants, and bioactive peptides. All these bioactive compounds represent a potential source of raw material for nutraceuticals and food industries. In order to get efficient gene delivery from chitosan-based nanoparticles, their properties (i.e., chitosan derivatives, N/P ratio, chitosan molecular weight, and deacetylation degree) should be optimized and well understood. The present status for clinical trials of chitosan-based gene delivery systems is not very satisfactory, and more research needs to be done in the future in order to tackle the challenges being faced. Macroalgae biomass as a third-generation feedstock is becoming an important source for bioenergy production; however, more research is needed to explore its potential to the maximum and scale up this practice at large scale. Future prospects for the use of marine-derived compounds depend on their bioavailability, development of eco-friendly techniques for a higher rate of extraction, preventing degradation, and identifying isolated components which can be purified for use in specific forms.

REFERENCES

Abedian, Z., Moghadamnia, A. A., Zabihi, E., Pourbagher, R., Ghasemi, M., Nouri, H. R., . . . Jenabian, N. (2019). Anticancer properties of chitosan against osteosarcoma, breast cancer and cervical cancer cell lines. *Caspian Journal of Internal Medicine*, *10*(4), 439.

Abo Elsoud, M. M., & El Kady, E. M. (2019). Current trends in fungal biosynthesis of chitin and chitosan. *Bulletin of the National Research Centre*, *43*(1), 1–12.

Adams, J. M., Gallagher, J. A., & Donnison, I. S. (2009). Fermentation study on Saccharina latissima for bioethanol production considering variable pre-treatments. *Journal of applied Phycology*, *21*(5), 569–574.

Ahmadi, F., Oveisi, Z., Samani, S. M., & Amoozgar, Z. (2015). Chitosan based hydrogels: Characteristics and pharmaceutical applications. *Research in Pharmaceutical Sciences*, *10*(1), 1.

Ahmed, S., & Ikram, S. (2015). Chitosan & its derivatives: A review in recent innovations. *International Journal of Pharmaceutical Sciences and Research*, *6*(1), 14.

Aiba, S. I. (1989). Studies on chitosan: 2. Solution stability and reactivity of partially N-acetylated chitosan derivatives in aqueous media. *International Journal of Biological Macromolecules*, *11*(4), 249–252.

Alameh, M., Lavertu, M., Tran-Khanh, N., Chang, C. Y., Lesage, F., Bail, M., . . . Buschmann, M. D. (2018). siRNA delivery with chitosan: Influence of chitosan molecular weight, degree of deacetylation, and amine to phosphate ratio on in vitro silencing efficiency, hemocompatibility, biodistribution, and in vivo efficacy. *Biomacromolecules*, *19*(1), 112–131.

Ale, M. T., Maruyama, H., Tamauchi, H., Mikkelsen, J. D., & Meyer, A. S. (2011). Fucose-containing sulfated polysaccharides from brown seaweeds inhibit proliferation of melanoma cells and induce apoptosis by activation of caspase-3 in vitro. *Marine Drugs*, *9*, 2605–2621.

Amorim, R. V. D. S., Souza, W. D., Fukushima, K., & Campos-Takaki, G. M. D. (2001). Faster chitosan production by mucoralean strains in submerged culture. *Brazilian Journal of Microbiology*, *32*(1), 20–23.

Anaya, P., Cárdenas, G., Lavayen, V., García, A., & O'Dwyer, C. (2013). Chitosan gel film bandages: Correlating structure, composition, and antimicrobial properties. Journal of applied polymer science, 128(6), 3939-3948.

Anraku, M., Fujii, T., Kondo, Y., Kojima, E., Hata, T., Tabuchi, N., . . . Tomida, H. (2011). Antioxidant properties of high molecular weight dietary chitosan in vitro and in vivo. *Carbohydrate Polymers*, *83*(2), 501–505.

Anraku, M., Gebicki, J. M., Iohara, D., Tomida, H., Uekama, K., Maruyama, T., . . . Otagiri, M. (2018). Antioxidant activities of chitosans and its derivatives in in vitro and in vivo studies. *Carbohydrate Polymers*, *199*, 141–149.

Aquino, R. S., Landeira-Fernandez, A. M., Valente, A. P., Andrade, L. R., & Mourao, P. A. (2005). Occurrence of sulfated galactans in marine angiosperms: Evolutionary implications. *Glycobiology*, *15*(1), 11–20.

Arbia, W., Arbia, L., Adour, L., & Amrane, A. (2013). Chitin extraction from crustacean shells using biological methods—a review. *Food Technology and Biotechnology*, *51*(1), 12–25.

Atia, M. M. M., Buchenauer, H., Aly, A. Z., & Abou-Zaid, M. I. (2005). Antifungal activity of chitosan against Phytophthora infestans and activation of defence mechanisms in tomato to late blight. *Biological Agriculture & Horticulture*, *23*(2), 175–197.

Bai, R. K., Huang, M. E., & Jiang, Y. Y. (1988). Selective permeabilities of chitosanacetic acid complex membrane and chitosan-polymer complex membranes for oxygen and carbon dioxide. *Polymer Bulletin*, *20*, 83–88.

Bajaj, M., Winter, J., & Gallert, C. (2011). Effect of deproteination and deacetylation conditions on viscosity of chitin and chitosan extracted from Crangon crangon shrimp waste. *Biochemical Engineering Journal*, *56*(1–2), 51–62.

BeMiller, J. N. (2018). *Carbohydrate chemistry for food scientists*. Elsevier.

Ben-Shalom, N., Ardi, R., Pinto, R., Aki, C., & Fallik, E. (2003). Controlling gray mould caused by Botrytis cinerea in cucumber plants by means of chitosan. *Crop Protection*, *22*(2), 285–290.

Bhosle, N. B., Bhaskar, P. V., & Ramachandran, S. (1998). Abundance of dissolved polysaccharides in the oxygen minimum layer of the Northern Indian Ocean. Marine chemistry, 63(1-2), 171-182.

Borchard, G. (2001). Chitosans for gene delivery. *Advanced Drug Delivery Reviews*, *52*(2), 145–150.

Boskey, A. L., Gadaleta, S., Gundberg, C., Doty, S. B., Ducy, P., & Karsenty, G. (1998). Fourier transform infrared microspectroscopic analysis of bones of osteocalcin-deficient mice provides insight into the function of osteocalcin. *Bone*, *23*(3), 187–196.

Boyoglu, S., Vig, K., Pillai, S., Rangari, V., Dennis, V. A., Khazi, F., & Singh, S. R. (2009). Enhanced delivery and expression of a nanoencapsulated DNA vaccine vector for respiratory syncytial virus. *Nanomedicine: Nanotechnology, Biology and Medicine*, *5*(4), 463–472.

Bozkir, A., & Saka, O. M. (2004). Chitosan-DNA nanoparticles: Effect on DNA integrity, bacterial transformation and transfection efficiency. *Journal of Drug Targeting*, *12*(5), 281–288.

Cao, Ye, Yang, F. T., Yee, S. W., Melvin, W. J. L., & Subbu, V. (2019). Recent advances in chitosan-based carriers for gene delivery. *Marine Drugs*, *17*(6), 381.

Castro, R., Guerrero-Legarreta, I., & Bórquez, R. (2018). Chitin extraction from Allopetrolisthes punctatus crab using lactic fermentation. *Biotechnology Reports*, *20*, e00287.

Chades, T., Scully, S. M., Ingvadottir, E. M., & Orlygsson, J. (2018). Fermentation of mannitol extracts from brown macro algae by Thermophilic Clostridia. *Frontiers in Microbiology*, *9*, 1931.

Chandur, V. K., Badiger, A. M., & Shambashiva, R. K. (2011). Characterizing formulations containing derivatized chitosan with polymer blending. *International Journal of Chemistry and Pharmaceutical*, *4*(1), 950–967.

Chandy, T., & Sharma, C. P. (1990). Chitosan-as a biomaterial. *Biomaterials, Artificial Cells and Artificial Organs*, *18*(1), 1–24.

Chatterjee, S., Adhya, M., Guha, A. K., & Chatterjee, B. P. (2005). Chitosan from Mucor rouxii: Production and physico-chemical characterization. *Process Biochemistry*, *40*(1), 395–400. production and physico-chemical characterization. *Process Biochemistry*, *40*(1), 395–400.

Chatterjee, S., Chatterjee, S., Chatterjee, B. P., & Guha, A. K. (2008). Enhancement of growth and chitosan production by Rhizopus oryzae in whey medium by plant growth hormones. *International Journal of Biological Macromolecules*, *42*(2), 120–126.

Chen, X., Fu, X., Huang, L., Xu, J., & Gao, X. (2021). Agar oligosaccharides: A review of preparation, structures, bioactivities and application. *Carbohydrate Polymers*, 265.

Cheung, R. C. F., Ng, T. B., Wong, J. H., & Chan, W. Y. (2015). Chitosan: An update on potential biomedical and pharmaceutical applications. *Marine Drugs*, *13*(8), 5156–5186.

Ciancia, M., Fernández, P. V., & Leliaert, F. (2020). Diversity of sulfated polysaccharides from cell walls of coenocytic green algae and their structural relationships in view of green algal evolution. *Frontiers in Plant Science*, *11*, 1452.

Cuong, H. N., Minh, N. C., Van Hoa, N., & Trung, T. S. (2016). Preparation and characterization of high purity β-chitin from squid pens (Loligo chenisis). *International Journal of Biological Macromolecules*, *93*, 442–447.

Dash, M., Chiellini, F., Ottenbrite, R. M., & Chiellini, E. (2011). Progress in polymer science. *Chitosan—A Versatile Semi-Synthetic Polymer in Biomedical Applications. Special Issue on Biomaterials*, *36*(8), 981–1014.

Antonino, R. S. C. M. D. Q., Fook, B. R. P. L., Lima, V. A. D. O., Rached, R. D. F., Lima, E. P. N., Lima, R. S., . . . & Fook, M. L. (2017). Preparation and characterization of chitosan obtained from shells of shrimp (Litopenaeus vannamei Boone). Mar. Drugs, 15(5), 1-12.

Denkbas, E. B., & Ottenbrite, R. M. (2006). Perspectives on: Chitosan drug delivery systems based on their geometries. *Journal of Bioactive and Compatible Polymers*, *21*(4), 351–368.

do-Amaral, C. C. F., Pacheco, B. S., Seixas, F. K., Pereira, C. M. P., & Collares, T. (2020). Antitumoral effects of fucoidan on bladder cancer. *Algal Research*, *47*.

Dong, X. D., Yu, J., Meng, F. Q., Feng, Y. Y., Ji, H. Y., & Liu, A. (2019). Antitumor effects of seleno-short-chain chitosan (SSCC) against human gastric cancer BGC-823 cells. *Cytotechnology*, *71*(6), 1095–1108.

Dong, Y., Wei, Z., & Xue, C. (2021). Recent advances in carrageenan-based delivery systems for bioactive ingredients: A review. *Trends in Food Science & Technology*, *112*, 348–361.

Englyst, K. N., Liu, S., & Englyst, H. N. (2007). Nutritional characterization and measurement of dietary carbohydrates. *European Journal of Clinical Nutrition*, *61*(1), S19–S39.

Fao Aquaculture Network, FAO Aquaculture Newsletter. (2019). (Vol. 60). Retrieved from www.fao.org/3/ca5223en/CA5223EN.pdf.

Felt, O., Buri, P., & Gurny, R. (1998). Chitosan: A unique polysaccharide for drug delivery. *Drug Development and Industrial Pharmacy*, *24*(11), 979–993.

Ferrier, D. R. (2014). *Biochemistry*. Lippincott Williams & Wilkins.

Fitton, J. H., Stringer, D. N., & Karpiniec, S. S. (2015). Therapies from Fucoidan: An update. *Marine Drugs*, *13*, 5920–5946.

Gibot, L., Chabaud, S., Bouhout, S., Bolduc, S., Auger, F. A., & Moulin, V. J. (2015). Anticancer properties of chitosan on human melanoma are cell line dependent. *International Journal of Biological Macromolecules*, *72*, 370–379.

Gonil, P., & Sajomsang, W. (2012). Applications of magnetic resonance spectroscopy to chitin from insect cuticles. *International Journal of Biological Macromolecules*, *51*(4), 514–522.

Goyal, M. R., Suleria, H. A. R., & Kirubanandan, S. (Eds.). (2019). *Technological processes for marine foods, from water to fork: Bioactive compounds, industrial applications, and genomics*. CRC Press.

Guan, J., Li, L., & Mao, S. (2017). Applications of carrageenan in advanced drug delivery. In *Seaweed polysaccharides* (pp. 283–303). Elsevier.

Hamed, I., Özogul, F., & Regenstein, J. M. (2016). Industrial applications of crustacean by-products (chitin, chitosan, and chitooligosaccharides): A review. *Trends in Food Science & Technology*, *48*, 40–50.

Hassan, M. I., Taher, F. A., Mohamed, A. F., & Kamel, M. R. (2016). Chitosan nanoparticles prepared from Lucilia cuprina maggots as antibacterial agent. *Journal of the Egyptian Society of Parasitology*, *46*(3), 519–526.

He, X., Xue, M., Jiang, S., Li, W., Yu, J., & Xiang, S. (2019). Fucoidan promotes apoptosis and inhibits EMT of breast cancer cells. *Biological and Pharmaceutical Bulletin*, *42*(3), 442–447.

Hidai, C., & Kitano, H. (2018). Nonviral gene therapy for cancer: A review. *Diseases*, *6*(3), 57.

Horn, S. J., Aasen, I. M., & Østgaard, K. (2000). Production of ethanol from mannitol by Zymobacter palmae. *Journal of Industrial Microbiology and Biotechnology*, *24*(1), 51–57.

Hossain, M. S., & Iqbal, A. (2014). Production and characterization of chitosan from shrimp waste. *Journal of the Bangladesh Agricultural University*, *12*(1), 153–160.

Hosseinnejad, M., & Jafari, S. M. (2016). Evaluation of different factors affecting antimicrobial properties of chitosan. *International Journal of Biological Macromolecules*, *85*, 467–475.

Hou, X., From, N., Angelidaki, I., Huijgen, W. J., & Bjerre, A. B. (2017). Butanol fermentation of the brown seaweed Laminaria digitata by Clostridium beijerinckii DSM-6422. *Bioresource Technology*, *238*, 16–21.

Huang, T. W., Ho, Y. C., Tsai, T. N., Tseng, C. L., Lin, C., & Mi, F. L. (2020). Enhancement of the permeability and activities of epigallocatechin gallate by quaternary ammonium chitosan/fucoidan nanoparticles. *Carbohydrate Polymers*, *242*, 116312.

Ibrahim, H. M., & El-Zairy, E. M. R. (2015). Chitosan as a biomaterial—structure, properties, and electrospun nanofibers. *Concepts, Compounds and the Alternatives of Antibacterials*, 81–101.

Illum, L., Jabbal-Gill, I., Hinchcliffe, M., Fisher, A. N., & Davis, S. S. (2001). Chitosan as a novel nasal delivery system for vaccines. *Advanced Drug Delivery Reviews*, *51*(1–3), 81–96.

Inamdar, N., & Mourya, V. K. (2011). Chitosan and anionic polymers—Complex formation and applications. In *Polysaccharides: Development, properties and applications* (pp. 333–377). Nova Science Publishers.

Ishii, T., Okahata, Y., & Sato, T. (2001). Mechanism of cell transfection with plasmid/chitosan complexes. *Biochimica et Biophysica Acta (BBA)-Biomembranes*, *1514*(1), 51–64.

Ivanova, D. G., & Yaneva, Z. L. (2020). Antioxidant properties and redox-modulating activity of chitosan and its derivatives: Biomaterials with application in cancer therapy. *BioResearch Open Access*, *9*(1), 64–72.

Jacob, J., Haponiuk, J. T., Thomas, S., & Gopi, S. (2018). Biopolymer based nanomaterials in drug delivery systems: A review. *Materials Today Chemistry*, *9*, 43–55.

Jensen, A., & Haug, A. (1956). *Geographical and seasonal variation in the chemical composition of Laminaria Hyperborea and Laminaria digitata from the Norwegian coast*. Akademisk trykningssentral.

Jo, G. H., Jung, W. J., Kuk, J. H., Oh, K. T., Kim, Y. J., & Park, R. D. (2008). Screening of protease-producing Serratia marcescens FS-3 and its application to deproteinization of crab shell waste for chitin extraction. *Carbohydrate Polymers*, *74*(3), 504–508.

John, R. P., Anisha, G. S., Nampoothiri, K. M., & Pandey, A. (2011). Micro and macroalgal biomass: A renewable source for bioethanol. *Bioresource Technology*, *102*(1), 186–193.

Joseph, S. M., Krishnamoorthy, S., Paranthaman, R., Moses, J. A., & Anandharamakrishnan, C. (2021). A review on source-specific chemistry, functionality, and applications of chitin and chitosan. *Carbohydrate Polymer Technologies and Applications*, *2*, 100036.

Juenet, M., Aid-Launais, R., Li, B., Berger, A., Aerts, J., Ollivier, V., . . . Chauvierre, C. (2018). Thrombolytic therapy based on fucoidan-functionalized polymer nanoparticles targeting P-selectin. *Biomaterials*, *156*, 204–216.

Jung, W. J., Jo, G. H., Kuk, J. H., Kim, K. Y., & Park, R. D. (2006). Extraction of chitin from red crab shell waste by cofermentation with Lactobacillus paracasei subsp. tolerans KCTC-3074 and Serratia marcescens FS-3. *Applied Microbiology and Biotechnology*, *71*(2), 234–237.

Kadam, S. U., Tiwari, B. K., & O'donnell, C. P. (2014). Extraction, structure and biofunctional activities of laminarin from brown algae. *International Journal of Food Science & Technology*, *50*, 24–31.

Kang, J., Jia, X., Wang, N., Xiao, M., Song, S., Wu, S., . . . Guo, Q. (2021). Insights into the structure-bioactivity relationships of marine sulfated polysaccharides: A review. *Food Hydrocolloids*, 107049.

Karagozlu, M. Z., & Kim, S. K. (2014). Anticancer effects of chitin and chitosan derivatives. *Advances in Food and Nutrition Research*, *72*, 215–225.

Kaur, S., & Dhillon, G. S. (2015). Recent trends in biological extraction of chitin from marine shell wastes: A review. *Critical Reviews in Biotechnology*, *35*(1), 44–61.

Kawashima, T., Murakami, K., Nishimura, I., Nakano, T., & Obata, A. (2012). A sulfated polysaccharide, fucoidan, enhances the immunomodulatory effects of lactic acid bacteria. *International Journal of Molecular Medicine*, *29*(3), 447–453.

Kim, B. S., Park, J. Y., Kang, H. J., Kim, H. J., & Lee, J. (2014). Fucoidan/FGF-2 induces angiogenesis through JNK-and p38-mediated activation of AKT/MMP-2 signalling. *Biochemical and Biophysical Research Communications*, *450*(4), 1333–1338.

Kim, S. K., & Venkatesan, J. (2015). Introduction to marine biotechnology. In *Springer handbook of marine biotechnology* (pp. 1–10). Springer.

Kim, S. Y., & Joo, H. G. (2015). Evaluation of adjuvant effects of fucoidan for improving vaccine efficacy. *Journal of Veterinary Science*, *16*(2), 145–150.

Klongthong, W., Muangsin, V., Gowanit, C., & Muangsin, N. (2020). Chitosan biomedical applications for the treatment of viral disease: A data mining model using bibliometric predictive intelligence. *Journal of Chemistry*, *2020*.

Knudsen, K. E. B., Lærke, H. N., & Jørgensen, H. (2013). 5 carbohydrates and carbohydrate utilization in Swine. *Sustainable Swine Nutrition*, 109–137.

Kraan, S. (2012). Algal polysaccharides, novel applications and outlook. In *Carbohydrates-comprehensive studies on glycobiology and glycotechnology*. InTech.

Kumar, M., Brar, A., Vivekanand, V., & Pareek, N. (2017). Production of chitinase from thermophilic Humicola grisea and its application in production of bioactive chitooligosaccharides. *International Journal of Biological Macromolecules*, *104*, 1641–1647.

Kumar, S., Mukherjee, A., & Dutta, J. (2020). Chitosan based nanocomposite films and coatings: Emerging antimicrobial food packaging alternatives. *Trends in Food Science & Technology*, *97*, 196–209.

Kumari, S., Rath, P., Kumar, A. S. H., & Tiwari, T. N. (2015). Extraction and characterization of chitin and chitosan from fishery waste by chemical method. *Environmental Technology & Innovation*, *3*, 77–85.

Kurita, K., & Goosen, M. F. A. (1997). *Applications of chitin and chitosan* (p. 297). CRC Press.

Lee, S. M., & Lee, J. H. (2012). Ethanol fermentation for main sugar components of brown algae using various yeasts. *Journal of Industrial and Engineering Chemistry*, *18*(1), 16–18.

Li, B., Lu, F., Wei, X., & Zhao, R. (2008). Fucoidan: Structure and bioactivity. *Molecules*, *13*, 1671–1695.

Li, K., e al. (1999). Reduction in fat storage during 4G-fi-D~ galactosylsucrose (lactosucrose) treatment in mice fed a high~ fat diet. *Journal of Traditional Medicines*, *16*, 66–71.

Liang, T. W., Huang, C. T., Dzung, N. A., & Wang, S. L. (2015). Squid pen chitin chitooligomers as food colorants absorbers. *Marine Drugs*, *13*(1), 681–696.

Lim, S. J., & Wan Aida, W. M. (2017). Extraction of sulfated polysaccharides (Fucoidan) from brown seaweed. In *Seaweed polysaccharides* (pp. 27–46). Elsevier.

Liu, S., Sun, J., Yu, L., Zhang, C., Bi, J., Zhu, F., . . . Yang, Q. (2012). Extraction and characterization of chitin from the beetle Holotrichia parallela Motschulsky. *Molecules, 17*(4), 4604–4611.

Liu, Y., Yuan, Y., Duan, S., Li, C., Hu, B., Liu, A., . . . Wu, W. (2020). Preparation and characterization of chitosan films with three kinds of molecular weight for food packaging. *International Journal of Biological Macromolecules, 155*, 249–259.

Lo, W. H., Deng, F. S., Chang, C. J., & Lin, C. H. (2020). Synergistic antifungal activity of chitosan with fluconazole against Candida albicans, Candida tropicalis, and fluconazole-resistant strains. *Molecules, 25*(21), 5114.

Ma, Z., Garrido-Maestu, A., & Jeong, K. C. (2017). Application, mode of action, and in vivo activity of chitosan and its micro-and nanoparticles as antimicrobial agents: A review. *Carbohydrate Polymers, 176*, 257–265.

Mane, S. R., Pathan, E. K., Kale, D., Ghormade, V., Gadre, R. V., Rajamohanan, P. R., . . . Deshpande, M. V. (2017). Optimization for the production of mycelial biomass from Benjaminiella poitrasii to isolate highly deacetylated chitosan. *Journal of Polymer Materials, 34*(1), 145–156.

Manni, L., Ghorbel-Bellaaj, O., Jellouli, K., Younes, I., & Nasri, M. (2010). Extraction and characterization of chitin, chitosan, and protein hydrolysates prepared from shrimp waste by treatment with crude protease from Bacillus cereus SV1. *Applied Biochemistry and Biotechnology, 162*(2), 345–357.

Mansouri, S., Lavigne, P., Corsi, K., Benderdour, M., Beaumont, E., & Fernandes, J. C. (2004). Chitosan-DNA nanoparticles as non-viral vectors in gene therapy: Strategies to improve transfection efficacy. *European Journal of Pharmaceutics and Biopharmaceutics, 57*(1), 1–8.

Mao, H. Q., Roy, K., Troung-Le, V. L., Janes, K. A., Lin, K. Y., Wang, Y., . . . Leong, K. W. (2001). Chitosan-DNA nanoparticles as gene carriers: Synthesis, characterization and transfection efficiency. *Journal of Controlled Release, 70*(3), 399–421.

Martins, A. F., Facchi, S. P., Follmann, H. D., Pereira, A. G., Rubira, A. F., & Muniz, E. C. (2014). Antimicrobial activity of chitosan derivatives containing N-quaternized moieties in its backbone: A review. *International Journal of Molecular Sciences, 15*(11), 20800–20832.

Mati-Baouche, N., Elchinger, P. H., de Baynast, H., Pierre, G., Delattre, C., & Michaud, P. (2014). Chitosan as an adhesive. *European Polymer Journal, 60*, 198–212.

McClean, S., Prosser, E., Meehan, E., O'Malley, D., Clarke, N., Ramtoola, Z., & Brayden, D. (1998). Binding and uptake of biodegradable poly-DL-lactide micro-and nanoparticles in intestinal epithelia. *European Journal of Pharmaceutical Sciences, 6*(2), 153–163.

Meng, X., Yang, L., Kennedy, J. F., & Tian, S. (2010). Effects of chitosan and oligochitosan on growth of two fungal pathogens and physiological properties in pear fruit. *Carbohydrate Polymers, 81*(1), 70–75.

Min, S. K., Kwon, O. C., Lee, S., Park, K. H., & Kim, J. K. (2012). An antithrombotic fucoidan, unlike heparin, does not prolong bleeding time in a murine arterial thrombosis model: A comparative study of Undaria pinnatifida sporophylls and Fucus vesiculosus. *Phytotherapy Research, 26*(5), 752–757.

Moskovitz, J., Yim, M. B., & Chock, P. B. (2002). Free radicals and disease. *Archives of Biochemistry and Biophysics, 397*(2), 354–359.

Mukhamejanov, E., & Kurilenko, V. (2020). Fucoidan: A nutraceutical for metabolic and regulatory systems homeostasis maintenance. *World Journal of Advanced Research and Reviews, 6*(1), 255–264.

Muthukumar, J., Chidambaram, R., & Sukumaran, S. (2021). Sulfated polysaccharides and its commercial applications in food industries—A review. *Journal of Food Science and Technology, 58*(7), 2453–2466.

Muzzarelli, R. A., Boudrant, J., Meyer, D., Manno, N., DeMarchis, M., & Paoletti, M. G. (2012). Current views on fungal chitin/chitosan, human chitinases, food preservation, glucans, pectins and inulin: A tribute to Henri Braconnot, precursor of the carbohydrate polymers science, on the chitin bicentennial. *Carbohydrate Polymers, 87*(2), 995–1012.

Negm, N. A., Hefni, H. H., Abd-Elaal, A. A., Badr, E. A., & Abou Kana, M. T. (2020). Advancement on modification of chitosan biopolymer and its potential applications. *International Journal of Biological Macromolecules, 152*, 681–702.

Ngo, D. H., & Kim, S. K. (2013). Sulfated polysaccharides as bioactive agents from marine algae. *International Journal of Biological Macromolecules, 62*, 70–75.

Ngo, D. H., & Kim, S. K. (2014). Antioxidant effects of chitin, chitosan, and their derivatives. *Advances in Food and Nutrition Research, 73*, 15–31.

Ngo, D. H., Vo, T. S., Ngo, D. N., Kang, K. H., Je, J. Y., Pham, H. N. D., . . . Kim, S. K. (2015). Biological effects of chitosan and its derivatives. *Food Hydrocolloids, 51*, 200–216.

Nigam, P. S., & Singh, A. (2011). Production of liquid biofuels from renewable resources. *Progress in Energy and Combustion Science, 37*(1), 52–68.

Nishinari, K., & Fang, Y. (2017). Relation between structure and rheological/thermal properties of agar. A mini review on the effect of alkali treatment and the role of agaropectin. *Food Structure, 13*, 24–34.

Nishinari, K., & Fang, Y. (2021). Molar mass effect in food and health. *Food Hydrocolloids, 112*, 106110.

Nitta, S. K., & Numata, K. (2013). Biopolymer-based nanoparticles for drug/gene delivery and tissue engineering. *International Journal of Molecular Sciences, 14*(1), 1629–1654.

Nobe, R., Sakakibara, Y., Fukuda, N., Yoshida, N., Ogawa, K., & Suiko, M. (2003). Purification and characterization of laminaran hydrolases from Trichoderma viride. *Bioscience, Biotechnology, and Biochemistry, 67*(6), 1349–1357.

Nwe, N., & Stevens, W. F. (2004). Effect of urea on fungal chitosan production in solid substrate fermentation. *Process Biochemistry, 39*(11), 1639–1642.

Offei, F., Mensah, M., Thygesen, A., & Kemausuor, F. (2018). Seaweed bioethanol production: A process selection review on hydrolysis and fermentation. *Fermentation, 4*(4), 99.

Ogle, W. O., Speisman, R. B., & Ormerod, B. K. (2013). Potential of treating age-related depression and cognitive decline with nutraceutical approaches: A mini review. *Gerontology, 59*(1), 23–31.

Okamoto, H., Shiraki, K., Yasuda, R., Danjo, K., & Watanabe, Y. (2011). Chitosan-interferon-β gene complex powder for inhalation treatment of lung metastasis in mice. *Journal of Controlled Release, 150*(2), 187–195.

Onsøyen, E., & Skaugrud, O. (1990). Metal recovery using chitosan. *Journal of Chemical Technology and Biotechnology (Oxford, Oxfordshire: 1986), 49*(4), 395–404.

Park, H. Y., Park, S. H., Jeong, J. W., Yoon, D., Han, M. H., Lee, D. S., . . . Choi, Y. H. (2017). Induction of p53-independent apoptosis and G1 cell cycle arrest by fucoidan in HCT116 human colorectal carcinoma cells. *Marine Drugs, 15*(6), 154.

Park, S. C., Nam, J. P., Kim, J. H., Kim, Y. M., Nah, J. W., & Jang, M. K. (2015). Antimicrobial action of water-soluble β-chitosan against clinical multi-drug resistant bacteria. *International Journal of Molecular Sciences, 16*(4), 7995–8007.

Percival, E., & McDowell, R. H. (1967). *Chemistry and enzymology of marine algal polysaccharides*. Academic Press.

Percot, A., Viton, C., & Domard, A. (2003). Optimization of chitin extraction from shrimp shells. *Biomacromolecules, 4*(1), 12–18.

Pomin, V. H. (2014). Marine medicinal glycomics. *Frontiers in Cellular and Infection Microbiology, 4*, 5.

Pomin, V. H., & Mourão, P. A. (2008). Structure, biology, evolution, and medical importance of sulfated fucans and galactans. *Glycobiology*, *18*, 1016–1027.

Pradhan, B., Patra, S., Nayak, R., Behera, C., Dash, S. R., Nayak, S., et al. (2020). Multifunctional role of fucoidan, sulfated polysaccharides in human health and disease: A journey under the sea in pursuit of potent therapeutic agents. *International Journal of Biological Macromolecules*, *164*, 4263–4278.

Qureshi, D., Nayak, S. K., Maji, S., Kim, D., Banerjee, I., & Pal, K. (2019). Carrageenan: A wonder polymer from marine algae for potential drug delivery applications. *Current Pharmaceutical Design*, *25*, 1172–1186.

Raftery, R. M., Castaño, I. M., Chen, G., Cavanagh, B., Quinn, B., Curtin, C. M., . . . O'Brien, F. J. (2017). Translating the role of osteogenic-angiogenic coupling in bone formation: Highly efficient chitosan-pDNA activated scaffolds can accelerate bone regeneration in critical-sized bone defects. *Biomaterials*, *149*, 116–127.

Raftery, R. M., et al. (2015). Development of a gene-activated scaffold platform for tissue engineering applications using chitosan-pDNA nanoparticles on collagen-based scaffolds. *Journal of Controlled Release*, *210*, 84–94.

Resmi, R., Yoonus, J., & Beena, B. (2021). Anticancer and antibacterial activity of chitosan extracted from shrimp shell waste. *Materials Today: Proceedings*, *41*, 570–576.

Rhoades, J., & Roller, S. (2000). Antimicrobial actions of degraded and native chitosan against spoilage organisms in laboratory media and foods. *Applied and Environmental Microbiology*, *66*(1), 80–86.

Rioux, L., Turgeon, S. L., & Beaulieu, M. (2007). Characterization of polysaccharides extracted from brown seaweeds. *Carbohydrate Polymers*, *3*, 530–553.

Roberfroid, M., & Slavin, J. (2000). Nondigestible oligosaccharides. *Critical Reviews in Food Science and Nutrition*, *40*(6), 461–480.

Saharan, V., Sharma, G., Yadav, M., Choudhary, M. K., Sharma, S. S., Pal, A., . . . Biswas, P. (2015). Synthesis and in vitro antifungal efficacy of Cu-chitosan nanoparticles against pathogenic fungi of tomato. *International Journal of Biological Macromolecules*, *75*, 346–353.

Sanjeewa, K. A., Lee, J. S., Kim, W. S., & Jeon, Y. J. (2017). The potential of brown-algae polysaccharides for the development of anticancer agents: An update on anticancer effects reported for fucoidan and laminarin. *Carbohydrate Polymers*, *177*, 451–459.

Santos, V. P., Maia, P., Alencar, N. D. S., Farias, L., Andrade, R. F. S., Souza, D., . . . Campos-Takaki, G. M. (2019). Recovery of chitin and chitosan from shrimp waste with microwave technique and versatile application. *Arquivos do Instituto Biológico*, *86*.

Saravana, P. S., & Chun, B. S. (2017). Seaweed polysaccharide isolation using subcritical water hydrolysis. In *Seaweed polysaccharides* (pp. 47–73). Elsevier.

Saravanakumar, G., Jo, D. G., & H Park, J. (2012). Polysaccharide-based nanoparticles: A versatile platform for drug delivery and biomedical imaging. *Current Medicinal Chemistry*, *19*(19), 3212–3229.

Savin, S., Craciunescu, O., Oancea, A., Ilie, D., Ciucan, T., Antohi, L. S., . . . Oancea, F. (2020). Antioxidant, cytotoxic and antimicrobial activity of chitosan preparations extracted from Ganoderma lucidum mushroom. *Chemistry & Biodiversity*, *17*(7), e2000175.

Sharma, K., Somavarapu, S., Colombani, A., Govind, N., & Taylor, K. M. (2013). Nebulised siRNA encapsulated crosslinked chitosan nanoparticles for pulmonary delivery. *International Journal of Pharmaceutics*, *455*(1–2), 241–247.

Singh, P., Sivanandam, T. M., Konar, A., & Thakur, M. K. (2020). Role of nutraceuticals in cognition during aging and related disorders. *Neurochemistry International*, 104928.

Sithranga Boopathy, N., & Kathiresan, K. (2010). Anticancer drugs from marine flora: An overview. *Journal of Oncology*, *2010*.

Slavin, J. L. (2012). Structure, nomenclature, and properties of carbohydrates. In *Biochemical, physiological & molecular aspects of human nutrition*. Elsevier.

Souza, B. W., Cerqueira, M. A., Bourbon, A. I., Pinheiro, A. C., Martins, J. T., Teixeira, J. A., . . . Vicente, A. A. (2012). Chemical characterization and antioxidant activity of sulfated polysaccharide from the red seaweed Gracilaria birdiae. *Food Hydrocolloids*, *27*(2), 287–292.

Sudarshan, N. R., Hoover, D. G., & Knorr, D. (1992). Antibacterial action of chitosan. Food Biotechnology, 6(3), 257-272.

Suntornsuk, W., Pochanavanich, P., & Suntornsuk, L. (2002). Fungal chitosan production on food processing by-products. *Process Biochemistry*, *37*(7), 727–729.

Szymańska, E., & Winnicka, K. (2015). Stability of chitosan—a challenge for pharmaceutical and biomedical applications. *Marine Drugs*, *13*(4), 1819–1846.

Tanino, Y., Hashimoto, T., Ojima, T., & Mizuno, M. (2016). F-fucoidan from Saccharina japonica is a novel inducer of galectin-9 and exhibits anti-allergic activity. *Journal of Clinical Biochemistry and Nutrition*, *59*(1), 25–30.

Tanna, B., & Mishra, A. (2019). Nutraceutical potential of seaweed polysaccharides: Structure, bioactivity, safety, and toxicity. *Comprehensive Reviews in Food Science and Food Safety*, *18*(3), 817–831.

Tayel, A. A., Moussa, S. H., Salem, M. F., Mazrou, K. E., & El-Tras, W. F. (2016). Control of citrus molds using bioactive coatings incorporated with fungal chitosan/plant extracts composite. *Journal of the Science of Food and Agriculture*, *96*(4), 1306–1312.

Thakur, M. (2019). Marine bioactive components: Sources, health benefits, and future prospects. In *Technological processes for marine foods, from water to fork* (pp. 61–72). Apple Academic Press.

Tian, T., Chang, H., He, K., Ni, Y., Li, C., Hou, M., . . . Ji, M. (2019). Fucoidan from seaweed Fucus vesiculosus inhibits 2, 4-dinitrochlorobenzene-induced atopic dermatitis. *International Immunopharmacology*, *75*, 105823.

Tsvetkov, Y. E., Paulovičová, E., Paulovičová, L., Farkaš, P., & Nifantiev, N. E. (2020). Synthesis of biotin-tagged chitosan oligosaccharides and assessment of their immunomodulatory activity. *Frontiers in Chemistry*, *8*, 1083.

Tungtong, S., Okonogi, S., Chowwanapoonpohn, S., Phutdhawong, W., & Yotsawimonwat, S. (2012). Solubility, viscosity and rheological properties of water-soluble chitosan derivatives. *Maejo International Journal of Science and Technology*, *6*(2), 315.

Udayangani, R. M. A. C., Somasiri, G. D. P., Wickramasinghe, I., & Kim, S. K. (2020). Potential health benefits of sulfated polysaccharides from marine algae. *Encyclopedia of Marine Biotechnology*, *1*, 629–635.

Vaclavik, V. A., Christian, E. W., & Campbell, T. (2008). *Essentials of food science* (Vol. 42). Springer.

Van Dijken, J. P., & Scheffers, W. A. (1986). Redox balances in the metabolism of sugars by yeasts. *FEMS Microbiology Reviews*, *1*(3–4), 199–224.

van Hal, J. W., Huijgen, W. J. J., & López-Contreras, A. M. (2014). Opportunities and challenges for seaweed in the biobased economy. *Trends in Biotechnology*, *32*(5), 231–233.

Venkatesan, J., & Kim, S. K. (2010). Chitosan composites for bone tissue engineering—an overview. *Marine Drugs*, *8*(8), 2252–2266.

Vo, T. S., Ngo, D. H., & Kim, S. K. (2012). Potential targets for anti-inflammatory and anti-allergic activities of marine algae: An overview. *Inflammation & Allergy-Drug Targets (Formerly Current Drug Targets-Inflammation & Allergy) (Discontinued)*, *11*(2), 90–101.

Vongchan, P., Sajomsang, W., Kasinrerk, W., Subyen, D., & Kongtawelert, P. (2003). Anticoagulant activities of the chitosan polysulfate synthesized from marine crab shell by semi-heterogeneous conditions. *Science Asia*, *29*, 115–120.

Wang, B., He, C., Tang, C., & Yin, C. (2011). Effects of hydrophobic and hydrophilic modifications on gene delivery of amphiphilic chitosan based nanocarriers. *Biomaterials*, *32*(20), 4630–4638.

Wang, X., Yi, K., & Zhao, Y. (2018). Fucoidan inhibits amyloid-β-induced toxicity in transgenic Caenorhabditis elegans by reducing the accumulation of amyloid-β and decreasing the production of reactive oxygen species. *Food & Function*, *9*(1), 552–560.

Wen, Y., Gao, L., Zhou, H., Ai, C., Huang, X., Wang, M., et al. (2021). Opportunities and challenges of algal fucoidan for diabetes management. *Trends in Food Science & Technology*, *111*, 628–641.

Yegappan, R., Selvaprithiviraj, V., Amirthalingam, S., & Jayakumar, R. (2018). Carrageenan based hydrogels for drug delivery, tissue engineering and wound healing. *Carbohydrate Polymers*, *198*, 385–400.

Younes, I., & Rinaudo, M. (2015). Chitin and chitosan preparation from marine sources. Structure, properties and applications. *Marine Drugs*, *13*(3), 1133–1174.

Younes, I., Sellimi, S., Rinaudo, M., Jellouli, K., & Nasri, M. (2014). Influence of acetylation degree and molecular weight of homogeneous chitosans on antibacterial and antifungal activities. *International Journal of Food Microbiology*, *185*, 57–63.

Zhao, F., Liu, Q., Cao, J., Xu, Y., Pei, Z., Fan, H., et al. (2020). A sea cucumber (Holothuria leucospilota) polysaccharide improves the gut microbiome to alleviate the symptoms of type 2 diabetes mellitus in Goto-Kakizaki rats. *Food and Chemical Toxicology*, *135*, 110886.

Zheng, L. Y., & Zhu, J. F. (2003). Study on antimicrobial activity of chitosan with different molecular weights. Carbohydrate polymers, 54(4), 527-530.

2 Marine Algae in Diabetes and Its Complications

Suman Samaddar

CONTENTS

2.1 INTRODUCTION

Diabetes mellitus (DM) is a metabolic disorder linked with chronic hyperglycemia due to the relative or absolute deficiency in insulin hormone (Kooti et al., 2016). It is the third leading cause of death worldwide and associated with major complications such as diabetic nephropathy, neuropathy, retinopathy leading to adult blindness, and amputations due to diabetic foot ulcers (Bhattacharjee et al., 2014). With changes in global diet and lifestyle, the incidence and prevalence of diabetes mellitus have risen rapidly all over the world, and it is one of the major chronic diseases that cause human disability (Saeedi et al., 2019). According to International Diabetes Federation's (IDF's) latest (ninth edition) Global Diabetes Map, statistics suggest that the number of currently confirmed diabetes cases is over 463 million, and the average global growth rate is 51%. It is estimated that by 2045, the number of diabetics will reach 700.2 million (Lin et al., 2021). Both types of DM have different clinical and pathological features. However, type 2 DM is mostly prevalent among the world population, and it is associated with insulin resistance due to overproduction of glucose by the liver and underutilization of glucose by muscle and adipose tissues (Wang

DOI: 10.1201/9781003303916-2

et al., 2013). Insulin resistance is the primary cause of type 2 diabetes, according to the pathophysiology of the disease. Insulin resistance is more common in obese people than in healthy people, and it can be compensated for by the β-cells producing too much insulin. The activities of β-cells will be disrupted as a result of insulin overproduction, leading to chronic postprandial hyperglycemia and fasting hyperglycemia. Chronic hyperglycemia impairs the function of β-cells and worsens the state of insulin resistance (Kahn et al., 2006). In addition to chronic hyperglycemia, type 2 diabetes is linked to dyslipidemia, which disrupts the normal lipid metabolism process (Mooradian, 2009). Hence, to restore glucose metabolism and prevent long-term consequences, treatment methods for type 2 diabetes should be established.

Marine weeds are photosynthetic algae that are rich in nutrients and phytochemicals with a wide range of biological activity (Jiménez-Escrig and Goñi, 1999; Bocanegra, 2009) and are abundant in every ocean. Traditional Asian diets rich in algae have been linked to a lower prevalence of chronic diseases such as cancer, cardiovascular disease, and heart disease (Brown et al., 2014; Xu et al., 2017). There are three main classes, or phyla, of seaweed: Phaeophyceae (brown algae), Rhodophyta (red algae), and Chlorophyta (green algae). The majority of marine algae are red varieties, which make up 6000 species of organisms. Brown algae accounts for another 2000 species. Green varieties include 1200 species of the seaweed (Guiry and Guiry, 2019; Kim, 2011). The marine algae have been studied for biologically active components that include dietary fiber, minerals, lipids, proteins, omega-3 fatty acids, essential amino acids, terpenoids, sterols, polysaccharides, carotenoids, phycobilins, vitamins, tocopherol, and phycocyanins. In particular, the brown algae have a variety of biological compounds such as fucoidans, phycocolloids, and phlorotannins. These natural bioactive compounds possess antiviral, antifungal, antibacterial, antioxidant, anti-inflammatory, hypercholesterolemic, and hypolipidemic and antineoplastic properties (Cerna, 2011; Misurcova et al., 2012; Tabarsa et al., 2012). This chapter aims at discussing the potentials of marine algae in diabetes and its complications.

2.2 MARINE ALGAE AND DIABETES

Insulin and several oral antidiabetic medications such as sulfonylureas (glimepiride, glipizide), non-sulfonylureas (metformin), α-glucosidase inhibitors (acarbose, miglitol), thiazolidinediones (rosiglitazone, pioglitazone), dipeptidyl peptidase-4 inhibitors (saxagliptin, sitagliptin), incretin mimetics (albiglutide, dulaglutide), meglitinides (nateglinide, repaglinide), and SGLT-2 inhibitors (canagliflozin, dapagliflozin) are currently accessible treatments for type 2 diabetes. These medicines, on the other hand, have severe mechanism-based adverse effects such as hypoglycemia, flatulence, and worsening of gastrointestinal disorders. Marine algae, as alternatives, are known to possess a large number of bioactive chemicals with promising pharmacological and biological applications. Most of the brown seaweeds which belong to the genus *Ecklonia* and family Lessoniaceae have been reported to exhibit antidiabetic activities mainly through the inhibitory action of α-amylase and α-glucosidase enzymes due to the presence of different phlorotannins such as eckol, dieckol, 6,6′-bieckol, phlorofucofuroeckol-A, phloroglucinol, and 7-phloroeckol. The methanol extract of *Ecklonia cava* exhibited potent antidiabetic activity through the inhibitory action on the α-glucosidase enzyme (IC_{50} 10.7 μM) compared to the standard acarbose due to the presence of phlorotannin compounds, namely, dieckol, fucodiphloroethol G, 6,6′-bieckol, 7-phloroeckol, and phlorofucofuroeckol-A. Similar phlorotannins isolated from the methanol extract of *Ecklonia stolonifera* were reported to possess an inhibitory activity against α-glucosidase enzymes, which was attributed to the presence of phlorofucofuroeckol-A, dieckol, and 7-phloroeckol (Lee and Jeon, 2013). The phlorotannin eckol (IC_{50} 11.16 μM) isolated from *Ecklonia maxima* exhibited potent α-glucosidase inhibitory activity compared to the phloroglucinol (IC_{50} 1991 μM). Similarly, fucofuroeckol-A and dioxinodehydroeckol isolated from *Ecklonia bicyclis* were also reported to have an inhibitory effect on α-amylase and α-glucosidase enzymes (Abdelsalam et al., 2019). Eckol, dieckol, and 7-phloroeckol isolated from *Eisenia bicyclis* also strongly inhibit the α-amylase enzymes by 87% at 1 mM of concentration in addition to the inhibitory activity on α-glucosidase enzyme and formation of advanced glycation end products (Moon et al., 2014). *Ishige okamurae* is a marine brown algae that belongs to the family Ishigeaceae and is reported to have potent antidiabetic activity through the mechanism of α-amylase and α-glucosidase inhibition. Recent studies have found that *I. okamurae* are rich in phlorotannins such as phloroglucinol, diphlorethohydroxycarmalol, 6–6-bieckol, octaphlorethol A, and ishophloroglucin, which are responsible for the antioxidant and antidiabetic activities (Yang et al., 2019). Brown algae *Ascophyllum nodosum* exhibited hypoglycemic activity by inhibiting α-amylase and α-glucosidase. Fractionation and liquid chromatography-mass spectrometry (LC-MS) analysis of the extract revealed the presence of a series of phlorotannin structures which exhibit the inhibitory activity on α-amylase and α-glucosidase enzymes (Nwosu et al., 2011). The phlorotannins present in marine brown algae *Sargassum hystrix* have also been found to possess α-amylase and α-glucosidase inhibition properties. Further, the antidiabetic activity of *S. hystrix* has been confirmed using streptozotocin-induced rats. *S. hystrix* was found to reduce preprandial (186.4 mg/ml) and postprandial (186.9 mg/ml) blood glucose levels significantly in diabetic rat groups (Gotama et al., 2018). Similarly, the edible seaweed *Sargassum polycystum* has been found to possess a hypoglycemic effect in streptozotocin-induced type 2 diabetic rats. The extracts of *S. polycystum* significantly reduce the blood glucose level in diabetic rats by 27.8% and 35.2% compared to the diabetic rats treated with 250 mg/kg of metformin (84.76%). Moreover, compared to the histopathology of the pancreas in the diabetic control group, atrophy and abnormalities of the nuclei and the cytoplasm of pancreatic cells have been significantly suppressed with the treatment of ethanol (150 mg/kg), aqueous (300 mg/kg) extracts, and metformin (Motshakeri et al., 2014). The brown seaweeds *Padina boergesenii* and *Padina tetrastromatica* have exhibited α-amylase and α-glucosidase inhibition properties (Senthilkumar et al., 2014). *Fucus vesiculosus* is a brown algae with prominent antidiabetic activities.

Ultra-high-pressure liquid chromatography coupled to mass spectrometry (UHPLC-MS) analysis revealed that the ethyl acetate fraction of *Fucus vesiculosus* has phlorotannins such as fucols, fucophlorethols, fuhalols, fucofurodiphlorethol, fucofurotriphlorethol, and fucofuropentaphlorethol showed promising antidiabetic activity through the inhibitory activity of α-glucosidase (IC_{50} 0:82 ± 0.05 μg/ml; acarbose IC_{50}206.6 ± 25.1 μg/ml) and α-amylase (IC_{50}2.8 ± 0.3 μg/ml; acarbose IC_{50}0.7 ± 0.2 μg/ml) enzymes (Catarino et al., 2019). Porous starch encapsulated fucoxanthin extracted and purified from *Sargassum angustifolium* displayed antidiabetic activity in streptozotocin and nicotinamide-induced type 2 diabetic mice. In addition, total cholesterol, triglyceride, and low-density lipoprotein were also lowered and high-density lipoprotein (HDL) was elevated in the treated groups (NajmeOliyaei et al., 2021).

As the main component of algae, bromophenols may be responsible for the reported antidiabetic activity of many marine organisms. These compounds arise from the tendency of the phenol moiety to undergo electrophilic bromination to varying degrees. Bromophenols isolated from the red algae *Rhodomela confervoides* have been found to possess potent in vitro protein-tyrosine phosphatase 1B (PTP1B) inhibitory effects, with IC_{50} values fluctuating between 0.8 μM and 4.5 μM. This change in potencies could be attributed to the bromine content of these compounds or to their side chains (Shi, Guo et al., 2013; Shi, Li et al., 2012). Other brominated phenols isolated from the red algae *Symphyocladia latiuscula* also exerted positive inhibitory activity (Liu et al., 2011).

Fucoidans are commonly sulfated marine polysaccharides that have a wide spectrum of bioactivities. *Ecklonia cava, Fucus vesiculosus*, and *Cladosiphon okamuranus* were the most widely studied species for fucoidans (Wijesinghe and Jeon, 2012). Fucoidans are a group of fucans, i.e., sulfated polysaccharides extracted from brown seaweeds and are characterized by fucose-rich sulfated groups. Other examples of fucans are ascophyllans (xylofucoglycuronan and xylofucomanuronan) and sargassans (glycuronofucogalactan). The position of sulfate groups in marine sulfated polysaccharides has a significant impact on their beneficial biological effects. Fucoidans isolated from various seaweed species have diverse bioactive capabilities depending on their compositional structure, charge density, distribution, and bonding of the sulfate substitution (Ale et al., 2011). Fucoidans isolated from brown seaweeds had potential beneficial effects on diabetes. Fucoidans from *Ascophyllum nodosum* and *Turbinaria ornate* displayed α-amylase inhibition activities (Kim et al., 2015; Lakshmanasenthil et al., 2014). In addition, fucoidans have been verified to provide pancreatic protection. Fucoidan from *Acaudina molpadioides* protected pancreatic islet cells against apoptosis via inhibition of inflammation in type 2 diabetic mice (Wang et al., 2016). In streptozotocin-treated β-cells and mice, fucoidan ameliorated pancreatic β-cell death and impaired insulin synthesis via the Sirt-1-dependent pathway (Yu et al., 2017). It stimulated insulin secretion and provided pancreatic protection via the cyclic adenosine monophosphate (cAMP) signaling pathway (Jiang et al., 2015). Additionally, insulin sensitivity was enhanced by increasing the expression of diabetes-related genes in 3T3-L1 adipocytes (Kim et al., 2007). Low-molecular-weight fucoidan also improved the action of insulin via adenosine-activated protein kinase (AMPK) stimulation (Jeong et al., 2013).

MicroRNAs (miRNAs) and other small non-coding RNAs serve as endogenous gene regulators through binding to specific sequences in RNA and modifying gene expression toward up- or down-regulation. miRNAs have become compelling therapeutic targets and play crucial roles in regulating the process of insulin resistance. Polysaccharides or oligosaccharides from marine seaweed or resources are known to affect the levels of miRNAs. In subjects with metabolic syndrome, specifically obesity, prediabetes, and type 2 DM, a large amount of miRNAs are deregulated (e.g. miR-21, miR-24.1, miR-27a, miR-28–3p, miR-29b, miR-30d, miR-34a, miR-93, miR-126, miR-146a, miR-148, miR-150, miR-155, and miR-223) (Nunez et al., 2016). The miR-29 family has emerged as important regulators of glucose metabolism. It was among the most abundantly expressed miRNAs in the pancreas and liver and showed strong regulatory functions in obesity and diabetes (Dooley et al., 2016). The miR-29 family of miRNAs in humans comprises three mature members, miR-29a, miR-29b, and miR-29c. MiR-29 has been shown to affect glucose metabolism, lipid metabolism, and insulin responsiveness in skeletal muscle, while both miR-29a and miR-29c could regulate glucose uptake and insulin-stimulated glucose metabolism (Massart et al., 2017). It has been shown that fucoidan inhibited breast cancer progression by dual regulation of the miR-29c/ADAM-12 and miR-17–5p/PTEN axes (Wu et al., 2016). Furthermore, fucoidan has been confirmed to markedly up-regulate miR-29b in human hepatocellular carcinoma cells to regulate the DNA methyltransferase 3B-metastasis suppressor 1 (DNMT3B-MTSS1) axis and inhibit epithelial-mesenchymal transition (EMT) (increased E-cadherin and decreased N-cadherin). DNMT3B is an important downstream target of miR-29b. Induction of miR-29b results in suppression of DNMT3B and a consequent increase in MTSS1, which is usually repressed in human hepatocellular carcinoma. Fucoidan is a novel target of DNMT3B (Yan et al., 2015). Based on the reported studies, it was confirmed that fucoidan could up-regulate the level of miR-29b andmiR-29c and down-regulate the level of miR-17–5p, which would contribute to the treatment of type 2 diabetes. Several clinical trials conducted on the potential health benefits of regular algae consumption in T2DM are referenced and detailed in Table 2.1.

2.3 DIABETIC COMPLICATIONS

Complications involving the small and large blood vessels are likely to occur in uncontrolled diabetes. The danger of developing diabetic complications depends upon the period and harshness of diabetes. Complications of the microvasculature in diabetes include retinopathy, nephropathy, and neuropathy. Retinopathy is possibly the most common complication of chronic diabetes, known to affect around 10,000 people and producing fresh incidences of blindness every year. Aldose reductase participates in the development of microvascular and macrovascular impediments. Hyperactivity of aldose reductase in hyperglycemic episode results in the accumulation of sorbitol

TABLE 2.1
Clinical Trials on the Health Benefits of Algae Consumption in Type 2 Diabetes Mellitus (T2DM)

Study	Study Characteristics	Algae and Content	Biological Activity	Reference
Intervention trial	Diet with large amount of seaweed 417 male Japanese T2DM 65 years or older	Total vegetable intake >150 g of daily total vegetable >200 g of total vegetable intake Green vegetable intake	↓ HbA1c, TG, waist circumference ↓ HbA1c ↓ Serum TG ↓ HbA1c ↓ Body mass index, TG, waist circumference	(Takahashi et al., 2012)
Double-blind, randomized, placebo-controlled crossover study	23 participants 19–59 years old	Blend of *Ascophyllum nodosum* and *Fucus vesiculosus*	↓ Insulin concentrations ↑ Insulin sensitive	(Murray et al., 2018)
Randomized crossover trial	26 participants	*Undaria pinnatifida* (4 g dry alga) + Rice (200 g)	↓ Postprandial glycemia ↓ Insulin levels	(Yoshinaga and Mitamura, 2019)
Randomized crossover study	12 participants	70 g Mekabu (*sporophylls* of *Undaria pinnatifida*)	↓ Postprandial glycemia ↓ Glucose area under the curve	(Tanemura et al., 2014)
Randomized controlled trial	12 overweight, healthy males Aged 40 year	*Ascophyllum nodosum* enriched bread (4%)	↓ Energy intake	(Hall et al., 2012)
Randomized controlled trial	10 healthy male volunteers, studied on three occasions	Agar (2.0 g)	↓ Delay gastric emptying No effect on the postprandial glucose response	(Sanaka et al., 2007)
Randomized controlled trial	76 obese patients with type 2 diabetes 12 weeks	Agar (180 g) + Traditional Japanese food	↓ HbA1c, ↓ Visceral fat area, subcutaneous fat area, total body fat, ↓ Insulin area under the curve after oral glucose tolerance test ↓ Total cholesterol	(Maeda et al., 2004)
Randomized controlled trial	48 overweight or obese participants 10 days	Sodium alginate from *Laminaria digitata*	No effect on gastric motor functions, satiation, appetite, or gut hormones	(Odunsi et al., 2010)

(Continued)

TABLE 2.1 *(Continued)*
Clinical Trials on the Health Benefits of Algae Consumption in Type 2 Diabetes Mellitus (T2DM)

Study	Study Characteristics	Algae and Content	Biological Activity	Reference
Randomized placebo-controlled trial	176 participants 5 weeks	Fiber supplements of alginate + balanced 1200 Kcal diet	↓ Body weight	(Birketvedt et al., 2005)
Double-blind, placebo-controlled, randomized cross-over trial	38 healthy adults (Asian and non-Asian)	Polyphenol-rich *Fucus vesiculosus* extract (500–200 mg)	↑ Risk of insulin resistance among Asian populations	(Murray et al., 2018)

in cells. Sorbitol, being impermeable, exerts hyperosmotic stress and causes pain. Non-enzymatic glycosylation of important plasma proteins attributes to the formation of advanced glycated end (AGE) products. Oxidative stress also contributes to cell injury in hyperglycemia. Antioxidant treatment may ease some vascular complications connected to diabetes. Diabetic retinopathy can be background or proliferative. Background retinopathy typifies tiny hemorrhages in the retinal layer at the center appearing as "dots", hence, called "dot hemorrhages". Proliferative retinopathy is symbolized by neoangiogenesis on the retinal surface causing vitreous hemorrhage, sometimes identified by "cotton wool spots". Diabetic nephropathy is one of the main etiologies of renal failure, described by proteinuria above 500 mg within 24 hours of diabetes onset, but this is followed by hypoproteinuria, or "microalbuminuria". This phenomenon is frequent to both types of diabetes. The renal pathological changes include increased thickness of the basement membrane of glomeruli, formation of mesangial nodules (Kimmelsteil-Wilson bodies), and microaneurysm, besides others. The core mechanism of damage probably involves a few or all of the etiologies as diabetic retinopathy (Fowler, 2008).

2.3.1 Diabetic Neuropathy

Diabetic neuropathy is the earliest and most frequent complication of the three major complications associated with diabetes, but it is often not diagnosed until the disease has seriously progressed (Feldman et al., 2019). Usually more than 50% of patients with 25 years of diabetes or more are affected that makes it a very commonly occurring disease affecting the nervous system. Diabetic neuropathy has an effect on all peripheral nerves, including motor neurons, pain fibers, and the autonomic nervous system. Hence, it can influence all systems and organs innervated by the nerves. Neuropathy due to diabetes is classified as autonomic, peripheral, focal, and proximal, out of which peripheral neuropathy is most commonly occurring.

Peripheral neuropathy leads to predominant sensory impairment, and its symptoms are hyperesthesia, pain, and a gradual loss of sensation due to the loss of nerve fibers. When hypoalgesia occurs, trauma and mechanical irritation cannot be noticed; as a result, foot ulcers and gangrene may occur, leading to amputation of the lower limbs (Pop-Busui et al., 2017). Diabetic neuropathy is a multifactorial disease. The probable etiology comprises hyperglycemia, non-enzymatic glycation of proteins, activation of a polyol pathway, free radicals and oxidative stress, a decrease in nitric oxide (NO) levels, and activation of protein kinase C-β (PKC-β) (Feldman et al., 2019; Pop-Busui et al., 2017; Martin et al., 2006; Galer et al., 2000). Studies show that these factors perform synergistically (Feldman et al., 1997).

2.3.1.1 Polyol Pathway and Advanced Glycated End Products

The polyol pathway of glucose metabolism plays a crucial function in the development of neuropathy (Gabbay, 1973). It is an alternative route of glucose metabolism in which the enzyme aldose reductase catalyzes the reduction of glucose to sorbitol, then to fructose by sorbitol dehydrogenase. Aldose reductase (AR) requires NADPH as a co-factor, and sorbitol dehydrogenase (SDH) needs NAD^+. During hyperglycemia, sorbitol accumulates in AR-containing tissues, as it is impermeable to the cell membranes and cannot diffuse out and, hence, creates hyperosmotic stress on the cell, thereby inducing neuropathic pain (Niimi et al., 2021; Kinoshita and Nishimura, 1988). Treatment with inhibitors of aldose reductase has been shown to prevent various complications, including nephropathy, neuropathy, and cataractin animal models (Oates and Mylari, 1999). It is reported that the AR inhibitor epalrestat prevents high glucose–induced smooth muscle cell proliferation and hypertrophy (Yasunari et al., 1995), thereby preventing their dysfunction and remodeling (Tawata et al., 1992). Accumulation of intracellular sorbitol and fructose due to polyol activation leads to diminution of other organic electrolytes like taurine and *myo*-inositol that regulate cellular osmolality (Stevens et al., 1993). Lessening of *myo*-inositol in the peripheral nerves gets in the way of phosphoinositide production, leading to inadequate diacylglycerol to sustain the content of protein kinase C (PKC) essential for Na+/K+-ATPase activation (Zhu and Eichberg, 1990; Greene et al., 1987). Amendments in PKC activation also interfere with an important myelin protein's (PO) phosphorylation of peripheral nerves and play an important pathogenetic role in primary segmental demyelination (Row-Rendleman and Eichberg, 1994). Enhanced activity of vascular PKC-β is thought to play a noteworthy role in microvascular complications.

Proteins or lipids are modified by glycosylation non-enzymatically in hyperglycemia to form AGEs and are subsequently oxidized. Schiff bases and Amadori products are formed due to early oxidation and glycation processes. Persistent glycation leads to molecular rearrangements that aids in AGE formation (Schmidt et al., 1994). Glycated products generate reactive oxygen species (ROS), engage receptors on the cell surface, and cross-link them. Important factors critical to AGE formation incorporate the rate of turnover of proteins for glycoxidation, the extent of hyperglycemia, and oxidative stress in the

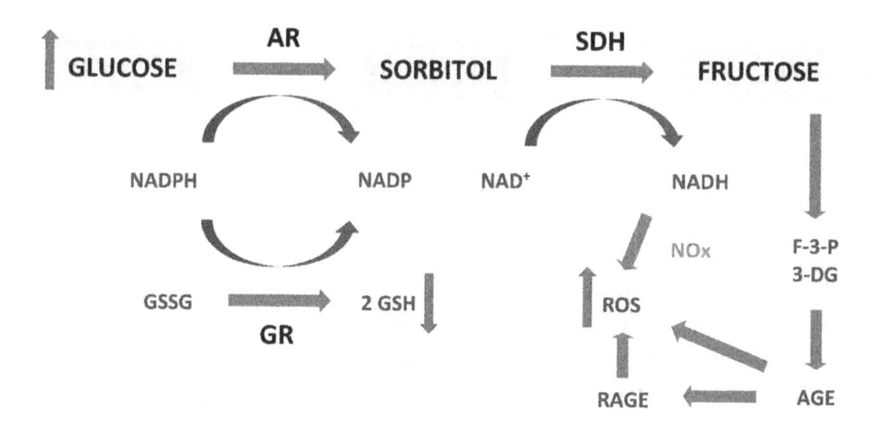

FIGURE 2.1 Polyol pathway and formation of AGEs.

milieu (Figure 2.1). If these conditions prevail, glycation and oxidation of both intracellular and extracellular proteins are certain. The AGE formation (by the Maillard reaction) happens due to the reaction between Schiff bases and Amadori product with the -NH_2 moieties of biomolecules (Giovino et al., 2020). During Amadori reorganization, these intermediary carbonyl groups that are highly reactive accumulate and are known as α-dicarbonyls or oxoaldehyde. These are the products of 3-deoxyglucosone and methylglyoxal. Such accumulation is called "carbonyl stress". The α-dicarbonyls react with -SH, -NH_2, and guanidine functional groups resulting in browning, cross-linking, and denaturation of the target proteins. They also react with arginine and lysine to form stable, non-fluorescent AGE products such as N-α-(carboxymethyl)lysine (CML). Protein glycation increases free radical activity that brings forth bimolecular damage in diabetes. AGEs act as initiators of a lot of abnormal cellular and tissue responses, such as the illicit expression of growth factors, extracellular matrix accumulation, and induction of cell death (Suzuki et al., 1999).[

2.3.1.2 Oxidative Stress and Free Radicals

Non-enzymatic glycation of proteins by glucose may form oxidants (Baynes and Thorpe, 1999). The activation of the polyol pathway brings about oxidative stress by reduction of NADPH that leads to depletion of reduced glutathione reductase and an increase in its oxidized counterpart, as NADPH is a common co-factor to both GSH and AR (De Mattia et al., 1994). Hyperglyaemia-induced increased PKC activation enhances oxidative stress due to activation of mitochondrial NADPH oxidase (Inoguchi et al., 2000). This leads to the formation of hydrogen peroxide (H_2O_2) and hydroxyl radical ($OH^{\cdot-}$) from the conversion of the superoxide anion ($O_2^{\cdot-}$) (Nishikawa et al., 2000). Polyol pathway activation also leads to decreased NADPH/NADP+ ratios, increasing oxidative stress. This happens as regeneration of the cellular antioxidant GSH is reduced and also reducing the activity of catalase breaking down H_2O_2 to H_2O as NADPH gets depleted. Additionally, inflamed vascular tissue stimulates inducible nitric oxide synthase (NOS) expression in smooth muscle cells and macrophages, consequently forming the free radical NO that reacts with superoxide ions to form highly reactive peroxynitrite. Lipid peroxidation, oxidation of low-density lipoprotein (LDL), and protein nitration are the lethal effects of peroxynitrite (Griendling and FitzGerald, 2003). Oxidative-reductive changes of transcription factors and enzymes and acutely, as well as fluctuating NADPH/NADP+ ratios, mediate cellular signaling under normal physiological conditions. Hydrogen peroxide originating from metabolism in mitochondrial activates the c-Jun N-terminal kinase (JNK) and inhibits the synthesis from glycogen synthase (Nemoto et al., 2000; Sundaresan et al., 1995). Oxidant production persistently may adversely affect cellular function and physiology. Enhanced PKC or inducible NO (iNO) activities may alter the normal functioning of affected tissues. Production of H_2O_2 in normal physiological levels encourages cell cycle progression responding to growth factors, while its exorbitant production may halt DNA synthesis through the induction of Cdc25c (Savitsky and Finkel, 2002). Excessive oxidative stress can decrease the neural conductivity of nerves affected by diabetes (Hounsom et al., 2001). Breaks in DNA strands may also be induced by oxidative stress in individual cells, thereby inducing death (Du et al., 2003). The mechanism of diabetic neuropathy contributed by various pathways is summarized in Figure 2.2.

2.3.1.3 Marine Algae in Diabetic Neuropathy

Inflammation has been found to be the pathophysiological mechanism underlying many chronic diseases such as cardiovascular disease, diabetes, certain cancers, arthritis, and neurodegenerative diseases (Allen and Barres, 2009). Numerous studies have documented anti-inflammatory activities of marine algae in vitro and in vivo (Abad et al., 2008). However, scientific analysis of anti-neuroinflammatory activity of marine algae has been poorly carried out and until now only a few studies were reported. *Ecklonia cava* has been reported to possess anti-inflammatory activity (Jung, Ahn et al., 2009). *E. cava* was able to suppress the levels of pro-inflammatory mediators such as NO, prostaglandine-E_2 (PGE_2), and pro-inflammatory cytokines (tumor necrosis factor-α [TNF-α], interleukin-6 [IL-6], and interleukin-1β [IL-1β]) in lipopolysaccharides (LPS)-stimulated BV2 cells by blocking nuclear factor-κB (NF-κB) and mitogen-activated protein kinases (MAPKs) activation (Jung, Ahn et al., 2009; Jung, Heo et al., 2009). Furthermore, *Neorhodomela aculeate* decreased NO production and inhibited iNOS synthase expression in interferon-gamma (IFN-γ)–stimulated BV2 cells (Lim et al., 2006). Another study conducted by Cui et al. provide the first evidence that fucoidan isolated from *Laminaria japonica* has a potent inhibitory effect against LPS-induced NO production in BV2 cells. Fucoidan at a concentration of 125 μg/mL significantly inhibited NO production to 75% (Cui et al., 2010). NO is a cytotoxic, short-lived, highly diffusible signaling molecule (Heales et al., 1999). A number of studies demonstrated that NO generated by iNOS causes injury and cell death of neurons and oligodendrocytes in the central nervous

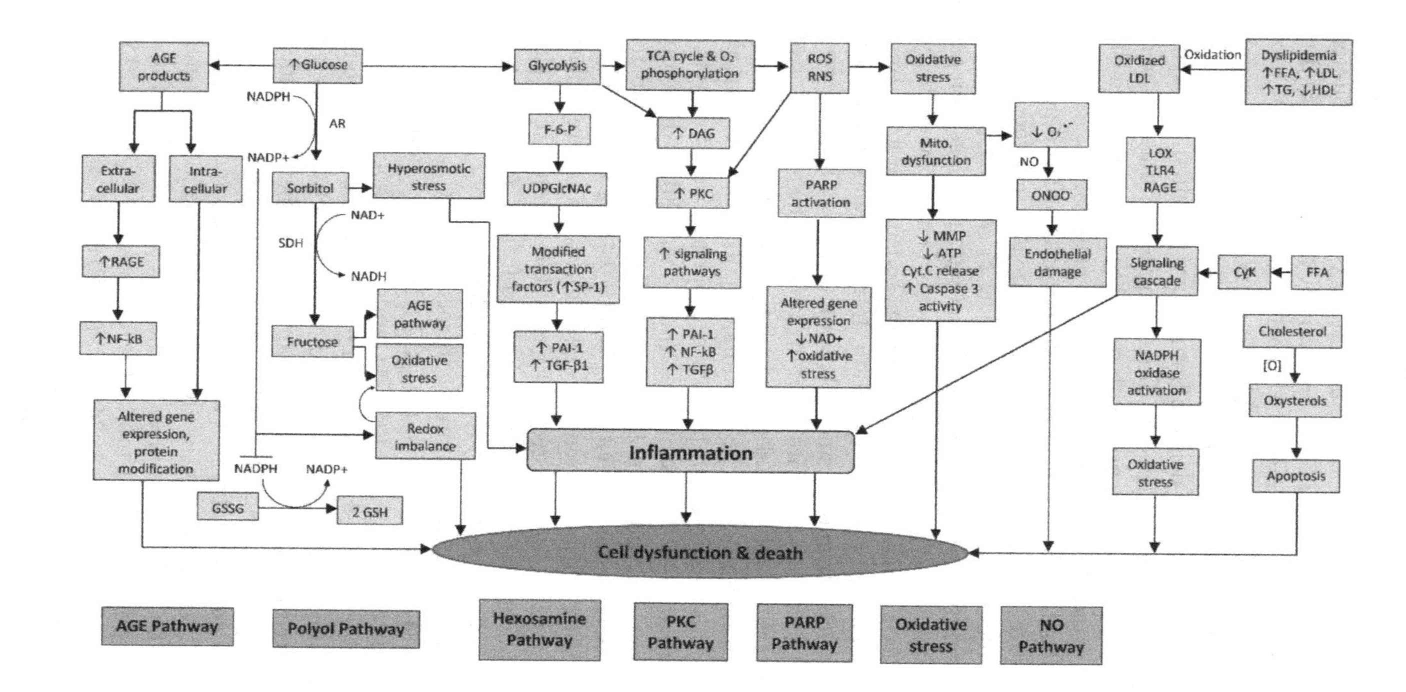

FIGURE 2.2 Mechanism of diabetic neuropathy.

system (CNS); hence, NO is implicated in the pathogenesis of various neurodegenerative diseases (Heales et al., 1999; Lee et al., 2000). *Ulva conglobate* methanolic extracts were able to suppress the expression of pro-inflammatory enzymes, iNOS, and cyclooxygenase-2 (COX-2), which accounted for the large production of NO and PGE2, respectively (Jin et al., 2006). Fucoidan, isolated from *Fucus vesiculosus*, has been found to protect rat cholinergic neuronal death induced by $A\beta_{1-42}$ by blocking the activation of caspase-9 and caspase-3, thereby inhibiting apoptosis (Jhamandas et al., 2005). Though these studies depict the anti-neuroinflammatory potential of marine algae, they were not conducted on specific disease models of diabetic neuropathy, and hence, there exists a gray area with regard to information on the neuroprotective potential of marine algae in diabetic neuropathy.

The polyphenolic fraction of *E. cava* has been reported to possess considerable neuroprotective activity in diabetic neuropathy induced in experimental animals (Samaddar and Koneri, 2019a). The *E. cava* polyphenols (ECPPs) were found to inhibit aldose reductase (AR) activity, as well as their expression in diabetic animals, thereby improving the nerve conduction velocity (NCV), compound muscle action potential (CMAP), and muscle grip strength and reduced the threshold of pain nociception. Improvements in the sciatic nerve Na^+K^+-ATPase activity and intraneural accumulation of sorbitol, an index of AR overactivity, were evident with ECPP treatment. The production of pro-inflammatory cytokines (IL-6, IL-1 β, and TNF-α) and expression of PKC were also diminished significantly. The brominated polyphenols of *Symphyocladia latiuscula* were also reported to possess neuroprotective activity in diabetic neuropathy similar to ECPPs (Samaddar and Koneri, 2019b). Five brominated polyphenols were isolated from *S. latiuscula* that were reported to possess aldose reductase inhibitory activity (IC_{50} is 0.11, 0.4, 0.4, 1.15, and 0.25 μg/mL) (Xiukun and Ming, 2012) and were suggested for use in the treatment of complications of diabetes, such as eye and nerve damage in T2DM patients. In another experiment, five bromophenols isolated from *S. latiuscula* have shown significant AR inhibitory activity (Wei et al., 2005).

The phlorotannins such as phlorofucofuroeckol-A (IC_{50} 2.4×10^3 μM), eckol (IC_{50} 1.6×10^3 μM), phloroglucinol (IC_{50} 2.4×10^3 μM), fucofuroeckol A (IC_{50} 7.4×10^2 μM), dieckol (IC_{50} 7.4×10^2 μM), and 8,8′-bieckol (IC_{50} 6.9×10^2 μM) isolated from *E. cava* have been reported to exhibit a potent inhibitory effect on the formation of fluorescence-bound advanced glycation end products (AGEs) compared to the reference drug aminoguanidine hydrochloride (IC_{50} 8.1×10^3 μM) (Sugiura et al., 2017). Phlorotannins in the methanol extract of brown algae *Padina pavonica*, *Sargassum polycystum*, and *Turbinaria ornata* inhibited glucose-induced protein glycation and formation of protein-bound fluorescent AGEs. The results revealed that the phlorotannins present in *P. pavonica* have a potent ability to inhibit the formation of AGEs (IC_{50} 15.16 ± 0.26 μg/ml) than *S. polycystum* (IC_{50} 35.245 ± 2.3 μg/ml) and *T. ornata* (IC_{50} 22.7 ± 0.3 μg/ml) compared to the standard thiamine (IC_{50} 263 μg/ml) and phloroglucinol (IC_{50} 222.33 μg/ml). In addition, further in vivo studies confirmed the hypoglycemic effect of phlorotannins present in brown algae *P. pavonica, S. polycystum*, and *T. ornate*, which exhibited inhibitory activity on the formation of AGEs in *Caenorhabditis elegans* (a nematode) with induced hyperglycamia (Shakambari et al., 2015).

2.3.2 Diabetic Retinopathy

Diabetic retinopathy is classically described by progressive alterations in the microvasculature that lead to retinal ischemia, neovascularization, altered retinal permeability, and macular edema. Diabetic retinopathy is now recognized to be an inflammatory, neurovascular complication, with neuronal injury/dysfunction preceding clinical microvascular damage. Importantly, the same pathophysiologic mechanisms that damage the pancreatic β-cell (e.g., inflammation, epigenetic changes, insulin resistance, fuel excess, and abnormal metabolic environment) also lead to cell and tissue damage, causing organ dysfunction and elevating the risk of all complications, including diabetic retinopathy (Duh et al., 2017).

The retina is the most metabolically active tissue in the body, making it very susceptible to oxidative stress both from light-induced electron injury and oxygen free radical production leading to increased inflammation. Indeed, diabetes and its associated hyperglycemia, insulin resistance, dyslipidemia, etc., all lead to altered biochemical pathways (polyol, AGEs, PKC, hexosamine, and renin-angiotensin system) that stimulate glial cell dysfunction (Kusuhara et al., 2018; Xu et al., 2018). This dysfunction leads to increased inflammatory cytokines and chemokine production, aberrant growth factor signaling and ROS, resulting in neuroglial degeneration and vascular dysfunction and associated alteration of the blood-retinal barrier, hypoxia, and vascular permeability resulting in edema and angiogenesis. This ultimately leads to the development and progression of diabetic retinopathy (Kusuhara et al., 2018).

Glucose oxidation makes the retina extremely susceptible to the generation of oxidized and NO_2 species (ROS/RNS). Further, the high tissue content of polyunsaturated fatty acids, oxygen uptake, and glucose oxidation make the retina extremely susceptible to the generation of ROS/RNS that results in modification of proteins, peroxidation of lipids, and DNA injury, mainly from mitochondrial dysfunction in the involved neurovascular unit cells (Pan et al., 2008). Furthermore, impaired antioxidant defense systems, including reduced enzymes such as catalase and glutathione peroxidase and superoxide dismutase, also lead to generation of retinal ROS/RNS that are exacerbated by variations in tissue exposure to glucose (Al-Shabrawey and Smith, 2010.). Inflammatory cytokines (e.g., TNF-α, IL-6, IL-8, IL-1β, etc.) and chemokines (e.g., monocyte chemoattractant protein-1, intercellular adhesion molecule [ICAMs]) are up-regulated in the serum, as well as ocular samples of diabetes patients, and are correlated with retinopathy severity (Gouliopoulos et al., 2018). Multiple interconnected pathways (e.g., polyol, AGEs, PKC, renin-angiotensin system, and hexosamine pathway) that are activated by diabetes increase the expression of inflammatory and angiogenic mediators, thereby inducing aberrant growth factor signaling. This is directedly linked to the neurodegeneration of vascular dysfunction (Rübsam et al., 2018). Balance between up-regulated and down-regulated neuroprotective factors in the diabetic retina plays a key role in the health of the retinal neurons. In early stages of diabetic retinopathy, down-regulation of key factors, including pigment epithelium-derived factor, somatostatin, glucagon-like peptide 1 (GLP-1), and other neurotrophic factors, is counterbalanced by an up-regulation of VEGF (Simó et al., 2018). Integrity of the blood-retinal barrier (BRB) is compromised by alterations in the neurovascular unit, leading to changes in permeability of endothelial cells and surrounding pericytes of retinal capillaries that leads to increased secretion of cytokines and growth factors that result in low-grade intravascular and extravascular inflammation with capillary basement membrane thickening, ICAM secretion, and white blood cell obstruction and degradation (Simó et al., 2018). Indeed, the disruption of the BRB is one of the most important events in the early stages of diabetic retinopathy (Kusuhara et al., 2018). The breakdown of the inner BRB results in the recognized microvascular leakage of the inner retinal vessels, producing macular edema and lipid exudates, but there appears also to be an important role of the outer BRB as well (Rübsam et al., 2018).

2.3.2.1 Current Treatments and Approaches

Current treatments for diabetic retinopathy include local treatments: 1) laser treatments (panretinal laser photocoagulation therapy, focal/grid laser) indicated in peripheral ischemia associated with neovascular proliferation (Simó and Hernández, 2014); 2) vitrectomy for vitreous hemorrhage and cryotherapy for retinal detachment (Bressler et al., 2018); 3) intravitreal steroids (fluocinolone acetonide insert [Retiser, Iluvien], dexamethasone [Ozurdex], triamcinolone acetonide [Kenacort Retard], and anti-TNF-α [infliximab, adalimumab]) for inflammation (Reddy et al., 2018), intravitreal injections of anti–vascular endothelial growth factor (VEGF) antibodies (aflibercept [Eylea], ranibizumab [Lucentis], pegaptanib [Macugen]), bevacizumab [Avastin]) against VEGF-mediated retinopathy (Wang and Lo, 2018). Systemic treatments include traditional approaches to the treatment of hyperglycemia (Wang and Lo, 2018).

2.3.2.2 Marine Algae in Diabetic Retinopathy

The ethyl acetate fraction of a brown algae *Ecklonia stolonifera* was found to inhibit rat lens AR more effectively than other fractions. Further isolation revealed that the inhibitory activity on rat lens AR was mainly due to the presence of phlorotannins such as 7-phloroeckol and 2-phloroeckol in the ethyl acetate fraction (Jung et al., 2008). The dichloromethane fraction of brown seaweed *Saccharina japonica* exhibited the inhibitory activity of rat lens AR enzyme due to the presence of active porphyrin derivatives such as pheophorbide-A and pheophytin-A (Jung et al., 2013). A series of bromophenols have been obtained from the red alga *Symphyocladia latiuscula* that have been reported to possess AR inhibition (Samaddar and Koneri, 2019b; Xiukun and Ming, 2012; Wei et al., 2005), anti-PKC (Samaddar and Koneri, 2019b), and free radical scavenging activities (Choi et al., 2000; Duan et al., 2007; Xu et al., 2013). *E. cava* has also been reported to possess AR inhibition and anti-PKC activities (Samaddar and Koneri, 2019a). It could be extrapolated that these marine algae might be beneficial in diabetic retinopathy too. However, information is lacking on the protective activities of marine algae on specific disease models of diabetic retinopathy, which calls for further research to be conducted.

2.3.3 Diabetic Nephropathy

Diabetic nephropathy is a clinical syndrome characterized by persistent albuminuria and a progressive decline in renal function, and the term infers the presence of a typical pattern of glomerular disease. The clinical features characterizing diabetic nephropathy include thickening of the basement membrane, extracellular matrix accumulation, and tubular and glomerular hypertrophy (Porrini et al., 2015; Xie et al., 2017). Many epidemiological studies demonstrate that ethnicity, family history, gestational diabetes, elevated blood pressure, dyslipidemia, obesity and insulin resistance are the major risk factors for diabetic nephropathy. Other putative risk factors include elevated glycosylated hemoglobin level (HbA1c), elevated systolic pressure, proteinuria, and smoking (Eberhard, 2006). Genome-wide transcriptome studies (Hameed et al., 2018) and high-throughput technologies (Kato and Natarajan, 2014) indicate the activation of inflammatory signaling pathways and oxidative stress, highlighting the role of genetic factors. Evidence suggests that epigenetic mechanisms such as DNA methylation, noncoding RNAs, and histone modifications can also play a pivotal role in the pathogenesis of diabetic nephropathy. Accordingly, cytokine TNF-α, IL-6, and IL-1β gene promoter polymorphisms and modulation in expression have been linked to diabetic nephropathy susceptibility in subjects. Dysregulation of the local metabolic environment triggered by inflammation and subsequent tissue remodeling may initiate kidney damage (Zheng and Zheng, 2016). Excess intracellular glucose has been shown to activate cellular signaling pathways such as the diacylglycerol (DAG)-PKC pathway, AGEs, polyol pathway, hexosamine pathway, and oxidative stress (Ni et al., 2015). Many studies have linked these pathways to key steps in the development of glomerulosclerosis. In addition to these metabolic pathways, Rho-kinase, an effector of small-GTPase binding protein Rho, has been linked to various steps in the ultrastructural damage of diabetic nephropathy by inducing endothelial dysfunction, mesangial excessive extracellular matrix (ECM) production, podocyte abnormality, and tubulointerstitial fibrosis (Kawanami et al., 2016).

2.3.3.1 Marine Algae in Diabetic Nephropathy

The polysaccharides isolated from the green algae *Caulerpa racemosa* (sea grapes) have been found to possess profound protective effects against high fructose-streptozotocin–induced nephropathy in diabetic animals (Meng et al., 2021). There was a significant decrease in fasting blood glucose, urinary albumin, serum creatinine, blood urea nitrogen (BUN), total cholesterol (TC), triglycerides (TG), and LDL levels after treatment with *C. racemose.*

Levels of renal superoxide dismutase (SOD), catalase (CAT), and glutathione peroxidase (GSH-Px) activities were also prominently increased, whereas the renal malondialdehyde (MDA) level was markedly decreased. Furthermore, renal inflammatory cytokines, IL-1β, IL-6, TNF-α, and TGF-β, were considerably reduced compared to diabetic control rats. Additionally, the treated rats showed a significant decrease in renal fibrosis, as evidenced by decreased expression of TGF-β1, collagen-1, fibronectin, and alpha-smooth muscle actin (α-SMA) in the kidneys. Histopathological lesions showed reduced hyperglycemia, dyslipidemia, and renal oxidative stress indicators with tubule thickening and glomerular hypertrophy. The increased renal levels of IL-1β, IL-6, TNF-α, and TGF-β were also notably reversed dose-dependently with alleviation of nephropathic histology. Furthermore, the polysaccharides reduced the expression of α-SMA, fibronectin, collagen-1, and TGF-β1 in the renal tissues.

Seaweeds are one of many sources of important minerals such as zinc, magnesium, potassium, and calcium (Lordan et al., 2011; Gomez-Gutierrez et al., 2011). Zinc supplementation has been found to improve β-cell function, insulin resistance, glucose intolerance, and hyperglycemia in both type 1 and type 2 diabetics through its potent insulinomimetic action and could prevent secondary complications such as diabetic retinopathy, renal failure, and macrovascular complications (Miao, Sun et al., 2013; Chimienti, 2013). Zinc supplementation as zinc sulfate (5 mg/kg of body weight) to OVE26 mice up-regulated the expression of nuclear factor (erythroid-derived 2)–like 2 (Nrf2), a key regulator of antioxidant defense mechanisms, and metallothionein, a potent antioxidant, with zinc treatment attenuating aortic inflammation, fibrosis, and increased wall thickness (Miao, Wang et al., 2013).

Low-molecular-weight (8.177 kDa) fucoidan (LMWF) isolated from *S. japonica* was studied as a novel target and method for preventing and curing diabetic nephropathy. The LMWF reduced levels of urea and glycated serum protein level in the blood; decreased micrototal protein and albumin levels in the urine; maintained the structural integrity of the glomerular basement membranes and glomerulus; protected glycosaminoglycans from abnormal degradation; inhibited AGE generation and accumulation; and suppressed the expression of inflammatory cytokines, such as JNK, IL-6, VEGF, and connective tissue growth factor (CTGF). Thus, LMWF is found to improve the glomerular filtration rate (GFR) and modulate inflammatory response in the kidney followed by the amelioration and slowing down of the development and progression of diabetic nephropathy in rats (Yingjie et al., 2017). Another study with LMWF on STZ-induced diabetic nephropathy in rats showed that LMWF down-regulates the expression of proinflammatory cytokines: TGF-β, TNF-α, ICAM-1, JNK, and MAPK at transcriptional and protein levels. LMWF relieves inflammation in the kidneys of diabetic nephropathy rats by inhibiting P-selectin and selectin-dependent inflammatory cytokines, thereby protecting renal tissue from damage caused by leukocytic and macrophage infiltration (Yingjie et al., 2016).

The protective effect of diphlorethohydroxycarmalol (DPHC), a polyphenol isolated from an edible seaweed, *Ishige okamurae*, was investigated on methylglyoxal-induced oxidative stress in human embryonic kidney (HEK) cells. DPHC treatment inhibited methylglyoxal-induced cytotoxicity and ROS production. It also activated the Nrf2 transcription factor and increased the mRNA expression of antioxidant and detoxification enzymes: superoxide dismutase 1 (SOD-1), catalase (CAT), γ-glutamyl cysteine synthetase (GCL), heme oxygenase-(HO-) 1, and NAD(P)H quinone dehydrogenase (NQO)1, consequently reducing MGO-induced advanced glycation end-product formation. In addition, DPHC increased glyoxalase-1 mRNA expression and attenuated MGO-induced advanced glycation end-product formation in HEK cells. These results suggest that DPHC possesses a protective activity against MGO-induced cytotoxicity in human kidney cells by preventing oxidative stress and advanced glycation end-product formation. Therefore, it could be used as a potential therapeutic agent for the prevention of diabetic nephropathy (Cha et al., 2018).

2.4 CONCLUSION

Neuropathy, nephropathy, and retinopathy are the three most common and prominent complications of chronic and uncontrolled diabetes, which, unfortunately, are the major reasons for diabetes-related mortality. Though glycemic control still remains the heart of therapy, organ-specific treatment strategy at the tissue and cellular levels is indispensable today. Marine natural products provide an excellent opportunity to study diverse and unique compounds capable of being used as therapeutic agents. Marine algae are known to possess a large number of bioactive chemicals with promising pharmacological and biological applications. Many of them have been found to possess promising antidiabetic activities, and the majority of such research studies have been restricted to preliminary screening of α-amylase, α-glucosidase, PTP1B inhibitory, and insulinomimetic activities; the information database lacks in-depth mechanism-based investigations. The effects of marine algae on diabetic complications need to be researched extensively, as the existing information is lacking. Investigations need to be carried out on disease-specific models at the level of etiological targets of each complication. Even though a decent number of investigations on the influence of marine algae on diabetic neuropathy and nephropathy has been conducted thus far, the number of studies specifically targeting the etiological factors in diabetic retinopathy is lean. As nature continues to provide undiminished quantities of marine algae, on one hand, on the other hand, it places a significant burden on us to investigate and interpret the precise function of their constituents utilizing cutting-edge technologies.

REFERENCES

Abad, M., Bedoya, L., and Bermejo, P. 2008. Natural marine anti-inflammatory products. *Mini Rev. Med. Chem.* 8:740–754.

Abdelsalam, S.S., Korashy, H.M., Zeidan, A., and Agouni, A. 2019. The role of protein tyrosine phosphatase (PTP)-1B in cardiovascular disease and its interplay with insulin resistance. *Biomolecules*. 9(286):1–23.

Al-Shabrawey, M., and Smith, S. 2010. Prediction of diabetic retinopathy: Role of oxidative stress and relevance of apoptotic biomarkers. *EPMA J.* 1:56–72.

Ale, M.T., Mikkelsen, J.D., and Meyer, A.S. 2011. Important determinants for fucoidan bioactivity: A critical review of structure-function relations and extraction methods for fucose-containing sulfated polysaccharides from brown seaweeds. *Mar. Drugs.* 9:2106–2130.

Allen, N., and Barres, B. 2009. Neuroscience: Glia—More than just brain glue. *Nature.* 457:675–677.

Baynes, J.W., and Thorpe, S.R. 1999. Role of oxidative stress in diabetic complications: A new perspective on an old paradigm. *Diabetes.* 48:1–9.

Bhattacharjee, R., Mitra, A., Dey, B., and Pal, A. 2014. Exploration of anti-diabetic potentials amongst marine species-a minireview. *Indo. Glob. J. Pharm. Sci.* 4(2):65–73.

Birketvedt, G.S., Shimshi, M., Erling, T., and Florholmen, J. 2005. Experiences with three different fiber supplements in weight reduction. *Med. Sci. Monit.* 11:5–8.

Bocanegra, A., Bastida, S., Benedí, J., Ródenas, S., and Sánchez-Muniz, F.J. 2009. Characteristics and nutritional and cardiovascular-health properties of seaweeds. *J. Med. Food.* 12:236–258.

Bressler, S.B., Odia, I., Glassman, A.R., Danis, R.P., Grover, S., Hampton, G.R., et al. 2018. Changes in diabetic retinopathy severity when treating diabetic macular edema with ranibizumab: DRCR.net protocol I 5-year report. *Retina.* 38:1896–1904.

Brown, E.M., Allsopp, P.J., Magee, P.J., Gill, C.I.R., Nitecki, S., Strain, C.R., and McSorley, E.M. 2014. Seaweed and human health. *Nutr. Rev.* 72:205–216.

Catarino, M.D., Silva, A.M.S., Mateus, N., and Cardoso, S.M. 2019. Optimization of phlorotannins extraction from *Fucus vesiculosus* and evaluation of their potential to prevent metabolic disorders. *Mar. Drugs.* 17(3):162–169.

Cerna, M. 2011. Seaweed proteins and amino acids as nutraceuticals. *Adv. Food Nutr. Res.* 64:297312.

Cha, S.H., Hwang, Y, Heo, S.J., and Jun, H.S. 2018. Diphlorethohydroxycarmalol attenuates methylglyoxal-induced oxidative stress and advanced glycation end product formation in human kidney cells. *Oxid. Med. Cell. Longev.* 2018(3654095):1–14.

Chimienti, F. 2013. Zinc, pancreatic islet cell function and diabetes: New insights into an old story. *Nutr. Res. Rev.* 26:1–11.

Choi, J.S., Park, H.J., Jung, H.A., Chung, H.Y., Jung, J.H., and Choi, W.C. 2000. A cyclohexanonyl bromophenol from the red alga *Symphyocladia latiuscula. J. Nat. Prod.* 63:1705–1706.

Cui, Y., Zhang, L., Zhang, T., Luo, D., Jia, Y., Guo, Z., et al. 2010. Inhibitory effect of fucoidan on nitric oxide production in lipopolysaccharide activated primary microglia. *Clin. Exp. Pharmacol. Physiol.* 37:422–428.

De Mattia, G., Laurenti, O., Bravi, C., Ghiselli, A., Iuliano, L., and Balsano, F. 1994. Effect of aldose reductase inhibition on glutathione redox status in erythrocytes of diabetic patients. *Metabolism.* 43:965–968.

Dooley, J., Garcia-Perez, J.E., Sreenivasan, J., Schlenner, S.M., Vangoitsenhoven, R., Papadopoulou, A.S., et al. 2016. The microRNA-29 family dictates the balance between homeostatic and pathological glucose handling in diabetes and obesity. *Diabetes.* 65:53–61.

Du, X., Matsumura, T., Edelstein, D., Rossetti, L., Zsengeller, Z., Szabo, C., and Brownlee, M. 2003. Inhibition of GAPDH activity by poly(ADP-ribose) polymerase activates three major pathways of hyperglycemic damage in endothelial cells. *J. Clin. Invest.* 112:1049–1057.

Duan, X.J., Li, X.M., and Wang, B.G. 2007. Highly brominated mono- and bis-phenols from the marine red alga *Symphyocladia latiuscula* with radical-scavenging activity. *J. Nat. Prod.* 70:1210–1213.

Duh, E.J., Sun, J.K., and Stitt, A.W. 2017. Diabetic retinopathy: Current understanding, mechanisms, and treatment strategies. *J. Clin. Invest. Insight*. 2:e93751.
Eberhard, R. 2006. Diabetic nephropathy. *Saudi J. Kidney Dis. Transplant*. 6(17):481–490.
Feldman, E.L., Callaghan, B.C., Pop-Busui, R., Zochodne, D.W., Wright, D.E., Bennett, D.L., et al. 2019. Diabetic neuropathy. *Nat. Rev. Dis. Primers*. 5:42–48.
Feldman, E.L., Stevens, M.J., and Greene, D.A. 1997. Pathogenesis of diabetic neuropathy. *Clin. Neuro. Sci*. 4:365–370.
Fowler, M.J. 2008. Microvascular and macrovascular complications of diabetes. *Clin. Diab*. 26(2):77–82.
Gabbay, K.H. 1973. The sorbitol pathway and the complications of diabetes. *N. Engl. J. Med*. 288:831–836.
Galer, B.S., Gianas, A., and Jensen, M.P. 2000. Painful diabetic polyneuropathy: Epidemiology, pain description, and quality of life. *Diabetes Res. Clin. Pract*. 47:123–128.
Giovino, A., Benny, J., and Martinelli, F. 2020. Advanced glycation end products (AGEs): Biochemistry, signaling, analytical methods, and epigenetic effects. *Oxid. Med. Cell. Longev*. 2020:1–18.
Gomez-Gutierrez, C.M., Guerra-Rivas, G., Soria-Mercado, I.E., Ayala-Sánchez, N.E. 2011. Marine edible algae as disease preventers. *Adv. Food Nutr. Res*. 64:29–39.
Gotama, T.L., Husni, A., and Ustadi. 2018. Antidiabetic activity of *Sargassum hystrix* extracts in streptozotocin-induced diabetic rats. *Prev. Nutr. Food Sci*. 23(3):189–195.
Gouliopoulos, N.S., Kalogeropoulos, C., Lavaris, A., Rouvas, A., Asproudis, I., Garmpi, A., et al. 2018. Association of serum inflammatory markers and diabetic retinopathy: A review of literature. *Eur. Rev. Med. Pharmacol. Sci*. 22:7113–7128.
Greene, D.A., Lattimer, S.A., and Sima, A.A.F. 1987. Sorbitol, phosphoinositides and sodium-potassium ATPase in the pathogenesis of diabetic complications. *N. Engl. J. Med*. 316:599–606.
Griendling, K.K., and FitzGerald, G.A. 2003. Oxidative stress and cardiovascular injury. I. basic mechanisms and *in vivo* monitoring of ROS. *Circulation*. 108:1912–1916.
Guiry, M.D., and Guiry, G.M. 2019. *AlgaeBase*. World-Wide Electronic Publication, National University of Ireland. www.algaebase.org.
Hall, A., Fairclough, A., Mahadevan, K., and Paxman, J. 2012. *Ascophyllum nodosum* enriched bread reduces subsequent energy intake with no effect on post-prandial glucose and cholesterol in healthy, overweight males. A pilot study. *Appetite*. 58:379–386.
Hameed, I., Masoodi, S.R., Malik, P.A., Mir, S.A., Ghazanfar, K., and Ganai, B.A. 2018. Genetic variations in key inflammatory cytokines exacerbates the risk of diabetic nephropathy by influencing the gene expression. *Gene*. 661:51–59.
Heales, S., Bolaños, J., Stewart, V., Brookes, P., Land, J., and Clark, J. 1999. Nitric oxide, mitochondria and neurological disease. *Biochim. Biophys. Acta*. 1410:215–228.
Hounsom, L., Corder, R., Patel, J., and Tomlinson, D.R. 2001. Oxidative stress participates in the breakdown of neuronal phenotype in experimental diabetic neuropathy. *Diabetologia*. 44:424–428.
Inoguchi, T., Li, P., Umeda, F., Yu, H.Y., Kakimoto, M., Imamura, M., et al. 2000. High glucose level and free fatty acid stimulate reactive oxygen species production through protein kinase C-dependent activation of NAD(P)H oxidase in cultured vascular cells. *Diabetes*. 49:1939–1945.
Jeong, Y.T., Kim, Y.D., Jung, Y.M., Park, D.C., Lee, D.S., Ku, S.K., et al. 2013. Low molecular weight fucoidan improves endoplasmic reticulum stress-reduced insulin sensitivity through AMP-activated protein kinase activation in L6 myotubes and restores lipid homeostasis in a mouse model of type 2 diabetes. *Mol. Pharmacol*. 84:147–157.
Jhamandas, J.H., Wie, M.B., Harris, K., MacTavish, D., and Kar, S. 2005. Fucoidan inhibits cellular and neurotoxic effects of β-amyloid (Aβ) in rat cholinergic basal forebrain neurons. *Eur. J. Neurosci*. 21:2649–2659.

Jiang, X., Yu, J., Ma, Z., Zhang, H., and Xie, F. 2015. Effects of fucoidan on insulin stimulation and pancreatic protection via the cAMP signaling pathway *in vivo* and *in vitro*. *Mol. Med. Rep.* 12:4501–4507.

Jiménez-Escrig, A., and Goñi, I. 1999. Nutritional evaluation and physiological effects of edible seaweeds. *Arch. Latinoam. Nutr.* 49:114–120.

Jin, D., Lim, C., Sung, J., Choi, H., Ha, I., and Han, J. 2006. *Ulva conglobata*, a marine algae, has neuroprotective and anti-inflammatory effects in murine hippocampal and microglial cells. *Neurosci. Lett.* 402:154–158.

Jung, H.A., Islam, M.N., Lee, C.M., Oh, S.H., Lee, S., Jung, J.H., and Choi, J.S. 2013. Kinetics and molecular docking studies of an anti-diabetic complication inhibitor fucosterol from edible brown algae *Eisenia bicyclis* and *Ecklonia stolonifera*. *Chem. Biol. Interac.* 206(1):55–62.

Jung, H.A., Yoon, N.Y., Woo, M.H., and Choi, J.S. 2008. Inhibitory activities of extracts from several kinds of seaweeds and phlorotannins from the brown alga *Ecklonia stolonifera* on glucose mediated protein damage and rat lens aldose reductase. *Fisheries Sci.* 74(6):1363–1365.

Jung, W.K., Ahn, Y.W., Lee, S.H., Choi, Y.H., Kim, S.K., Yea, S.S., et al. 2009. *Ecklonia cava* ethanolic extracts inhibit lipopolysaccharide-induced cyclooxygenase-2 and inducible nitric oxide synthase expression in BV2 microglia via the MAP kinase and NF-[kappa]B pathways. *Food Chem. Toxicol.* 47:410–417.

Jung, W.K., Heo, S., Jeon, Y., Lee, C., Park, Y., and Byun, H. 2009. Inhibitory effects and molecular mechanism of dieckol isolated from marine brown alga on COX-2 and iNOS in microglial cells. *J. Agr. Food Chem.* 57:4439–4446.

Kahn, S.E., Hull, R.L., and Utzschneider, K.M. 2006. Mechanisms linking obesity to insulin resistance and type 2 diabetes. *Nature*. 444(7121):840–846.

Kato, M., and Natarajan, R. 2014. Diabetic nephropathy—emerging epigenetic mechanisms. *Nat. Rev. Nephrol.* 10(9):517–530.

Kawanami, D., Matoba, K., and Utsunomiya, K. 2016. Signaling pathways in diabetic nephropathy. *Histol Histopathol.* 31(10):1059–1067.

Kim, K.J., Lee, O.H., Lee, H.C., Kim, Y.C., and Lee, B.Y. 2007. Effect of fucoidan on expression of diabetes mellitus related genes in mouse adipocytes. *Food Sci. Biotechnol.* 16:212–217.

Kim, K.T., Rioux, L.E., and Turgeon, S.L. 2015. Molecular weight and sulfate content modulate the inhibition of a-amylase by fucoidan relevant for type 2 diabetes management. *Pharmanutrition*. 3:108–114.

Kim, S.K. 2011. Application of seaweeds in the food industry. In *Handbook of Marine Macroalgae: Biotechnology and Applied Phycology*, Chapter 34, pp. 522–529. Wiley-Blackwell.

Kinoshita, J.H., and Nishimura, C. 1988. The involvement of aldose reductase in diabetic complications. *Diabetes Metab. Rev.* 4:323–327.

Kooti, W., Farokhipour, M., Asadzadeh, Z., Larky, D.A., and Samani, A. 2016. The role of medicinal plants in thetreatment of diabetes: A systematic review. *Electron Physician*. 8(1):1832–1842.

Kusuhara, S., Fukushima, Y., Oguram, S., Inoue, N., and Uemura, A. 2018. Pathophysiology of diabetic retinopathy: The old and the new. *Diabetes Metab. J.* 42:364–376.

Lakshmanasenthil, S., Vinothkumar, T., Geetharamani, D., Marudhupandi, T., Suja, G., and Sindhu, N.S. 2014. Fucoidan—a novel a-amylase inhibitor from *Turbinaria ornata* with relevance to NIDDM therapy. *Biocatal. Agric. Biotechnol.* 3:66–70.

Lee, J., Grabb, M., Zipfel, G., and Choi, D. 2000. Brain tissue responses to ischemia. *J. Clin. Invest.* 106:723–731.

Lee, S., and Jeon, Y. 2013. Anti-diabetic effects of brown algae derived phlorotannins, marine polyphenols through diverse mechanisms. *Fitoterapia*. 86:129–136.

Lim, C., Jin, D., Sung, J., Lee, J., Choi, H., Ha, I., and Han, J. 2006. Antioxidant and anti-inflammatory activities of the methanolic extract of *Neorhodomela aculeate* in hippocampal and microglial cells. *Biol. Pharm. Bull.* 29:1212–1216.

Lin, G., Wan, X., Liu, D., Wen, Y., Yang, C., and Zhao, C. 2021. COL1A1 as a potential new biomarker and therapeutic target for type 2 diabetes. *Pharmacol. Res.* 165:105436.

Liu, X., Li, X., Gao, L., Cui, C., Li, C., Li, J., and Wang, B. 2011. Extraction and PTP1B inhibitory activity of bromophenols from the marine red alga *Symphyocladia latiuscula*. *Chin. J. Oceanol. Limnol.* 29:686–690.

Lordan, S., Ross, R.P., and Stanton, C. 2011. Marine bioactives as functional food ingredients: Potential to reduce the incidence of chronic diseases. *Mar. Drugs*. 9:1056–1100.

Maeda, H., Yamamoto, R., Hirao, K., and Tochikubo, O. 2004. Effects of agar (kanten) diet on obese patients with impaired glucose tolerance and type 2 diabetes. *Diabetes Obes. Metab*. 7:40–46.

Martin, C.L., Albers, J., and Herman, W.H. 2006. Neuropathy among the diabetes control and complications trial cohort 8 years after trial completion. *Diabetes Care*. 29:340–344.

Massart, J., Sjögren, R.J.O., Lundell, L.S., Mudry, J.M., Franck, N., O'Gorman, D.J., et al. 2017. Altered miR-29 expression in type 2 diabetes influences glucose and lipid metabolism in skeletal muscle. *Diabetes*. 66:1807–1818.

Meng, C., Li, Y., Famurewa, A.C., and Olatunji, O.J. 2021. Antidiabetic and nephroprotective effects of polysaccharide extract from the seaweed *Caulerpa racemosa* in high fructose-streptozotocin induced diabetic nephropathy. *Diabetes Metab. Syndr. Obes*. 14:2121–2131.

Miao, X., Sun, W., Miao, L., Fu, Y., Wang, Y., Su, G., et al. 2013. Zinc and diabetic retinopathy. *J Diabetes Res*. 2013:425854.

Miao, X., Wang, Y., Sun, J., Sun, W., Tan, Y., Cai, L., et al. 2013. Zinc protects against diabetes-induced pathogenic changes in the aorta: Roles of metallothionein and nuclear factor (erythroid-derived 2)-like 2. *Cardiovasc Diabetol*. 12:54–61.

Misurcova, L., Skrovankova, S., Samek, D., Ambrozova, J., and Machu, L. 2012. Health benefits of algal polysaccharides in human nutrition. *Adv. Food. Nutr. Res*. 66:75–145.

Moon, H.E., Islam, M.N., and Ahn, B.R. 2014. Protein tyrosine phosphatase 1B and α-glucosidase inhibitory phlorotannins from edible brown algae, *Ecklonia stolonifera* and *Eisenia bicyclis*. *Biosci. Biotechnol. Biochem*. 75(8):1472–1480.

Mooradian, A.D. 2009. Dyslipidemia in type 2 diabetes mellitus. *Nat. Clin. Pract. Endocrinol. Metab*. 5(3):150–159.

Motshakeri, M., Ebrahimi, M., Goh, Y.M., Othman, H.H., Hair-Bejo, M., and Mohamed, S. 2014. Effects of brown seaweed (*Sargassum polycystum*) extracts on kidney, liver, and pancreas of type 2 diabetic rat model. *Evid. Based Complement. Alternat. Med*. 2014(379407):1–11.

Murray, M., Dordevic, A.L., Ryan, L., and Bonham, M.P. 2018. The impact of a single dose of a polyphenol-rich seaweed extract on postprandial glycaemic control in healthy adults: A randomised cross-over trial. *Nutrients*. 10:270–278.

NajmeOliyaei, N., Moosavi-Nasab, M., Tamaddon, A.M., and Tanideh, N. 2021. Antidiabetic effect of fucoxanthin extracted from *Sargassum angustifolium* on streptozotocin nicotinamide-induced type 2 diabetic mice. *Food Sci. Nutr*. 9:3521–3529.

Nemoto, S., Takeda, K., Yu, Z.X., Ferrans, V.J., and Finkel, T. 2000. Role for mitochondrial oxidants as regulators of cellular metabolism. *Mol. Cell Biol*. 20:7311–7318.

Ni, W.J., Tang, L.Q., and Wei, W. 2015. Research progress in signaling pathway in diabetic nephropathy. *Diabetes Metab. Res. Rev*. 31(3):221–233.

Niimi, N., Yako, H., Takaku, S., Chung, S.K., and Sango, K. 2021. Aldose reductase and the polyol pathway in Schwann cells: Old and new problems. *Int. J. Mol. Sci*. 22(1031):1–14.

Nishikawa, T., Edelstein, D., Du, X.L., Yamagishi, S.I., Matsumura, T., and Kaneda, Y. 2000. Normalizing mitochondrial superoxide production blocks three pathways of hyperglycaemic damage. *Nature*. 404:787–790.

Nunez, L.Y.O., Garufi, G., and Seyhan, A.A. 2016. Altered levels of circulating cytokines and microRNAs in lean and obese individuals with prediabetes and type 2 diabetes. *Mol. Biosyst*. 13:106–1021.

Nwosu, F., Morris, J., Lund, V.A., Stewart, D., Ross, H.A., and McDougall, G.J. 2011. Anti-proliferative and potential antidiabetic effects of phenolic-rich extracts from edible marine algae. *Food Chem*. 126(3):1006–1012.

Oates, P.J., and Mylari, B.L. 1999. Aldose reductase inhibitors: Therapeutic implications for diabetic complications. *Expert Opin. Investig. Drugs*. 8:2095–2119.

Odunsi, S.T., Vázquez-Roque, M.I., Camilleri, M., Papathanasopoulos, A., Clark, M.M., Wodrich, L., et al. 2010. Effect of alginate on satiation, appetite, gastric function, and selected gut satiety hormones in overweight and obesity. *Obesity*. 18:1579–1584.

Pan, H.Z., Zhang, H., Chang, D., Li, H., and Sui, H. 2008. The change of oxidative stress products in diabetes mellitus and diabetic retinopathy. *Br. J. Ophthalmol*. 92:548–5451.

Pop-Busui, R., Boulton, A.J., Feldman, E.L., Bril, V., Freeman, R., Malik, R.A., et al. 2017. Diabetic neuropathy: A position statement by the American diabetes association. *Diabetes Care*. 40:136–154.

Porrini, E., Ruggenenti, P., Mogensen, C.E., Barlovic, D.P., Praga, M., Cruzado, J.M., et al. 2015. Non-proteinuric pathways in loss of renal function in patients with type 2 diabetes. *Lancet Diabetes Endocrinol*. 3:382–391.

Reddy, R.K., Pieramici, D.J., Gune, S., Ghanekar, A., Lu, N., Quezada-Ruiz, C., et al. 2018. Efficacy of ranibizumab in eyes with diabetic macular edema and macular nonperfusion in RIDE and RISE. *Ophthalmology*. 125:1568–1574.

Row-Rendleman, C.L., and Eichberg, J. 1994. PO phosphorylation in nerves from normal and diabetic rats: Role of protein kinase C and turnover of phosphate groups. *Neurochem. Res*. 19:1023–1031.

Rübsam, A., Parikh, S., and Fort, P.E. 2018. Role of inflammation in diabetic retinopathy. *Int. J. Mol. Sci*. 19:E942.

Saeedi, P., Petersohn, I., Salpea, P., Malanda, B., Karuranga, S., Unwin, N., et al. 2019. *Global and regional diabetes prevalence estimates for 2019 and projections for 2030 and2045: Results from the international diabetes federation diabetes atlas* (9th ed., Vol. 157). Diabetes Research and Clinical Practice.

Samaddar, S., and Koneri, R. 2019a. Neuroprotective efficacy of polyphenols of marine brown macroalga *Ecklonia cava* in diabetic peripheral neuropathy. *Phcog. Mag*. 15:S468–S475.

Samaddar, S., and Koneri, R. 2019b. Polyphenols of marine red macroalga *Symphyocladia latiuscula* ameliorate diabetic peripheral neuropathy in experimental animals. *Heliyon*. 5(e01781):1–8.

Sanaka, M., Yamamoto, T., Anjiki, H., Nagasawa, K., and Kuyama, Y. 2007. Effects of agar and pectin on gastric emptying and post-prandial glycaemic profiles in healthy human volunteers. *Clin. Exp. Pharmacol. Physiol*. 34:1151–1155.

Savitsky, P.A., and Finkel, T. 2002. Redox regulation of Cdc25C. *J. Biol. Chem*. 277: 20535–20540.

Schmidt, A.M., Hori, O., Brett, J., Yan, S.D., Wautier, J.L., and Stern, D. 1994. Cellular receptors for advanced glycation end products: Implications for induction of oxidant stress and cellular dysfunction in the pathogenesis of vascular lesions. *Arterioscler. Thromb*. 14:1521–1528.

Senthilkumar, P., Sudha, S., and S. Prakash. 2014. Antidiabetic activity of aqueous extract of *Padina boergesenii* in streptozotocin-induced diabetic rats. *Int. J. Pharm. Pharmaceut. Scie*. 6(5):418–422.

Shakambari, G., Ashokkumar, B., and Varalakshmi, P. 2015. Phlorotannins from brown algae: Inhibition of advanced glycation end products formation in high glucose induced *Caenorhabditis elegans. Ind. J. Expt. Biol.* 53(6):371–379.

Shi, D., Guo, S., Jiang, B., Guo, C., Wang, T., Zhang, L., and Li, J. 2013. HPN, a synthetic analogue of bromophenol from red alga *Rhodomela confervoides*: Synthesis and anti-diabetic effects in C57BL/KsJ-db/db mice. *Mar. Drugs.* 11:350–362.

Shi, D., Li, J., Jiang, B., Guo, S., Su, H., and Wang, T. 2012. Bromophenols as inhibitors of protein tyrosine phosphatase 1B with antidiabetic properties. *Bioorgan. Med. Chem. Lett.* 22:2827–2832.

Simó, R., and Hernández, C. 2014. European consortium for the early treatment of diabetic retinopathy (EUROCONDOR). Neurodegeneration in the diabetic eye: New insights and therapeutic perspectives. *Trends Endocrinol. Metab.* 25:23–33.

Simó, R., Stitt, A.W., and Gardner, T.W. 2018. Neurodegeneration in diabetic retinopathy: Does it really matter? *Diabetologia.* 61:1902–1912.

Stevens, M.J., Lattimer, S.A., Kamijo, M., Huysen, C.V., Sima, A.A.F., and Greene, D.A. 1993. Osmotically-induced nerve taurine depletion and the compatible osmolyte hypothesis in experimental diabetic neuropathy in the rat. *Diabetologia.* 36:608–614.

Sugiura, S., Minami, Y., Taniguchi, R., Tanaka, R., Miyake, H., Mori, T., et al. 2017. Evaluation of antiglycation activities of phlorotannins in human and bovine serum albumin-methylglyoxal models. *Nat. Prod. Comm.* 12(11):1793–1796.

Sundaresan, M., Yu, Z.X., Ferrans, V.J., Irani, K., and Finkel, T. 1995. Requirement for generation of H_2O_2 for platelet-derived growth factor signal transduction. *Science.* 270:296–299.

Suzuki, D., Miyata, T., Saotome, N., Horie, K., Inagi, R., Yasuda, Y., et al. 1999. Immunohistochemical evidence for an increased oxidative stress and carbonyl modification of proteins in diabetic glomerular lesions. *J. Am. Soc. Nephrol.* 10:822–832.

Tabarsa, M., Rezaei, M., Ramezanpour, Z., and Waaland, J.R. 2012. Chemical compositions of the marine algae *Gracilaria salicornia* (Rhodophyta) and *Ulva lactuca* (Chlorophyta) as a potential food source. *J. Sci. Food Agric.* 92:2500–2506.

Takahashi, K., Kamada, C., Yoshimura, H., Okumura, R., Iimuro, S., Ohashi, Y., et al. 2012. Effects of total and green vegetable intakes on glycated hemoglobin A1c and triglycerides in elderly patients with type 2 diabetes mellitus: The Japanese elderly intervention trial. *Geriatr. Gerontol. Int.* 12:50–58.

Tanemura, Y., Yamanaka-Okumura, H., Sakuma, M., Nii, Y., Taketani, Y., and Takeda, E. 2014. Effects of the intake of *Undaria pinnatifida* (Wakame) and its sporophylls (Mekabu) on postprandial glucose and insulin metabolism. *J. Med. Investig.* 61:291–297.

Tawata, M., Ohtaka, M., Hosaka, Y., and Onaya, T. 1992. Aldose reductase mRNA expression and its activity are induced by glucose in fetal rat aortic smooth muscle (A10) cells. *Life Sci.* 51:719–726.

Wang, J., Hu, S., Wang, J., Li, S., and Jiang, W. 2016. Fucoidan from *Acaudina molpadioides* protects pancreatic islet against cell apoptosis via inhibition of inflammation in type 2 diabetic mice. *Food Sci. Biotechnol.* 25:293–300.

Wang, W., and Lo, A.C.Y. 2018. Diabetic retinopathy: Pathophysiology and treatments. *Int. J. Mol. Sci.* 19(1816):1–14.

Wang, Z., Wang, J., and Chan, P. 2013. Treating type 2 diabetes mellitus with traditional Chinese and Indian medicinal herbs. *Evid. Based. Complement. Alternat. Med.* 2013:1–17.

Wei, W., Yoshihito, O., Haibo, S., Yongqi, W., and Toru, O. 2005. Structures and aldose reductase inhibitory effects of bromophenols from the red alga *Symphyocladia latiuscula. J. Nat. Prod.* 68:620–622.

Wijesinghe, W.A.J.P., and Jeon, Y.J. 2012. Biological activities and potential industrial applications of fucose rich sulfated polysaccharides and fucoidans isolated from brown seaweeds: A review. *Carbohydr. Polym.* 88:13–20.

Wu, S.Y., Yan, M.D., Wu, A.T.H., Yuan, K.S., and Liu, S.H. 2016. Brown seaweed fucoidan inhibits cancer progression by dual regulation of mir-29c/ADAM12 and miR-17–5p/PTEN axes in human breast cancer cells. *J. Cancer*. 7:2408–2419.

Xie, R., Zhang, H., Wang, X.Z., Yang, X.Z., Wu, S.N., Wang, H.G., et al. 2017. The protective effect of betulinic acid (BA) diabetic nephropathy on streptozotocin (STZ)-induced diabetic rats. *Food Funct*. 8:299–306.

Xiukun, L., and Ming, L. 2012. Bromophenols from marine algae with potential anti-diabetic activities. *J. Ocean Univ. China*. 11(4):533–538.

Xu, J., Chen, L.J., Yu, J., Wang, H.J., Zhang, F., Liu, Q., et al. 2018. Involvement of advanced glycation end products in the pathogenesis of diabetic retinopathy. *Cell Physiol. Biochem*. 48:705–717.

Xu, S.Y., Huang, X., and Cheong, K.L. 2017. Recent advances in marine algae polysaccharides: Isolation, structure, and activities. *Mar. Drugs*. 15:388–396.

Xu, X., Yin, L., Gao, L., Gao, J., Chen, J., Li, J., and Song, F. 2013. Two new bromophenols with radical scavenging activity from marine red alga *Symphyocladia latiuscula*. *Mar. Drugs*. 11:842–847.

Yan, M.D., Yao, C.J., Chow, J.M., Chang, C.L., Hwang, P.A., Chuang, S.E., et al. 2015. Fucoidan elevates microRNA-29b to regulate DNMT3B-MTSS1 axis and inhibit EMT in human hepatocellular carcinoma cells. *Mar. Drugs*. 13:6099–6116.

Yang, H.W., Fernando, K.H.N., Oh, J.Y., Li, X., Jeon, Y.J., and Ryu, B.M. 2019. Anti-obesity and anti-diabetic effects of *Ishige okamurae*. *Mar. Drugs*. 17(4):202–209.

Yasunari, K., Kohno, M., Kano, H., Yokokawa, H., Horio, T., and Yoshikawa, J. 1995. Aldose reductase inhibitor prevents hyperproliferation and hypertrophy of cultured rat vascular smooth muscle cells induced by high glucose. *Arterioscler. Thromb. Vasc. Biol*. 15:2207–2212.

Yingjie, X., Zhang, Q., Luo, D., Wang, J., and Duan, D. 2016. Low molecular weight fucoidan modulates P-selectin and alleviates diabetic nephropathy. *Int. J. Biol. Macromol*. 91:233–240.

Yingjie, X., Zhang, Q., Luo, D., Wang, J., and Duan, D. 2017. Low molecular weight fucoidan ameliorates the inflammation and glomerular filtration function of diabetic nephropathy. *J. Appl. Phycol*. 29(1):531–542.

Yoshinaga, K., and Mitamura, R. 2019. Effects of *Undaria pinnatifida* (Wakame) on postprandial glycemia and insulin levels in humans: A randomized crossover trial. *Plant Foods Hum. Nutr*. 74:461–467.

Yu, W.C., Chen, Y.L., Hwang, P.A., Chen, T.H., and Chou, T.C. 2017. Fucoidan ameliorates pancreatic β-cell death and impaired insulin synthesis in streptozotocin-treated β-cells and mice via a Sirt-1-dependent manner. *Mol. Nutr. Food Res*. 61:1700136.

Zheng, Z., and Zheng, F. 2016. Immune cells and inflammation in diabetic nephropathy. *J. Diabetes Res*. 2016(1841690):1–10.

Zhu, X., and Eichberg, J. 1990. 1,2-Diacylglycerol content and its arachidonyl-containing molecular species are reduced in sciatic nerve from streptozotocin-induced diabetic rats. *J. Neurochem*. 55:1087–1090.

3 Marine Biopolymers
Diversity in Structure and Application

Ngo Dang Nghia

CONTENTS

DOI: 10.1201/9781003303916-3

3.1 INTRODUCTION

Existing in many organisms as the constituents for mainly supporting the strength of tissues, the biopolymers from the marine beings exhibit the diversity not only in the molecular levels but in the way they aggregate, in the behaviors they react to the environment in which they function. All of this makes the marine biopolymers difficult to master in understanding and dealing with their structures, properties, and applications. Among the polymers, including phytocolloids as agar, alginate, carrageenan, and polymers from marine animals as chitin, chitosan, collagen, and gelatin, we would like to select two typical biopolymers to go further in analyzing the structures and properties related to the applications for alginate and chitin/chitosan.

Alginate is manufactured from brown algae and is used in a large range of applications, from ice cream, textiles, paper, cosmetics, therapeutic drugs, and wound dressing to advanced research in medicine for drug delivery, cell culture and tissue engineering.

Chitin and its impressive derivative chitosan, produced commercially from marine crustaceans, mainly crab and shrimp shell, have shown many value properties and were found useful in many cases. Alginate and chitosan have the properties that can complement each other, and both are often combined in many applications. Chitin/chitosan and their derivatives get more interest and dominate the use of biopolymers.

The studies on the two biopolymers are developing with an increasing rate. If in the 1960s, the scientists focused on exploring the properties and structures, then nowadays, with advanced techniques, the research ranges from molecular structural aspects to complicated configurations related to applications in biomedicine. Many strange behaviors of alginate and chitin/chitosan have been elucidated and help us in going further to the sophisticated structures and properties.

This chapter is limited in the analysis of the chemical structure, the relationship between structural properties of alginate and chitin/chitosan, that are helpful in understanding and applying in biomedicine.

3.2 ALGINATE

3.2.1 Source

Alginate is the biopolymers, commercially produced from seaweed, that has the large range of applications, from paper, texture printing, and the food industry to cosmetics, pharmaceuticals, medicine, biotechnology, and tissue engineering. It is hard to list all applications of alginate. The alginate market is valued at 759.8 million USD with the volume of 44,480 tons in 2021. The revenue forecast in 2028 is over 1 billion USD with the volume of 59,139 tons, corresponding to the compound annual growth rate (CAGR) of 5% from 2021 to 2028 (*Alginate Market Size, Share & Trends Analysis Report 2021–2028*, 2021).

The diversity in applications of alginate is based on the variation on structure, which leads to flexible properties that can adopt to different requirements in technology.

Alginate is the constructed polymer for seaweed and bacteria and is found in the cell wall. It may contribute to the strength of seaweed in respond to the flow and wave of seawater in the tidal areas. Although many species of brown algae exist, up to 40% in dry weight, the industry uses the most abundant species, including *Ascophyllum nodosum, Macrocystis pyrifera, Laminaria hyperborea, Laminaria digitata, Laminaria japonica,* and *Sargassum. Ascophyllum* is found in the cold waters of Northern Hemisphere and grows in the eulittoral zone. *Macrocystis pyrifera*, distributed in Baja California, is the largest brown seaweed with the holdfast fixed in the bottom of the sea and rising to the surface water. The length of this seaweed can get up to 20 meters. *Laminaria hyperborea* is located in the cold water, in the mid-sublittoral zone. They have a strong stipe and can survive to 15 years. *Laminaria digitata* and *Laminaria japonica* grow in the cold water, mainly in Japan and China. *Laminaria japonica* is mainly used for food and only the surplus production is used for alginate (McHugh, 2003). Located in tropical areas, *Sargassum* is used for alginate production but not much because of the poor quality of alginate. In bacteria, alginate is produced with a more defined chemical structure and so that changes the properties of alginate. *Azotobacter* and *Pseudomonas* are used for obtaining the bacteria alginate, which get more interest in biomedical applications for controlled structure by bacteria modification and are tailor-made for meeting the specific application (Hay et al., 2013).

3.2.2 Structure and Diversity

In terms of chemical structure, alginate is a family of linear copolymers composed of the two uronates in blocks of (1,4)-linked β-D-mannuronic (M) and α-L-guluronic (G) residues (Fischer, & Dörfel, 1955) (Figure 3.1). The two units in the figure show the difference in configuration with $^4C_1$ for mannuronate and $^1C_4$ for guluronate, which plays the significant role in building the features in relationship between structural properties of the alginate. Since the different configurations in the M and G unit, with the same number of units, the length of every block with be different, as in Figure 3.2, and the longest in polyM and shortest in polyG. Figure 3.2 also shows the feature in the spatial arrangement of every block, which can explain the ability of the gelation

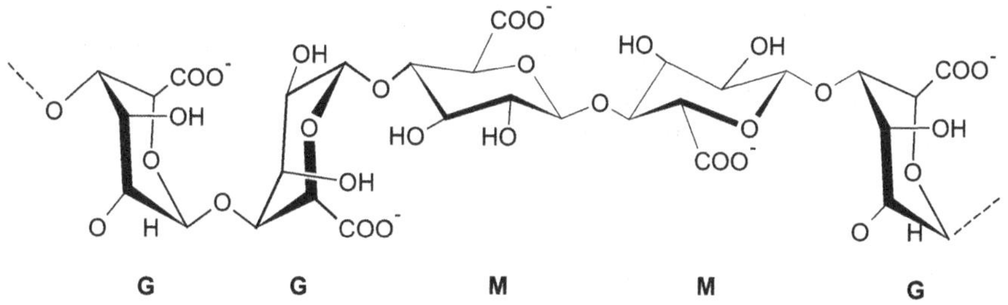

FIGURE 3.1 Molecular structure of alginate. G: guluronate, M: mannuronate.

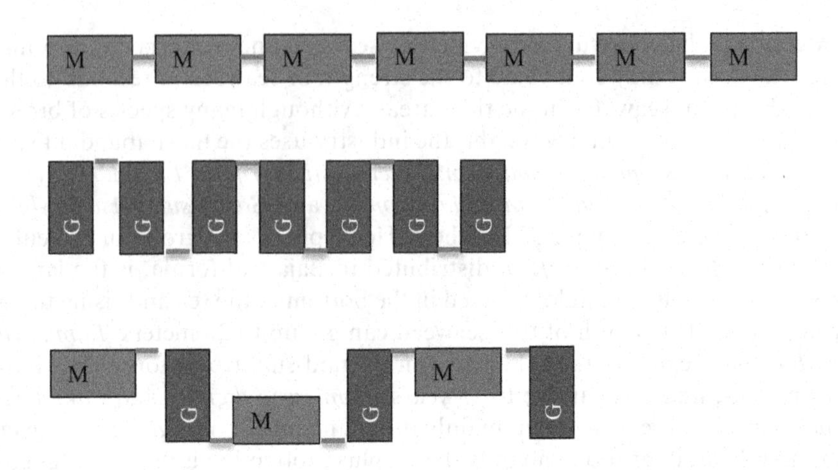

FIGURE 3.2 Relative spatial arrangement of a/ block polyM, b/ polyG and c/ poly MG.

of alginate. The two uronate residues are linked not randomly but into three kinds of blocks: polyM(-MMMMM-), polyG (-GGGGGG-), and alternative MG (-MGMGMG-). The diversity of the chemical structure of alginate is based on the difference in the length of each block and their distribution in the chain of polymers. For characterization of alginate structure, the mole ratio of mannuronate and guluronate M/G is used (Haug & Larsen, 1962). However, the gelation ability of alginate depends on the block polyG, so the dyad (group of two monomers) and triad (group of three monomers) relative frequencies are used to show many guluronate residues consecutively. Although in many papers, to express the probability of the occurrence of an event of a monad, dyad, and triad, the term frequency was used, but relative frequency is more precise. The set of parameters for describing the chemical structure of alginate includes M/G ratio; four dyad frequencies of MM, GG, MG, GM; and eight triad relative frequencies MMM, GMM, MMG, GMG (M center), GGG, MGG, GGM, MGM (G center) (Grasdalen et al., 1981).

F_M means the relative frequency of the occurrence of the event M in the polymer chain, similar for F_G. F_{MG} is the relative frequency of the event of M beside G. Because the events of M and G built up a sample space, the total of their relatives frequencies is 1. In similar, the sample space of dyads includes four events of MM, MG, GM, GG, and the sample space of triads built up from eight events of MMM, MMG, GMM, GMG, GGG, GGM, MGG, MGM, and the total of relative frequencies for every sample space is 1.

In probabilities and statistics, we have (Grasdalen et al., 1979):

$$F_M + F_G = 1 \tag{1}$$

$$F_{MM} + F_{MG} + F_{GM} + F_{GG} = 1 \tag{2}$$

$$F_{MMM} + F_{GMM} + F_{MMG} + F_{GMG} + F_{GGG} + F_{MGG} + F_{GGM} + F_{MGM} = 1 \tag{3}$$

For long chains, more than 20 monomers, the correction of the reducing end is not necessary, so:

$$F_{MG} = F_{GM} \tag{4}$$

We also have the relationship between monad and dyad relative frequencies as:

$$F_{MM} + F_{MG} = F_M \tag{5}$$

$$F_{GG} + F_{GM} = F_G \tag{6}$$

Then between dyad and triad (Grasdalen et al., 1981):

$$F_{MMM} + F_{GMM} = F_{MM} \tag{7}$$

$$F_{MMG} + F_{GMG} = F_{MG} \tag{8}$$

These frequencies can be acquired by ^{13}C or ^{1}H nuclear magnetic resonance (NMR) spectra. With the higher resolution of NMR spectrometer in the future, the tetrad relative frequencies may be achieved.

The ^{1}H NMR spectra can be used for obtaining the data for calculating the dyad frequencies and G-centered triad frequencies in high-field NMR (Grasdalen, 1983). For getting all the triad relative frequencies, we have to acquire the ^{13}C spectra with high-resolution NMR and many scans. In the position of carbon C1 in the spectra, the peak is split into eight small peaks by the influence of the neighbors that express eight triads (Grasdalen et al., 1981).

The data of relative frequencies of alginate (monad and dyad) from different species of brown algae, acquired by proton NMR spectra, is presented in Table 3.1. In the table, we can see the M/G and dyad frequencies are different from species to species of seaweed. The old stipe of *Laminaria hyperborea* gave the highest G content, up to 70% and F_{GG} of 66%. Alginate made from old tissue of *Laminaria hyperborea* shows the strong gelation. The alginates from *Sargassum* have the relative lower G content, in which *Turbinaria ornata* produced the highest G content. In nature this brown seaweed has a strong and rigid thallus. For comparation, alginate from bacteria *Azotobacter vinelandii* is very rich in mannuronic content with F_M up to 85%. The alternative block MG occurred with the low relative frequency.

For obtaining more information about the structure of alginate, the ^{13}C NMR spectra was used. The composition of monad, dyad, and triad frequencies of alginate from several brown algae are shown in Table 3.2. In this table, the triad frequency F_{GGG} of *Laminaria hyperborea* was highest, 0.51 and lowest is alginate from *Ascophylum nodosum*, 0.12. Among *Sargassum*, the alginate from the stipe of *Sargassum mcclurei* gave the F_{GGG} 0.41, higher than other *Sargassum*. Conversely, the alginate from *Laminaria digitata* and *Sargassum kjellmanianum* gave the high F_{MMM}, 0.38 and 0.39, respectively. The mixed triad frequencies were relatively low in all the seaweed.

TABLE 3.1
Composition of Uronate of Alginate from Different Sources by 1H NMR Spectra

Alginate Sources	Monad Frequencies		Dyad Frequencies			M/G	References
	F_M	F_G	F_{MM}	F_{MG} F_{GM}	F_{GG}		
Ascophylum nodosum	0.60	0.40	0.4	0.2	0.2	1.50	(Grasdalen et al., 1979)
Laminaria digitata	0.62	0.38	0.49	0.13	0.25	1.63	(Grasdalen et al., 1979)
Laminaria hyperborea (Old stipe)	0.30	0.70	0.26	0.04	0.66	0.43	(Grasdalen et al., 1979)
Sargassum mcclurei	0.54	0.46	0.44	0.1	0.36	1.20	(Nghia, 2000)
Sargassum polycystum	0.51	0.49	0.43	0.08	0.41	1.04	(Nghia, 2000)
Sargassum kjellmanianum	0.56	0.44	0.45	0.11	0.33	1.27	(Nghia, 2000)
Turbinaria ornata	0.43	0.57	0.41	0.02	0.55	0.75	(Nghia, 2000)
Azotobacter vinelandii	0.85	0.15	0.70	0.15	0	5.7	(Grasdalen et al., 1979)

The dyad and triad frequencies from different species of seaweed show that the distribution of M and G did not follow the Bernoulli and first-order Markov principles.

The chemical structure of alginate varies from species to species of seaweed. In the same species, the alginate structure depends on the part as tip or thallus. In the hard tissue like cortex in the stipe, the content of G is higher than in the thallus, that needs to be tender for vibrating under the wave of seawater. In *Laminaria hyperborea* growing in Norway, the Guluronic content in stipe and in thallus are significantly different, 0.71 and 0.51, respectively (Grasdalen, 1983). Because of that, all the data of alginate chemical structure always ranges in the interval, which makes the properties diverse.

Beside the seaweed, alginate produced from bacteria via biosynthesis. The two genera of bacteria that can secrete alginate are *Pseudomonas* and *Azotobacter* in which the two species gaining more interest are the opportunistic human pathogen *Pseudomonas aeruginosa* (Linker & Jones, 1964) and the soil-dwelling *Azotobacter vinelandii* (Gorin & Spencer, 1966). The mechanism of alginate biosynthesis is similar in the two species but they produce alginate for different purposes. *Pseudomonas aeruginosa* secretes alginate to form the biofilm, whereas *Azotobacter vinelandii* produces alginate to build the resistant cysts. The alginate biosynthesis in bacteria is very complicated and includes many steps to produce mannuronic acid to build the polymannuronic chain. After that the acetylation in C2 and C3 of the pyranose rings, then the C5 epimerization changes L-mannuronic residues into D-glurunonic residues under the catalyst of C5-mannuronan epimerase. The alginate from *Azotobacter vinelandii* has the M/G ratio similar to alginate from seaweed, but the alginate from *Pseudomonas aeruginosa* shows a high content of guluronate (up to 40%) (Hay

TABLE 3.2
Composition and Distribution of Uronic Residues in Alginate from Different Brown Algae Acquired by ^{13}C NMR Spectra

Alginate Sources	Monad Frequencies		Dyad Frequencies			Triad Frequencies						Ref.
	F_M	F_G	F_{MM}	F_{GM} F_{MG}	F_{GG}	F_{MMM}	F_{MMG} F_{GMM}	F_{GMG}	F_{GGG}	F_{GGM} F_{MGG}	F_{MGM}	
Ascophylum nodosum	.57	0.43	0.34	0.23	0.20	0.23	0.11	0.09	0.12	0.08	0.15	aa
Macrocystis pyrifera	0.58	0.42	0.38	0.20	0.22	0.32	0.06	0.14	0.17	0.05	0.15	aa
Laminaria digitata	0.60	0.40	0.46	0.14	0.26	0.38	0.08	0.06	0.22	0.04	0.10	bb
Laminaria hyperborea	0.32	0.68	0.20	0.12	0.56	0.15	0.05	0.07	0.51	0.05	0.07	ca
Sargassum mcclurei (stipe)	0.49	0.51	0.41	0.08	0.43	0.35	0.06	0.02	0.41	0.02	0.06	cc
Sargassum kjellmanianum	.58	0.42	0.46	0.12	0.30	0.39	0.07	0.05	0.27	0.03	0.09	cc
Sargassum polycystum	0.50	0.50	0.40	0.10	0.40	0.34	0.06	0.04	0.36	0.03	0.07	cc

a. (Stokke et al., 1991)
b: (Grasdalen et al., 1981)
c: (Nghia, 2000)

et al., 2013). However, when using the C5 epimerase to catalyze the epimerization of polymannuronate in vitro, the equilibrium content of guluronate can reach the value of 75% (Schürks et al., 2002).

Even with the set of parameters noted earlier, it is not enough to characterize the chemical structure of alginate because there is no information of the length of every block and the distribution of blocks in the whole polymer chain. The full information of every alginate chain is only obtained by sequencing, for example, by cleaving the linkage with alginate lyase (Aarstad et al., 2012).

The molecular weight of commercial alginate ranges from 10^4 to 10^5 g/mol. It depends on the source of alginate and the extraction technology (Clementi et al., 1998; Fourest & Volesky, 1997; Fertah et al., 2017). The high temperature and acidic conditions can break the glycosidic linkage and reduce the degree of polymerization. In addition, during the storage, the dP of alginate gradually decreases (McHugh, 1987).

3.2.3 Properties of Alginate

3.2.3.1 Rheology Properties

Alginate has two main properties: the thickening and the gelation ability. Based on the long chain of rich OH group uronate residues, the sodium alginate dissolves in water and makes a high-viscosity solution. The higher dP alginate is, the higher viscosity of solution is obtained. This feature is applied in industry for stabilizing and thickening the solution. In rheological properties, sodium alginate solution is non-Newtonian (Haug & Smidsrød, 1962), and its behavior follows the pseudoplastic law (Mancini et al., 1996), which means the viscosity decreases with the increase of the shear stress. Related to this application, the high mannuronate alginate is preferred. With the flat chain of block polyM, the alginate chain is more flexible in configuration.

The rheological properties of alginate solution are still controversial. In some papers, the reduced viscosity η_{sp} / c increases with the concentration of sodium alginate (Mancini et al., 1996). Cristina et al. recognized the relationship between reduced viscosity and concentration separated into two parts: with the very low concentration until 0.25 g/dL, the reduced viscosity decreases with the increase of concentration but in the higher concentration, the reduced viscosity increased (Cristina et al., 2014). This strange behavior of sodium alginate solution may be the influence of ionic strength to the conformation of alginate molecules (Smidsrød, 1970).

For measuring the molecular weight of alginate, the Mark-Houwink-Sakurada equation, expressing the relationship between the intrinsic viscosity and the average molecular weight is used: $[\eta] = KM_w^a$, in which K and a are the experiment constants. Alginate chains composed of many negative discharged groups (OH^- and COO^-), and they may have the extended configuration by their repulsions and cause the increase of the viscosity. If cations are added to the solution (salts or acid), being characterized by ionic strength I, alginate will be neutralized either partially or totally; therefore the configuration of alginate will change from a relative rigid rod to a flexible coil that leads to the reduction of viscosity. Olav Smidsrød has proved the linear relationship between intrinsic viscosity $[\eta]$ and $\frac{1}{\sqrt{I}}$ (Smidsrød, 1970). With the different salts, the influence of ionic strength to the intrinsic viscosity is different, too. But in the infinite ionic strength, all the salts have the same effect on viscosity. Based on these results, the constants K and a in the Mark-Houwint-Sakurada equation have the different values depending on the ionic strength. For instance, when $I = 0.01$; $I = 0.1$ then $K = 4.8 \times 10^{-6}$; 2.0×10^{-5} and $a = 1.15$; 1.0, respectively. In infinite ionic strength, $I = \infty$, the constant $a = 0.83$, which means the configuration of the alginate molecule is still quite extended (Smidsrød, 1970).

This feature in configuration of alginate makes it a high-viscosity agent in many applications.

3.2.3.2 Gelation of Alginate

The gelation ability of alginate with the cations is more interesting and found in so many applications, especially Ca^{2+} gelation. For a long time, the model egg-box with the blocks polyG was believed to build the junction zone in the gel (Grant et al., 1973). With the configuration of guluronate, the polyG blocks make the ribbon chain and create the hole for trapping the cation Ca^{2+} following the 2/1 helical conformation. The polyG block with Ca^{2+} will be piled up with another polyG block to make the junction zone. The polyM and MG blocks, having no significant role in building up the junction, will make an interrupt zone with the 3D crosslink network to keep the solution inside. However, based on recent studies, there is more evidence for the role of the MG block and polyM block that can take part in the junction zone with the polyG block (Donati et al., 2005). In addition, x-ray scattering examined the gel rising for the proof for a 3/1 helical conformation, although it has not excluded the 2/1 helical conformation in case of the fast gelation process (Li et al., 2007).

In the fast gelation with Ca^{2+}, as illustrated in Figure 3.3, when the sodium alginate is dropped into the Ca^{2+} solution ($CaCl_2$), the process of gelation happens immediately in the surface of the drop. In the meantime, when the cations of Ca^{2+} penetrate into the drop, the gelation will pull the alginate molecules from the center of the drop to the surface. Because of that, the content of alginate near the surface of the drop (becoming the gel bead) is higher in the center and we obtain the heterogenous gel. Figure 3.3 shows that the center of the gel bead has the very low alginate content and becomes the hole of solution. This feature is convenient for immobilization of cells since the cells will accumulate with high density near the surface, the mass transfer into and out of the gel is faster, and the effectiveness of the biocatalyst is better. The mechanism of fast gelation is proved in Figure 3.4. In the cross-section of a calcium alginate gel bead after the immobilization of yeast cells, Figure 3.4 clearly shows the hollow area (zero zone) in the center of the gel bead with only aqueous solution, with the gel with cells concentrated near the surface of the gel bead.

When the homogenous gelation is required, the ion Ca^{2+} will be released gradually. At first, the alginate is mixed to $CaCO_3$ that is insoluble in water at neutral pH. After that the D-glucono-δ-lactone will be added to the mixture

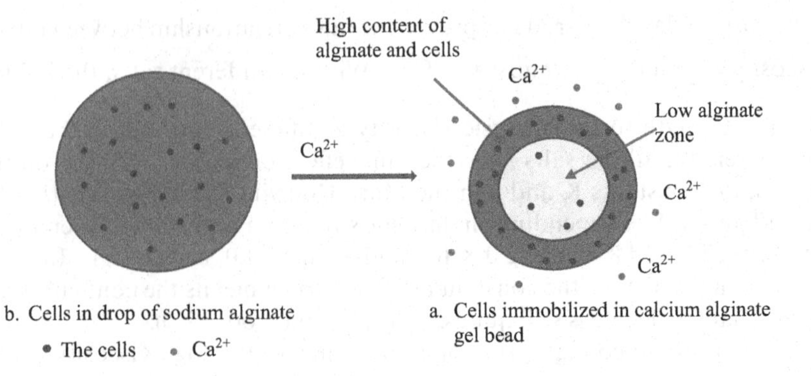

FIGURE 3.3 The shrinkage of the gel bead in the gelation of alginate with a calcium ion.

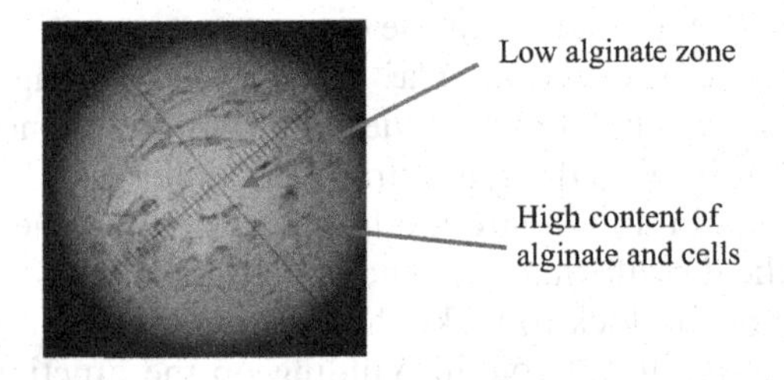

FIGURE 3.4 The cross-section of a hollow calcium alginate gel bead encapsuled in a yeast cell.

of Ca^{2+} and alginate for reducing the pH in order to dissociate the ion Ca^{2+} into the solution, and the gelation occurs gradually (Draget et al., 1990).

In the process of gelation, the alginate molecules are reorganized, more compact, and shrink in volume. In the high content of polyG alginate, the shrinkage, measured by the ratio between the weight of the gel bead and the weight of drop w/w_0, is much more than the low polyG alginate (Yotsuyanagi et al., 1987). This effect can be explained by the 3D network built in the case of high polyG taking up a larger volume than the low polyG alginate.

3.3 CHITIN/CHITOSAN

3.3.1 Chitin

3.3.1.1 Source

Chitin is the polysaccharide composed of β -(1 $\rightarrow$ 4)-*N*-acetyl-D-glucosamine and is the second biosynthesis biopolymer in quantity in the world, just after cellulose. Chitin is synthesized in nature to form the structural components such as in the exoskeleton of arthropods or in the cell wall of fungi and yeast. The main source of commercial production of chitin is only from crab and shrimp shells. The chitin production has two main steps: demineralization by acid and deproteinization by alkaline. The remaining component in the crustacean shells after these operations is mainly chitin. The yield of chitin in commercial sources as lobster, crab, and shrimp shell waste is about 17.5% (lobster shell waste), 20% (shrimp shell waste), and 23.75% (*Squilla* shell waste) (Mohan et al., 2021). Under the conditions of concentrated alkaline and high temperature, the chitin is partially deacetylated, but can be soluble in a dilute acid solution, referred to as chitosan.

3.3.1.2 Structure

The chitin appears in two main allomorphs, α and β forms (Rudall & Kenchington, 1973; Hackman & Goldberg, 1965) depending on the source. The third allomorph γ is seen as the variant of α . The specific arrangements of the chitin chains in allomorphs, as in Figure 3.5, make the chitin from different sources with different in physical and crystal properties. The α allomorph is the most popular in fungi, yeast, krill, lobsters, crabs, and shrimp shell. The β form is found in squid pens and pogonophoran and vestimetiferan worms (Blackwell et al., 1965; Rudall, 1969). The crystallographic parameters of α and β chitin show that the two molecular chains are antiparallel in α form, while in β form, there is only the one parallel orientation of all the molecular chains. In the γ structure, every two parallel chains have one opposite chain (Figure 3.5) (Hackman & Goldberg, 1965).

Chitin exists in the crustacean cuticle under hierarchical levels of structures. Firstly, narrow and long crystalline units are constructed, based on the assemblage of chitin chains. The chains arrange according to the allomorphs. Secondly, these parallel chitin units are wrapped in the periphery by the globular proteins to make the chitin-protein units, which group into fibrils with the different diameters. Thirdly, the fibrils arrange parallelly to make the sheet, including successive planes, with the direction rotating progressively from plane to higher-level plane (Giraud-Guille, 1984).

The protein-chitin sheets also include minerals (mainly $CaCO_3$), a small quantity of lipid and pigment to make the shells of crustaceans (Rødde et al., 2008). The chains of chitin have many hydrogen bonds in the inner chains that cause the insolubility. In the form, there are inter sheets of hydro bonds but not in the form. Chitin therefore is in practice insoluble in all normal solvents. For estimating the molecular weight of chitin through viscosity, the mix of chitin and LiCl in dimethylacetamide solvent is used for dissolving it.

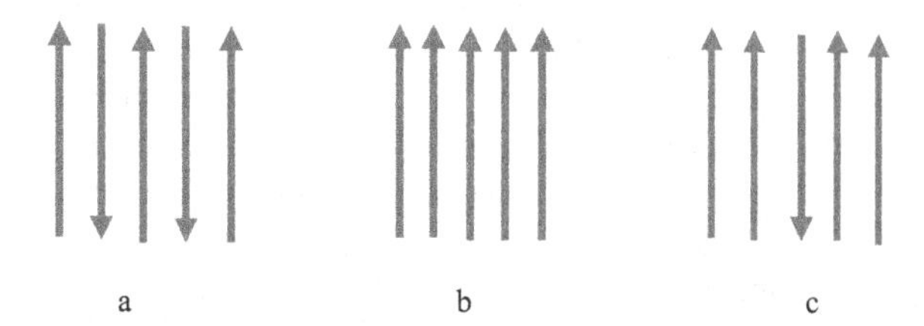

FIGURE 3.5 The three allomorphs of chitin a: α , b: β and c: γ .

Although chitin is referred to poly -(14)-*N*-acetyl-D-glucosamine, a small part of residues is deacetylated and characterized by degree of acetylation (DA) or degree of deacetylation (DD), with DA + DD = 1(or 100%). In marine crustacean shell waste, the DA of chitin is about 65% to 90% (Mohan et al., 2021).

Chitin can dissolve in specific solutions. Weimarn found that chitin can dissolve or swell in inorganic salt aqueous solutions in order of LiSCN > $Ca(SCN)_2$ > CaI_2 > $CaBr_2$ > $CaCl_2$; LiSCN > LiI > LiBr > LiCl; NaSCN > NaI (Weimarn, 1927). Austin reported that dimethylacetamide (DMAc) and N-methylpyrrolidone (NMP) mixed with LiCl can be used as a chitin solvent (US4059457A., 1977). Einbu, Vårum et al. found that chitin could be dissolved in NaOH aqueous solutions, by first immersing chitin powder in a concentrated NaOH at 4 °C for 72 h then diluted to 10 wt% NaOH using crushed ice (Einbu et al., 2004). Although the chitin can be dissolved in concentrated acid, it is degraded and not used in the analysis of molecular weight.

The insolubility of chitin in water and normal solvents makes its application limited. However, its derivative, chitosan, obtained by the deacetylation of chitin that induces the solubilization in dilute acid solution, has a large range of applications.

For characterization of the chemical structure of chitin and chitosan, the molecular weight (MW) and DD are the most important. By adjusting the extraction parameters, we can obtain different chitin and chitosan in terms of MW and DD. The positions in the chitin chain where the deacetylation happens under chemical reaction is random. That means the structure of chitin/chitosan from every batch will be different. Even with the same MW and DD, the distribution of deacetyl glucosamine in the chain can be significantly different. The diverse structure makes the large application of chitosan possible, but it is also a big challenge in controlling the relationship structure-properties of chitosan, especially in biomedicine.

For estimating the MW, the intrinsic viscosity related to the Mark-Houwink-Sakurada equation is used with constants $K = 2.4 \times 10^{-1}$ and $a = 0.69$ at 25°C (Terbojevich et al., 1988).

In case of measuring the DD, the spectra as shown by Fourier-transform infrared spectroscopy and NMR spectroscopy is very useful. Many studies related to measuring the DD have been published and the results were different depending on the sources of chitin and the extraction method and then the deacetylation in the case of chitosan.

3.3.1.3 Preparation

In chitin and chitosan processing, under the reaction of an acid or alkaline, degradation of polymers is hard to avoid because the glycosidic linkage is sensible with acid and alkaline conditions. In case of conserving the whole chitin molecules, in terms of keeping the degree of polymerization, the dilute acid (or organic acid) and dilute alkaline under room temperature are used with the longer time of reaction, and the operation is repeated many times.

In demineralization, many different acids were used, including strong and weak acids as HCl, HNO_3, H_2SO_4, CH_3COOH, and HCOOH, in which the dilute HCl were preferred (No & Hur, 1998).

Although the commercial resources for chitin/chitosan production is limited in crab and shrimp shells, the differences in conditions (temperature, pH and time) for treatment and extraction in acid and alkaline agents resulted in the chitin/chitosan being very diverse in structure and properties.

If the demineralization can be conducted successfully with very low mineral residue, the deproteinization still needs more efforts. The little the protein residue that remains, the harder the deproteinization. It is explained by the close linkage between chitin chains and the protein chain to make the elemental fibers for the chitin sheets. The rest of the protein is a big problem of chitin/chitosan in medical applications because the protein can cause allergy or immunogenicity. Therefore, in tissue engineering, the purity of chitosan in terms of protein residue is very important. There is a wide range of agents for the deproteinization step such as NaOH, Na_2CO_3, $NaHCO_3$, KOH, K_2CO_3, $Ca(OH)_2$, Na_2SO_3, $NaHSO_3$, $CaHSO_3$, Na_3PO_4, and Na_2S, but NaOH is preferred in most papers (Roberts, 1992; Percot et al., 2003).

3.3.2 Chitosan

3.3.2.1 Preparation

The chitosan is a derivative of chitin through the process of de *N*-acetylation that is carried out effectively by concentrated NaOH at high temperature as illustrated in the Figure 3.6. Through this process, a quantity of acetyl groups is removed, and the chain of polymer reveals the amino groups $-NH_3^+$. The deacetylation can be conducted in an acid or base but the strong acid can break out the glycosidic linkage and reduces the degree of polymerization (Sannan

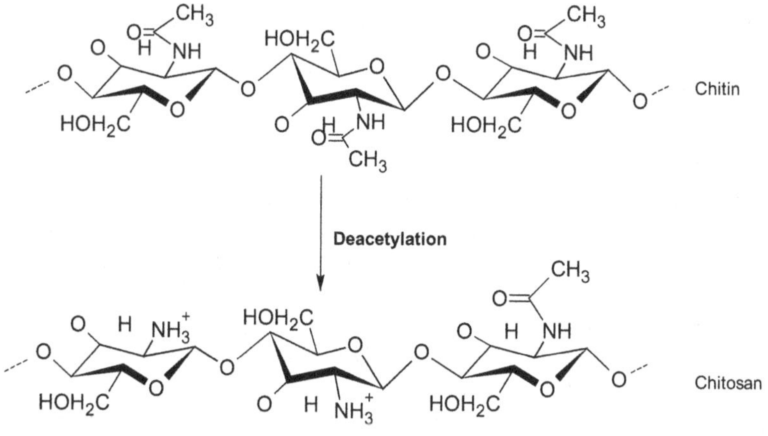

FIGURE 3.6 Molecular structure of chitin and chitosan

et al., 1976; Vårum et al., 1991). Since there is a degree of deacetylation in the natural chitin, the main thing for distinguishing between chitin and chitosan is the ability of solubility of chitosan in dilute acid while the chitin is insoluble. The solubility of chitosan is achieved when the degree of deacetylation is high enough, about 50% (Rinaudo, 2006).

3.3.2.2 Conformation

The intrinsic of chitosan solution depends not only the MW but also the degree of acetylation (F_A) and the ionic strength. When studying the intrinsic viscosity of a chitosan solution with F_A ranging from 0.11 to 0.52 and the ionic strength from 0.12 M to 0.3 M, Berth and Dautzenberg recognized the best fit of the MHS equation to data in the form (Berth & Dautzenberg, 2002):

$$[\eta]\,(\text{mL/g}) = 8.43\times10^{-3}\,M_W^{0.93} \quad (9)$$

The influence of DD on the intrinsic viscosity and so on the constant K and a in the MHS equation was examined in the study of Wei Wang et al. The association of K and a to DD were expressed in the regression equations (Wang et al., 1991):

$$K = 1.64\times10^{-30}.DD^{14.0} \quad (10)$$

$$a = -1.02\times10^{-2}.DD + 1.82 \quad (11)$$

3.3.2.3 Ion Binding

Chitosan can bind to metal ions to form the complexes. The order of selectivity of chitosan to some metal ions is $Cu^{2+} \gg Hg^{2+} > Zn^{2+} > Cd^{2+} > Ni^{2+} > Co^{2+} \sim Ca^{2+}$ (Rhazi et al., 2002). The symbol $\gg$ means the affirmative of chitosan to Cu^{2+} is stronger significantly in comparing to others ions. Chitosan with different fractions of acetylated units (F_A of 0.01 and 0.49) shows the similar selective behavior toward both negative and positive ions: no selectivity to chloride and nitrate ions, but strong selectivity towards molybdate polyoxyanions. Especially, the selectivity to Cu^{2+} is large excessive to other ions such as Zn^{2+}, Cd^{2+} and Ni^{2+} (Inger et al., 2003).

3.3.2.4 Solubility and Viscosity

The solubility is an interesting property in commercial chitosan. This property depends on three parameters. The pH of the solution is the parameter that links to the charge of D units in the polymer chain. The ionic strength is important and relates to the salting out effect. Finally, the existence of some ions as copper or molypdate, with strong selectivity, influence the solubility of chitosan. All chitosan is soluble in the pH below 6.5 and precipitate when the pH increases from 6.5 to 8. With the same pH, the solubility increases with F_A (Vårum & Smidsrød, 2004) to alginate, the relationship between intrinsic viscosity $[\eta]$ and $\frac{1}{\sqrt{I}}$ of chitosan is linear (Anthonsen et al.,1993).

When we examine the influence of chitosan structure on the hydrodynamics, there are the two opposite effects. In principle, the longer the polymer chain, the higher the resistance to the flow of the fluid, which means higher viscosity. However, the rate of the viscosity increase can be influenced by the degree of acetylation of chitosan. In case of high F_A, the chitosan chain possesses more *N*-acetyl groups that promote the hydrogen bonds between the polymers and water molecules, that induces the increase in resistance of chitosan solution under shear stress. On the other hand, in the low F_A chains, the electrostatic effect of positive charged amino groups makes the chain more extended and also increases the resistance for shear strain and so the viscosity. Anthonsen et al. obtained the regression equations between intrinsic viscosity and the MW of chitosan with different F_A in linear form $Y = a + bX$, in which Y is for $\log[\eta]$, X for $\log M_n$. The values of F_A are 0.6, 0.15 and 0.0, the values of b are 1.06, 0.778 and 0.583, respectively. That result reveals the tendency to increase the viscosity under the effects of N-acetyl groups. The F_A reveals the two opposite effects on viscosity in the K and a constant in MHS equation (Anthonsen et al., 1993):

$$logK = -0.427 - 3.821F_A \tag{12}$$

$$a = 0.6169 + 0.759F_A \tag{13}$$

3.3.2.5 Sequence and Distribution of Acetyl Groups

The NMR spectra is rich in information about the sequence and distribution of acetyl groups in the chain of chitin and chitosan. Through the NMR, the frequencies of dyads (F_{AA}, F_{AD}, F_{DA}, F_{DD}) and triads (F_{AAA}, F_{AAD}, F_{DAA}, F_{DAD}; F_{DDD}, F_{DDA}, F_{ADD}, F_{ADA}) are acquired. In the ^{13}C NMR spectra, the anomeric carbon C_1 reveals the influence of the nearest-neighbor residues, and we obtain the two groups of signals, one for A centered triads, and another for D centered triads. The signals of D centered triads dominate in the chemical shift ranging from 99.5 to 100 ppm and correspond to low F_A (F_A = 0.06) chitosan, while the ones of M centered triads are strong in the zone of from 103.6 to 103.7 ppm with the high F_A (F_A = 0.54) chitosan. Based on the data of dyad frequencies, the distribution of acetyl groups is random and is consistent with the distribution of Bernoulli. The information from triad frequencies ruled out the block arrangement in chitosan (Vårum et al., 1991). This random distribution of acetyl group can be explained by the chaos movement of the molecules in the chemical reaction of deacetylation by alkaline. In chemical thermodynamics, the reaction is carried out by the random collisions between the reagents. In this case, the alkaline molecules collide randomly to the acetyl linkage along the chitin chain that results in the deacetylation and leaves the chain of chitosan with a random distribution of acetyl groups. The random collision also reveals that although the same F_A is in a chitosan sample, the distribution of acetyl groups in every chain is not identical. This also explains that the distribution of the acetyl group in different batches may not be the same.

3.4 BIOMEDICAL APPLICATIONS

Marine polysaccharides have the large range of applications in food industry, cosmetics, industry, pharmaceuticals, and biomedicine and are explored in many review papers. In this chapter, we limit the area of application in biomedicine in relation to gelation ability, including drug delivery system, wound dressing, cell culture, and tissue engineering. These applications require the sophisticated structure and state of the art in preparation. In some decades, there have been so many studies of alginate and chitin/chitosan in biomedicine. The reason for that is the two biopolymers have the same properties, including non-toxicity, biocompatibility, bioadhesion, biodegradability, and nonimmunogenicity, as shown in Figure 3.7 (Gombotz & Wee, 1998; Dash et al., 2011), stimulating the process of recover or wound healing (Aderibigbe & Buyana, 2018; Yu et al., 2016; Feng et al., 2021), ability of preparing in different forms as hydrogel, film/membrane, sponge, foam, and nanofiber (Dash et al., 2011; Kumar et al., 2020).

In many applications, we often meet the combination of alginate and chitosan in making hydrogels with the functional properties better than only alginate or chitosan gel. Chitosan is the polycation with the pKa-values of the amino-groups of 6.5, while alginate is a polyanion with the pKa-values carboxyl-groups of ca. 3.5. They therefore will link by an ionic interaction stronger by than hydrogen bonds in the pH interval from 4 to 6. When analyzing the structure of alginate and chitosan, Khong et al., under supervision

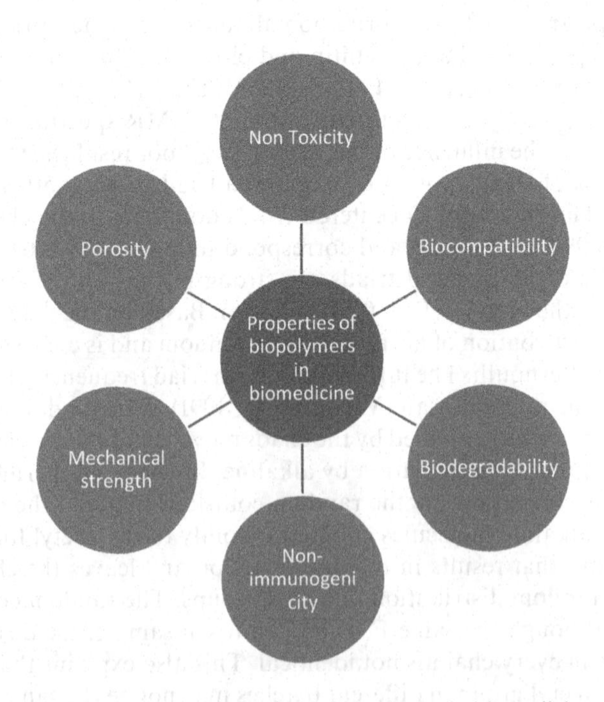

FIGURE 3.7 The main properties of marine biopolymers in biomedicine.

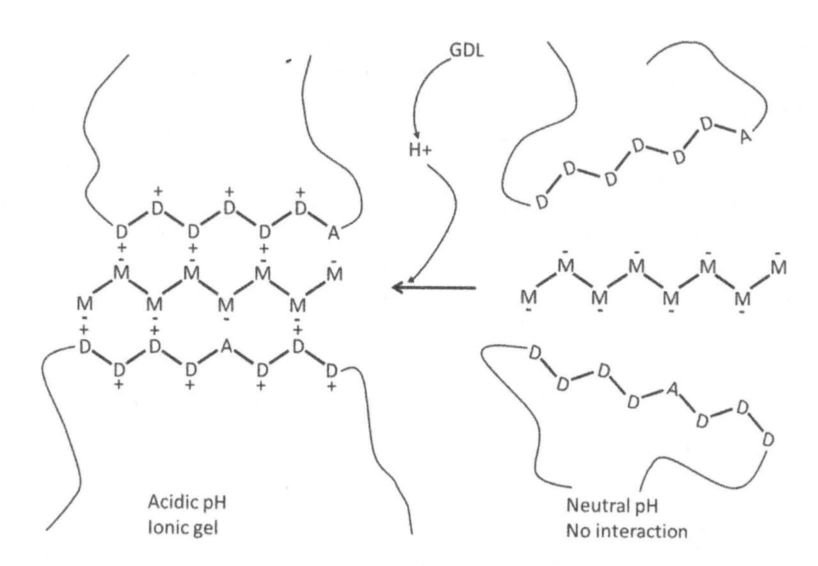

FIGURE 3.8 Schematic illustration of the two gelling systems. In the first system (a) the oligomers are derived from chitosan and the polymer is alginate, while in the second system (b) the oligomers are derived from alginate and the polymer is chitosan.

Source: Khong et al., 2013, with permission.

of Prof Vårum, recognized that there was a similar configuration between the block polyM in alginate and chitosan. The mannuronic units have the $^4C_1$ configuration that linked through β-(1 − 4) diequatorial glycosidic linkages with the length of disaccharides about 10.3 to 10.4 Å while it is shorter, about 8.7 Å, between the two guluronic units with the $^1C_4$ configuration. The same distance of the mannuronic units and a fraction of deacetyl N-glucosamine with the same positions of a pair of opposite charges induces the gelation, as illustrated in Figure 3.8. In this experiment, the partially acetylated chitosan is mixed with polymannuronic alginate oligomers at a neutral pH to chitosan, and there is no ionic interaction with alginate. The proton donor glucono-δ-lactone (GDL) is added for controlled release of protons that lowers the pH and protonates the amino groups of the chitosan inducing the gelation. Figure 3.8 shows the linkage between every three monomers in a chain (Khong et al., 2013).

3.4.1 Drug Delivery

The controlled-release technology is getting more interest for many advantages comparing to the conventional medical drugs. In normal medicine (as a tablet or pill), after administration, the drugs dissolve quickly in stomach or intestine into a liquid and then release the active substance into plasma and distribute throughout the body. However, once entering the body, the process of elimination begins. The drug is metabolized in the liver and passes out the body through urine, breath, sweat or with small quantities in saliva and breast milk. This process causes the strong fluctuation of the drug content in plasma: the very high drug content may be up to a toxic level in the early time after administration and may cause side effects to the patients; the duration of effective drug content is short, then the content decreased to an inappropriate value (Moodley et al., 2012). This situation introduces to increase the frequency of administration, makes the patients uncomfortable by short time of effectiveness and more side effects. The controlled-release technology is based on the absorption or entrapping of the drug into the matrix gel of biopolymer as alginate of chitosan. When the drug moves into the gastric intestinal tract, the gel is degraded and that releases the drug gradually. This technology can eliminate the limits of the conventional drugs (Kumar, 2000).

3.4.1.1 Alginate in Drug Delivery

Alginate can be used in drug delivery under different forms as tablets, capsule, hydrogels, films, or nanofibers. The simple way used sodium alginate as an excipient that is mixed into a drug and pressed into tablet. When going through the gastrointestinal (GI) tract, the tablet will absorb the water in the stomach at a low pH (pH = 1.2 to 2), swells, transforms to alginic acid gel and then releases the drugs. After that when the tablet moves to the intestine, the pH will increase to 6.5 to 7. With the high Na^+ concentration in the GI tract, the alginic acid will exchange ions and return the sodium alginate and dissolve. This process depends on many factors such as the MW of alginate, the M/G ratio, the particle size, the ratio of alginate/drug, and the porosity of the tablet. Liew et al. have carried out the impressive examination on the influence of these factors on the release of chlorpheniramine maleate from sodium alginate excipient tablet. The result showed that in low pH, the small sodium alginate particles rehydrated faster than the large particles, transformed to acid gel, and reduce the burst release. The alginate with high mannuronate slows down the release better than high guluronate. The high concentration of alginate also controls the release better than low concentration. In general, the alginate excipient can keep the release of drug within eight hours (Liew et al., 2006). The influence of various grades of alginate (with different M/G ratio, size and viscosity), sodium/calcium alginate and ammonium/calcium alginate, pH of medium on swelling, erosion and drug (metronidazole) and release from alginate-based matrix tablets were investigated in the study of Sriamornsak et al. The influence of the M/G ratio of sodium alginate is different depending on the pH. The high M alginate tablet released the drug faster in a buffer of pH 6.8 than in pH 1.2, while the high G alginate tablet delivered drug faster in acid solution with pH 1.2 and slowest in pH 6.8. Especially, the two mixtures of salt alginate, sodium/calcium alginate and ammonium/calcium alginate, were disintegrated very quickly in pH 1.2, released up to 90% in 30 minutes (Sriamornsak et al., 2007). Beside the solid form in tablet drug, alginate can be prepared under hydrogel, film, sponge, and scaffold forms that can contain and release the drugs in different applications in medicine as discussed in the next sections.

3.4.1.2 Chitosan in Drug Delivery

The solubility/insolubility of chitosan when changing the pH of the solution is determined by the pKa value of the D-units of approximately 6.5, the degree of acetylation F_A, and the MW. A polymeric chitosan with $F_A$will not precipitate even with pH values well above the pKa-value (Vårum et al., 1994).

The chitin is non-toxic and easy to make in a gel by adjusting the pH to a low value. The matrix of chitosan can absorb or entrap the drug then release it slowly into body. In addition, chitosan shows antacid and antiulcer activities which is of benefit in prevent the stomach from drug irritation (Hillyard et al., 1964). In the acid conditions of the stomach, the chitosan gel will float and swell gradually and release the drug.

Miyazaki has compared the solution time under powder form and the release time from chitin and chitosan gel of the two drugs indomethacin and papaverine. The results showed the very fast dissolution of the powder form, about 20 minutes, while the sustainable release of drug from chitin and chitosan gel can last to 5 hours (Miyazaki et al., 1981).

The drug release from the gel occurs by the degradation (chemistry or enzyme), swelling or erode of the gel. After contact with the liquid in the body, the tablet will absorb water. The matrix will swell and increase the dimension of the pores that favors the diffusion of the small molecule drugs. In the low pH, the chitosan will be soluble; the crosslink matrix is degraded and releases the drug.

Many studies examine the release in vitro with different buffers and static conditions (Miyazaki et al., 1981; Felt et al., 1998). However, in the GI tract, the pH changes through the digestion system. Starting with the highly acidic gastric environment, pH is 1.5 to 2, the pH rises rapidly to pH 6 in the duodenum and increases along the small intestine to pH 7.4 at the terminal ileum. After that the pH in the cecum drops below pH 6 and rises in the colon, reaching pH 6.7 at the rectum. In the GI tract, under peristalsis of the GI tract and churning of the stomach (Hua, 2020), the tablet may be abrasive that can influence the release the drugs. The change of pH along the GI tract is the challenge for controlled drug delivery (Lin et al., 2017). In case of chitosan gel, the fluctuation of pH can change the association of the matrix. For improving the chitosan gel properties, some anions or polymers like carrageenan and alginate are used (Leng, 2009).

3.4.2 Wound Healing

The skin takes more responsibilities in physiological functions of the body. It protects the internal organs from invasions of pathogens like bacteria and viruses; regulates the heat exchange through sensible heat and latent heat (through sweat); receives the signals due to exposure to chemical and physical stresses as toxins, pressure, or temperature (cold or hot); keeps the fluid homeostasis, and works as an active immune organ (Salmon et al., 1994). Naturally, the skin can restore the integrity after injury through the complicated process with the interaction of many factors. In acute injury, the healing process includes four phases: hemostasis, inflammation, proliferation, and remodeling. The first phase of wound healing, hemostasis, starts immediately just after wounding and includes vasoconstriction to limit the bleeding, temporary blockage by the platelet plug to seal the break in vessel wall, and coagulation of blood to form the fibrin clot (Daniel, 1993; Periayah et al., 2017). The inflammation is the second phase of wound healing and begins after injured blood vessels leak the transudate. Inflammation controls the bleeding and prevents the wound from infection. The platelets not only form the hemostasis plug but secrete several mediators, such as platelet-derived growth factor, to attract and activate macrophages and fibroblasts. Beside cleaning the wound from foreign particles and bacteria by neutrophils, many different growth factors, the cytokines, and cells initiate the formation of granule tissues. In the phase of proliferation, the epidermal cells separate dermal cells and migrate between collagenous dermis and fibrin eschar. One or two days after injury, the epiderma cells at the wound margin proliferate. For supporting materials to the wound in order to build the new tissues, the new blood vessels are formed under the complex process that are based on the extracellular matrix in the wound bed and several angiogenesis factors. Finally, in the remodeling phase, the wound contraction and the extracellular matrix reorganization recover the wound and leave the scar (Singer & Clark, 1999).

The wound dressing can be classified as the traditional, biomaterial-based, interactive, and bioactive dressings (Dhivyaa et al., 2015) but overlap. The traditional dressing (or passive) has the two main targets: to help in eliminate the bleeding and to protect the wound from surrounding. This dressing, made by cotton, often adheres to the wound and damages the newly formed epithelium on removal. The biomaterial-based dressings are categorized into allografts, tissue derivatives, and xenografts. The interactive wound dressings are prepared by biopolymers and synthetic polymers in different forms as gels, foams, films, spray, and composites. The most common biopolymers used are alginate, chitosan, and gelatin. The bioactive wound dressing in which growth factors and antimicrobials are added improve the wound healing process (Sezer & Cevher, 2011; Aderibigbe & Buyana, 2018).

The wound dressing plays the role of controlling the bleeding and protecting the wound from infection. Further, it helps to set up the humid conditions for healing, absorb the blood and body fluid (exudate), and gas transfer. Beside the protection role, the wound dressing can release some drug for limiting pathogens and facilitating the wound healing. Nowadays, many new wound dressings with different materials and conformations are developed (Shakeel Ahmed & Ikram, 2016; Feng et al., 2021; Jayakumar et al., 2011). Every kind of wound dressing has the specific advantages and disadvantages. Among them, hydrogels from biopolymers as chitosan, collagen and alginate are possible. The chitosan hydrogel bandages have the properties of antibacterial based on the positive charge of the molecules, moisture conditions, biocompatible, and promotion of the healing process. The multiple biopolymer dressing seems to integrate the advantages of different biopolymers and become the current use.

3.4.2.1 Alginate in Wound Healing

Alginate has valuable properties such as biocompatible, water retention, porosity, and ability to load and release bioactive agents. To improve the mechanical characteristics, alginate is often used with other polymers and chitosan, gelatin, and poly(vinyl alcohol). Alginate wound dressings can be prepared under forms of hydrogels, film/membranes, nanofibers, foams, and topical formulations (Aderibigbe & Buyana, 2018). Beside the alginate hydrogel base, the drugs are loaded into the matrix and released in the progress of treatment.

In combination of alginate hydrogel and simvastatin in mesoporous hydroxyapatite microspheres, Yu et al. recognized the dressing enhanced the formation of new vascular and re-epithelialization of the cutaneous wounds. The simvastatin released from the gel has an effect on the angiogenic differentiation of human umbilical vein endothelial cells (HUVECs) (Yu et al., 2016).

Alginate hydrogels were mixed with zine oxide and revealed re-epithelialization after 48 h, revealing that the zinc oxide nanoparticles enhanced the proliferation and migration of keratinocyte cells to the wound site (Mohandas et al., 2015); with polyvinyl pyrrolidone containing nanosilver exhibiting good antimicrobial activity (Singh & Singh, 2012); with chitin/chitosan and fucoidan providing a moist environment for an enhanced healing process in full-thickness skin defects on rats, forming granulation tissue and capillary formation (Singh & Singh, 2012).

Alginate wound dressing under nanofibers form mimics the extracellular matrix, thereby setting up the conditions for the mass transfer of growth factors, motive of the cells, which enhances the proliferation of epithelial cells and the formation of new tissue. The alginate nanofibers are made in a combination of alginate and synthetic polymers as poly(vinyl alcohol) and poly(ethylene oxide). For making the nanofiber of alginate by electrospinning, the combination of the poly(vinyl alcohol)/sodium alginate was studied by Üstündağ et al. This nanofiber mat was examined as a wound dressing in vivo (rabbit) in comparing with cotton gauze, Bactigras, and Suprasorb-A (made from calcium alginate). The results showed that after 12 days, the wound healing performance, in terms of wound constriction, was nearly the same, in which the Suprasorb-A was the best (Üstündağ et al., 2010). The similar nanofiber with moxifloxacin hydrochloride also was found in study of Fu et al. This dressing was applied in rats with full-thickness round wounds and revealed significant wound healing. It also exhibited good antibacterial activity against *S. aureus* and *P. aeruginosa* (Fu et al., 2016).

Park et al. prepared sodium alginate nanofibers by blending with poly(ethylene oxide) and lecithin. Lecithin, a natural surfactant, was combined to fabricate the uniform nanofiber. This electrospun nanofiber exhibited good water absorption, biocompatibility, and promoted the wound healing (Park et al., 2010).

3.4.2.2 Chitosan in Wound Healing

Chitosan dressings can be prepared under different forms in wound healing technology, including chitosan with synthetic/natural polymer blend scaffolds, chitosan-based sponges, chitosan-based oil immobilized scaffolds, chitosan-based extract immobilized scaffolds, and chitosan-based drug loaded scaffolds. These chitosan-based dressings can adopt different characteristics of the wound as dry or more exudate, acute or chronic, area and depth of injury, pains, and protection from bacteria (Shakeel Ahmed & Ikram, 2016).

Zang et al. prepared the CCS sponge made from 3% chitosan and 1% collagen for studying the wound healing in comparing chitosan and collagen dressings. The results demonstrated the new dressing as safe towards NIH3T3 cells, inhibiting *E. coli* and *S. aureus,* holding the moisture in vitro. When testing in vivo the CCS dressing on rats, the wound healed faster, the injure area constricted and was smaller in comparing to only chitosan and only collagen sponge (Zhang et al., 2021). It can be explained by that the hydrogels possessed properties of safety, biocompatible, antibacterial, moisture retention from chitosan and facilitate the proliferation of epiderma cells from collagen.

Although there are many studies on the chitosan-based dressing in wound healing, the papers did not support the properties of chitosan molecules in which the leak of information about the degree of acetylation and its distribution is very important. On the other hand, most studies concentrate on the total effect of the chitosan dressing based on the images of the wound appearance and tissue anatomy but not going further to the biochemical aspect to show clearly the role of chitosan in every phase of the wound healing process. If more available data about the relationship between molecular properties and the function of chitosan were available, we could approximate the best design of chitosan-based dressing fitting to the specific kind of wound.

In the future, with the development of therapeutics and 3D bioprinting, we hope to see more complex wound dressings containing bioactive compounds and drugs that can be released in a program to treat actively the specific wound.

3.4.3 Tissue Engineering

Tissue engineering creates functional constructs for tissue repair, studies on stem cell behavior, and the models for studying diseases. Tissue engineering includes three main elements: three-set porous scaffold, the cells, and the growth factors. The scaffold builds up the appropriate environment for the cells proliferating, differentiating, and synthesizing the tissue. The growth factors initiate and promote cells in the complex process to the regeneration of the new tissue. For specific purposes, the triad should be selected and tailored to meet the requirement (Akter, 2016).

The objective of the scaffolds is to make the convenient environment for the formation of the tissue. They play a role of artificial extracellular matrices that support the conditions for the cells' functioning, including proliferation, differentiation, and synthesis (Chan & Leong, 2008). To meet the requirement, the scaffold should have properties including water absorption, structural stability, porosity, pore size, interconnectivity, mechanical properties, biodegradation, cell adhesion, cell proliferation, non-toxicity, and biocompatibility (Ahmed et al., 2018).

3.4.3.1 Alginate in Tissue Engineering

Alginate after purification, mainly for eliminating the protein residue, reveals the biocompatibility (Ménard et al., 2010). In cell transplantation, the high mannuronate alginate is preferred because it is low viscous and can make a gel with higher alginate content. The high alginate content gel increases its stability and reduces the diffusion permeability that leads to protect immobilized cells effectively against the host immune system. In many early studies, the high mannuronate alginate seemed to provoke the immune system, but after purification, it revealed the non-immunogenicity (Klöck et al., 1997).

Alginate has been used in tissue engineering including islet cell transplantation (Lim & Sun, 1980), bone (Man et al., 2012), cartilage (Farokhi et al., 2020), and vascularization (Gandhi et al., 2013).

Islet cells is the kind of cell in the pancreas, including alpha cells and beta cells, the last one producing the insulin, a hormone for controlling the glucose in blood. In the type I diabetic patient, a kind of autoimmune disease, the immune system does not recognize the islet cells and kills them as foreign substances, which makes the body not produce insulin. With this disease, the patients must inject insulin every day, which is inconvenient. Therefore, islet cell transplantation is a promising therapy. However, this therapy needs the pancreas from at least two deceased donors. In this situation, the development of tissue engineering is very necessary (Matsumoto, 2010). In islet cell transplantation, the hydrogels are made from crosslinks of alginate and cation Ca^{2+} or Ba^{2+} by the extrusion or electrostatic spraying technique. The last one has the advantage of making the small size gel beads but requiring not a high viscosity of alginate solution. The alginate gel used in many studies of islet cell transplantation showed good compatibility, avoided the attack of lymphocytes, controlled the glucose in plasma just one day after transplantation, and increased the survivor time (Mallett & Korbutt, 2008). Beside finding the source of beta cells with high bioactive like the nature islet cell, the technique in cell encapsulation is developed, from microencapsulation to micro-capsule then nanoencapsulation with only one islet cell for one nanoencapsule. With the 3D bioprinting, the cell scaffolds can get the desired forms. The islet cell nanoencapsule revealed many advantage as enhance the exchange mass by high area/volume ratio, protect islet from the immune system better, can distribute the nanoencapsules to the target organ (Abadpour et al., 2021).

Bone fractures are caused mainly by diseases, trauma, or accident. For treatment of bone defects, the normal therapies such as autograft bone, allograft bone, and xenograft bone are used. However, these treatments have disadvantages such as more chance for infections, reaction of the immune system, and limits in availability. Bone healing is a complicated process that requires mechanical stability and revascularization along with osteoinduction, osteoconduction, and osseointegaration. Scaffolds made by calcium-alginate showed the ability to facilitate the growth and differentiation of human osteoblast cell clusters with maintained cell viability, up-regulated bone-related gene expression, and biological apatite crystals formation (Chen et al., 2015). The bone substitute should be the appropriate scaffold for promoting the osteoconductivity (bone grows on a surface), containing the growth factors to enhancing the osteoinductivity (the stimulation of the immature cells to develop into preosteoblasts), and making the matrix for osseointegration (Albrektsson & Johansson, 2001). Alginate by itself can not satisfy all the requirements of the bone substitute; therefore, it must be combined with other compounds such as calcium silicate, calcium phosphate, hydroxyapatite, and other polymers such as collagen, gelatin, and chitosan. Sathain et al. examined the properties of the scaffold made from alginate, carrageenan, and calcium silicate. The scaffolds showed good mechanic properties, nontoxicity to human living cells, and proper diclofenac release from the scaffold, promoting the formation of hydroxyapatite in the surface of scaffold that is suitable for the treatment of acute inflammation after surgery (Sathain et al., 2021).

3.4.3.2 Chitosan in Tissue Engineering

Among the materials for making scaffolds, chitosan hydrogel possesses all the requirements and gets more interest. Chitosan produces electrostatic complexes when mixing with small compounds or negatively charged polymers such as alginate, hyaluronic acid, and proteins. Chitosan scaffolds can be prepared in different shapes and sizes, including hydrogels, films, nano/microfibers, tubes, microspheres, nano/microparticles, or membranes.

Physical chitosan hydrogels should have two features: the network must be strong enough based on interchain interactions of hydrogel, and the networks of the hydrogel must be able to exchange and diffuse water molecules through their hydrated structure (Dash et al., 2011). The main physical interactions involved in forming chitosan hydrogels include, for example, ionic, polyelectrolyte, interpolymer complex, and hydrophobic associations (Pita-Lopez et al., 2021).

The thermosensitive chitosan hydrogels have been applied in many in vitro and in vivo models including fibroblasts, chondrocytes, osteoblasts, stem cells, nucleus pulposus cells, neural cell blood of rabbit or human, mice, rat, and rabbit. In these studies, the chitosan hydrogel expressed the properties of non-toxicity, cell viability, cell proliferation, cell attachment, expression osteogenic markers, chondrogenic markers, and cardiac markers.

One of the important requirements of the scaffold is the architecture for how to mimic the natural hierarchy of the natural tissue. Nowadays, with the 3D printing technique, the scientists can obtain the microstructure following the design (Alison et al., 2019). In addition, by integrating with the sacrificial template technique, the porosity and pore size of scaffold can be controlled. In this method, the sacrificial template is made by polymer like polylacid polymer (PLA) via 3D printing technique for obtaining fiber template. The solution of chitosan in acetic acid was inhaled to fill the void space in the template. After freeze-drying, the chitosan will cover the fiber of PLA. The dichloromethane (DMC) is used to dissolve the PLA, which means all the template is removed (scarified) and leaves the porous chitosan scaffold (Jiang et al., 2021). For high activities in biomedicine, the chitosan used in these studies has a high degree of deacetylation (>90%).

3.5 CONCLUSION

The diversity in the molecular structure of alginate and chitin/chitosan serves versatile requirements in different areas, especially in biomedicine. The sophisticate microstructural features of these biopolymers help us to understand the properties and their behaviors in interacting to the environment where they are used. The diversity in structure and in the properties also poses challenges to manufacturers in keeping the sustainable technical characteristics of the products. With the difference in natural molecular structures from different sources, changes in state-of-the-art technology are expected. Beside the large studies experimenting on this, we hope the research in modelling the process of these polymers will help in building the theory, which is helpful in terms of general and fundamental knowledge. With more information about the structure and properties of the biopolymer products, scientists and technologist have the great opportunity in selecting one that can meet the requirements in specific applications and increase the effectiveness.

ACKNOWLEDGMENT

We appreciate the NORAD and the Norwegian University of Science and Technology NTNU for supporting research in marine biopolymers.

REFERENCES

Aarstad, O. A., Tøndervik, A., Sletta, H., & Gudmund, S.-B. (2012). Alginate Sequencing: An Analysis of Block Distribution in Alginates Using Specific Alginate Degrading Enzymes. *Biomacromolecules*, *13*, 106–116.

Abadpour, S., Wang, C., Niemi, E. M., & Scholz, H. (2021). Tissue Engineering Strategies for Improving Beta Cell Transplantation Outcome. *Current Transplantation Reports*, *8*, 205–219.

Aderibigbe, B. A., & Buyana, B. (2018). Alginate in Wound Dressings. *Pharmaceutics*, *10*(2), 42. https://doi.org/10.3390/pharmaceutics10020042

Ahmed, S., Annu, A. A., & Sheikh, J. (2018). A Review on Chitosan Centred Scaffolds and Their Applications in Tissue Engineering. *International Journal of Biological Macromolecules*, *116*, 849–862.

Ahmed, S., & Ikram, S. (2016). Chitosan Based Scaffolds and Their Applications in Wound Healing. *Achievements in the Life Sciences*, *10*, 27–37.

Akter, F. (2016). Principles of Tissue Engineering. *Tissue Engineering Made Easy*. http://dx.doi.org/10.1016/B978-0-12-805361-4.00002-3 Elsevier Inc.

Albrektsson, T., & Johansson, C. (2001). Osteoinduction, Osteoconduction and Osseointegration. *European Spines*, *2*, S96–101. doi: 10.1007/s005860100282.

Alginate Market Size, Share & Trends Analysis Report 2021–2028. (2021). https://doi.org/GVR-2-68038-244-0.

Alison, L., Menasce, S., Bouville, F., Tervoort, E., Mattich, I., Ofner, A., & Studart, A. R. (2019). 3D printing of sacrificial templates into hierarchical porous materials. *Scientific Reports*, *9*(409).

Anthonsen, M. W., Kjell, M. V., & Smidsrod, O. (1993). Solution Properties of Chitosans: Conformation and Chain Stiffness of Chitosans with Different Degrees of N-Acetylation. *Carbohydrate Polymers*, *22*, 193–201.

Austin, P. R. (1977). *Chitin Solution*, Patent US4059457A. University of Delaware, GD Searle LLC, United States.

Berth, G., & Dautzenberg, H. (2002). The Degree of Acetylation of Chitosans and Its Effect on the Chain Conformation in Aqueous Solution. *Carbohydrate Polymers*, *47*, 39–51.

Blackwell, J., Parker, K. D., Rudall, K. M. (1965). Chitin in Pogonophore Tubes. *Journal of Marine Biology Association UK*, *45*, 659–661.

Chan, B. P., & Leong, K. W. (2008). Scaffolding in Tissue Engineering: General Approaches and Tissue-Specific Considerations. *European Spine Journal*, *17*, 467–479.

Chen, C.-Y., Ke, C.-J., Yen, K.-C., Hsieh, H.-C., Sun, J.-S., & Lin, F.-H. (2015). 3D Porous Calcium-Alginate Scaffolds Cell Culture System Improved Human Osteoblast Cell Clusters for Cell Therapy. *Theranostics*, *5*, 643–655.

Clementi, F., Mancini, M., & Mauro Moresi. (1998). Rheology of Alginate from Azotobacter Vinelandii in Aqueous Dispersions. *Journal of Food Engineering*, *36*, 51–62.

Cristina, R.-R., Loic, H., Valle, E. M. M. del, & Galán, M. A. (2014). Rheological Characterization of Commercial Highly Viscous Alginate Solutions in Shear and Extensional Flows. *Rheologica Acta*, *53*, 559–570.

Daniel, G. B. (1993). An Overview of Hemostasis. *Toxicologic Pathology*, *21*, 170–179.

Dash, M., Chiellini, F., Ottenbrite, R. M., & Chiellini, E. (2011). Chitosan—A Versatile Semisynthetic Polymer in Biomedical Applications. *Progress in Polymer Science*, *36*, 981–1014.

Dhivyaa, S., Padma, V. V., & Santhinia, E. (2015). Wound Dressings—A Review. *BioMedicine*, *5*, 24–28.

Donati, I., Holtan, S., Mørch, Y. A., Borgogna, M., Dentini, M., & Skjåk-Bræk, G. (2005). New Hypothesis on the Role of Alternating Sequences in Calcium-Alginate Gels. *Biomacromolecules*, *6*, 1031–1040.

Draget, K. I., Østgaard, K., & Smidsrød, O. (1990). Homogeneous Alginate Gels: A Technical Approach. *Carbohydrate Polymers*, *14*, 159–178.

Einbu, A., Naess, S. N., Elgsaeter, A., & Vårum, K. M. (2004). Solution Properties of Chitin in Alkali. *Biomacromolecules, 5,* 2048–2054. doi: 10.1021/bm049710d

Farokhi, M., Shariatzadeh, F. J., Solouk, A., & Mirzadeh, H. (2020). Alginate Based Scaffolds for Cartilage Tissue Engineering: A Review. *International Journal of Polymeric Materials and Polymeric Biomaterials*, *69*.

Felt, O., Buri, P., & Gurny, R. (1998). Chitosan: A Unique Polysaccharide for Drug Delivery. *Drug Development and Industrial Pharmacy*, *24*, 979–993.

Feng, P., Luo, Y., Ke, C., Qiu, H., Wang, W., Zhu, Y., Ruixia Houl, L. X., & Wu, S. (2021). Chitosan-Based Functional Materials for Skin Wound Repair: Mechanisms and Applications. *Frontiers in Bioengineering and Biotechnology*, *9*, 650598, 1–15.

Fertah, M., Belfkira, A., Dahmane, E. M., Taourirte, M., & Brouillette, F. (2017). Extraction and Characterization of Sodium Alginate from Moroccan Laminaria Digitata Brown Seaweed. *Arabian Journal of Chemistry*, *10*, S3707–S3714.

Fischer, F., & Dörfel, H. (1955). Die polyuronsauren der braunalgen-(kohlenhydrate der algen-I). *hoppe-seyler's zeitschrift für physiologische chemie*, *302*, 186–203.

Fourest, E., & Volesky, B. (1997). Alginate Properties and Heavy Metal Biosorption by Marine Algae. *Applied Biochemistry and Biotechnology*, *67*, 215–226.

Fu, R., Li, C., Yu, C., Xie, H., Shi, S., Li, Z., . . . Lu, L. (2016). A Novel Electrospun Membrane Based on Moxifloxacin Hydrochloride/Poly (Vinyl Alcohol)/Sodium Alginate for Antibacterial Wound Dressings in Practical Application. *Drug Delivery*, *23*, 818–829.

Gandhi, J. K., Opara, E. C., & Brey, E. M. (2013). Alginate-Based Strategies for Therapeutic Vascularization. *Therapeutic Delevery*, *4*, 327–341.

Giraud-Guille, M. M. (1984). Fine Structure of the Chitin-Protein System in the Crab Cuticle. *Tissue Cell*, *16*, 75–92.

Gombotz, W. R., & Wee, S. F. (1998). Protein release from alginate matrices. *Advanced Drug Delivery Reviews*, *31*, 267–285.
Gorin, P. A. J., & Spencer, J. F. T. (1966). Exocellular Alginic Acid from Azotobacter Vinelandii. *Canadian Journal of Chemistry*, *44*, 993–998.
Grant, G. T., Morris., E. R., Rees, D. A., Smith, P. J. C., & Thom, D. (1973). Biological Interactions Between Polysacchartdes and Divalent Cations: The Egg-Box Model. *Febs Letters*, *32*, 195–198.
Grasdalen, H. (1983). High-Field, 'H-n.m.r. Spectroscopy of Alginate: Sequential Structure and Linkage Conformations. *Carbohydrate Research*, *118*, 255–260.
Grasdalen, H., Larsen, B., & Smidsrod, O. (1979). A P.M.R. Study of the Composition and Sequence of Uronate Residues in Alginates. *Carbohydrate Research*, *68*, 23–31.
Grasdalen, H., Larsen, B., & Smidsrod, O. (1981). 13C-N M-R. Studies of Monomeric Composition and Sequence in Alglnate. *Carbohydrate Research*, *89*, 179–191.
Hackman, R. H., & Goldberg, M. (1965). Studies on Chitin. VI. The Nature of Alpha- and Beta-Chitins. *Australian Journal of Biological Sciences*, *18*, 935–946.
Haug, A., & Larsen, B. (1962). Quantitative Determination of the Uronic acid Composition of Alginates. *Acta Chemica Scandinavica*, *16*, 1908–1918.
Haug, A., & Smidsrød, O. (1962). Determination of Intrinsic Viscosity of Alginates. *Acta Chemica Scandinavica*, *16*, 1569–1578.
Hay, I. D., Rehman, Z. U., Moradali, M. F., Wang, Y., & Bernd, H. A. R. (2013). Microbial Alginate Production, Modification and Its Applications. *Microbial Biotechnology*, *6*, 637–650.
Hillyard, I. W., Doczi, J., & Kiernan, P. B. (1964). Antacid and Antiulcer Properties of the Polysaccharide Chitosan in the Rat. *Proceedings of the Society for Experimental Biology and Medicine*, *115*, 1108–1112.
Hua, S. (2020). Advances in Oral Drug Delivery for Regional Targeting in the Gastrointestinal Tract—Influence of Physiological, Pathophysiological and Pharmaceutical Factors. *Frontiers in Pharmacology*, *11*, 524. doi: 10.3389/fphar.2020.00524.
Inger, M. N., Vold, M., Våruma, K., Guibal, E., & Smidsrød, O. (2003). Binding of Ions to Chitosan—Selectivity Studies. *Carbohydrate Polymers*, *54*, 471–477.
Jayakumar, R., Prabaharan, M., Sudheesh, K. P., Nair, S. V., & Tamura, H. (2011). Biomaterials Based on Chitin and Chitosan in Wound Dressing Applications. *Biotechnology Advances*, *29*, 322–337.
Jiang, U., Zhang, K., Du, L., Cheng, Z., Zhang, T., Ding, J., Li, W., Xu, B., Zhu, M. (2021). Construction of Chitosan Scaffolds with Controllable Microchannel for Tissue Engineering and Regenerative Medicine. *Materials Science & Engineering C*, *126*, 112178.
Khong, T. T., Aarstad, O. A., Skjåk-Bræk, G., Draget, K. I., & Vårum, K. M. (2013). Gelling Concept Combining Chitosan and Alginate—Proof of Principle. *Biomacromolecules*, *14*, 2765–2771.
Klöck, G., Pfeffermann, A., Ryser, C., Gröhn, P., Kuttler, B., Hahn, H. J., & Zimmermann, U. (1997). Biocompatibility of Mannuronic Acid-Rich Alginates. *Biomaterials*, *18*, 707–713.
Kumar, M. N. V. R. (2000). A Review of Chitin and Chitosan Applications. *Reactive & Functional Polymers*, *46*, 1–27.
Kumar Nayak, A., Chandra Mohanta, B., Saquib Hasnain, M., NiyazHoda, M., & Tripathi, G. (2020). Alginate-Based Scaffolds for Drug Delivery in Tissue Engineering. In A. K. Nayak & M. S. Hasnain (Eds.), *Alginates in Drug Delivery* (pp. 359–386). London: Academic Press.
Larsen, B., Salem, D. M. S. A., Sallam, M. A. E., Beltagy, A., & Mishrikey, M. M. (2003). Characterization of the Alginates from Algae Harvested at the Egyptian Red Sea Coast. *Carbohydrate Research*, *338*, 2325–2336.
Leng, A. S. J. (2009). *Oral Absorption Enhancement of Hydrophobic Drugs by Chitosan Based Amphiphilic Polymers*. Doctoral thesis. London: University College London.

Li, L., Fang, Y., Vreeker, R., & Appelqvist, I. (2007). Reexamining the Egg-Box Model in Calcium-Alginate Gels with X-ray Diffraction. *Biomacromolecules*, *8*, 464–468.

Liew, C. V., Chan, L. W., Ching, A. L., & Heng, P. W. S. (2006). Evaluation of Sodium Alginate as Drug Release Modifier in Matrix Tablets. *International Journal of Pharmaceutics*, *309*, 25–37.

Lim, F., & Sun, A. M. (1980). Microencapsulated Islets as Bioartificial Endocrine Pancreas. *Science*, *210*, 908–910.

Lin, L., Yao, W., Rao, Y., Lu, X., & Jian, Q. G. (2017). pH-Responsive Carriers for Oral Drug Delivery: Challenges and Opportunities of Current Platforms. *Drug Delivery*, *24*, 569–581.

Linker, A., & Jones, R. S. (1964). A Polysaccharide Resembling Alginic Acid from a Pseudomonas Micro-organism. *Nature*, *204*, 187–188.

Mallett, A. G., & Korbutt, G. S. (2008). Alginate Modification Improves Long-Term Survival and Function of Transplanted Encapsulated Islets. *Tissue Engineering Part A*, *15*(6). doi: 10.1089/ten.tea.2008.0118.

Man, Y., Wang, P., Xiang, Y. G. L., Yang, Y., Qu, Y., Gong, P., & Deng, L. (2012). Angiogenic and Osteogenic Potential of Platelet-Rich Plasma and Adipose-Derived Stem Cell Laden Alginate Microspheres. *Biomaterials*, *33*, 8802–8811.

Mancini, M., Moresi, M., & Sappino, F. (1996). Rheological Behaviour of Aqueous Dispersions of Algal Sodium Alginates. *Journal of Food Engineering*, 283–295.

Matsumoto, S. (2010). Islet Cell Transplantation for Type 1 Diabetes. *Journal of Diabetes*, *2*, 16–22.

McHugh, D. J. (1987). *Production and Utilization of Products from Commercial Seaweeds* (pp. 1–189). FAO Fisheries Technical Paper No. 288. Rome: Food and Agriculture Organization of the United Nations.

McHugh, D. J. (2003). *A Guide to the Seaweed Industry* (105 p). FAO Fisheries Technical Paper 441. Rome: Food and Agriculture Organization of the United Nations.

Ménard, M., Dusseault, J., Langlois, G., Baille, W. E., Tam, S. K., Yahia, L., . . . Hallé, J.-P. (2010). Role of Protein Contaminants in the Immunogenicity of Alginates. *Journal of Biomedical Materials Research B*, *93*, 333–340.

Miyazaki, S., Ishii, K., & Nadai, T. (1981). The Use of Chitin and Chitosan as Drug Carriers. *Chemical Pharmaceutical Bulletin*, *29*, 3067–3069.

Mohan, K., Muralisankar, T., Jayakumar, R., & Rajeevgandhi, C. (2021). A Study on Structural Comparisons of Alpha-Chitin Extracted from Marine Crustacean Shell Waste. *Carbohydrate Polymer Technologies and Applications*, *2*, 100037. doi.org/10.1016/j.carpta.2021.100037

Mohandas, A., Sudheesh Kumar, P. T., Raja, B., Lakshmanan, V. K., & Jayakumar, R. (2015). Exploration of Alginate Hydrogel/Nano Zinc Oxide Composite Bandages for Infected Wounds. *International Journal of Nanomedicine*, *10*, 53–66.

Moodley, K., Pillay, V., Choonara, Y. E., Toit, L. C., Du, K., Ndesendo, V. M., . . . Bawa, P. (2012). Oral Drug Delivery Systems Comprising Altered Geometric Configurations for Controlled Drug Delivery. *International Journal of Molecular Sciences*, *13*, 18–43.

Nghia, N. D. (2000). *Optimization of Sodium Alginate Extraction from Sargassum in Vietnam and Applications*. Nha Trang: University of Fisheries.

No, H. K., & Hur, E. Y. (1998). Control of Foam Formation by Antifoam During Demineralization of Crustacean Shell in Preparation of Chitin. *Journal of Agricultural and Food Chemistry*, *46*, 3844–3846.

Park, S. A., Park, K. E., & Kim, W. (2010). Preparation of Sodium Alginate/Poly(Ethylene Oxide) Blend Nanofibers with Lecithin. *Macromolecular Research*, *18*, 891–896.

Percot, A., Viton, C., & Domard, A. (2003). Characterization of Shrimp Shell Deproteinization. *Biomacromolecules*, *4*, 1380–1385.

Periayah, M. H., Halim, A. S., & Saad, A. Z. M. (2017). Mechanism Action of Platelets and Crucial Blood Coagulation Pathways in Hemostasis. *International Journal of Hematology-Oncology and Stem Cell Research, 11*, 319–327.

Pita-Lopez, M. L., Fletes-Vargas, G., Espinosa-Andrews, H., & Rodríguez-Rodríguez, R. (2021). Physically Cross-Linked Chitosan-Based Hydrogels for Tissue Engineering Applications: A State-of-the-Art Review. *European Polymer Journal, 145*, 110176.

Rhazi, M., Desbrières, J., Tolaimate, A., Rinaudo, M., Votterod, P., & Alaguic, A. (2002). Contribution to the Study of the Complexation of Copper by Chitosan and Oligomers. *Polymers, 43*, 1267–1276.

Rinaudo, M. (2006). Chitin and Chitosan: Properties and Applications. *Progress in Polymer Science, 31*, 603–632.

Roberts, G. A. F. (1992). Preparation of Chitin and Chitosan. In: *Chitin Chemistry*. London: Palgrave. https://doi.org/10.1007/978-1-349-11545-7_2

Rødde, R. H., Einbu, A., & Vårum, K. M. (2008). A Seasonal Study of the Chemical Composition and Chitin Quality of Shrimp Shells Obtained from Northern Shrimp (Pandalus Borealis). *Carbohydrate Polymers, 71*, 388–393.

Rudall, K. M. (1969). Chitin and Its Association with Other Molecules. *Journal of Polymer Science Part C, 28*, 83–102.

Rudall, K. M., & Kenchington, K. (1973). The Chitin System. *Biological Reviews, 48*, 597–633.

Salmon, J. K., Armstrong, C. A., & Ansel, J. C. (1994). The Skin as an Immune Organ. *Western Journal of Medicine, 160*, 146–152.

Sannan, T., Kurita, K., & Yoshio, I. (1976). Studies on Chitin, 2†. Effect of Deacetylation on Solubility. *Die Makromolekulare Chemie, 177*, 3589–3600.

Sathain, A., Monvisade, P., & Siriphannon, P. (2021). Bioactive Alginate/Carrageenan/Calcium Silicate Porous Scaffolds for Bone Tissue Engineering. *Materials Today Communications, 26*, 102165. doi.org/10.1016/j.mtcomm.2021.102165

Schürks, N., Wingender, J., Flemming, H.-C., & Mayer, C. (2002). Monomer Composition and Sequence of Alginates from Pseudomonas Aeruginosa. *International Journal of Biological Macromolecules, 30*, 105–111.

Sezer, A. D., & Cevher, E. (2011). Biopolymers as Wound Healing Materials: Challenges and New Strategies. In R. Pignatello (Ed.), *Biomaterials Applications for Nanomedicine* (pp. 383–414). Rijeka: InTech.

Singer, A. J., & Clark, R. A. F. (1999). Mechanisms of Disease: Cutaneous Wound Healing. *New England Journal of Medicine, 341*, 738–746. doi: 10.1056/NEJM199909023411006.

Singh, R., & Singh, D. (2012). Radiation Synthesis of PVP/Alginate Hydrogel Containing Nanosilver as Wound Dressing. *Journal of Materials Science: Materials in Medicine, 23*, 2649–2658.

Sinha, V. R., Singla, A. K., Wadhawan, S., Kaushik, R., Kumria, R., Bansal, K., & Dhawan, S. (2004). Chitosan Microspheres as a Potential Carrier for Drugs. *International Journal of Pharmaceutics, 274*, 1–33.

Smidsrød, O. (1970). Solution Properties of Alginate. *Carbohydrate Research, 13*, 359–372.

Sriamornsak, P., Thirawong, N., & Korkerd, K. (2007). Swelling, Erosion and Release Behavior of Alginate-Based Matrix Tablets. *European Journal of Pharmaceutics and Biopharmaceutics, 66*, 435–450.

Stokke, B. T., Smidsrod, O., Bruheim, P., & Skjak-Braek, G. (1991). Distribution of Uronate Residues in Alginate Chains in Relation to Alginate Gelling Properties. *Macromolecules, 24*, 4637–4645.

Terbojevich, M., Carraro, C., & Cosani, A. (1988). Solution Studies of the Chitin-Lithium Chloride-N, N-Dimethylacetamide System. *Carbohydrate Research, 180*, 73–86.

Üstündağ, G. C., Karaka, E., Özbek, S., & İlkin Çavuşoğlu. (2010). In Vivo Evaluation of Electrospun Poly (Vinyl Alcohol)/Sodium Alginate Nanofibrous Mat as Wound Dressing. *Tekstil ve Konfeksiyon*, *20*, 290–298.

Vårum, K. M., Anthonsen, M. W., Hans, G., & Olav, S. (1991). 13C-N.m.r. Studies of the Acetylation Sequences in Partially N-Deacetylated Chitins (Chitosans). *Carbohydrate Research*, *217*, 19–27.

Vårum, K. M., Mette, O. H., & Smidsrød, O. (1994). Water-Solubility of Partially N-Acetylated Chitosans as a Function of pH: Effect of Chemical Composition and Depolymerisation. *Carbohydrate Polymers*, *25*, 65–70.

Vårum, K. M., & Smidsrød, O. (2004). Structure-Property Relationship in Chitosans. In S. Dumitriu (Ed.), *Polysaccharides Structural Diversity and Functional Versatility* (Second). Boca Raton, FL: CRC Press.

Wang, W., Bo, S., Li, S., & Qin, W. (1991). Determination of the Mark-Houwink Equation for Chitosans with Different Degrees of Deacetylation. *International Journal of Biological Macromolecules*, *13*, 281–285.

Weimarn, P. P. V. (1927). Conversion of Fibroin, Chitin, Casein, and Similar Substances into the Ropy-Plastic State and Colloidal Solution. *Industrial & Engineering Chemistry*, *19*, 109–110.

Yotsuyanagi, T., Ohkubo, T., Ohhashi, T., & Ikeda, K. (1987). Calcium-Induced Gelation of Alginic Acid and pH-Sensitive Reswelling of Dried Gels. *Chemical Pharmaceutical Bulletin*, *35*, 1555–1563.

Yu, W., Jiang, Y.-Y., Sun, T.-W., Chao Qi, B. H. Z., Chen, F., Shi, Z., . . . He, Y. (2016). Design of a Novel Wound Dressing Consisting of Alginate Hydrogel and Simvastatin-Incorporated Mesoporous Hydroxyapatite Microspheres for Cutaneous Wound Healing. *RSC Advances*, *6*, 104375–104387.

Zhang, M.-X., Zhao, W.-Y., Fang, Q.-Q., Wang, X.-F., Chen, C.-Y., Shi, B.-H., . . . Shou-Jie, W., Wei-Qiang, T., & Li-Hong, W. (2021). Effects of Chitosan-Collagen Dressing on Wound Healing in Vitro and in Vivo Assays. *Journal of Applied Biomaterials & Functional Materials*, 1–10.

4 Marine Polysaccharides from Algae
Bioactivities and Application in Drug Research

Wen-Yu Lu, Hui-Jing Li, and Yan-Chao Wu

CONTENTS

4.1 INTRODUCTION

The marine world has a unique ecological environment, providing extraordinary biodiversity and related chemical diversity. The structure and chemical properties of bioactive natural products in the ocean are usually not possessed by terrestrial natural products (Moghadamtousi et al., 2015). These bioactive substances can be divided into many structural types, including polysaccharides, glycosides, peptides, terpenes, sterols, polyethers, alkaloids, macrolides and unsaturated fatty acids. Marine natural products have obvious curative effects on various human diseases. Therefore, many new compounds isolated from marine organisms have been applied to biochemical and pharmaceutical research (Laurienzo,

DOI: 10.1201/9781003303916-4

2010). At the end of 1950, Bergmann officially reported the first marine natural product with biological activity (Malve, 2015). Humans initially regarded carbohydrates as a source of energy and cellulose in food. Polysaccharides and their conjugates (glycoproteins, glycolipids, etc.) are involved in the regulation of various cell life phenomena. They are immunomodulators. It can activate immune cells and improve the immune function of the body without side effects on normal cells. Therefore, polysaccharides, as antineoplastic drugs, treatment of AIDS and other antiviral drugs and anti-aging drugs, have shown more and more application prospects in clinic. Marine polysaccharides have been found to be the most important biological macromolecules in a large number of marine extracts. Agar polysaccharide sulfate, carrageenan polysaccharide sulfate and kelp polysaccharide sulfate have been used in clinic for their anticoagulant, lipid-lowering and hemostatic effects. Marine polysaccharides with different structural characteristics widely exist in various marine species (Wang et al., 2012), such as marine animals (Fonseca et al., 2009; Laurienzo, 2010; Singh et al., 2000; Zierer and Mourão, 2000), marine plants and marine microorganisms (Debbab et al., 2010; Thomas et al., 2010; Waters et al., 2010). Among various marine species, agar, carrageenan and algin from seaweed are used as culture medium, suspending agent and emulsifier in biochemistry and medicine.

Seaweed is the most abundant resource in the ocean, which is rich in functional metabolites, such as polysaccharides, proteins, peptides, lipids, vitamins, polyphenols and mineral salts (Figure 4.1) (Brown et al., 2014). Marine algae or

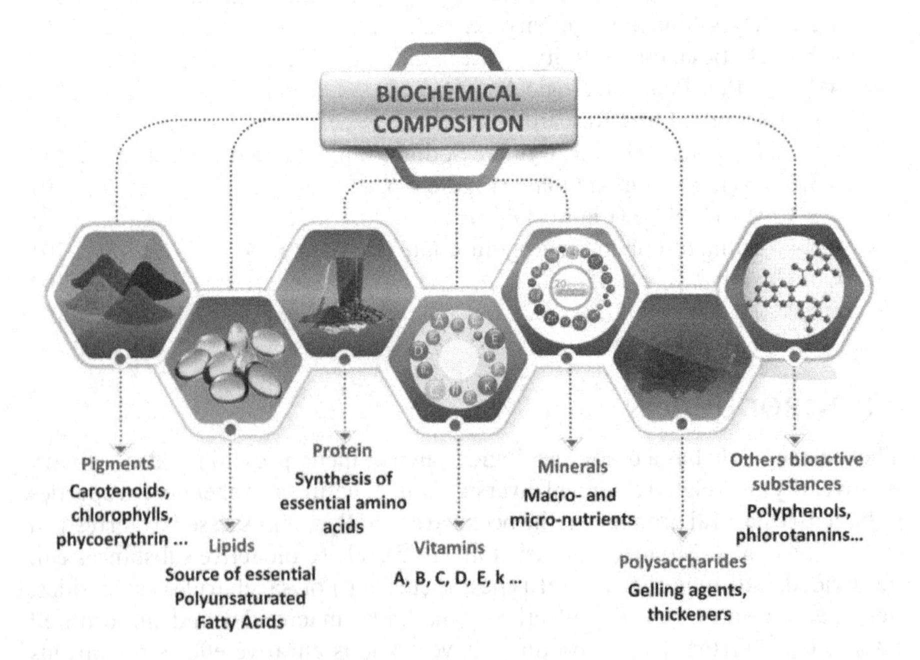

FIGURE 4.1 Bioactive compounds from marine seaweeds.

Source: Hentati et al., 2020.

seaweed can be divided into three categories according to their pigmentation and chemical composition: (1) brown seaweed (Phaeophyceae) (2) red seaweed (Rhodophyceae) and (3) green seaweed (Chlorophyceae). A variety of seaweeds are widely used in Asia as dishes or foods such as soups, condiments and salads (Rioux et al., 2017). In Asian countries, seaweed-rich diet has been associated with a decrease in the incidence rate of chronic diseases such as cancer, heart diseases and cardiovascular diseases (Brown et al., 2014; Mohamed et al., 2012a). Polysaccharides from seaweed are usually closely related to pharmacological activities such as antiviral, antioxidant, antitumor and immune regulation (de Jesus Raposo et al., 2015; Hamed et al., 2015).

4.2 BIOACTIVITIES

4.2.1 Antitumor Activity

Cancer, also known as malignant tumor, is part of a group of diseases involving abnormal cell growth. It is possible to kill normal cells and quickly spread to all parts of the body. Cancer can be classified according to cell types, such as cancer derived from epithelial cells (most common in the elderly population), sarcoma derived from connective tissues, cancer derived from hematopoietic cells (such as blood cancer), cancer derived from pluripotent cells (such as dysgerminoma and seminoma), and cancer derived from immature embryos (which is the most common type in children) (Weinberg, 2007). So far, about 100 known cancers have been found to affect humans. There are many causes of cancer: for example, obesity, tobacco, poor diet and excessive, drinking, genetic defects and following radiation (King and Robins, 2006). Cancer is a global medical problem. For a long time, clinical scientists have been looking for less toxic and more effective anti-cancer methods. Natural compounds derived from the ocean, with low cost and availability, have great potential in treating cancer. Seaweed consumption is associated with its health benefits. Seaweed is considered to be a potential source for the development of anticancer drugs, functional foods and pharmacological products (Cardozo et al., 2007; Chu, 2011; Lordan et al., 2011; Pangestuti and Kim, 2015; Smit, 2004; Thomas and Kim, 2011).

Recent studies have shown that marine polysaccharides have obvious antitumor activity in vitro and in vivo, which have been validated with a variety of tumor cell lines (Table 4.1). Compounds containing 3,6-anhydro-L-galactose and D-galactose, and a linear structure of repeated disaccharide agar disaccharide units (isolated from *Gracilariopsis lemaneiformis*) inhibited the viability of A549, B16 and MKN-28 cell lines (Kang et al., 2017). The antitumor mechanism of marine polysaccharides is considered to be due to the stimulation of cell-mediated immune responses (Kang et al., 2016). For example, polysaccharides from *Sargassum fusiforme* directly attacks cancer cells or enhances the immune function of the host, significantly inhibiting the growth of human HepG2 cell-transplanted tumor in nude mice (Fan et al., 2017). Palanisamy et al. (Palanisamy et al., 2017) reported that fucoidans extracted from *Sargassum*

TABLE 4.1
Marine Algae Sources with Anti-Cancer Activity

Type of PS	Source	Biological Properties	References
Polysaccharides	*Gracilariopsis lemaneiformis*	Anticancer (A549, B16 and MKN-28 cell lines)	Kang et al., 2017
Polysaccharides	*Sargassum fusiforme*	Anticancer (HepG2 cells)	Fan et al., 2017
Fucoidan	*Sargassum polycystum*	Anticancer (MCF-7 cells)	Palanisamy et al., 2017
Fucoidan	*Sargassum cinereum*	Anticancer (Caco-2 cells)	Narayani et al., 2019
Fucoidan	*Fucus vesiculosus*	Anticancer (HCT-116 cells)	Park et al., 2017
Fucoidan	*Sargassum hemiphyllum*	Anticancer (HCC cells)	Yan et al., 2015
Polysaccharides	*Ulva lactuca*	Anticancer	Abd-Ellatef et al., 2017

polycystum exhibited anti-proliferative at concentrations of 50 g/mL, and they could induce apoptosis in breast cancer cell line (MCF-7) through activation of caspase-8. The results of Narayani et al. (Narayani et al., 2019) showed that fucoidans isolated from brown seaweed *Sargassum cinereum* played an effective anticancer and apoptotic role in human colon adenocarcinoma cell line (Caco-2) by preventing metastasis. Fucoidan isolated from *Fucus vesiculosus* reduced the viability of human colorectal cancer cell lines (HCT-116) by about 60% in a dose- and time-dependent manner. The compound blocks the cell cycle in G1 phase and induces p53 independent apoptosis (Park et al., 2017). Yan et al. (Yan et al., 2015) showed that fucoidan obtained from *Sargassum hemiphyllum* can induce the expression of miR-29b, which helps to inhibit the expression of DNA methyltransferase 3B in human hepatocellular carcinoma (HCC) cells. The sea lettuce *Ulva lactuca* is a widely distributed green algae. Its polysaccharide has anticancer activity against many types of cells. For example, polysaccharides isolated from *U. lactuca* have shown in vitro and in vivo activity against breast cancer. The report also showed that polysaccharides derived from *U. lactuca* had chemopreventive effects, because the use of these compounds for 10 weeks prevented the histological changes and carcinogenic lesions of breast tissue in rat DMBA induced breast cancer models (Abd-Ellatef et al., 2017).

4.2.2 Antioxidant Activity

Oxygen is the key substance in the normal metabolic activities of aerobic organisms. In the environment of high redox potential, organisms inevitably produce reactive oxygen species (ROS), including hydrogen peroxide (H_2O_2), hydroxyl radical (•OH), peroxyl radical (ROO^-), superoxide anion ($•O_2^-$) and nitroxide radical (NO•) (Wu et al., 2015). ROS play an important role in various physiological activities of organisms. Low or moderate concentrations of ROS prevent infectious agents from infecting host cells and interfering with cell mitosis (Valko et al., 2007). High concentration of ROS may destroy the balance of pro-oxidants/antioxidants in organisms and cause oxidative stress (Valko et al., 2006).

Oxidative stress is caused by the generation of free radicals and the unfairness between neutralization, resulting in various diseases (Ul-Haq et al., 2019), such as diabetes, cancer, neurodegenerative diseases, inflammatory diseases, aging and immune system damage (Gandhi and Abramov, 2012; Pisoschi and Pop, 2015). Supplementing exogenous antioxidants is the most effective and widely used strategy to reduce oxidative stress (Poljsak et al., 2013). Natural products, such as carotenoids, which display strong antioxidant activity in scavenging free radicals and alleviating cell damage caused by oxidation, have been added to food additives, medicines and health care products (Massini et al., 2016; Nimse and Pal, 2015; Xu et al., 2017).

In recent years, with the continuous development and utilization of marine resources, many polysaccharides with antioxidant activity have been found from seaweed (Table 4.2) (Kadam et al., 2013). The chelating, FRAP and anti DPPH activities of algal sulfated polysaccharides depend on the availability of many functional groups. In addition, more than one of -O-, -C=O, -COOH, -OH, $-PO_3H_2$, $-NR_2$, -S- and -SH is conducive to antioxidant capacity (Hentati et al., 2018). Ulvan is a sulfated polysaccharide extracted from green algae (*Ulva lactuca* and *Ulva rigida*). The main components are rhamnose (12.73%–45%), sulfate (12.80%–23%), uronic acids (6.50%–25.96%) and xylose (2%–12%). The antioxidant activity of ulvan depends on the concentration of sulfated polysaccharides (Yaich et al., 2017). Polysaccharides from *Sargassum* usually exhibit high DPPH radical scavenging ability. For example, fucoidan isolated from *Sargassum cinereum* showed a DPPH scavenging activity of 51.99% at a concentration of 80 μg/ml (Somasundaram et al., 2016). Fucoidan can not only reduce the concentration of MDA and NO in liver, upregulates the level of GSH but also reduce the mRNA expressions of TNF-α, IL-1β and MMP-2 and inhibit the production of ROS in liver, thereby inhibiting insulin resistance

TABLE 4.2
Marine Algae Sources with Antioxidant Activity

Type of PS	Source	Biological Properties	References
SPs	*Ulva lactuca* and *Ulva rigida*	Antioxidant	Yaich et al., 2017
Fucoidan	*Sargassum cinereum*	Antioxidant	Somasundaram et al., 2016
SPs	*Laurencia papillosa*	Antioxidant/anticancer	Murad et al., 2016

induced by high-fat diet and alleviating the development of non-alcoholic fatty liver disease (NAFLD) (Heeba and Morsy, 2015). A study showed that ASPE, a sulfated polysaccharides extracted from *Laurencia papillosa*, inhibits the proliferation of MDA-MB-231 human breast cancer cells in vitro by inhibiting the ratio of ROS and Bax/Bcl-2 protein in cells (Murad et al., 2016).

4.2.3 Antiviral Activity

At present, an outbreak of coronavirus called coronavirus disease 19 (COVID-19) has spread to > 210 countries, which is rare in acute infectious diseases in recent years, and posed a great threat to global public health (Liu et al., 2020). In addition, different types of infectious diseases caused by emerging or re-emerging viruses still pose a threat to human health. Therefore, great efforts need to be made in the research and development of antiviral drugs (Clercq, 2004). Despite the increasing number of antiviral drugs approved for clinical use, there are still problems in the treatment of infectious diseases due to the insufficient efficacy, high toxicity and high cost of current antiviral drugs (Scully and Samaranayake, 2016; Wang et al., 2012a). Therefore, there is an urgent need to develop new antiviral drugs as safe and effective alternative drugs or supplementary drugs. Some polysaccharides isolated from natural sources have been found to have antiviral and immunomodulatory activities and are suitable for the development of antiviral reagents (Ivanova et al., 1994). Although the life cycle of virus varies from species to species, there are six basic stages: attachment, penetration (also known as virus entry), uncoating, replication, assembly and release, which may become the target of inhibitory reagents. Marine polysaccharides, especially those from seaweed, have a unique structure, which can interfere with different stages of virus infection process and play a role in killing various viruses (Figure 4.2) (Bouhlal et al., 2011; Wang et al., 2012a), so they have attracted extensive attention (Damonte et al., 2004). A study conducted by Gerber and his colleagues in 1958 showed that polysaccharides from seaweed could inhibit mumps and influenza B virus (Gerber et al., 1958). Subsequently, the antiviral activities of other polysaccharides isolated from red algae against HSV and other viruses have been reported (Burkholder and Sharma, 1970; Deig et al., 1974; Ehresmann et al., 1977; Richards et al., 1978). In addition, the high polysaccharide content of seaweed provides rich resources for drug discovery and development.

The mechanism of action of seaweed polysaccharide is diverse, which can play a role in different stages of virus cycle, from preventing virus attachment to intracellular antiviral activity in different ways (Besednova et al., 2019; Liu et al., 2016). The reported antiviral activity of polysaccharides is mainly focused on SARS-CoV-2 (Table 4.3). It is reported that iota-carrageenan containing lozenges can inhibit the viral activity of hCoV-OC43 (Morokutti-Kurz et al., 2017). This is because the polyanionic property of polysaccharide can inhibit virus adsorption. Kwon et al. showed that due to the multivalent interaction between polysaccharides and virus particles, the fucoidan components from brown algae *Sacharina japonica* can effectively inhibit the entry of SARS-CoV-2

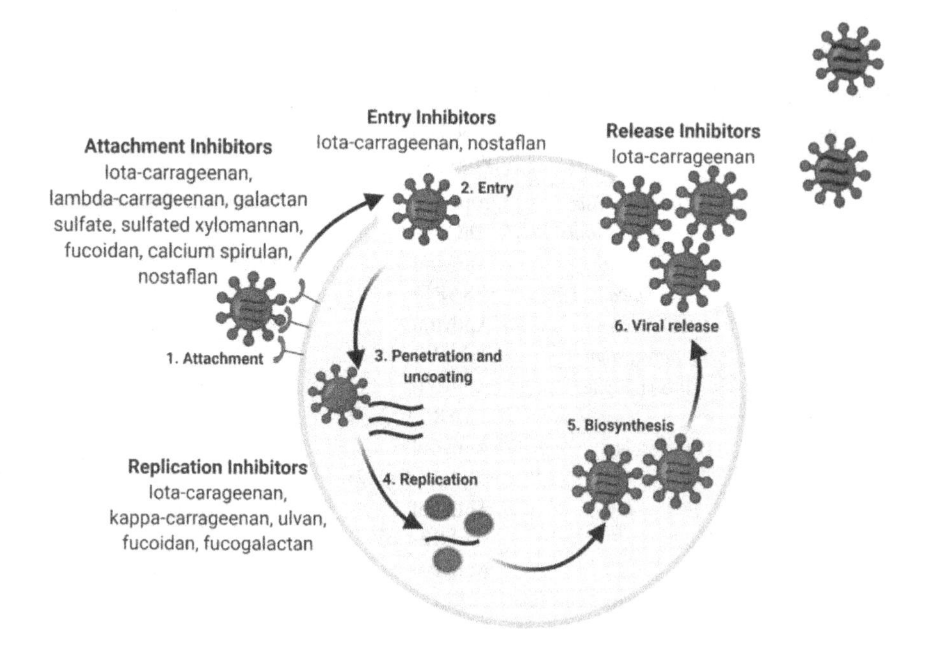

FIGURE 4.2 Mechanism of action of seaweed sulfated polysaccharides at various stages of the viral cycle.

Source: Jabeen et al., 2021.

into Vero cells in vitro (Kwon et al., 2020). λ-carrageenan showed effective inhibition of SARS-CoV-2 in Vero cells in vitro, preventing the virus from adhering to cell receptors (Jang et al., 2021). Fucoidan and iota-carrageenan from brown and red algae, respectively, inhibit viral infection by preventing host cells from entering Vero E6 cells, showing the antiviral potential against SARS-CoV-2 (Song et al., 2020). In a study, the antiviral activities of different sulfated polysaccharides from seaweed were compared, in which iota-carrageenan had the greatest antiviral activity against SARS-CoV-2 (Morokutti-Kurz et al., 2021). Bansal et al. used Vero E6 cells to confirm the antiviral activity of iota-carrageenan nasal spray in vitro (2021).

4.2.4 Immunomodulating Activity

Immunomodulation is a therapeutic method that regulate the balance of cytokines in the human body by limiting inflammation and controlling immune response or by stimulating the defective immune system. Macrophages are immune cells of the innate immune system. They play an important role in maintaining homeostasis by changing their functions according to tissues. In addition, macrophages are the main source of pro-inflammatory factors (Wijesekara et al., 2011). A variety of cytokines regulate the activation,

TABLE 4.3
Marine Algae Source with Antiviral Activity

Type of PS	Source	Biological Properties	References
Iota-carrageenan	*Gracilariopsis lemaneiformis*	Antiviral (hCoV-OC43)	Morokutti-Kurz et al., 2017
Fucoidan	*Sacharina japonica*	Antiviral (SARS-CoV-2)	Kwon et al., 2020
λ-carrageenan	*Sargassum polycystum*	Antiviral (SARS-CoV-2)	Jang et al., 2021
Fucoidan	brown algae	Antiviral (SARS-CoV-2)	Song et al., 2020
Iota-carrageenan	red algae	Antiviral (SARS-CoV-2)	Song et al., 2020
Iota-carrageenan	–	Antiviral (SARS-CoV-2)	Morokutti-Kurz et al., 2021
Iota-carrageenan	–	Antiviral	Bansal et al., 2021

development, proliferation, killing of natural killer cells (NK cells) and chemotaxis. Raulet's study showed that interleukin-2 and IL-15 can stimulate the proliferation of NK cells and the secretion of a variety of cytokines (Raulet, 2006). In fact, activated NK cells can secrete soluble cytokines such as IFN and tumor necrosis factor (TNF) to enhance the body's immune response. Some polysaccharides and glycosides obtained from natural sources are considered as biological response regulators, which can enhance various immune responses. They can maintain homeostasis by regulating T/B lymphocytes, NK cells (Figure 4.3), macrophages (Figure 4.4) and complement system (Huang et al., 2019).

The activation of macrophages is one of the indicators of immunomodulatory activity, which is one of the greatest pharmacological contributions of seaweed polysaccharides. The effects of polysaccharides found in different seaweeds on RAW 264.7 macrophages and NK cells were studied (Table 4.4). Their immunomodulatory properties can be further studied to enhance the defense ability against pathogens and invasive cells in vivo. All polysaccharide components of *Cystoseira indica* were tested on RAW 264.7 cells at concentrations ranging from 10 to 50 μg/ml. CIF2 polysaccharide is a proliferation stimulator of RAW264.7 cells, which can increase cell growth to about 25% (Han et al., 2016). Brown algae *Dictyopteris divaricata* showed similar immunostimulatory activity. A significant increase in RAW 264.7 macrophages proliferation and nitric oxide (NO) production were observed. A significant increase in proliferation and no production of RAW 264.7 macrophages was observed. The sulfate polysaccharides of *Dictyopteris divaricata* were a proliferation stimulant of RAW 264.7 cells at the concentration range of 50–400 μg/mL (Cui et al., 2018). Fucoidan could significantly increase

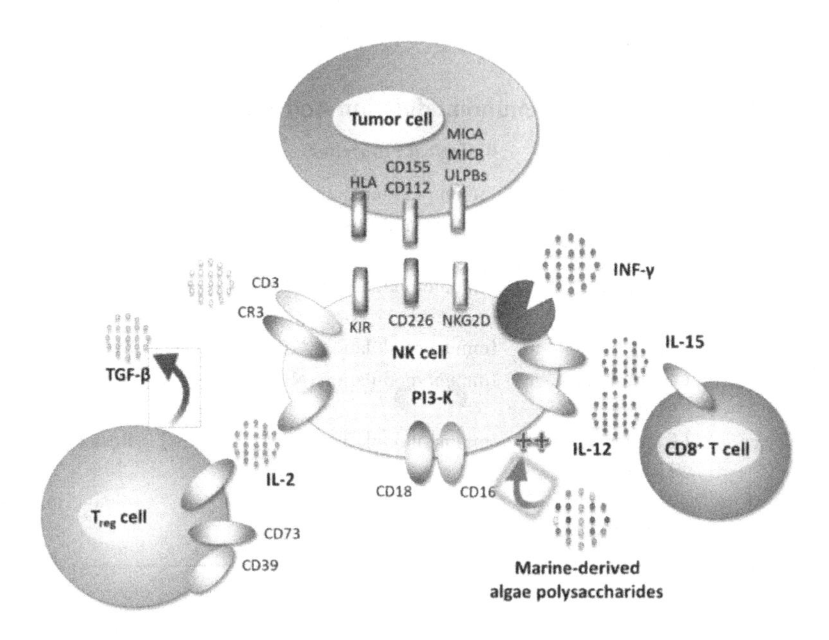

FIGURE 4.3 Signaling pathways involved in natural killer cell (NK cells) activation by bioactive algal polysaccharides.

Source: Hentati et al., 2020.

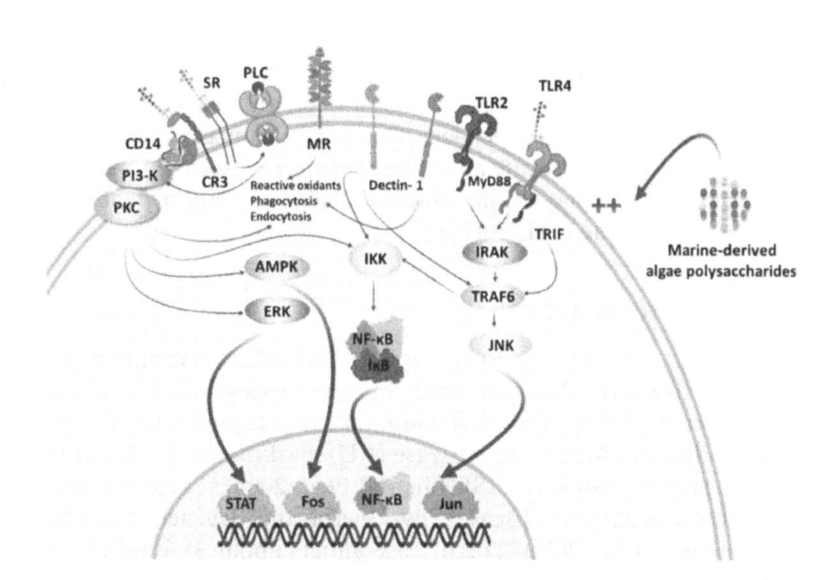

FIGURE 4.4 Signaling pathways involved in macrophage activation by algal-sulfated polysaccharides.

Source: Hentati et al., 2020.

TABLE 4.4
Marine Algae Source with Immunomodulation Activity

Type of PS	Source	Biological Properties	References
Polysaccharide	*Cystoseira indica*	Immunomodulation (RAW264.7 cells)	Han et al., 2016
CIF2 Polysaccharide	*Sacharina japonica*	Immunomodulation (RAW264.7 cells)	Han et al., 2016
SPs	*Dictyopteris divaricata*	Immunomodulation (RAW264.7 cells)	Cui et al., 2018
Fucoidan		Immunomodulation (NK cells)	Shen et al., 2018
SPs	Chlorophyceae *Codium fragile*	Immunomodulation (NK cells)	Surayot and You, 2017
Carrageenans	Gigartinaceae and Tichocarpaceae	Immunomodulation (RAW264.7 cells)	Rostami et al., 2017
Sulfated agarose	*Polysiphonia senticulosa*	Immunomodulation (NK cells)	Zhao et al., 2017

the cytolytic activity of NK cells by stimulating macrophage mediated immune response signal molecules (such as interleukin-2 and IL-12 and IFN-γ) (Shen et al., 2018). Sulfated polysaccharides obtained from Chlorophyceae *Codium fragile* can improve the activation of NK cells by inducing the expression of activation receptors, cytokine secretion and the release of lysing proteins, perforin and granzyme B (Surayot and You, 2017). Five types of carrageenan extracted from red seaweed (belonging to Gigartinaceae and Tichocarpaceae) have been reported to increase pro-inflammatory IL-6 and TNF-α levels and induce anti-inflammatory IL-10 secretion in a dose-dependent manner (Rostami et al., 2017). Sulfated agarose, which is rich in pyruvate and xylose substitutes isolated from red algae *Polysiphonia senticulosa*, showed immunomodulatory activity by increasing the activity of NK cells (Zhao et al., 2017).

4.2.5 Neuroprotective Activity

Neurodegenerative diseases are progressive damage of neurons, which are mainly related to the death of neurons. This leads to a gradual loss of cognitive and physical function. The most common neurodegenerative diseases are Huntington's disease, Alzheimer's disease (AD), Parkinson's disease (PD) and amyotrophic lateral sclerosis (ALS) (Lin and Beal, 2006). There are two main types of neurodegenerative diseases: dyskinesia and degeneration/dementia disorders (Huang et al., 2018). The disease afflicts about 35.6 million people around the world. The number is expected to double by 2030 and triple by 2050. Even in most high-income countries, it is difficult for dementia patients to obtain adequate medical care, where only about 50% of dementia patients are correctly diagnosed, while in low-income and middle-income countries, less

TABLE 4.5
Marine Algae Sources with Neuroprotective Activity

Type of PS	Source	Biological Properties	References
Fucoidans	*Sargassum fusiforme*	Neuroprotective	Hu et al., 2016
Polysaccharide	*Sargassum muticum*	Neuroprotective	Huang et al., 2018
SPs	*Gracilaria cornea*	Neuroprotective	Souza et al., 2017
Polysaccharide	*Ecklonia cava*	Neuroprotective	Park et al., 2018
Polysaccharide	*Ecklonia radiata*	Neuroprotective	Alghazwi et al., 2020

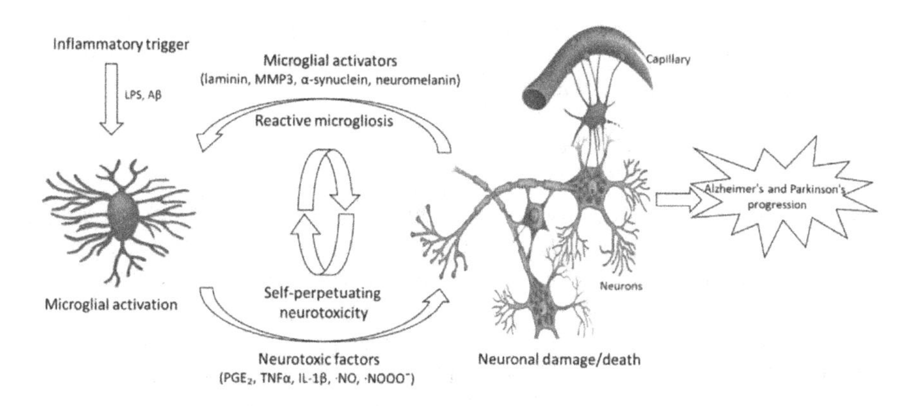

FIGURE 4.5 Microglia-mediated neurotoxicity in Al Microglia-mediated neurotoxicity in Alzheimer's and Parkinson's diseases.

Source: Pereira and Valado, 2021.

than 10% of cases are diagnosed. With the aging of the population, the number of patients with dementia is also increasing (Jager et al., 2014).

Recent studies have shown that microglia activation and excessive release of pro-inflammatory mediators by microglia were observed in the pathogenesis of AD, PD, dementia and neuronal death in stroke and traumatic brain injuries (Figure 4.5). Therefore, the mechanism of controlling the inflammatory response released by microglia may have important therapeutic significance for the treatment of neurodegenerative diseases. Some relevant reports have pointed out that seaweed ingredients are the relevant source of neuroprotective agents, especially suitable for preventive treatment (Table 4.5) (Pereira, 2018). AD is associated with failure to clear β-amyloid peptide (Aβ) from the extracellular blood vessel walls and plaques. Located around neurons Aβ aggregation has toxic effects, making neurons vulnerable to free radicals (Li et al., 2019; Masters et al., 2015). Hu et al. reported that fucoidans isolated from ethanol purification of *Sargassum fusiforme* has therapeutic potential in improving cognitive impairment and enhancing cognitive ability in mice (Hu et al., 2016). In one study, polysaccharide extracts from *Sargassum muticum*, a brown seaweed native to Japan, were found to be added to cells containing 6-hydroxydopamine (6-OHDA). 6-OHDA has been used to study neurodegenerative injury, because it increases the concentration of hydrogen peroxide, resulting in oxidative stress and decreased cell viability (Huang et al., 2018). Sulfated polysaccharide (SA-Gc) extracted from red seaweed *Gracilaria cornea* promoted neuroprotection in vivo by reducing oxidative/nitro activity stress and the change of monoamine content induced by 6-OHDA. In addition, SA-Gc regulates the transcription of neuroprotective and inflammatory genes and restores rats behavioral activity and weight gain to normal (Souza et al., 2017). *Ecklonia cava* polysaccharide extracts reduced mitochondrial-mediated protein expression and protein aggregation (Park et al., 2018). An study found that polysaccharide extracts from *Ecklonia radiata* (a brown seaweed) can prevent apoptosis and Aβ toxicity and recovery of damaged cells after oxidative damage in vitro (Alghazwi et al., 2020).

4.2.6 Hypolipidemic Activity

Nowadays, abnormal blood lipids pose a threat to the health of young people, so they have attracted more and more attention (Li et al., 2012). Hyperlipidemia is one of the major causes of cardiovascular diseases (such as type 2 diabetes, hypertension, cerebral thrombosis and atherosclerosis) (Tangvarasittichai, 2017). Chemical drugs are used in clinical treatment of hyperlipidemia due to their therapeutic efficacy (Weintraub, 2013). However, different degrees of adverse reactions were observed during and after treatment (Egom and Hafeez, 2016). Considering the existing evidence that algae based diet can prevent hyperlipidemic atherosclerosis by changing cholesterol absorption and metabolism, so as to reduce the levels of plasma triglyceride (TG), total cholesterol (TC) and low-density lipoprotein (LDL) cholesterol in hyperlipidemic mammals (Mohamed et al., 2012b).

The types of polysaccharides with hypolipidemic activity extracted from seaweed are fucoidan, SPs and ulvans (Table 4.6). The use of fucoidan can reduce the levels of serum TC, TG and LDL-C (low density lipoprotein cholesterol). while it is found that the level of serum HDL-C (high-density lipoprotein cholesterol) is increased in Kunming mice (Chen et al., 2017). Fucoidan (KF), derived from the first cultivated *Kjellmaniella crassifolia* in Dalian, China, is 71.68% carbohydrate and 20.04% sulfate. The monosaccharide of KF is mainly composed of fucose (31.89%) and galactose (23.54%), and sulfate is bound at the equatorial C-4 position and the axial C-2/3 position of fucpyranose. KF can significantly improve hyperlipidemia and antioxidant stress in animals, and the effect of KF is time and dose-dependent. The molecular mechanism of KF is to regulate hyperlipidemia by inhibiting cholesterol and fatty acid synthesis and accelerating mitochondrial β -oxidation or peroxisome oxidative degradation of fatty acids in vivo (Peng et al., 2018). Porphyran is a sulfated polysaccharide extracted from *Pyropia*, a marine red alga. At 28 days after administration, porphyran reduced TG, TC and the ratio of LDL-C/HDL-C. Porphyron at the dose

TABLE 4.6
Marine Algae Sources with Hypolipidemic Activity

Type of PS	Source	Biological Properties	References
Fucoidan	*Kjellmaniella crassifolia*	Hypolipidemic Activity	Peng et al., 2018
SPs	*Pyropia*	Hypolipidemic Activity	Cao et al., 2016
SPs	*Monostroma nitidum*	Hypolipidemic Activity	Hoang et al., 2015
Ulvans	*Ulva. pertusa*	Hypolipidemic Activity	Yu et al., 2017

of 200 mg $kg^{-1}d^{-1}$ can significantly reduce the percentage of weight gain (BWG) and serum lipid in mice, similar to Zhibituo (Cao et al., 2016). Two types of sulfated polysaccharides (SPs) purified from *Monostroma nitidum* (MF1 and MF2) significantly reduced cellular lipid concentrations. The decrease of cell lipid concentrations is accompanied by the decrease of cholesterol synthesis gene expressions, and induces cholesterol degradation, LDL uptake and peroxisome β-oxidative gene expression (Hoang et al., 2015). The Acetylated derivatives of ulvans from *U. pertusa* showed higher anti-hyperlipidemic activity than natural ulvans, especially in reducing TG and LDL-C levels (Yu et al., 2017).

4.2.7 Antidiabetic Activity

Globalization, industrialization and changes in human environment, behavior and lifestyle have led to a rising incidence rate of obesity and diabetes (Xiao and Högger, 2015). Diabetes is a serious chronic disease characterized by hyperglycemia caused by insulin action, insulin secretion or both defects (Ramachandran et al., 2016). Its main characteristic symptoms are polyuria, thirst and overeating (ADA, 2005). Different levels of insulin resistance (Pontiroli, 2004) and postprandial hyperglycemia play an important role in the development of type 2 diabetes and related complications (Lee et al., 2012). Effective control of postprandial blood glucose level is the key to diabetes care, and can improve the quality of life of patients with type 2 diabetes. At present, some anti-diabetic drugs are either ineffective or have serious side effects (Lee et al., 2014). Therefore, it is still an important issue to search for more effective and safer anti-diabetic drugs from natural sources (Vinayagam et al., 2017).

The functional components of large algae have been found to have antidiabetic properties and are commonly used as food supplements (Pangestuti and Kim, 2011). Polysaccharides with antidiabetic activity are mainly derived from brown seaweed (Table 4.7). Fucoidans are a complex and heterogeneous sulfated polysaccharide, which usually exists in brown macroalgae. Fucoidans extracted from *S. wightii Greville ex J. Agardh* can inhibit α-glucosidase (Vinoth et al., 2015). In vitro, the polysaccharides extracted from *G. lemaneiformis* inhibited α-glucosidase activity. After giving polysaccharide 21 days, the blood glucose level of diabetic mice decreased significantly (Liao et al., 2015). Jia et al. extracted three kinds of algal polysaccharides (APs) from *Scagassum*

TABLE 4.7
Marine Algae Sources with Antidiabetic Activity

Type of PS	Source	Biological Properties	References
Fucoidans	*S. wightii Greville ex J. Agardh*	Antidiabetic	vinoth et al., 2015
Polysaccharides	*G. lemaneiformis*	Antidiabetic	Liao et al., 2015
Polysaccharides	*Scagassum* (SCP), *Sargassum fusiforme (Harv.) Setch.* (SFP) and *Macrocystis pyrifera (L.) Ag* (MAP)	Antidiabetic	Jia et al., 2020
SPs	*Sargassum vulgare*	Antidiabetic	Ben Abdallah Kolsi et al., 2017
SPs	*Ascophyllum nodosum*	Antidiabetic	Okimura et al., 2019
SPs	*Sargassum pallidum*	Antidiabetic	Xiao et al., 2019

(SCP), *Sargassum fusiforme (Harv.) Setch.* (SFP) and *Macrocystis pyrifera (L.) Ag* (MAP) by hot water extraction. Oral administration of APs significantly suppressed weight loss and increased water intake, and significantly controlled elevated blood glucose, TG and TC levels in diabetic rats (Jia et al., 2020). SPs isolated from seaweed *Sargassum vulgare* significantly corrected fasting and postprandial blood glucose in diabetic rats after 30 days of intraperitoneal injection of SPs. SPs could significantly reduce the serum concentration of T-Ch, TG and LDL-Ch in diabetic animal models (Ben Abdallah Kolsi et al., 2017). SPs from *Ascophyllum nodosum* (Okimura et al., 2019) and *Sargassum pallidum* (Xiao et al., 2019) showed inhibitory effect on α-Glucosidase. *A. nodosum* SP can increase the secretion of glucagon-like peptide-1 (GLP-1). GLP-1 promotes the secretion of insulin from pancreatic b-cells. At week 8 after ingestion, SP resulted in a decrease in glucose in the treatment group compared with the placebo group.

4.2.8 Other Pharmacologic Activities

Other pharmacological activities attributed to the presence of polysaccharides in seaweed have been reported in the literature (Table 4.8). Cui et al. showed that fucoidan component NP2 extracted from brown algae Pheophyceae *Nemacytus decipiens* can increase the percentage of plasma t-PA/PAI-1 levels, indicating that it has high fibrinolytic activity, which means that it may be used as a new antithrombotic compound (Cui et al., 2018). The sulfated ramified polysaccharide (CP2–1) isolated from green algae *Codium divaricatum* has high dose-dependent anticoagulant activity (Li et al., 2015). The low-molecular-weight (LMW) fucoidan (Mn = 7.3 kDa and Mw = 7.6 kDa) isolated from brown algae *Laminaria japonica* showed better oral absorption and stronger antithrombotic activity (Zhao et al., 2016). The analysis of activated partial

TABLE 4.8
Marine Algae Sources with Other Pharmacologic Activities

Type of PS	Source	Biological Properties	References
Fucoidan	Pheophyceae *Nemacytus decipiens*	antithrombotic	Cui et al., 2018
SPs	*Codium divaricatum*	antithrombotic	Li et al., 2015
Fucoidan	*Laminaria japonica*	antithrombotic	Zhao et al., 2016
SPs	*Gracilaria debilis*	antithrombotic	Sadhasivam et al., 2015
SPs	*Sargassum wightii*	Antibacterial	Pugazhendhi et al., 2019
Ulvans	*Ulva armoricana*	Antibacterial	Massironi et al., 2019
Ulvans	*Ulva rigida*	Antibacterial	Tziveleka et al., 2018
polysaccharides	*Sargassum fusiforme*	Skin protective	Ye et al., 2018
SPs	*Laurencia obtuse*	*Gastroprotective*	Lajili et al., 2018
SPs	*Hypnea musciformis*	antidiarrheal	Sousa et al., 2016
SPs	*Gracilaria cervicornis*	antidiarrheal	Bezerra et al., 2018

thromboplastin time (APTT) and prothrombin time (PT) (14.11 and 8.23 IU/mg) showed that the sulfated polysaccharide isolated from the red marine algae *Gracilaria debilis* had important anticoagulant ability (Sadhasivam et al., 2015).

Magnesium oxide nanoparticles (MgONPs) were synthesized from sulfated polysaccharides extracted from *Sargassum wightii.* The nanoparticles showed bactericidal activity against gram-negative bacteria, especially in MRSA 56 and gram-positive bacteria, *Pseudomonas aeruginosa* had the highest inhibitory activity (Pugazhendhi et al., 2019). Ulvans, which was found in *Ulva armoricana* for the synthesis of silver nanoparticles (AgNP), observed further rapid bactericidal activity against gram-negative bacteria. All tests showed that the synthesized AgNP showed stronger antibacterial properties than free ulvan (Massironi et al., 2019). The same effect was also observed in *Ulva rigida.* Compared with free ulvan, ulvan loaded nanocomposite carrier showed higher inhibitory activity in Staphylococcus aureus (Tziveleka et al., 2018).

Sargassum fusiforme alleviated the oxidative stress induced by UVB in Kunming mice and is expected to be used as a cosmetic skin protector (Ye et al., 2018). The sulfated polysaccharide in *Laurencia obtuse* has high-molecular-weight sulfated polysaccharide. It has significant gastric protective effect. In ethanol/HCl-induced gastric mucosa of rats, SPs showed a significant increase in glutathione (GSH) level and a decrease in thiobarbituric acid reactive substances in gastric tissue, which are usually regarded as by-products of lipid peroxidation. This suggests that the possible mechanism of gastric protection is through the reduction of oxidative stress (Lajili et al., 2018). Sousa et al. (Sousa et al., 2016) and Bezerra et al. (Bezerra et al., 2018) tested the efficacy of sulfated polysaccharides from *Hypnea musciformis* and *Gracilaria cervicornis* as an antidiarrheal in rats and mice, respectively. In both studies, pre-treatment of castor oil–induced and cholera toxin–induced animals with sulfated polysaccharide showed decreased fecal and diarrhea fecal excretion, increased Na^+/K^+ ATPase activity in the small intestine and decreased cholera toxin-induced fluid secretion and chloride excretion in the small intestine. This may be attributed to the reduction of toxins that may bind to GM1 receptors. However, compared with *H. musciformis*, sulfated polysaccharides isolated from *G. cervicornis* had no effect on gastrointestinal motility, which may attribute its activity on cholinergic receptor activation to the mechanism of action, similar to the drug loperamide.

4.3 APPLICATION IN DRUG RESEARCH

4.3.1 Propylene Glycol Alginate Sodium Sulfate (Figure 4.6, 1)

Propylene glycol sodium alginate sodium sulfate (PSS), a sulfated polysaccharide derivative of alginate, was isolated from Marine brown algae. PSS is an oral heparin-like drug. It is the first drug approved by China food and Drug Administration (CFDA) in 1987 for the treatment of hyperlipidemia and ischemic cardiovascular and cerebrovascular diseases (Wu et al., 2016). The basic carbohydrate chain of PSS is composed of 1→4 linked β-$_D$-mannituronic acid (M) and α-$_L$-guluronic acid (G), including the homooligomeric regions of M-blocks, G-blocks, and MG-blocks (Figure 4.6). In general, the relative molecular weight (Mw) of PSS is 10–20 kDa and the organic sulfur content is 9.0–14.0%. PSS could be obtained by hydrolysis, esterification and sulfation of sodium alginate. The biological activity of PSS is related to its Mw, uronic acid composition (M/G ratio) and degree of sulfate substitution (DS) (Xue et al., 2018).

An important safety issue for PSS is bleeding. Studies have shown that PSS components with a lower M/G ratio and a higher Mw could over inhibit the activity of coagulation factors and platelet functions, which is the main cause of bleeding (Ma et al., 2019; Xin et al., 2016). But PSS also has many important

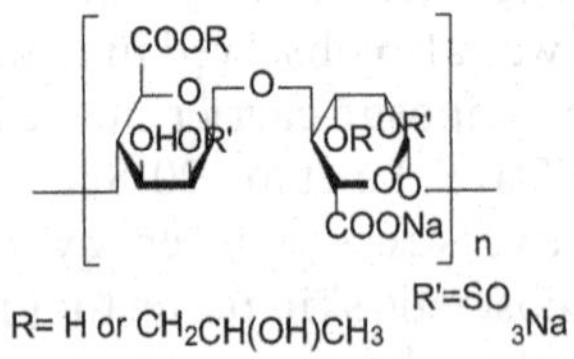

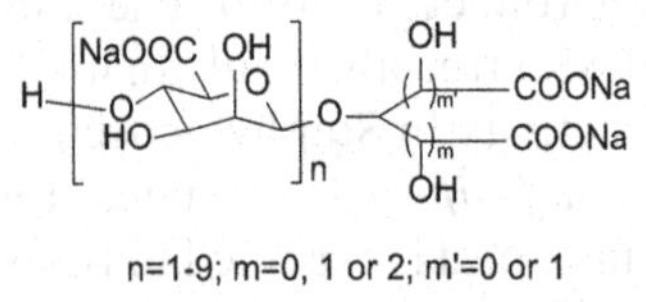

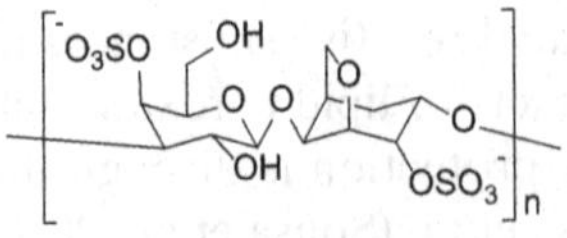

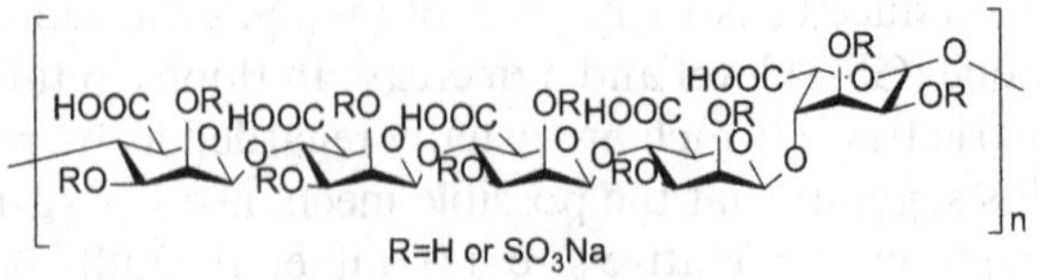

FIGURE 4.6 The structures of polysaccharides.

biological activities, such as treating diabetes, lowering blood viscosity, anti-hypertension, skin diseases and kidney diseases. Therefore, it still has research values. Nowadays, PSS has been redeveloped to improve its effectiveness and security (Xu et al., 2017).

4.3.2 Sodium Oligomannate (Figure 4.6, 2)

Sodium oligomannate (GV-971) is a mixture of acidic linear oligosaccharides from marine brown algae. It ranges from dimer to decamer, and its molecular weight could reach ~1 kDa. GV-971 was prepared by depolymerizing propylene glycol sodium alginate sodium sulfate and then oxidizing, leaving carboxyl groups at the reducing end. In 2019, it was approved by the China Food and Drug Administration (CFDA) for the treatment of AD. The drug fills the gap that there are no novel drugs to treat AD in the past 17 years, which provides patients with new drug options. A phase II trial showed that the treatment of GV-971 was associated with elevated levels of Aβ1–42 in cerebrospinal fluid (CSF) (Syed, 2020). A phase III clinical trial demonstrated that GV-971 could inhibit intestinal microbiome dysregulation and related phenylalanine/isoleucine accumulation, control neuroinflammation and reverse cognitive impairment (Wang et al., 2019).

4.3.3 Carragelose (Figure 4.6, 3)

The main ingredient of Carragelose is carrageenan. Carrageenan is an extract of the Rhodophyceae seaweeds. Carrageenan is a sulfated galactose polymer, which is composed of galactose and 3,6-ether galactose. The three principal copolymers are designated as Iota (ι), Kappa (κ), and Lamda (λ) based on the amount and location of the ester sulfate (S) and the presence of the 3,6-anhydro bridge (DA) in the 4-coupled residues. Among them, Carragelose (Iota carrageenan) has extensive inhibitory effect on respiratory virus in vitro, and its safety is also very good. The drug could inhibit the attachment and entry of virus into cells, reduce the replication of virus, and then reduce the symptoms caused by virus (Eccles et al., 2010; Genicot et al., 2014). At present, carragelose can be used as an over-the-counter drug in the European Union and parts of Asia and Australia.

4.3.4 Sulfated Polymannuroguluronate (Figure 4.6, 4)

Sulfated polymannuroguluronate (SPMG) is a novel heparin-like sulfated polysaccharide rich in 1,4-linked β-D-mannituronate, which was extracted from marine brown algae. Structurally, each sugar residue contains an average of 1.5 sulfates and 1.0 carboxyl groups, with an average molecular weight of 10 kDa. In vitro and in vivo studies have shown that SPMG could inhibit HIV replication and interfere with HIV entry into host T lymphocytes. Studies have shown that the basic domain of SPMG has high binding affinity to the trans-activator (Tat) of HIV-1 virus, thereby inhibiting Tat induced AIDS-KS. Another binding domain of SPMG is the KKR domain of high affinity heparin binding site. SPMG could inhibit Tat mediated-SLK cell adhesion by directly binding to the KKR region (Wu et al., 2011). A study confirmed that SPMG could enhance the resistance of PC12 cells to Tat protein in HIV and prevent apoptosis. This is because SPMG reduces Tat-induced calcium overload (Hui et al., 2010). In China, SPMG has entered the II phase clinical trials, and it is expected to become the first marine sulfated polysaccharide approved as an anti-AIDS drug.

4.4 CONCLUSION

Seaweed has the advantages of strong adaptability and high yield. It is the most important source of many bioactive compounds. Therefore, it is of great significance to make better use of seaweed resources. The bioactive polysaccharides extracted from seaweed often have complex components and great changes in molecular weight. Because of the multiple hydroxyl groups of sugars and the α and β configurations at the reduction terminal, it is possible for sugars to have multiple isomers when joining. Therefore, it is very difficult to study the relationship between its structure and function. However, it also brings many opportunities to scientists.

Polysaccharides are one of the main components of seaweed and have significant biological activity. Studies have proved that seaweed polysaccharides play an important role in human health and nutrition. The by-products of seaweed processing with bioactive polysaccharides can be easily used to produce functional components. Considering the valuable biological functions and beneficial effects of seaweed polysaccharides, these components may be used to prepare and design new functional foods and drugs to support the reduction or regulation of diet-related chronic dysfunction. However, since the structural and pharmacological characteristics of these compounds may vary depending on species, location and harvest time, the synthesis of standardized commercial products based on algal polysaccharide components will be a major difficulty.

REFERENCES

Abd-Ellatef, G.E.F., Ahmed, O.M., Abdel-Reheim, E.S. and Abdel-Hamid, A.H.Z. 2017. Ulva lactuca polysaccharides prevent Wistar rat breast carcinogenesis through the augmentation of apoptosis, enhancement of antioxidant defense system, and suppression of inflammation. *Breast Cancer*. 9: 67–83.

ADA. 2005. Diagnosis and classification of diabetes mellitus. *Diabetes Care*. 28: S37–S42.

Alghazwi, M., Charoensiddhi, S., Smid, S. and Zhang, W. 2020. Impact of Ecklonia radiata extracts on the neuroprotective activities against amyloid beta (Aβ1–42) toxicity and aggregation. *J. Funct Foods*. 68: 103893–103901.

Bansal, S., Jonsson, C.B., Taylor, S.L., et al. 2021. Iota-carrageenan and xylitol inhibit SARS-CoV-2 in Vero cell culture. *PLoS One*. 16: e0259943.

Ben Abdallah Kolsi, R., Bkhairia, I., Gargouri, L., Ktari, N., Chaaben, R., El Feki, A., et al. 2017. Protective effect of Sargussum vulgare sulfated polysaccharide against molecular, biochemical and histopathological damage caused by alloxan in experimental diabetic rats. *Int. J. Biol. Macromol*. 105: 598–607.

Besednova, N., Zaporozhets, T., Kuznetsova, T., Makarenkova, I. and Ermakova, S. 2019. Metabolites of seaweeds as potential agents for the prevention and therapy of influenza infection. *Mar. Drugs*. 17: 373–393.

Bezerra, F.F., Lima, G.C., de Sousa, N.A., et al. 2018. Antidiarrheal activity of a novel sulfated polysaccharide from the red seaweed Gracilaria cervicornis. *J. Ethnopharmacol.* 224: 27–35.

Bouhlal, R., Haslin, C., Chermann, J.C., et al. 2011. Antiviral activities of sulfated polysaccharides isolated from Sphaerococcus coronopifolius (Rhodophytha, Gigartinales) and Boergeseniella thuyoides (Rhodophyta, Ceramiales). *Mar. Drugs*. 9: 1187–1290.

Brown, E.S., Allsopp, P.J., Magee, P.J., et al. 2014. Seaweed and human health. *Nutr. Rev.* 72: 205–216.

Burkholder, P.R. and Sharma, G.M. 1970. Antimicrobial agents from the sea. *Lloydia*. 32: 466–483.

Cao, J., Wang, S., Yao, C., Xu, Z. and Xu, X. 2016. Hypolipidemic effect of porphyran extracted fromPyropia yezoensisin ICR mice with high fatty diet. *J. Appl. Phycol.* 28: 1315–1322.

Cardozo, K.H., Guaratini, T., Barros, M.P., et al. 2007. Metabolites from algae with economical impact. *Comp. Biochem. Phys. C*. 146: 60–78.

Chen, S., Wang, W., Cai, L., Zhong, S. and Xie, E. 2017. The regulatory effect of sargassum fucoidan on the lipid metabolism related enzymes and cholesterol synthesis key enzyme of hyperlipidemic mice. *J. Chin. Inst. Food Sci. Tech.* 17: 10–16.

Chu, W.-L. 2011. Potential applications of antioxidant compounds derived from algae. *Curr. Top. Nutraceutical Res.* 9: 83–98.

Clercq, E.D. 2004. Antiviral drugs in current clinical use. *J. Clin. Virol.* 30: 115–133.

Cui, K., Tai, W., Shan, X., Hao, J., Li, G. and Yu, G. 2018. Structural characterization and anti-thrombotic properties of fucoidan from Nemacystus decipiens. *Int. J. Biol. Macromol.* 120: 1817–1822.

Cui, Y., Liu, X., Li, S., et al. 2018. Extraction, characterization and biological activity of sulfated polysaccharides from seaweed Dictyopteris divaricata. *Int. J. Biol. Macromol.* 117: 256–263.

Damonte, E.B., Matulewicz, M.C. and Cerezo, A.S. 2004. Sulfated seaweed polysaccharides as antiviral agents. *Curr. Med. Chem.* 11: 2399–2419.

de Jesus Raposo, M.F., de Morais, A.M. and de Morais, R.M.S.C. 2015. Marine polysaccharides from algae with potential biomedical applications. *Mar. Drugs*. 13: 2967–3028.

Debbab, A., Aly, A.H., Lin, W.H. and Proksch, P. 2010. Bioactive compounds from marine bacteria and fungi. *Microb. Biotechnol.* 3: 544–563.

Deig, E.F., Ehresmann, D.W., Hatch, M.T. and Riedlinger, D.J. 1974. Inhibition of herpesvirus replication by marine algae extracts. *Antimicrob. Agents and Ch.* 6: 524–525.

Eccles, R., Meier, C., Jawad, M., Weinmüllner, R. and Prieschl-Grassauer, E. 2010. Efficacy and safety of an antiviral iota-carrageenan nasal spray: A randomized, double-blind, placebo-controlled exploratory study in volunteers with early symptoms of the common cold. *Resp. Res.* 11: 108–108.

Egom, E.E.A. and Hafeez, H. 2016. Biochemistry of statins—sciencedirect. *Adv. Clin. Chem.* 73: 127–168.

Ehresmann, D.W., Deig, E.F., Hatch, M.T., DiSalvo, L.H. and Vedros, N.A. 1977. Antiviral substances from California marine algae. *J. Phycol.* 13: 37–40.

Fan, S.R., Zhang, J.F., Nie, W.J., Zhou, W.Y., Jin, L.Q., Chen, X.M. and Lu, J.X. 2017. Antitumor effects of polysaccharide from Sargassum fusiforme against human hepatocellular carcinoma HepG2 cells. *Food Chem. Toxicol.* 102: 53–62.

Fonseca, R.J.C., Santos, G.R.C. and Mourao, P.A.S. 2009. Effects of polysaccharides enriched in 2,4-disulfated fucose units on coagulation, thrombosis and bleeding. *Practical and Conceptual Implications. Thrombosis & Haemostasis*. 102: 829–836.
Gandhi, S. and Abramov, A.Y. 2012. Mechanism of oxidative stress in neurodegeneration. *Oxid. Med. Cell Longev*. 2012: 428010–428020.
Genicot, S. M., Groisillier, A., Rogniaux, H., Meslet-Cladière, L., Barbeyron, T. and Helbert, W. 2014. Discovery of a novel iota carrageenan sulfatase isolated from the marine bacterium Pseudoalteromonas carrageenovora. *Front. Chem*. 2: 1–15.
Gerber, P., Dutcher, J.D., Adams, E.V. and Sherman, J.H. 1958. Protective effect of seaweed extracts for chicken embryos infected with influenza B or mumps virus. *Exp. Biol. Med*. 99: 590–593.
Hamed, I., Özogul, F., Özogul, Y. and Regenstein, J.M. 2015. Marine bioactive compounds and their health benefits: A review. *Compr. Rev. Food Sci. F*. 14: 446–465.
Han, Y., Wu, J., Liu, T., et al. 2016. Separation, characterization and anticancer activities of a sulfated polysaccharide from Undaria pinnatifida. *Int. J. Biol. Macromol*. 83: 42–49.
Heeba, G.H. and Morsy, M.A. 2015. Fucoidan ameliorates steatohepatitis and insulin resistance by suppressing oxidative stress and inflammatory cytokines in experimental non-alcoholic fatty liver disease. *Environ. Toxicol. Phar*. 40: 907–914.
Hentati, F., Delattre, C., Ursu, A.V., et al. 2018. Structural characterization and antioxidant activity of water-soluble polysaccharides from the Tunisian brown seaweed Cystoseira compressa. *Carbohydr. Polym*. 198: 589–600.
Hentati, F., Tounsi, L., Djomdi, D., et al. 2020. Bioactive polysaccharides from seaweeds. *Molecules* 25: 3152.
Hoang, M.H., Kim, J.Y., Lee, J.H., You, S.G. and Lee, S.J. 2015. Antioxidative, hypolipidemic, and anti-inflammatory activities of sulfated polysaccharides from Monostroma nitidum. *Food Sci. Biotechnol*. 24: 199–205.
Hu, P., Li, Z., Chen, M., et al. 2016. Structural elucidation and protective role of a polysaccharide from Sargassum fusiforme on ameliorating learning and memory deficiencies in mice. *Carbohydr. Polym*. 139: 150–158.
Huang, C.Y., Kuo, C.H. and Chen, P.W. 2018. Compressional-puffing pretreatment enhances neuroprotective effects of fucoidans from the brown seaweed sargassum hemiphyllum on 6-hydroxydopamine-induced apoptosis in SH-SY5Y cells. *Molecules*. 23: 78–109.
Huang, L., Shen, M. and Morris, G.A. 2019. Sulfated polysaccharides: Immunomodulation and signaling mechanisms. *Trends. Food Sci. Tech*. 92: 1–11.
Hui, B., Li, J. and Geng, M.Y. 2010. Sulfated polymannuroguluronate, a novel anti-acquired immune deficiency syndrome drug candidate, decreased vulnerability of PC12 cells to human immunodeficiency virus tat protein through attenuating calcium overload. *J. Neurosci. Res*. 86: 1169–1177.
Ivanova, V., Rouseva, R., Kolarova, M., Serkedjieva, J., Rachev, R. and Manolova, N. 1994. Isolation of a polysaccharide with antiviral effect from Ulva Lactuca. *Prep. Biochem. Biotech*. 24: 83–97.
Jabeen, M., Dutot, M., Fagon, R., Verrier, B. and Monge, C. 2021. Seaweed sulfated polysaccharides against respiratory viral infections. *Pharmaceutics* 13: 733.
Jager, P.D., Srivastava, G., Lunnon, K., et al. 2014. Alzheimer's disease. *Nat. Neurosci*. 17: 1156–1163.
Jang, Y., Shin, H., Lee, M.K., et al. 2021. Antiviral activity of lambda-carrageenan against influenza viruses and severe acute respiratory syndrome coronavirus 2. *Sci. Rep-UK*. 11: 1–26.
Jia, R.-B., Wu, J., Li, Z.-R., et al. 2020. Structural characterization of polysaccharides from three seaweed species and their hypoglycemic and hypolipidemic activities in type 2 diabetic rats. *Int. J. Biol. Macromol*. 155: 1040–1049.

Kadam, S.U., Tiwari, B.K. and O'Donnell, C.P. 2013. Application of novel extraction technologies for bioactives from marine algae. *J. Agric. Food Chem.* 60: 4667–4675.

Kang, Y.N., Li, H., Wu, J., et al. 2016. Transcriptome profiling reveals the antitumor mechanism of polysaccharide from marine algae gracilariopsis lemaneiformis. *Plos One*. 11: e0158279.

Kang, Y.N., Wang, Z.J., Xie, D., Sun, X., Yang, W., Zhao, X. and Xu, N. 2017. Characterization and potential antitumor activity of polysaccharide from Gracilariopsis lemaneiformis. *Mar. Drugs*. 15: 100–113.

King, R.J.B. and Robins, M.W. 2006. *Cancer Biology*. Pearson Education.

Kwon, P.S., Oh, H., Kwon, S.J., Jin, W. and Dordick, J.S. 2020. Sulfated polysaccharides effectively inhibit SARS-CoV-2 in vitro. *Cell Discov*. 6: 4–7.

Lajili, S., Ammar, H.H., Mzoughi, Z., et al. 2018. Characterization of sulfated polysaccharide from Laurencia obtusa and its apoptotic, gastroprotective and antioxidant activities. *Int. J. Biol. Macromol.* 126: 326–336.

Laurienzo, P. 2010. Marine polysaccharides in pharmaceutical applications: An overview. *Mar. Drugs*. 8: 2435–2465.

Lee, S.H., Kang, N., Kim, E.A., et al. 2014. Antidiabetogenic and antioxidative effects of octaphlorethol a isolated from the brown algae Ishige foliacea in streptozotocin-induced diabetic mice. *Food Sci. Biotechnol.* 23: 1261–1266.

Lee, S.H., Min, K.H., Han, J.S., et al. 2012. Effects of brown algae, Ecklonia cava on glucose and lipid metabolism in C57BL/KsJ-db/db mice, a model of type 2 diabetes mellitus. *Food Chem. Toxicol.* 50: 575–582.

Li, J.H., Wang, L.M., Li, Y.C., Bi, Y.F. and Zhao, W.H. 2012. Pidemiologic characteristics of dyslipidemia in Chinese adults 2010. *Chin. J. Prev. Med.* 46: 414–418.

Li, N., Mao, W., Yan, M., et al. 2015. Structural characterization and anticoagulant activity of a sulfated polysaccharide from the green alga Codium divaricatum. *Carbohydr. Polym.* 121: 175–182.

Li, Z., Chen, X., Zhang, Y., Liu, X., Wang, C. and Teng, L. 2019. Protective roles of Amanita caesarea polysaccharides against Alzheimer's disease via Nrf2 pathway. *Int. J. Biol. Macromol.* 121: 29–37.

Liao, X., Yang, L., Chen, M., Yu, J., Zhang, S. and Ju, Y. 2015. The hypoglycemic effect of a polysaccharide (GLP) from Gracilaria lemaneiformis and its degradation products in diabetic mice. *Food Funct*. 6: 2542–2549.

Lin, M.T. and Beal, M.F. 2006. Mitochondrial dysfunction and oxidative stress in neurodegenerative diseases. *Nature*. 443: 787–795.

Liu, C., Zhou, Q., Li, Y., Garner, L.V. and Albaiu, D. 2020. Research and development on therapeutic agents and vaccines for COVID-19 and related human coronavirus diseases. *ACS Central Sci.* 6: 315–331.

Liu, Q., Zhou, Y.-H., Ye, F. and Yang, Z.-Q. 2016. Antivirals for respiratory viral infections: Problems and prospects. *Semin. Resp. Crit. Care* 37: 640–646.

Lordan, S., Ross, R.P. and Stanton, C. 2011. Marine bioactives as functional food ingredients: Potential to reduce the incidence of chronic diseases. *Mar. Drugs*. 9: 1056–1100.

Ma, H., Qiu, P., Xin, M., et al. 2019. Structure-activity relationship of propylene glycol alginate sodium sulfate derivatives for blockade of selectins binding to tumor cells. *Carbohydr. Polym.* 210: 225–233.

Malve, H.J. 2015. Exploring the ocean for new drug developments: Marine pharmacology. *Pharm. Bioallied Sci.* 8: 83–91.

Massini, L., Rico, D., Martin-Diana, A.B. and Barry-Ryan, C. 2016. Apple peel flavonoids as natural antioxidants for vegetable juice applications. *Eur. Food Res. Technol.* 242: 1459–1469.

Massironi, A., Morelli, A., Grassi, L., et al. 2019. Ulvan as novel reducing and stabilizing agent from renewable algal biomass: Application to green synthesis of silver nanoparticles. *Carbohydr. Polym.* 203: 310–321.

Masters, C.L., Bateman, R., Blennow, K., Rowe, C.C., Sperling, R.A. and Cummings, J.L. 2015. Alzheimer's disease. *Nat. Rev. Dis. Primers*. 1: 15056–15073.

Moghadamtousi, S.Z., Nikzad, S., Kadir, H., Abubakar, S. and Zandi, K. 2015. Potential antiviral agents from marine fungi: An overview. *Mar. Drugs*. 13: 4520–4538.

Mohamed, S., Hashim, S.N. and Rahman, H.A. 2012a. Seaweeds: A sustainable functional food for complementary and alternative therapy. *Trends Food Sci. Tech*. 23: 83–96.

Morokutti-Kurz, M., Frba, M., Graf, P., Groe, M. and Prieschl-Grassauer, E. 2021. Iota-carrageenan neutralizes SARS-CoV-2 and inhibits viral replication in vitro. *Plos One*. 16: 1–13.

Morokutti-Kurz, M., Graf, C. and Prieschl-Grassauer, E. 2017. Amylmetacresol/2,4-dichlorobenzyl alcohol, hexylresorcinol, or carrageenan lozenges as active treatments for sore throat. *Int. J. Gen. Med*. 10: 53–60.

Murad, H., Hawat, M., Ekhtiar, A., et al. 2016. Induction of G1-phase cell cycle arrest and apoptosis pathway in MDA-MB-231 human breast cancer cells by sulfated polysaccharide extracted from Laurencia papillosa. *Cancer Cell Int*. 16: 39–49.

Narayani, S.S., Saravanan, S., Ravindran, J., Ramasamy, M.S. and Chitra, J. 2019. In vitro anticancer activity of fucoidan extracted from Sargassum cinereum against Caco-2 cells. *Int. J. Biol. Macromol*. 138: 618–628.

Nimse, S.B. and Pal, D. 2015. Free radicals, natural antioxidants, and their reaction mechanisms. *RSC Adv*. 5: 27986–28006.

Okimura, T., Jiang, Z., Liang, Y., Yamaguchi, K. and Oda, T. 2019. Suppressive effect of ascophyllan HS on postprandial blood sugar level through the inhibition of α-glucosidase and stimulation of glucagon-like peptide-1 (GLP-1) secretion. *Int. J. Biol. Macromol*. 125: 453–458.

Palanisamy, S., Vinosha, M., Marudhupandi, T., Rajasekar, P. and Prabhu, N.M. 2017. Isolation of fucoidan from Sargassum polycystum brown algae: Structural characterization, in vitro antioxidant and anticancer activity. *Int. J. Biol. Macromol*. 102: 405–412.

Pangestuti, R. and Kim, S.K. 2011. Biological activities and health benefit effects of natural pigments derived from marine algae. *J. Funct. Foods*. 3: 255–266.

Pangestuti, R. and Kim, S.K. 2015. Seaweeds-derived bioactive materials for the prevention and treatment of female's cancer. In *Handbook of Anticancer Drugs from Marine Origin*. Springer International Publishing, 165–176.

Park, H.Y., Park, S.H., Jeong, J.W., Yoon, D. and Choi, Y.H. 2017. Induction of p53-independent apoptosis and G1 cell cycle arrest by fucoidan in HCT116 human colorectal carcinoma cells. *Mar. Drugs*. 15: 154–167.

Park, S.K., Kang, J.Y., Kim, J.M., et al. 2018. Protective effect of fucoidan extract from Ecklonia cava on hydrogen peroxide-induced neurotoxicity. *J. Microbiol. Biotech*. 28: 40–49.

Peng, Y., Wang, Y., Wang, Q., Luo, X. and Song, Y. 2018. Hypolipidemic effects of sulfated fucoidan from Kjellmaniella crassifolia through modulating the cholesterol and aliphatic metabolic pathways. *J. Funct. Foods*. 51: 8–15.

Pereira, L. 2018. Neurological activities of seaweeds and their extracts. In *Therapeutic and Nutritional Uses of Algae*. CRC Press, 485–502.

Pereira, L. and Valado, A. 2021. The seaweed diet in prevention and treatment of the neurodegenerative diseases. *Mar. Drugs* 19: 128.

Pisoschi, A.M. and Pop, A. 2015. The role of antioxidants in the chemistry of oxidative stress: A review. *Eur. J. Med. Chem*. 46: 55–74.

Poljsak, B., Suput, D. and Milisav, I. 2013. Achieving the balance between ROS and antioxidants: When to use the synthetic antioxidants. *Oxid. Med. Cell Longev*. 2013: 956792–956802.

Pontiroli, A.E. 2004. Type 2 diabetes mellitus is becoming the most common type of diabetes in school children. *Acta Diabetol.* 41: 85–90.
Pugazhendhi, A., Prabhu, R., Muruganantham, K., Shanmuganathan, R. and Natarajan, S. 2019. Anticancer, antimicrobial and photocatalytic activities of green synthesized magnesium oxide nanoparticles (MgONPs) using aqueous extract of Sargassum wightii. *J. Photoch. Photobio. B.* 190: 86–97.
Ramachandran, A., Snehalatha, C. and Nanditha, A. 2016. Classification and diagnosis of diabetes. In *Textbook of Diabetes.* 5th ed. John Wiley & Sons, Ltd., 23–28.
Raulet, D.H. 2006. Missing self recognition and self tolerance of natural killer (NK) cells. *Semin. Immunol.* 18: 145–150. Academic Press: London.
Richards, J.T., Kern, E.R., Glasgow, L.A., Overall, J.C., Deign, E.F. and Hatch, M.T. 1978. Antiviral activity of extracts from marine algae. *Antimicrob. Agents Ch.* 14: 24–30.
Rioux, L.E., Beaulieu, L. and Turgeon, S.L. 2017. Seaweeds: A traditional ingredients for new gastronomic sensation. *Food Hydrocolloid.* 68: 255–265.
Rostami, Z., Tabarsa, M., You, S. and Rezaei, M. 2017. Relationship between molecular weights and biological properties of alginates extracted under different methods from Colpomenia peregrina. *Process Biochem.* 58: 289–297.
Sadhasivam, S., Sudharsan, N., Seedevi, P., et al. 2015. Antioxidant and anticoagulant activity of sulfated polysaccharide from Gracilaria debilis (Forsskal). *Int. J. Biol. Macromol.* 81: 1031–1038.
Scully, C. and Samaranayake, L. 2016. Emerging and changing viral diseases in the new millennium. *Oral Dis.* 22: 171–179.
Shen, P., Yin, Z., Qu, G. and Wang, C. 2018. Fucoidan and its health benefits. In *Bioactive Seaweeds for Food Applications.* Academic Press, 223–238.
Singh, D.K., Ray, A.R. and Macromol, J. 2000. Biomedical applications of chitin, chitosan, and their derivatives. *Sci. Rev. Macromol. Chem. Phys.* C40: 69–83.
Smit, A.J. 2004. Medicinal and pharmaceutical uses of seaweed natural products: A review. *J. Appl. Phycol.* 16: 245–262.
Somasundaram, S.N., Shanmugam, S., Subramanian, B. and Jaganathan, R. 2016. Cytotoxic effect of fucoidan extracted from Sargassum cinereum on colon cancer cell line HCT-15. *Int. J. Biol. Macromol.* 91: 1215–1223.
Song, S., Peng, H., Wang, Q., Liu, Z. and Zhu, B.W. 2020. Inhibitory activities of marine sulfated polysaccharides against SARS-CoV-2. *Food Funct.*11: 7415–7420.
Sousa, N.A., Barros, F.C.N., Araújo, T.S.L., Costa, D.S. and Medeiros, J.V.R. 2016. The efficacy of a sulphated polysaccharide fraction from Hypnea musciformis against diarrhea in rodents. *Int. J. Biol. Macromol.* 86: 865–875.
Souza, R.B., Frota, A.F., Sousa, R.S., et al. 2017. Neuroprotective effects of sulphated Agaran from marine alga Gracilaria cornea in rat 6-hydroxydopamine Parkinson's disease model: Behavioural, neurochemical and transcriptional alterations. *Basic Clin. Pharmacol.* 120: 159–170.
Surayot, U. and You, S.G. 2017. Structural effects of sulfated polysaccharides from Codium fragile on NK cell activation and cytotoxicity. *Int. J. Biol. Macromol.* 98: 117–124.
Syed, Y.Y. 2020. Sodium oligomannate: First approval. *Drugs.* 80: 445–446.
Tangvarasittichai, S. 2017. Atherogenic dyslipidemia: An important risk factor for cardiovascular disease in metabolic syndrome and type 2 diabetes mellitus patients. *Diabetes Obes Int. J.* 2: 1–19.
Thomas, N.V. and Kim, S.-K. 2011. Potential pharmacological applications of polyphenolic derivatives from marine brown algae. *Environ. Toxicol. Phar.* 32: 325–335.
Thomas, T.R.A., Kavlekar, D.P. and Lokabharathi, P.A. 2010. Marine drugs from sponge-microbe association—a review. *Mar. Drugs.* 8: 1417–1468.

Tziveleka, L.-A., Pippa, N., Georgantea, P., Ioannou, E., Demetzos, C. and Roussis, V. 2018. Marine sulfated polysaccharides as versatile polyelectrolytes for the development of drug delivery nanoplatforms: Complexation of ulvan with lysozyme. *Int. J. Biol. Macromol.* 118: 69–75.

Ul-Haq, I., Butt, M.S., Amjad, N., Yasmin, I. and Suleria, H.A.R. 2019. Marine-algal bioactive compounds: A comprehensive appraisal. In *Handbook of Algal Technologies and Phytochemicals*. CRC Press, 71–80.

Valko, M., Leibfritz, D., Moncol, J., Cronin, M.T.D., Mazur, M. and Telser, J. 2007. Free radicals and antioxidants in normal physiological functions and human diseases. *Int. J. Biochem. Cell Biol.* 39: 44–84.

Valko, M., Rhodes, C.J., Moncol, J., Izakovic, M. and Mazur, M. 2006. Free radicals, metals and antioxidants in oxidative stress-induced cancer. *Chem-Biol. Interact.* 160: 1–40.

Vinayagam, R., Xiao, J. and Xu, B. 2017. An insight into anti-diabetic properties of dietary phytochemicals. *Phytochem. Rev.* 16: 535–553.

Vinoth, K.T., Lakshmanasenthil, S., Geetharamani, D., Marudhupandi, T., Suja, G. and Suganya, P. 2015. Fucoidan: A α-D-glucosidase inhibitor from Sargassum wightii with relevance to type 2 diabetes mellitus therapy. *Int. J. Biol. Macromol.* 72C: 1044–1047.

Wang, W., Wang, S.X. and Guan, H.S. 2012. The antiviral activities and mechanisms of marine polysaccharides: An overview. *Mar. Drugs.* 10: 2795–2816.

Wang, X., Sun, G., Feng, T., et al. 2019. Sodium oligomannate therapeutically remodels gut microbiota and suppresses gut bacterial amino acids-shaped neuroinflammation to inhibit Alzheimer's disease progression. *Cell Res.* 29: 1–17.

Waters, A.L., Hill, R.T., Place, A.R. and Hamann, M.T. 2010. The expanding role of marine microbes in pharmaceutical development. *Curr. Opin. Biotech.* 21: 780–786.

Weinberg, R. 2007. *The Biology of Cancer*. 1st ed. W.W. Norton & Company. DOI: 10.1201/9780203852569.

Weintraub, H. 2013. Update on marine omega-3 fatty acids: Management of dyslipidemia and current omega-3 treatment options. *Atherosclerosis*. 230: 381–389.

Wijesekara, I., Pangestuti, R. and Kim, S.K. 2011. Biological activities and potential health benefits of sulfated polysaccharides derived from marine algae. *Carbohydr. Polym.* 84: 14–21.

Wu, J., Zhang, M., Zhang, Y., Zeng, Y., Zhang, L. and Xia, Z. 2016. Anticoagulant and FGF/FGFR signal activating activities of the heparinoid propylene glycol alginate sodium sulfate and its oligosaccharides. *Carbohydr. Polym.* 136: 641–648.

Wu, R.B., Wu, C.L., Liu, D., et al. 2015. Overview of antioxidant peptides derived from marine resources: The sources, characteristic, purification, and evaluation methods. *Appl. Biochem. Biotech.* 176: 1815–1833.

Wu, Y.L., Ai, J., Zhao, J.M., et al. 2011. Sulfated polymannuroguluronate inhibits Tat-induced SLK cell adhesion via a novel binding site, a KKR spatial triad. *Acta Pharmacol. Sin.* 32: 647–654.

Xiao, H., Fu, X., Cao, C., Li, C. and Chen, C. 2019. Sulfated modification, characterization, antioxidant and hypoglycemic activities of polysaccharides from Sargassum pallidum. *Int. J. Biol. Macromol.* 121: 407–414.

Xiao, J.B. and Högger, P. 2015. Dietary polyphenols and type 2 diabetes: Current insights and future perspectives. *Curr. Med. Chem.* 22: 23–38.

Xin, M., Ren, L., Sun, Y., Li, H.H. and Li, C.X. 2016. Anticoagulant and antithrombotic activities of low-molecular-weight propylene glycol alginate sodium sulfate (PSS). *Eur. J. Med. Chem.* 114: 33–40.

Xu, D.P., Li, Y., Meng, X., et al. 2017. Natural antioxidants in foods and medicinal plants: Extraction, assessment and resources. *Int. J. Mol. Sci.* 18: 96–127.

Xu, S., Wu, L., Zhang, Q., et al. 2017. Pretreatment with propylene glycol alginate sodium sulfate ameliorated concanavalin A-induced liver injury by regulating the PI3K/Akt pathway in mice. *Life Sci.* 185: 103–113.

Xue, Y.-T., Shuang, L., Liu, W.-J., et al. 2018. The mechanisms of sulfated polysaccharide drug of propylene glycol alginate sodium sulfate (PSS) on bleeding side effect. *Carbohydr. Polym.* 194: 365–374.

Yaich, H., Ben Amira, A., Abbes, F., et al. 2017. Effect of extraction procedures on structural, thermal and antioxidant properties of ulvan from Ulva lactuca collected in Monastir coast. *Int. J. Biol. Macromol.* 105: 1430–1439.

Yan, M.D., Yao, C.J., Chow, J.M., et al. 2015. Fucoidan elevates microRNA-29b to regulate DNMT3B-MTSS1 axis and inhibit EMT in human hepatocellular carcinoma cells. *Mar. Drugs.* 13: 6099–6116.

Ye, Y., Ji, D., You, L., Zhou, L., Zhao, Z. and Brennan, C. 2018. Structural properties and protective effect of Sargassum fusiforme polysaccharides against ultraviolet B radiation in hairless Kun Ming mice. *J. Funct. Foods.* 43: 8–16.

Yu, Y., Li, Y., Du, C., Mou, H. and Wang, P. 2017. Compositional and structural characteristics of sulfated polysaccharide from Enteromorpha prolifera. *Carbohydr. Polym.* 165: 221–228.

Zhao, X., Guo, F., Hu, J., et al. 2016. Antithrombotic activity of oral administered low molecular weight fucoidan from Laminaria Japonica. *Thromb. Res.* 114: 46–52.

Zhao, X., Jiao, G., Yang, Y., et al. 2017. Structure and immunomodulatory activity of a sulfated agarose with pyruvate and xylose substitutes from Polysiphonia senticulosa Harvey. *Carbohydr. Polym.* 176: 29–37.

Zierer, M.S. and Mourão, P.A.S. 2000. A wide diversity of sulfated polysaccharides are synthesized by different species of marine sponges. *Carbohydr. Res.* 328: 209–216.

5 Marine Polysaccharides in Pharmaceutical Applications

Riyasree Paul, Sourav Kabiraj, Sreejan Manna, and Sougata Jana

CONTENTS

DOI: 10.1201/9781003303916-5

5.1. INTRODUCTION

In recent times, biopolymers are considered one of the important excipients in the pharmaceutical industry. Naturally occurring gums are generated from plant sources or internal parts of plant tissues which can be disintegrated by a process known as gummosis. Through this process the plant tissue is transformed into cavities which further get distilled and transformed into carbohydrates, known to us as gums. Polysaccharides can also be generated from the plant stem parts when there is either any damage on the plant stem or the plant suffers bacterial, algal or fungal attack (Goswami and Naik 2014). These biopolymers are regarded as excellent candidates in developing conventional dosage forms, as well as in sustained-release drug delivery systems, including nanoparticles, microparticles and hydrogels-based systems (Avachat et al. 2011). These gums or polymers are chemically carbohydrates, consisting of long chains of monosaccharide known to us as polysaccharide. Over past few decades, polysaccharides have gained significant importance due to their inherent features, like their easy availability, non-toxic nature, flexibility, potential biodegradability and biocompatibility (Beneke et al. 2009). Polysaccharides can also be used as a binder in the formulation of solid dosage forms due to their adhesive property. The swelling ability of polysaccharides helps to disintegrate the solid dosage forms and also facilitates developing hydrogel-based systems (Deshmukh et al. 2011; Kumaran et al. 2010; Jian et al. 2008). They exhibit a protective colloidal property that helps to act as a suspending and emulsifying agent in the formulation of course dispersion system (Doharey et al. 2010). Apart from these, they can also be used as coating agent, developing microencapsulation and as gelling agent (Ogaji et al. 2013; Mankala et al. 2011; Dionísio and Grenha 2012).

Amongst various other polysaccharides, marine sources are considered one of the vast biological sources of natural polysaccharides. Marine polysaccharides are generally obtained from various marine algae (Brown et al. 2014). Marine algae or seaweeds are divided into three main classes based upon their chemical configuration and natural coloring (complexation): (1) Phaeophyceae (brown algae); (2) Rhodophyceae (red algae) and (3) Chlorophyceae (green algae) (Rioux et al. 2017). In the past few years, a huge number of polysaccharides segregated from marine seaweed show numerous applications in pharmaceutical, cosmetic and food industries. Polysaccharides are a kind of biomacromolecule which is present as structural constituents of cell wall of marine algae. Polysaccharides which are isolated from marine seaweed exhibit various pharmacological pursuits, i.e. antitumor, anticoagulant, antioxidant and immunomodulatory properties (de Jesus Raposo et al. 2015; Hamed et al. 2015). Generally, biological activities of polysaccharides are dependent upon their chemical characteristics including their glycosidic linkages, monosaccharide ratios, their molecular sizes, shapes and their types (Hu et al. 2013). Physicochemical characteristics and biological activities of marine algae polysaccharides help to enhance their multifunctional activities and application. Marine polysaccharides can also be used to develop super-porous hydrogels (Omidian et al. 2005). A few sulfated polysaccharides are reported to exhibit antineoplastic, anti-coagulant, anti-oxidant, anti-allergic, antiviral, anti-inflammatory, anti-angiogenic activity, etc. (Arata et al. 2017; Peasura et al. 2015; Ahmadi et al. 2015; Abdelahad et al. 2016; Newman and Cragg 2016; Fernando et al. 2016). In the field of drug delivery, marine polysaccharides are enormously investigated to controlled-release and site-specific drug targeting, including modified release tablets, microparticles, beads, nanocarriers, hydrogels, in situ gels, transdermal formulations, etc. (Jana, Saha et al. 2013; Jana, Maji et al. 2013; Sharma et al. 2019; Dong et al. 2020). This chapter describes the classification of marine polysaccharides according to their biological sources and their applications in delivering a wide range of therapeutic agents, including wound healing and tissue regeneration.

5.2. CLASSIFICATION OF MARINE POLYSACCHARIDES

5.2.1 Polysaccharides Derived from Marine Algae

The prime sources of marine polysaccharides are marine microalgae, including red macro algae. However, they can also be extracted from green or brown macro algae and can be employed in pharmaceutical as well as biomedical applications (Cardoso et al. 2016). Classification of marine polysaccharides depending on their biological sources is shown in Figure 5.1.

5.2.1.1 Alginates

Alginates are majorly extracted from brown seaweeds, i.e., *Laminaria japonica, Laminaria digitata, Laminaria hyperborean, Ascophyllum nodosum,* and *Macrocystis pyrifera* (Lee & Mooney 2012). Alginates can also be obtained

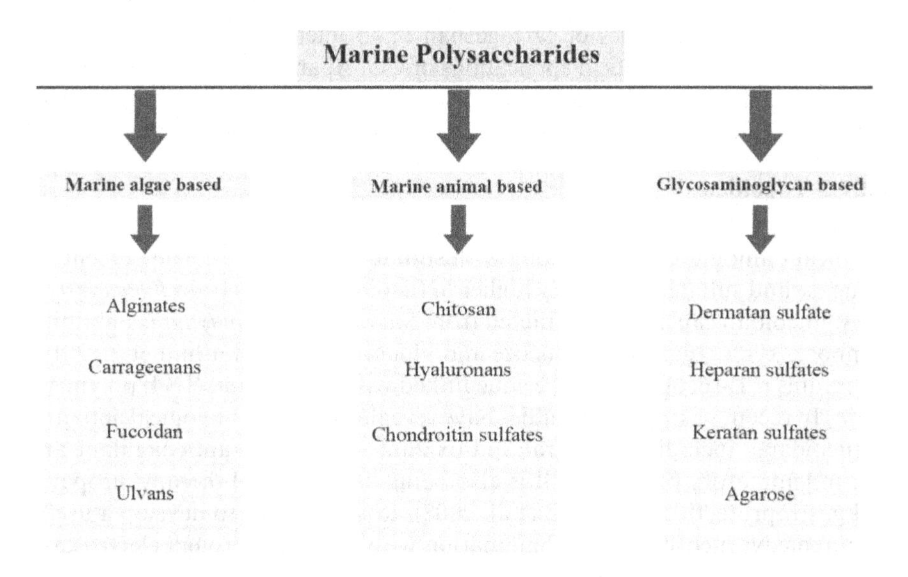

FIGURE 5.1 Classification of marine polysaccharides based on their sources.

from bacterial biosynthesis from the species of *Pseudomonas* and *Azobacter*. Structurally, it contains sequentially arranged two blocks including β-D-mannuronate (M) and (1–4)-linkages of α-L-guluronate (G) monomers (Jana et al. 2016). The M and G block ratio can be changed based on the geographical source variation. The ionic nature of alginate facilitates the formation of hydrogel matrices after cross-linking with bivalent or trivalent metal ions (Manna et al. 2021). The pH responsive nature of alginate is observed due to the presence of carboxyl group which helps to protect drug molecules in basic and neutral pH (Cardoso et al. 2016). The easy extraction and purification process of alginate has promoted its applications in the pharmaceutical and food industries. The excellent biodegradability and biocompatible nature of alginate combined with its low cytotoxic behavior has encouraged pharmaceutical researchers to thoroughly investigate it. In recent years, various applications of alginate have surfaced through several investigations focusing on the development of oral tablets, hydrogel matrices, in situ gels, nanocarriers, microparticulate, systems etc. (Manna et al. 2017; Jana, Sen et al. 2016; Jana, Das et al. 2013).

5.2.1.2 Carrageenans

Carrageenans (CG) are naturally occurring sulfated polysaccharides extracted from red seaweeds (Yermak et al. 2017). Depending on the sulfation degree carrageenans are divided into three different types, i.e. kappa carrageenan (κ-CG), lambda carrageenan (λ-CG) and iota carrageenan (τ-CG). κ CG, λ CG and τ CG are extracted from algae belonging to the Kappaphycus, Gigantinaceae and Eucheuma family, respectively (Cardoso et al. 2016; Manna and Jana 2021). Chemically it contains alternatively connected two units: unit A (α-(1Ñ3)) and unit B (β-(1Ñ4)) linked with glycosidic bonds. Unit A exhibits dextrorotatory properties and unit B shows levorotatory or dextrorotatory conformation. CG also possesses anticancer, anticoagulant, antiviral, antifungal and antiangiogenic properties. The gel-forming properties of carrageenan upon interaction with cations have enhanced the pharmaceutical applications of CG. Apart from being used in food industries as stabilizer and emulsifying agents, CG have found several applications in pharmaceutical and biomedical field (Cardoso et al. 2016; Salbach et al. 2012).

5.2.1.3 Fucoidan

Fucoidan is a polysaccharide, isolated from marine brown algae which contain significant amount of fucose along with sulfated ester groups. The percentage of fucose and sulfate is relatively higher in fucoidan isolated from *Fucus vesiculosus*. Fucoidan can also be extracted from *Sargassum stenophyllum* containing mannose, xylose, galactose, glucose and glucuronic acid (Luthuli et al. 2019). It contains α-L-fucopyranosyl residue linked with a (1Ñ3) and (1Ñ4) polymeric chain. In recent past, several studies have revealed different biological activities of fucoidans, including antiviral, antioxidant, antitumor, anticoagulant and anti-inflammatory properties. It is also being used in renal therapy, uropathy and gastroprotective effects (Li et al. 2008). It is also used to develop a useful burn remedy: fucospheres in combination with chitosan through electrostatic interaction (van Weelden et al. 2019).

5.2.1.4 Ulvans

Ulvans are sulfated marine polysaccharides extracted from marine microalgae belonging to the *Ulva* and *Enteromorpha* genus (Kaeffer et al. 1999). The extraction of ulvans involves depolymerization using organic solvent on feedstock (Melcher et al. 2017). It contains several sugar residues, including xylose, glucose, iduronic acid, glucuronic acid, etc. Ulvan exhibits variations in molecular weight, charge distribution and electron density depending on the percentage of sugar present in its structure (Marco et al. 2021; Fei et al. 2021).

5.2.2 Polysaccharides Derived from Marine Animals

Apart from algae, other marine sources are also rich in polysaccharide content. Amongst marine polysaccharides, chitin derivatives and glycosaminoglycans (GAG) are mostly obtained from marine animals (Claverie et al. 2020).

5.2.2.1 Chitosan

Chitosan is a derivative obtained from chitin which is extracted from the exoskeleton of several marine animals, i.e. shrimps, lobsters and crabs (Younes and Rinaudo 2015). The conversion of chitin to chitosan takes place through deacetylation reaction. The percentage of deacetylation may vary and result in a mutual presence of acetylated unit (N-acetyl-D-glucosamine) and deacetylated unit (D-glucosamine residues) (Acarturk et al. 1993; Cheung et al. 2015). Chitosan is soluble in acidic pH and exhibits cationic behavior in aqueous solution. This property of chitosan helps to develop electrostatic complexes with oppositely charged polymers (Jana et al. 2013). In recent past, chitosan is reported to have various biological properties, which include antitumor, antioxidant and antimicrobial behavior. Chitosan possesses excellent biodegradable and biocompatible profile which attributes towards the extensive application in pharmaceutical and biomedical field (Abd El-Hack et al. 2020; Jana and Sen 2017; Jana et al. 2015). Apart from these chitosan have found widespread applications in several food processing industries. It has also shown promises in the treatment of obesity, tissue engineering and wound healing scaffold (Kim et al. 2008).

5.2.2.2 Hyaluronan

Hyaluronan is glycosaminoglycans (GAG) based linear mucopolysaccharide containing alternative units of uronic acid and N-acetylated hexosamine in a repeating configuration. Hyaluronan is a non-sulfated polysaccharide, which is commonly known as hyaluronic acid. It is a negatively charged heteropolysaccharide. This marine polysaccharide consist of D-glucuronic acid unit and N-acetyl-D-glucosamine unit linked with glycosidic bonds (Gupta et al. 2019). Hyaluronic acid is commonly used in the cosmetic industry, tissue regeneration, wound healing, cell proliferation and as a polymer in drug delivery application (Goldberg and Buckwalter 2005). Apart from these, it is also used as a biological marker in determining the fraction of synovial fluids in different body joints for arthritis patients (Papakonstantinou et al. 2012).

5.2.2.3 Chondroitin Sulfates

This sulfated marine polysaccharide is primarily extracted from bovine and porcine cartilages. However, it can also be extracted from other marine animals, i.e. salmons, sea cucumbers, sharks, whales and mollusks. It consists of glucuronic acid disaccharide units attached with N-acetylgalactosamine (Henrotin et al. 2010; Mishra and Ganguli 2021). This sulfated GAG can also be used to prevent arthritis, tissue regeneration and as anticoagulant (Djerbal et al. 2017; Henrotin et al. 2010).

5.2.3 Polysaccharides Obtained from Glycosaminoglycan

Glycosaminoglycans are relatively less explored polysaccharides with respect to other marine polysaccharides. The complex extraction technique and lower bioavailability have restricted the pharmaceutical applications of glycosaminoglycans. However, a few polysaccharides, i.e. dermatan sulfate, keratan sulfate, heparin sulfate and agarose, have found their applications in the pharmaceutical industry (Merry. 2021).

5.2.3.1 Dermatan Sulfate

Dermatan sulfate is a linear polysaccharide containing a disaccharide unit of glucuronic acid or N-acetyl galactosamine (Westergren-Thorsson et al. 1991). It is also known as chondroitin sulfate B. Recent investigations revealed the efficacy of dermatan sulfate as an anticoagulant and as stabilizer for cytokine growth factors (Rostand and Esko 1997).

5.2.3.2 Heparan Sulfate

Heparan sulfate is a glycosaminoglycan-based polysaccharide which consists of a linear chain containing repeating units of D-glucuronic acid and iduronic acid in combination with either sulfated or acetylated D-glucosamine residue (Li et al. 2016). The biological activity of heparan sulfate may vary depending on the presence of sulfated residue (Wang, Dhurandhare et al. 2021).

5.2.3.3 Keratan Sulfates

This glycosaminoglycan-based marine polysaccharide consists of β-(1Ñ4) linked N-acetyl glucosamine unit and galactose residue. Based on the protein binding ability of keratan sulfate, it is classified in three different types, i.e. class I, class II and class III which are found in corneal cartilages, small cartilages and nerve tissue precipitates, respectively (Caterson et al. 2018; Tai et al. 1997).

5.2.3.4 Agarose

Agarose is extracted from the cell wall of red algae. The basic structural element of agarose is monosaccharide connected through repeating agarobiose units. It also contains galactose residue connected through the linkages of α-(1Ñ3) and β-(1Ñ4) (Hickson and Polson 1968). Agarose exhibits structural similarity with carrageenans except the L-conformation of unit A (α-(1Ñ3)) (Armisen 1995).

Several pharmaceutical applications of agarose have been reported in developing hydrogels and nanoparticulate delivery systems (Khodadadi et al. 2020).

5.3 PHARMACEUTICAL APPLICATIONS OF MARINE POLYSACCHARIDES

Several pharmaceutical applications of marine polysaccharides are listed in Table 5.1.

5.3.1 Delivery of Anticancer Agents

Ahmadi et al. investigated the efficacy of alginate-based nanocarriers developed with the combination of chitosan and sodium tripolyphosphate. The curcumin loaded nanoparticles were developed and subjected to in vitro evaluation. The average particle size was reported below 50 nm. The cytotoxic analysis indicated considerable antitumor efficacy in respect to free curcumin. The study findings indicated enhanced gene expression (Ahmadi et al. 2017). Ayub et al. employed a layer-by-layer synthesis approach to fabricate nanocarriers for colon-specific delivery. Oxidized sodium alginate was used in combination with cysteamine to form self-assembled nanospheres. Paclitaxel, a commonly used anticancer agent, was encapsulated in a nanoparticulate core. A pH-dependent swelling was reported when exposed to various gastrointestinal pH. Localization of NPs were reported in HT-29 cells which confirmed the efficacy of developed NPs in delivering colon-specific anticancer delivery (Ayub et al. 2019). Novel double-reactive hydrogels based on temperature-sensitive N-isopropylacrylamide and N-ethyl maleamic acid was employed with sodium alginate to synthesize free-radical polyamide. The stimulant-reactive characteristic of semi-isopropylacrylamide hydrogels shows significant sensitivity in both pH and temperature without restriction of N-ethyl containing malamic acid. In vitro studies have shown that the incorporated doxorubicin hydrochloride releases faster in simulation of tumor environment or endosome/lysosome than in simulated physiological conditions of humans (Dadfar et al. 2019).

pH-responsive novel metal nanoparticles of carrageenan oligosaccharide were developed by Chen et al. for tumor-specific delivery of epirubicin. The synthesized spherical shaped NPs demonstrated rapid release of epirubicin in compare to neutral pH. The flow cytometry method showed localization of epirubicin loaded NPs in the nucleus and induced relatively higher apoptosis of HepG2 and HCT-116 cells than free drug (Chen et al. 2019). In a recent study, Nogueira et al. assessed the efficacy of an iron oxide–based magnetic core containing a carrier system attached with a hybrid of κ-CG for delivering doxorubicin. The κ-CG containing hybrid shells showed high drug-loading capacity. The behavior of doxorubicin release depends on the mechanism of cation exchange within the ammonium group of doxorubicin of κ-CG. The effectiveness for both drug-loaded and free drug NPs was investigated in cancerous and non-cancerous cell lines, which indicated similar anti-proliferative

TABLE 5.1
Pharmaceutical Applications of Marine Polysaccharides

Sl. No.	Types of Therapeutic Agents	Marine Polysaccharides Used	Type of Drug Delivery Systems	Drug	References
1.	Anticancer agents	Alginate and chitosan	Nanoparticles	Curcumin	Ahmadi et al. 2017
		Sodium alginate	Nanoparticles	Paclitaxel	Ayub et al. 2019
		Sodium alginate	Hydrogels	Doxorubicin hydrochloride	Dadfar et al. 2019
		Carrageenan	Nanoparticles	Epirubicin	Chen et al. 2019
		κ-Carrageenan	Nanoparticles	Doxorubicin	Nogueira et al. 2020
		κ-CG and grapheme oxide conjugates	Nanocarriers	Doxorubicin	Vinothini et al. 2019
		Chitosan and chitosan-based	Hydrogels	Doxorubicin	Hosseinzadeh et al. 2019
		Chitosan	Nanoparticles	Daunorubicin	Cao et al. 2017
		Chitosan oligosaccharide	Nanomicelles	Genistein	Wang et al. 2019
2.	Antihypertensive agents	Alginate and chitosan	Nanocarriers	Amlodipine, captopril and valsartan	Niaz et al. 2016
		Alginate, chitosan and carrageenan	Matrix tablets	Diltiazem hydrochloride	Tapia et al. 2005
		Alginate	Mucoadhesive floating beads	Metoprolol tartrate	Biswas and Sahoo 2016
		Chitosan	Liposomes	Carvedilol	Chen et al. 2020
		Chitosan	Hydrogels	Latanoprost	Cheng et al. 2014
3.	Antiviral agents	Alginate	Microparticles	-	Meng et al. 2017
		Alginate	Microparticles	Acyclovir	Jain et al. 2016
		Alginate	Microparticles	Zidovudine	Roy et al. 2014
		Chitosan	Microbeads	Zidovudine	Szymanska et al. 2019
		Chitosan	Nanodroplets	Acyclovir	Donalisio et al. 2020

4.	NSAID agents	Alginate	Microbeads	Eudragit S-100	Chawla et al. 2012
		Alginate	Microspheres	Aceclofenac	Jana et al. 2013
		Alginate	Microspheres	Aceclofenac	Jana et al. 2015
		κ-CG and chitosan	Nanoparticles	Diclofenac sodium	Mahdavinia et al. 2015
		Chitosan and chondroitin sulfate	Nanoparticles	Ketoprofen.	Gul et al. 2018
		Boswellia resin-chitosan composites	Microparticles	Aceclofenac	Jana et al. 2015
		Chitosan	Nanocarriers	Aceclofenac	Jana and Sen 2017
		Chitosan	Nanogel	Aceclofenac	Jana et al. 2014
5.	Antibiotic agents	Alginate	Membranes	Ciprofloxacin hydrochloride	Sarwar et al. 2020
		Sodium alginate	Nanoparticles	Amoxicillin	Güncüm et al. 2018
		Alginate	In-situ gel	Metronidazole	Youssef et al. 2015
		Carrageenan and dextran sulfate	Nanodrugs	Ciprofloxacin	Cheow et al. 2015
		κ-CG	Nanoparticles	Ciprofloxacin	Mahdavinia et al. 2019
		Chitosan	Nanoparticles	Gentamicin, tetracycline and ciprofloxacin	El-Alfy et al. 2020
		Chitosan	Nanoparticles	Azithromycin	Mahaling et al. 2021
		N-trimethyl chitosan	Hydrogels	Ciprofloxacin	Hanna and Saad 2018
6.	Antidiabetic agents	Alginate	Floating and mucoadhesive core shell composite	Metformin HCl	Bera and Kumar 2018
		Alginate	Hydrogel	Metformin hydrochloride	Martínez-Gómez et al. 2017
		Alginate	Nanocarriers	Insulin	Collado-González et al. 2020
		Carboxymethylated τ-carrageenan	Nanoparticles	Insulin	Sahoo et al. 2017
		κ-CG	Carbon dots	Metformin	Das et al. 2019
		Chitosan	Nanocarriers	Insulin	Liu et al. 2019
		Chitosan	Nanoparticles	Insulin	Mumuni et al. 2020
		Chitosan	Nanoparticles	Metformin	Wang, Chin et al. 2021

(Continued)

TABLE 5.1 ***(Continued)***
Pharmaceutical Applications of Marine Polysaccharides

Sl. No.	Types of Therapeutic Agents	Marine Polysaccharides Used	Type of Drug Delivery Systems	Drug	References
7	Hormone delivery	Alginate	Nanocapsules	Testosterone	Jana et al. 2015
8.	Wound healing and tissue regeneration	Alginate	Microfibers	Curcumin	Sharma et al. 2020
		Alginate	Hydrogels	-	Karavasili et al. 2020
		Alginate	Dressing	Aloe vera	Koga et al. 2020
		κ-CG and τ-CG	Hydrogels	Manuka honey, eucalyptus oil and aloe vera gel	Pettinelli et al. 2020
		κ-CG	Hybrid hydrogels	Grape seed extract	Jaiswal et al. 2019
		κ-CG	Injectable hydrogels	-	Dev et al. 2020
		Methacrylated κ-carrageenan	3D bioprintable scaffold	-	Tytgat et al. 2019
		Chitosan	Nanoparticles incorporated hydrogels	-	Masood et al. 2019
		Chitosan	Hydrogel dressings	-	Zhao et al. 2017
		Chitosan	Composite membranes	-	Pereira et al. 2019

activity, especially with reduced cytotoxicity in non-cancerous cells (Nogueira et al. 2020). Vinothini et al. developed a nanocarrier-based delivery system containing doxorubicin for the treatment of cervical cancer. κ-CG and grapheme oxide conjugates were formed in combination with biotin to develop the nanocarriers system. The entrapment efficiency of doxorubicin was reported as 94%. HeLa cell lines were used to determine the cell viability which demonstrated a significant reduction ($p < 0.01$) in cellular viability (Vinothini et al. 2019).

An investigation performed by Hosseinzadeh et al. reported the use of chitosan in combination with acrylic acid and N-isopropyl acrylamide. Stimulus-reactive hydrogel nanocomposites were synthesized for delivering doxorubicin. The study results show the maximum loading capacity of doxorubicin was 89%. The new modified hydrogel showed excellent controlled-release ability of doxorubicin. The identical properties of chitosan and chitosan-based nanocomposites ensured the sustainable stimuli responsive delivery of anticancer drugs (Hosseinzadeh et al. 2019). Cao et al. developed carboxymethyl chitosan-based pH sensitive NPs for tumor targeting of daunorubicin. The developed nanocarriers showed negative zeta potential. Flow cytometry study involving HeLa cells confirmed the uptake and translocation of NPs within the cell nuclei (Cao et al. 2017). Chitosan oligosaccharide–based nanomicelles were developed by Wang et al. for delivering hydrophobic anticancer agent. pH responsive spherical shaped nanomicelles were loaded with genistein. The chitosan oligosaccharide complex and nanomicelles formation is shown in Figure 5.2. The protonation of the amino group of chitosan oligosaccharide facilitated the adsorption of nanomicelles onto tumor cells. At lesser intracellular pH, the dual pH responsive nanomicelles were disrupted to release the drug (Wang et al. 2019).

5.3.2 Delivery of Anti-Hypertensive Agents

Niaz et al. have investigated the efficacy of alginate- and chitosan-based nanocarriers loaded with commonly used antihypertensive drugs, i.e. amlodipine, captopril and valsartan. The NPs showed better stability and high drug loading capacity. The in vitro drug release study indicated excellent retention of the incorporated antihypertensive drugs in hybrid NPs. The study findings demonstrated <8% of drug release in the initial 24 hours (Niaz et al. 2016). Alginate-, chitosan- and carrageenan-based matrix tablets were developed by Tapia et al. for delivering diltiazem hydrochloride. The alginate and chitosan based tablets exhibited a slow erosion of the matrix. The drug release mechanisms were reported to be a combination of diffusion and polymeric relaxation. The alginate chitosan matrix exhibited ability for solvent uptake without disruption of the microstructure (Tapia et al. 2005). Alginate-based mucoadhesive floating beads were developed using the ionic gelation technique for delivering metoprolol tartrate. The study findings suggested the incorporation of tapioca starch has enhanced the efficacy and sustained-release behavior of metoprolol tartrate. Hydroxypropyl methylcellulose (HPMC) coating on developed beads has extended the metoprolol tartrate release up to 11 hours. Pharmacokinetic

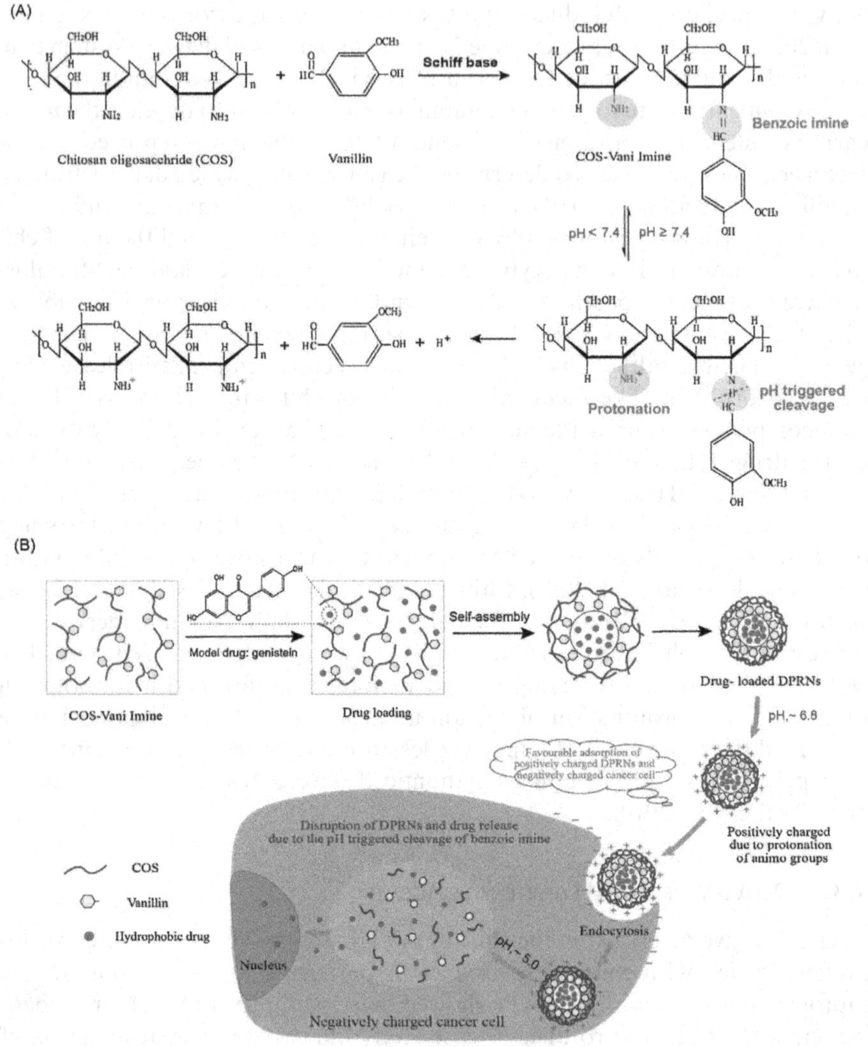

FIGURE 5.2 Schematic representation of (A) chitosan oligosaccharise-vanillin-imine complex exhibiting pH-responsive disruption and (B) drug loading and fabrication of self-assembled dual pH-responsive nanomicelles (DPRNs) indicating pH-specific drug release. (Reprinted from Wang et al. 2019; copyright 2019, with permission from Elsevier.)

analysis involving rabbit model indicated relatively higher oral bioavailability for floating beads (Biswas and Sahoo 2016).

Chitosan-coated carvedilol incorporated phospholipid layers were developed for enhancing oral retention period by Chen et al. The fiber diameter was determined using confocal laser scanning microscopic technique. Transmission electron microscopy has demonstrated the spherical shape of self-assembled liposomal structure after contacting with the aqueous phase. A permeation study using porcine buccal mucosa indicated higher permeation of carvedilol in respect to carvedilol suspension (Chen et al. 2020). Cheng et al. reported the synthesis of a chitosan- and gelatin-based hydrogel system for the treatment of ocular hypertension. Latanoprost was incorporated into the hydrogel system which demonstrated a sustained release in vitro. Intravitreal injection of developed formulation was administered on a rabbit model which exhibited considerable decrease of intraocular pressure (Cheng et al. 2014).

5.3.3 Delivery of Antiviral Agents

Meng et al. have investigated the application of novel alginate microparticles for vaginal delivery of antiviral drugs. Thiolated chitosan was employed for coating the synthesized microparticles. Optimization of several formulation batches was achieved using the alginate micro-particle core spray drying method. The average diameter was reported between 2 and 3 μm. The cytotoxic profile was evaluated using vaginal cell line of humans, which indicated no considerable cytotoxic activity. An increased mucoadhesion was reported for thiolated chitosan–coated microparticles (Meng et al. 2017). Another microparticulate delivery system for acyclovir was developed by Jain et al. Alginate in combination with carbopol was employed to develop microbeads. A drug-cholestyramine resin complex was used which resulted sustained release of acyclovir (Jain et al. 2016). A study conducted by Roy et al. reported the development of alginate-based zidovudine incorporated microparticles. The effect of hypromellose and polyacrylates wasalso investigated on the release profile of zidovudine. Complete release of drug was observed after 8 hours following Higuchi kinetics. The addition of HPMC K4M into the formulation resulted a sustained release of zidovudine (Roy et al. 2014).

The efficacy of chitosan glutamate was assessed by Szymanska et al. for zidovudine-loaded microbeads as a carrier for vaginal microbeads. Data from differential scanning calorimetry analysis and attenuated total reflectance – Fourier transform infrared spectroscopy confirmed the compatibility of the drug with chitosan glutamate after spray-drying. A 5:1 w/w polymer:drug ratio developed using an azeotropic mixture of ethanol and water was reported to spread easily in contact with simulated vaginal fluid. In vitro drug release study indicated an initial burst release of drug, followed by controlled release up to 4 hours (Szymanska et al. 2019). In a recent study, acyclovir-incorporated nanodroplets were developed using a sulfobutyl ether-β-cyclodextrin complex by Donalisio et al. The synthesized drug complex was incorporated using electrostatic interaction in a chitosan shell with an encapsulation efficiency of 97%. Increased antiviral efficacy was reported using the cell culture of HSV-2 with respect to free acyclovir (Donalisio et al. 2020).

5.3.4 Delivery of NSAIDs

Chawla et al. have investigated the efficacy of alginate-based mucoadhesive microbeads for delivering naproxen sodium. Emulsification method was employed to synthesize calcium chloride cross-linked alginate microbeads. Eudragit S-100 was used to coat the developed microbeads. The core microbeads showed an enhanced mucoadhesive property in respect to coated microbeads. The uncoated microspheres exhibited pH-dependent sustained drug release following Higuchi kinetics, where s the coated microspheres demonstrated Korsmeyer-Peppas kinetics (Chawla et al. 2012). Alginate-gellan gum based microspheres were synthesized for oral delivery of aceclofenac by Jana et al. The average size of the microparticles was reported between 270 and 490 μm. An in vitro drug release study revealed a sustained release of aceclofenac over 6 hours following Korsemeyer-Peppas kinetics. An in vivo study performed on a rabbit model revealed sustained absorption of drug with an excellent anti-inflammatory effect (Jana et al. 2013). Another study conducted by Jana et al. reported the development of alginate and locust bean gum based interpenetrating polymeric network (IPN) microspheres for oral delivery of aceclofenac. The ionic gelation technique was employed to develop calcium ion cross-linked microspheres. An in vitro study revealed sustained release of aceclofenac over 8 hours in pH 6.8 phosphate buffer. An increased polymer concentration has decreased the aceclofenac release percentage, as shown in Figure 5.3. Pharmacodynamic analysis exhibited a sustained anti-inflammatory effect after oral administration (Jana et al. 2015).

Mahdavinia et al. have developed κ-CG- and chitosan-based beads for oral delivery of diclofenac sodium. Incorporation of iron oxide magnetic NPs resulted in a decreased swelling ability. At pH 7.4, the maximum diclofenac sodium release was reported around 82% (Mahdavinia et al. 2015). Chitosan and chondroitin sulfate were employed to develop NPs for transdermal application of ketoprofen. The synthesized NPs were incorporated in an emulgel of argan oil. Sustained release of ketoprofen was observed up to 72 hours. A skin permeation study indicated higher permeation of ketoprofen than the marketed gel (Gul et al. 2018). Aceclofenac-loaded IPN nanocarriers were developed using glutaraldehyde cross-linked chitosan and locust bean gum. An increase in locust bean gum percentage decreased the drug entrapment. An in vitro study demonstrated restricted release of aceclofenac in acidic pH (Jana and Sen 2017). Jana et al. investigated the efficacy of chitosan-albumin based NPs for transdermal delivery of aceclofenac. The synthesized NPs were further incorporated into carbopol gel for easy topical application. The drug release study indicated sustained release pattern over 8 hours. A negative zeta potential of -22.10 mV was reported. The ex vivo skin permeation exhibited sustained penetration of aceclofenac through mice skin. Thein vivo study performed in carrageenan injected rats demonstrated higher swelling inhibition of rat paw compared to the marketed gel preparation (Jana et al. 2014).

5.3.5 Delivery of Antibiotics

Sarwar et al. have reported the synthesis of alginate- and polyethylene glycol–based membranes using a solvent casting method. A commonly used antibiotic,

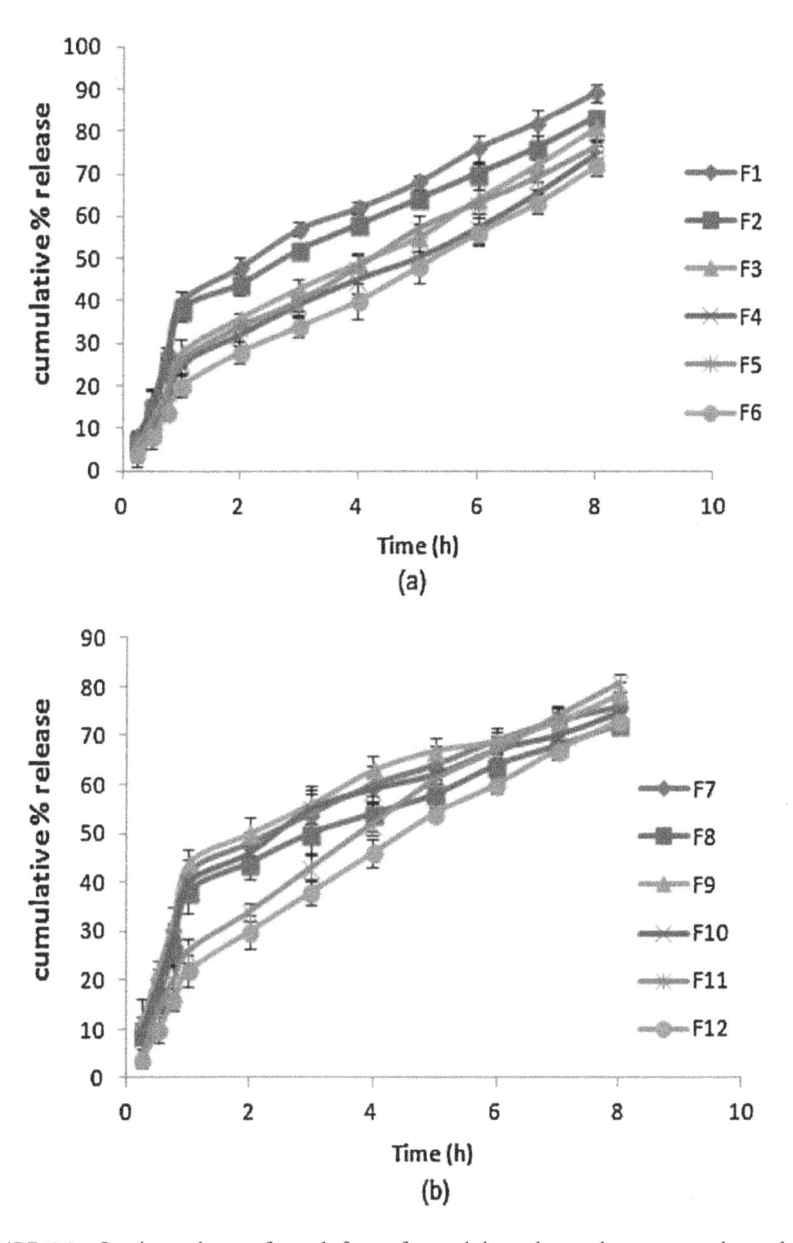

FIGURE 5.3 In vitro release of aceclofenac from alginate locust bean gum microspheres in pH 6.8 phosphate buffer for (a) F1-F6 and (b) F7-F12. (Reprinted from Jana et al. 2015; copyright 2015, with permission from Elsevier.)

ciprofloxacin hydrochloride, was incorporated in the developed membrane. Maximum ciprofloxacin hydrochloride reported was found around 80% in pH 1.2 within 70 to 120 minutes. At pH 7.4, 75% drug release was reported within 90 to 120 minutes (Sarwar et al. 2020). A study conducted by Güncüm reported the fabrication of sodium alginate and polyvinyl alcohol based NPs for controlled release of amoxicillin. The average size of developed NPs was reported between 336 and 558 nm. pH-responsive drug release was observed with an initial sudden release of amoxicillin. The minimum inhibitory concentration values indicated a comparable antimicrobial efficacy with pure amoxicillin (Güncüm et al. 2018). Youssef et al. have attempted to prolong the gastric residence time for metronidazole by developing a raft forming system using alginate and gellan gum based in in situ gel. The gellan gum based formulation exhibited less gelation ability. Floating lag time varied from few seconds to 1 minute, whereas floating duration was found to be more than 24 hours (Youssef et al. 2015).

The effect of carrageenan or dextran sulfate was investigated by Cheow et al. for developing amorphous nanodrugs through electrostatic complexation. Drug polysaccharide nanoplexes were synthesized to enhance the bioavailability of ciprofloxacin. The ciprofloxacin-dextran sulfate nanoplex showed higher permeation than the ciprofloxacin-carrageenan nanoplex. The developed nanoplexes were reported to be non-cytotoxic and to possess antimicrobial efficiency (Cheow et al. 2015). Mahdavinia et al. have synthesized hydroxyapatite nanoparticles for the sustained release of ciprofloxacin. A complex was formed by reacting high-molecular-weight chitosan with hydroxyapatite along with to κ-CG via electrostatic interaction for the development of a carrageenan-chitosan complex. Spherical NPs ranging between 20 and 30 nm was reported. The developed NPs were tested against bacterial efficacy for both gram-negative (*Salmonella typhi*) and gram-positive (*Staphylococcus pyogenes*) bacteria (Mahdavinia et al. 2019).

El-Alfy et al. have developed chitosan-based NPs for delivering commonly used antibiotics including gentamicin, tetracycline and ciprofloxacin. Chitosan NPs were prepared by using ionic gelation between chitosan and tripolyphosphate. Blended cotton fabric was treated with the developed composites to exhibit antibacterial efficacy. The study results indicated the inhibition of bacterial growth when treated with fabrics (El-Alfy et al. 2020). In a recent study, Mahaling et al. developed a nanoparticulate delivery system for ocular delivery of azithromycin. A chitosan and poly-L-lactide based nanoparticulate core is loaded with azithromycin. The synthesized nanoparticles are spherical in shape and lysozyme tolerant. It was also reported to be hemocompatible and cytocompatible in nature. An in vivo study involving mice demonstrated higher ocular drug permeation. A sustained release of drug for 300 hours was reported with considerable antibacterial and anti-inflammatory efficacy (Mahaling et al. 2021). A polyelectrolyte hydrogel-based drug carrier was developed by Hanna et al. for ciprofloxacin delivery. N-trimethyl chitosan in combination with modified xanthan gum was employed to synthesize the hydrogel drug carrier. In vitro drug release results indicated a maximum release of ciprofloxacin around 96% up to 150 minutes. Higher drug loading exhibited faster release of drug following zero-order release kinetics. The cytotoxic analysis performed on human lung cell lines exhibited $97 \pm 0.5\%$ cell viability up to 50 μg/ml concentration (Hanna et al. 2018).

5.3.6 Delivery of Antidiabetic Agents

Bera et al. have synthesized alginate-pectin based composites encapsulating a commonly used antidiabetic agent, metformin HCl. The stomach-specific controlled release of metformin HCl was achieved using the bioadhesive and floating properties of developed core-shell composites. Zinc acetate was used as a cross-linker for fabricating the core matrices through ionic gelation. The optimized formulation indicated zero-order release following case II transport of metformin HCl (Bera et al. 2018). An alginate- and PVA-based temperature and pH-sensitive hydrogel was developed using a green synthesis technique. Metformin hydrochloride was encapsulated in hydrogel matrix which demonstrated a minimal release at acidic pH and a considerable higher release at pH 8 after 6 hours (Martínez-Gómez et al. 2017). In a recent investigation, Collado-González et al. reported the synthesis of alginate based nanocarriers for oral delivery of insulin. Other polymers, including dextran sulfate, poloxamer 188, poly-(ethylene glycol), bovine serum albumin and chitosan, were also employed in the preparation of nanocomposites. The spherical-shaped nanocarriers were reported to range between 400 and 600 nm. Complete release of insulin was observed through high-performance liquid chromatography analysis (Collado-González et al. 2020).

Sahoo et al. have developed insulin encapsulated carrageenan-based NPs for oral delivery. Carboxymethylated τ-carrageenan issued in combination with chitosan to fabricate NPs. An in vitro study revealed pH-responsive insulin release from the nanoparticulate core. The gastric environment resulted a low release of insulin, whereas considerable higher release was observed in intestinal pH (Sahoo et al. 2017). Das et al. have reported the synthesis of carbon dots (C-dots) for oral delivery of metformin. κ-CG was employed with phenyl boronic acid to synthesize C-dots, having the ability of glucose-responsive drug release. Metformin encapsulated C-dots were subjected for drug release, which revealed pH-responsive release of metformin. 3-(4,5-dimethylthiazol-2-yl)-2,5-diphenyl-2H-tetrazolium bromide (MTT) assay has confirmed the biocompatible nature of the C-dots (Das et al. 2019).

Liu et al. have investigated the efficacy of insulin loaded chitosan nanocarriers. Polyelectrolyte complexes were synthesized using chitosan and polyethylene glycol monomethyl ether copolymers. The developed nano-formulation exhibited surface-dependent absorption of insulin maintaining a prolonged in vivo therapeutic effect (Liu et al. 2019). Another study performed by Mumuni et al. reported the development of chitosan-mucin NPs encapsulating insulin for oral delivery. A zeta potential of 31.2 mV was reported. An in vivo hypoglycemic study performed on diabetic rats indicated good efficacy of insulin loaded NPs after oral administration. The pharmacokinetic analysis exhibited low plasma clearance. Toxicological analysis revealed the non-toxic nature of NPs (Mumuni et al. 2020). In a recent investigation, Wang et al. described the synthesis of chitosan NPs for treating polycystic kidney disorder. Metformin was loaded in mucoadhesive nanocarriers. Ex vivo imaging indicated higher accumulation of NPs in the intestinal region. The area under the curve (AUC) was found to be 1.3 times greater than free drug after 24 hours of administration. Metformin-loaded NPs exhibited reduced cyst burden in respect to free metformin (Wang, Chin et al. 2021).

5.3.7 Wound Healing and Tissue Regeneration Applications

Sharma et al. employed alginate and gelatin to develop microfibers containing curcumin as a potential agent used for wound healing. The tensile strength of microfibers was reported between 1.08 and 3.53 N/mm^2. A cumulative curcumin release of 85% was observed after 72 hours. An in vivo study performed on cutaneous wound indicated higher drug accumulation in compared to the marketed formulation (Sharma et al. 2020). Karavasili et al. developed alginate and methyl cellulose based 3D printable hydrogels for wound healing application containing naturally obtained bioactive compounds. The developed carbohydrate inks exhibited excellent swelling behavior in hydrated conditions along with required antimicrobial efficacy. The formulations demonstrated good biocompatible profile with the human dermis layer. The in vivo wound healing study confirmed the cellular growth after application (Karavasili et al. 2020). In a recent investigation, Koga et al. reported the synthesis of zinc ion cross-linked alginate dressings for rapid healing of wounds. Aloe vera was also incorporated in the dressing and subjected tovarious evaluations. The developed films exhibited sufficient mechanical strength and malleability. An in vivo study performed in two groups of Wistar rats indicated enhancement in type I collagen fibers, quickening the healing process (Koga et al. 2020).

κ-CG and τ-CG based injectable hydrogel was developed by Pettinelli et al. for tissue repairing and wound healing applications. Gelatin and locust bean gum were also employed using physical cross-linking technique. The hydrogel exhibited a porous nature along with good mechanical stability. The rheological behavior along with the stained hydrogel is shown in Figure 5.4. In vitro evaluation performed on cell cultures indicated the ability of fibroblasts to grow and spread within the hydrogel which aids tissue repair (Pettinelli et al. 2020). Hydrogel films of κ-CG were fabricated by Jaiswal et al. containing grape seed extract and chitosan-linked sulfur NPs. The hybrid hydrogel film exhibited better mechanical properties and swelling ratio in compare to only κ-CG films. The developed films showed antimicrobial efficacy after application. The hybrid film demonstrated excellent efficacy in wound healing compared to the control group (Jaiswal et al. 2019). In a recent study, Dev et al. reported the development of injectable hydrogels of κ-CG and C-phycocyanin. The gelling ability of κ-CG was explored to obtain injectable gel matrix. The presence of a hydrogel matrix enhances dermal fibroblast proliferation. The study findings indicated the efficacy of the injectable hydrogel for tissue repairing and to initiate wound healing (Dev et al. 2020). Patient-specific scaffolding was fabricated using methacrylated κ-carrageenan and modified gelatin for tissue engineering. A 3D bioprintable scaffold was developed to enhance reconstruction of adipose tissue to avoid surgical complications in breast cancer. High water absorbing capacity was reported. Stem cells derived from adipose

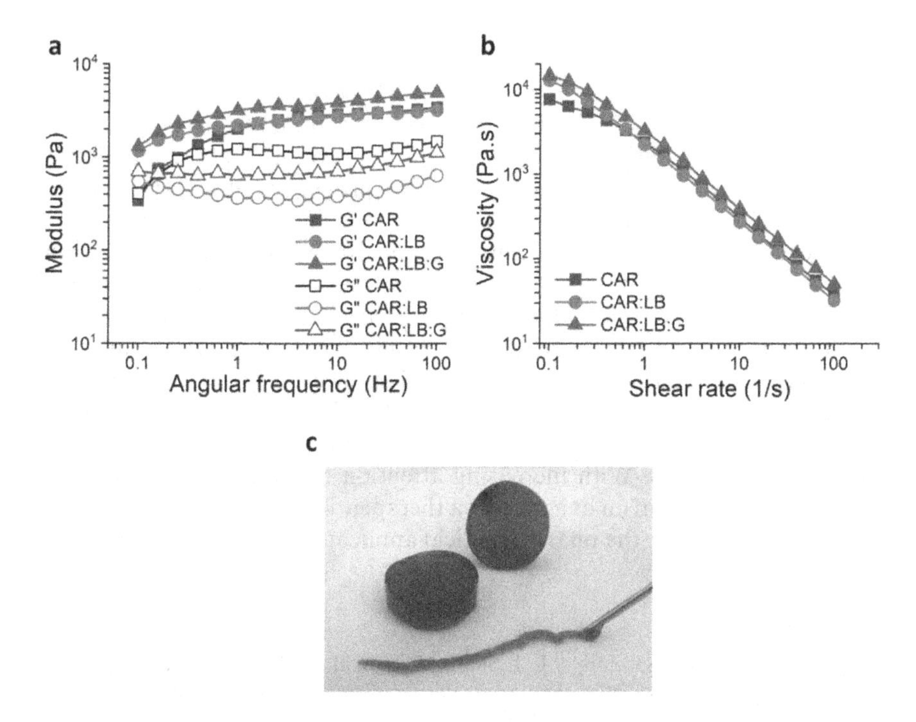

FIGURE 5.4 Rheological behavior of injectable hydrogel formulations indicating(a) the elastic modulus and viscous modulus of hydrogel preparation, (b) shear-thinning behavior of hydrogels and (c) bromophenol blue stained hydrogels for better visualization. (Reprinted from Pettinelli et al. 2020; copyright 2020, with permission from Elsevier.)

tissue indicated similar proliferation rate and cellular viability after 2 weeks (Tytgat et al. 2019).

Chitosan and PEG were cross-linked with silver NPs using glutaraldehyde to treat wounds in diabetes induced wounds. The wound healing effect was determined using diabetic rabbits. Compared to chitosan-PEG based hydrogels, the developed hydrogel exhibited higher swelling and higher porosity. It also reported enhanced microbial efficacy. Sustained release of silver NPs was observed over 7 days (Masood et al. 2019). Zhao et al. developed conductivity promoting hydrogel dressing to be used for self-healing application of wounds. The modified chitosan grafted polyaniline hydrogel formulation exhibited anti-oxidative and anti-infection properties. It also demonstrated blood clotting ability at a certain cross-linker ratio. The study reports suggested considerable wound healing ability (Zhao et al. 2017). Pereira et al. reported the fabrication of a chitosan-polyethylene glycol–based matrix containing zinc oxide, calcium phosphate and copper oxide. An improved mechanical property was also reported. The addition of chitosan resulted in a bacteriostatic effect without producing any cytotoxicity (Pereira et al. 2019).

5.4 CONCLUSION

Being abundantly available in nature and possessing favorable properties, marine polysaccharides are extensively exploited in several pharmaceutical fields. Amongst various polysaccharides, a few exhibit therapeutic efficacy, i.e. anti-inflammatory, anticancer and antiviral properties, etc. They are widely investigated as fundamental biomaterials in pharmaceutical industries and pharmaceutical research. The stimuli-responsive behavior combining with excellent biocompatibility and biodegradability makes marine-originated polysaccharides a suitable biopolymer for pharmaceutical applications. The classification of marine-originated polysaccharides is discussed in this chapter. We have also focused on pharmaceutical applications in delivering several classes of therapeutic agents, including wound healing and tissue repair potential. However, maximizing the potential of marine polysaccharides is still a challenge to scientists. With increasing attention on marine polysaccharides and their exact mechanism of action as a therapeutic agent and biodegradation pathways will broaden the pharmaceutical applications in the near future.

REFERENCES

Abd El-Hack, M. E. El-Saadony, M. T. Shafi, M. E. et al. 2020. Antimicrobial and antioxidant properties of chitosan and its derivatives and their applications: A review. *Int J Biol Macromol* 164:2726–2744.

Abdelahad, N. Barbato, F. O'Heir, S. et al. 2016. Reproduction of Sphaerococcus coronopifolius (Gigartinales, Rhodophyta) in natural populations of the Lazio coasts (central Italy) and in culture. *Cryptogamie Algol* 37:265–272.

Acarturk, F. Sencan, A. Celebi, N. 1993. Enhancement of the dissolution of spironolactone with chitosan and low molecular weight gelatin. *STP Pharma SCI* 3:369–373.

Ahmadi, A. Zorofchian, M. S. Abubakar, S. Zandi, K. 2015. Antiviral potential of algae polysaccharides isolated from marine sources: A review. *BioMed Res Int*:825203.

Ahmadi, F. Ghasemi-Kasman, M. Ghasemi, S. et al. 2017. Induction of apoptosis in HeLa cancer cells by an ultrasonic-mediated synthesis of curcumin-loaded chitosan-alginate-STPP nanoparticles. *Int J Nanomed* 12:8545–8456.

Arata, P. X. Genoud, V. Lauricella, A. M. Ciancia, M. Quintana, I. 2017. Alterations of fibrin networks mediated by sulfated polysaccharides from green seaweeds. *Thromb Res* 159:1–4.

Armisen R. 1995. World-wide use and importance of Gracilaria. *J Appl Phycol* 7:231–243.

Avachat, A. M. Dash, R. R. Shrotriya, S. N. 2011. Recent investigations of plant based natural gums, mucilages and resins in novel drug delivery systems. *Ind J Pharm Edu Res* 45:86–99.

Ayub, A. D. Chiu, H.I. Yusuf, S. N. A. M. Kadir, E. A. Ngalim, S. H. Lim, V. 2019. Biocompatible disulphide cross-linked sodium alginate derivative nanoparticles for oral colon-targeted drug delivery. *Artif Cells Nanomed Biotechnol* 47:353–369.

Beneke, C. E. Viljoen, A. M. Hamman, J. H. 2009. Polymeric plant-derived excipients in drug delivery. *Molecules* 14:2602–2620.

Bera, H. Kumar, S. 2018. Diethanolamine-modified pectin based core-shell composites as dual working gastroretentive drug-cargo. *Int J Biol Macromol* 108:1053–1062.

Biswas, N. Sahoo, R. K. 2016. Tapioca starch blended alginate mucoadhesive-floating beads for intragastric delivery of Metoprolol Tartrate. *Int J Biol Macromol* 83:61–70.

Brown, E. S. Allsopp, P. J. Magee, P. J. et al. 2014. Seaweed and human health. *Nutr Rev* 72:205–216.

Cao, J. Zheng, H. Hu, R. et al. 2017. pH-responsive nanoparticles based on covalently grafted conjugates of carboxymethyl chitosan and daunorubicin for the delivery of anti-cancer drugs. *J Biomed Nanotechnol* 13:1647–1659.

Cardoso, M. J. Costa, R. R. Mano, J. F. 2016. Marine origin polysaccharides in drug delivery systems. *Mar Drugs* 14:34.

Caterson, B. Melrose, J. 2018. Keratan sulfate, a complex glycosaminoglycan with unique functional capability. *Glycobiology* 28:182–206.

Chawla, A. Sharma, P. Pawar, P. 2012. Eudragit S-100 coated sodium alginate microspheres of naproxen sodium: Formulation, optimization and in vitro evaluation. *Acta Pharm* 62:529–545.

Chen, J. Pan, H. Duan, H. et al. 2020. Self-assembled liposome from core-sheath chitosan-based fibres for buccal delivery of carvedilol: Formulation, characterization and in vitro and ex vivo buccal absorption. *J Pharm Pharmacol* 72:343–355.

Chen, X. Han, W. Zhao, X. Tang, W. Wang, F. 2019. Epirubicin-loaded marine carrageenan oligosaccharide capped gold nanoparticle system for pH-triggered anticancer drug release. *Sci Rep* 9:6754.

Cheng, Y. H. Hung, K. H. Tsai, T. H. et al. 2014. Sustained delivery of latanoprost by thermosensitive chitosan-gelatin-based hydrogel for controlling ocular hypertension. *Acta Biomater* 10:4360–4366.

Cheow, W. S. Kiew, T. Y. Hadinoto, K. 2015. Amorphous nanodrugs prepared by complexation with polysaccharides: Carrageenan versus dextran sulfate. *Carbohydr Polym* 117:549–558.

Cheung, R. C. Ng, T. B. Wong, J. H. Chan, W. Y. 2015. Chitosan: An update on potential biomedical and pharmaceutical applications. *Mar Drugs* 13:5156–5186.

Claverie, M. McReynolds, C. Petitpas, A. Thomas, M. Fernandes, S. C. M. 2020. Marine-derived polymeric materials and biomimetics: An overview. *Polym (Basel)* 12:1002.

Collado-González, M. Ferreri, M. C. Freitas, A. R. et al. 2020. Complex polysaccharide-based nanocomposites for oral insulin delivery. *Mar Drugs* 18:55.

Dadfar, S. M. R. Pourmahdian, S. Tehranchi, M. M. Dadfar, S. M. 2019. Novel dual-responsive semi-interpenetrating polymer network hydrogels for controlled release of anticancer drugs. *J Biomed Mater Res A* 107:2327–2339.

Das, P. Maity, P. P. Ganguly, S. Ghosh, S. Baral, J. Bose, M. 2019. Biocompatible carbon dots derived from κ-carrageenan and phenyl boronic acid for dual modality sensing platform of sugar and its anti-diabetic drug release behavior. *Int J Biol Macromol* 132:316–329.

de Jesus Raposo, M. F. de Morais, A. M. de Morais, R. M. S. C. 2015. Marine polysaccharides from algae with potential biomedical applications. *Mar. Drugs* 13:2967–3028.

Deshmukh, T. Patil, P. Thakare, V. et al. 2011. Evaluation of binding properties of Butea monosperma Lam, gum in tablet formulation. *Int J Drug Discov Herb Res* 1:128–133.

Dev, A. Mohanbhai, S. J. Kushwaha, A. C. Sood, A. Sardoiwala, M. N. Choudhury, S. R. Karmakar, S. 2020. κ-carrageenan-C-phycocyanin based smart injectable hydrogels for accelerated wound recovery and real-time monitoring. *Acta Biomater* 109:121–131.

Dionísio, M. Grenha, A. 2012. Locust bean gum: Exploring its potential in biopharmaceutical application. *J Pharm Bioallied Sci* 4:175–185.

Djerbal, L. Lortat-Jacob, H. Kwok, J. 2017. Chondroitin sulfates and their binding molecules in the central nervous system. *Glycoconj J* 34:363–376.

Doharey, V. Sharma, N. Bindal, M. C. 2010. Assessment of the suspending properties of Cordia gheraf gum on paracetamol suspension. *Sch Res Lib* 2:510–517.

Donalisio, M. Argenziano, M. Rittà, M. et al. 2020. Acyclovir-loaded sulfobutyl ether-β-cyclodextrin decorated chitosan nanodroplets for the local treatment of HSV-2 infections. *Int J Pharm* 587:119676.

Dong, W. Ye, J. Wang, W. et al. 2020. Self-assembled lecithin/chitosan nanoparticles based on phospholipid complex: A feasible strategy to improve entrapment efficiency and transdermal delivery of poorly lipophilic drug. *Int J Nanomedicine* 15:5629–5643.

El-Alfy, E. A. El-Bisi, M. K. Taha, G. M. Ibrahim, H. M. 2020. Preparation of biocompatible chitosan nanoparticles loaded by tetracycline, gentamicin and ciprofloxacin as novel drug delivery systems for improvement of the antibacterial properties of cellulose based fabrics. *Int J Biol Macromol* 161:1247–1260.

Fei, J. Zhe, C. Yuanyuan, D. et al. 2021. Wound dressing hydrogel of enteromorpha prolifera polysaccharide-polyacrylamide composite: A facile transformation of marine blooming into biomedical material. *ACS Appl Mater Inter* 13: 14530–14542.

Fernando, I. S. Nah, J. W. Jeon, Y. J. 2016. Potential anti-inflammatory natural products from marine algae. *Environ Toxicol Pharmacol* 48:22–30.

Goldberg, V. M. Buckwalter, J. A. 2005. Hyaluronans in the treatment of osteoarthritis of the knee: Evidence for disease-modifying activity. *Osteoarthr Cartilage* 13:216–224.

Goswami, S. Naik, S. 2014. Natural gums and its pharmaceutical application. *J Sci Innov Res* 3:112–121.

Gul, R. Ahmed, N. Ullah, N. Khan, M. I. Elaissari, A. Rehman, A. U. 2018. Biodegradable ingredient-based emulgel loaded with ketoprofen nanoparticles. *AAPS Pharm Sci Tech* 19:1869–1881.

Güncüm, E. Işıklan, N. Anlaş, C. Ünal, N. Bulut, E. Bakırel, T. 2018. Development and characterization of polymeric-based nanoparticles for sustained release of amoxicillin—an antimicrobial drug. *Artif Cells Nanomed Biotechnol* 46:964–973.

Gupta, R. C. Lall, R. Srivastava, A. Sinha, A. 2019. Hyaluronic acid: Molecular mechanisms and therapeutic trajectory. *Front Vet Sci* 6:192.

Hamed, I. Özogul, F. Özogul, Y. Regenstein, J. M. 2015. Marine bioactive compounds and their health benefits: A review. *Compr Rev Food Sci Food Saf* 14:446–465.

Hanna, D. H. Saad, G. R. 2018. Encapsulation of ciprofloxacin within modified xanthan gum-chitosan based hydrogel for drug delivery. *Bioorg Chem* 84:115–124.

Henrotin, Y. Mathy, M. Sanchez, C. Lambert, C. 2010. Chondroitin sulfate in the treatment of osteoarthritis: From in vitro studies to clinical recommendations. *Ther Adv Musculoskelet Dis* 2:335–348.

Hickson, T. G. Polson, A. 1968. Some physical characteristics of the agarose molecule. *Biochim Biophys Acta* 165:43–58.

Hosseinzadeh, S. Hosseinzadeh, H. Pashaei, S. Khodaparast, Z. 2019. Synthesis of stimuli-responsive chitosan nanocomposites via RAFT copolymerization for doxorubicin delivery. *Int J Biol Macromol* 121:677–685.

Hu, D. J. Cheong, K. I. Zhao, J. Li, S. P. 2013. Chromatography in characterization of polysaccharides from medicinal plants and fungi. *J Sep Sci* 36:1–19.

Jain, S. K. Kumar, A. Kumar, A. Pandey, A. N. Rajpoot, K. 2016. Development and in vitro characterization of a multiparticulate delivery system for acyclovir-resinate complex. *Artif Cells Nanomed Biotechnol* 44:1266–1275.

Jaiswal, L. Shankar, S. Rhim, J. W. 2019. Carrageenan-based functional hydrogel film reinforced with sulfur nanoparticles and grapefruit seed extract for wound healing application. *Carbohydr Polym* 224:115191.

Jana, S. Bibek Laha, B. Maiti, S. 2015. Boswellia gum resin/chitosan polymer composites: Controlled delivery vehicles for aceclofenac. *Int J Biol Macromol* 77:303–306.

Jana, S. Das, A. Nayak, A. K. Sen, K. K. Basu, S. K. 2013. Aceclofenac-loaded unsaturated esterified alginate/gellan gum microspheres: In vitro and in vivo assessment. *Int J Biol Macromol* 57:129–137.

Jana, S. Gandhi, A. Sheet, S. Sen, K. K. 2015. Metal ion-induced alginate-locust bean gum IPN microspheres for sustained oral delivery of aceclofenac. *Int J Biol Macromol* 72:47–53.

Jana, S. Gangopadhaya, A. Bhowmik, B. B. Nayak, A. K. Mukherjee, A. 2015. Pharmacokinetic evaluation of testosterone-loaded nanocapsules in rats. *Int J Biol Macromol* 72:28–30.

Jana, S. Maji, N. Nayak, A. K. Sen, K. K. Basu, S. K. 2013. Development of chitosan-based nanoparticles through inter-polymeric complexation for oral drug delivery. *Carbohydr Polym* 98: 870–876.

Jana, S. Manna, S. Nayak, A. K. Sen, K. K. Basu, S. K. 2014. Carbopol gel containing chitosan-egg albumin nanoparticles for transdermal aceclofenac delivery. *Colloids Surf B Biointerfaces* 114:36–44.

Jana, S. Saha, A. Nayak, A. K. Sen, K. K. Basu, S. K. 2013. Aceclofenac-loaded chitosan-tamarind seed polysaccharide interpenetrating polymeric network microparticles. *Coll Surf B* 105:303–309.

Jana, S. Samanta, A. Nayak, A. K. Sen, K. K. Jana, S. 2015. Novel alginate hydrogel core-shell systems for combination delivery of ranitidine HCl and aceclofenac. *Int J Biol Macromol* 74:85–92.

Jana, S. Sen, K. K. 2017. Chitosan—Locust bean gum interpenetrating polymeric network nanocomposites for delivery of aceclofenac. *Int J Biol Macromol* 102:878–884.

Jana, S. Sen, K. K. Gandhi, A. 2016. Alginate based nanocarriers for drug delivery applications. *Curr Pharm Des* 22:3399–410.

Jana, S. Sharma, R. Maiti, S. Sen, K. K. 2016. Interpenetrating hydrogels of O-carboxymethyl Tamarind gum andalginate for monitoring delivery of acyclovir. *Int J Biol Macromol* 92:1034–1039.

Jian, S. Yadav, S. K. Patil, U. K. 2008. Preparation and evaluation of sustained release matrix tablet of furosemide. *Res J Pharm Tech* 1: 374–376.

Kaeffer, B. Bénard, C. Lahaye, M. Blottière, H. M. Cherbut, C. 1999. Biological properties of Ulvan, a new source of green seaweed sulfated polysaccharides, on cultured normal and cancerous colonic epithelial cells. *Planta Med* 65:527–531.

Karavasili, C. Tsongas, K. Andreadis, I. I. et al. 2020. Physico-mechanical and finite element analysis evaluation of 3D printable alginate-methylcellulose inks for wound healing applications. *Carbohydr Polym* 247:116666.

Khodadadi, M. Taghizadeh, A. Taghizadeh, M. et al. 2020. Agarose-based biomaterials for advanced drug delivery. *J Control Release* 326:523–543.

Kim, I. Y. Seo, S. J. Moon, H. S. et al. 2008. Chitosan and its derivatives for tissue engineering applications. *Biotechnol Adv* 26:1–21.

Koga, A. Y. Felix, J. C. Silvestre, R. G. M. et al. 2020. Evaluation of wound healing effect of alginate film containing Aloe vera gel and cross-linked with zinc chloride. *Acta Cir Bras* 35:e202000507.

Kumaran, A. K. S. G. Palanisamy, S. Rajasekaran, A. et al. 2010. Evaluation of Cassia roxburghii seed gum as binder in tablet formulations of selected drugs. *Int J Pharm Sci Nanotechnol* 2:726–732.

Lee, K. Y. Mooney, D. J. 2012. Alginate: Properties and biomedical applications. *Prog Polym Sci* 37:106–126.

Li, B. Lu, F. Wei, X. Zhao, R. 2008. Fucoidan: Structure and bioactivity. *Molecules* 13:1671–1695.

Li, J. P. Kusche-Gullberg, M. 2016. Heparan sulfate: Biosynthesis, structure, and function. *Int Rev Cell Mol Biol* 325:215–273.

Liu, C. Kou, Y. Zhang, X. Dong, W. Cheng, H. Mao, S. 2019. Enhanced oral insulin delivery via surface hydrophilic modification of chitosan copolymer based self-assembly polyelectrolyte nanocomplex. *Int J Pharm* 554:36–47.

Luthuli, S. Wu, S. Cheng, Y. Zheng, X. Wu, M. Tong, H. 2019. Therapeutic effects of fucoidan: A review on recent studies. *Mar Drugs* 17:487.

Mahaling, B. Baruah, N. Ahamad, N. Maisha, N. Lavik, E. Katti, D. S. 2021. A non-invasive nanoparticle-based sustained dual-drug delivery system as an eyedrop for endophthalmitis. *Int J Pharm* 606:120900.

Mahdavinia, G. R. Etemadi, H. Soleymani, F. 2015. Magnetic/pH-responsive beads based on caboxymethyl chitosan and κ-carrageenan and controlled drug release. *Carbohydr Polym* 128:112–121.

Mahdavinia, G. R. Karimi, M. H. Soltaniniya, M. Massoumi, B. 2019. In vitro evaluation of sustained ciprofloxacin release from κ-carrageenan-crosslinked chitosan/hydroxyapatite hydrogel nanocomposites. *Int J Biol Macromol* 126:443–453.

Mankala, S. K. Nagamalli, N. K. Raprla, R. Kommula, R. 2011. Preparation and characterization of mucoadhesive microcapsules of gliclazide with natural gums. *Stm J Pharm Sci* 4:38–48.

Manna, S. Jana, S. 2021. Carrageenan-based nanomaterials in drug delivery applications. In *Biopolymer-based nanomaterials in drug delivery and biomedical applications*, ed. H. Bera, C. M. Hossain, S. Saha, 365–382. Elsevier.

Manna, S. Jayasri, K. Annapurna, K. R. Kanthal, L. K. 2017. Alginate based gastroretentive raft forming tablets for enhanced bioavailability of tinidazole. *Int J Appl Pharm* 9:16–21.

Manna, S. Mal, M. Das, S. Mandal, D. Bhowmik, M. 2021. Ionically gelled alginates in drug delivery. In*Ionically gelled biopolysaccharide based systems in drug delivery, gels horizons: From science to smart materials*, ed. A. K. Nayak, M. S. Hasnain, D. Paul, 29–53. Springer.

Marco, B. Remy, T. Grace, V. Dennis, N. Aurelie, R. Ronan, P. V. Prasad, S. 2021. Aurelien Forget. Hydrogel-forming algae polysaccharides: From seaweed to biomedical applications. *Biomacromolecules* 22:1027–1052.

Martínez-Gómez, F. Guerrero, J. Matsuhiro, B. Pavez, J. 2017. In vitro release of metformin hydrochloride from sodium alginate/polyvinyl alcohol hydrogels. *Carbohydr Polym* 155:182–191.

Masood, N. Ahmed, R. Tariq, M. et al. 2019. Silver nanoparticle impregnated chitosan-PEG hydrogel enhances wound healing in diabetes induced rabbits. *Int J Pharm* 559:23–36.

Melcher, R. L. Neumann, M. Fuenzalida, W. J. P. Gröhn, F. Moerschbacher, B. M. 2017. Revised domain structure of Ulvan lyase and characterization of the first Ulvan binding domain. *Sci Rep* 7:44115.

Meng, J. Agrahari, V. Ezoulin, M. J. et al. 2017. Spray-dried thiolated chitosan-coated sodium alginate multilayer microparticles for vaginal HIV microbicide delivery. *AAPS J* 19:692–702.

Merry, C. L. R. 2021. Exciting new developments and emerging themes in glycosaminoglycan research. *J Histochem Cytochem* 69:9–11.

Mishra, S. Ganguli, M. 2021. Functions of, and replenishment strategies for, chondroitin sulfate in the human body. *Drug Discov Today* 26:1185–1199.

Mumuni, M. A. Kenechukwu, F. C. Ofokansi, K. C. Attama, A. A. Díaz, D. D. 2020. Insulin-loaded mucoadhesive nanoparticles based on mucin-chitosan complexes for oral delivery and diabetes treatment. *Carbohydr Polym* 229:115506.

Newman, D. J. Cragg, G. M. 2016. Natural products as sources of new drugs from 1981 to 2014. *J Nat Prod* 79:629–661.

Niaz, T. Nasir, H. Shabbir, S. Rehman, A. Imran, M. 2016. Polyionic hybrid nano-engineered systems comprising alginate and chitosan for antihypertensive therapeutics. *Int J Biol Macromol* 91:180–187.

Nogueira, J. Soares, S. F. Amorim, C. O. et al. 2020. Magnetic driven nanocarriers for pH-responsive doxorubicin release in cancer therapy. *Molecules* 25:333.

Ogaji, I. J. Okafor, I. S. Hoag, S. W. 2013. Grewia gum as a film coating agent in theophylline tablet formulation: Properties of theophylline tablets coated with grewia gum as a film coating agent. *J Pharm Bioallied Sci* 5:53–60.

Omidian, H. Rocca, J. G. Park, K. 2005. Advances in superporous hydrogels. *J. Control. Release* 102:3–12.

Papakonstantinou, E. Roth, M. Karakiulakis, G. 2012. Hyaluronic acid: A key molecule in skin aging. *Dermatoendocrinol* 4:253–258.

Peasura, N. Laohakunjit, N. Kerdchoechuen, O. Wanlapa, S. 2015. Characteristics and antioxidant of Ulvaintestinalis sulphated polysaccharides extracted with different solvents. *Int J Biol Macromol* 81:912–919.

Pereira, I. C. Duarte, A. S. Neto, A. S. Ferreira, J. M. F. 2019. Chitosan and polyethylene glycol based membranes with antibacterial properties for tissue regeneration. *Mater Sci Eng C Mater Biol Appl* 96:606–615.

Pettinelli, N. Rodríguez-Llamazares, S. Bouza, R. Barral, L. Feijoo-Bandín, S. Lago, F. 2020. Carrageenan-based physically crosslinked injectable hydrogel for wound healing and tissue repairing applications. *Int J Pharm* 589:119828.

Rioux, L. E. Beaulieu, L. Turgeon, S. L. 2017. Seaweeds: A traditional ingredients for new gastronomic sensation. *Food Hydrocoll* 68:255–265.

Rostand, K. S. Esko, J. D. 1997. Microbial adherence to and invasion through proteoglycans. *Infect Immun* 65: 1–8.

Roy, H. Rao, P. V. Panda, S. K. Biswal, A. K. Parida, K. R. Dash, J. 2014. Composite alginate hydrogel microparticulate delivery system of zidovudine hydrochloride based on counter ion induced aggregation. *Int J Appl Basic Med Res* 4:S31–S36.

Sahoo, P. Leong, K. H. Nyamathulla, S. Onuki, Y. Takayama, K. Chung, L. Y. 2017. Optimization of pH-responsive carboxymethylated iota-carrageenan/chitosan nanoparticles for oral insulin delivery using response surface methodology. *React Funct Polym* 119:145–155.

Salbach, J. Kliemt, S. Rauner, M. et al. 2012. The effect of the degree of sulfation of glycosaminoglycans on osteoclast function and signaling pathways. *Biomaterials* 33:8418–8429.

Sarwar, M. S. Ghaffar, A. Huang, Q. Zafar, M. S. Usman, M. Latif, M. 2020. Controlled-release behavior of ciprofloxacin from a biocompatible polymeric system based on sodium alginate/poly(ethylene glycol) mono methyl ether. *Int J Biol Macromol* 165:1047–1054.

Sharma, A. Mittal, A. Puri, V. Kumar, P. Singh, I. 2020. Curcumin-loaded, alginate-gelatin composite fibers for wound healing applications. *3 Biotech* 10:464.

Sharma, S. Sarkar, G. Srestha, B. Chattopadhyay, D. Bhowmik, M. 2019. In-situ fast gelling formulation for oral sustained drug delivery of paracetamol to dysphagic patients. *Int J Biol Macromol* 134:864–868.

Szymanska, E. Czarnomysy, R. Jacyna, J. et al. 2019. Could spray-dried microbeads with chitosan glutamate be considered as promising vaginal microbicide carriers? The effect of process variables on the in vitro functional and physicochemical characteristics. *Int J Pharm* 568:118558.

Tai, G. H. Nieduszynski, I. A. Fullwood, N. J. Huckerby, T. N. 1997. Human corneal keratan sulfates. *J Biol Chem* 272:28227–28231.

Tapia, C. Corbalán, V. Costa, E. Gai, M. N. Yazdani-Pedram, M. 2005. Study of the release mechanism of diltiazem hydrochloride from matrices based on chitosan-alginate and chitosan-carrageenan mixtures. *Biomacromolecules* 6:2389–2395.

Tytgat, L. Van, Damme, L. Ortega, Arevalo, M. D. P. et al. 2019. Extrusion-based 3D printing of photo-crosslinkable gelatin and κ-carrageenan hydrogel blends for adipose tissue regeneration. *Int J Biol Macromol* 140:929–938.

van Weelden, G. Bobiński, M. Okła, K. van Weelden, W. J. Romano, A. Pijnenborg, J. M. A. 2019. Fucoidan structure and activity in relation to anti-cancer mechanisms. *Mar Drugs* 17:32.

Vinothini, K. Rajendran, N. K. Munusamy, M. A. Alarfaj, A. A. Rajan, M. 2019. Development of biotin molecule targeted cancer cell drug delivery of doxorubicin loaded κ-carrageenan grafted graphene oxide nanocarrier. *Mater Sci Eng C Mater Biol Appl* 100:676–687.

Wang, J. Chin, D. Poon, C. et al. 2021. Oral delivery of metformin by chitosan nanoparticles for polycystic kidney disease. *J Control Release* 329:1198–1209.

Wang, Y. Khan, A. Liu, Y. et al. 2019. Chitosan oligosaccharide-based dual pH responsive nano-micelles for targeted delivery of hydrophobic drugs. *Carbohydr Polym* 223:115061.

Wang, Z. Dhurandhare, V. M. Mahung, C. A. et al. 2021. Improving the sensitivity for quantifying heparan sulfate from biological samples. *Anal Chem* 93:11191–11199.

Westergren-Thorsson, G. Onnervik, P. O. Fransson, L. A. Malmstrom, A. 1991. Proliferation of cultured fibroblasts is inhibited by L-iduronate-containing glycosaminoglycans. *J Cell Physiol* 147:523–530.

Yermak, I. M. Mischchenko, N. P. Davydova, V. N. et al. 2017. Carrageenans—sulfated polysaccharides from red seaweeds as matrices for the inclusion of echinochrome. *Mar Drugs* 15:337.

Younes, I. Rinaudo, M. 2015. Chitin and chitosan preparation from marine sources. Structure, properties and applications. *Mar Drugs* 13:1133–1174.

Youssef, N. A. H. A. Kassem, A. A. El-Massik, M. A. Boraie, N. A. 2015. Development of gastroretentive metronidazole floating raft system for targeting Helicobacter pylori. *Int J Pharm* 486:297–305.

Zhao, X. Wu, H. Guo, B. Dong, R. Qiu, Y. Ma, P. X. 2017. Antibacterial anti-oxidant electroactive injectable hydrogel as self-healing wound dressing with hemostasis and adhesiveness for cutaneous wound healing. *Biomaterials* 122:34–47.

6 Comparison of Healing Effect of DMSP in Green Sea Algae and Mesenchymal Stem Cells on Various Inflammatory Disorders

Kenji Nakajima

CONTENTS

DOI: 10.1201/9781003303916-6

6.1 SUMMARY

The healing effect of a single natural compound, dimethylsulfoniopropionate (DMSP), in green sea algae, and mesenchymal stem cells (MSCs) of varied tissues origins on various disorders in three germ layers and the neural crest of animals was parallelly described and compared in brief and in detail. DMSP completely healed the skin and gastric ulcers, type 1 diabetes mellitus, of great interest, free cell and sold cancers, dementia, Parkinson's and Alzheimer's diseases in rat and mouse, promoted the outgrowth and elongation of nerve cells, and prolonged significantly the life span of senescence accelerated mice and control mice. In these diseases, more and frequent usage of DMSP is able and preferable to the investigations in this line of work in the future. Whereas MSCs could not completely reinstate especially crucial cancers, the cells do exert the preferable healing effect on various diseases. In conclusion, the comparison indicates that the healing effect of DMSP is likely comparative to that of MSCs. Here the embryonic stem cells and induced pluripotent stem (iPS) cells were excluded for the ethical problem and the possibility of incidence of immune rejection and tumors. In contrast, the clinical trials with DMSP need to be rapidly performed in humans. Thereafter, clinical therapies with both DMSP and MSCs should also be attempted to eradicate these diseases, especially cancers.

6.2 INTRODUCTION

Population in the world is increasing year by year, and have rapidly elevated from about 1950 to date. In particular, the increase occurs in developing countries (The current US Census Bureau world population, 2019). The world population more than 60 years and older is expected to total 2 billion by 2050 from 900 million in 2015 and aged 80 years or older elevated from 125 million today to 434 million by 2050. In 2050, 80% of older people will be living in low- and middle-income countries (WHO, 2017). The causes of mortality are different between the developing countries and developed countries. The kinds of disorders in low-income countries differ from those in high-income countries

(WHO, 2017). Therefore, different clinic therapies are needed in both countries. Furthermore, the mortality of the population in the low-income countries suffered from communicable diseases (CDs) (infectious disorders) was 52% but those in high-income countries were 7%. In contrast, the mortality of the population caused by non-communicable disorders (NCDs) (non-infectious) was 70% in the world, but that in low-income countries was 37% but that in high-income countries was 88% (WHO, 2017). Therefore, various and effective clinical therapies need to be considered and performed in the world.

6.3 DIMETHYLSULFONIOPROPIONATE

6.3.1 Occurrence of DMSP

3-Dimethyl-2-carboxyethyl sulfonium (dimethylsulfoniopropionate [DMSP]) ($C_5H_{10}O_2S$, M.W. 134.04) is a tertiary sulfoniopropionate (sulfobetaine) zwitterion. The compound is synthesized by equimolar amounts of dimethylsulfide and propionate (halide) by refluxing under moderate temperature for about half a day. The compound is very soluble in water and methyl alcohol but insoluble in etylether. The products were identified by Nuclear magnetic resonance (NMR), Infrared absorption spectrometry (IR), elemental and mass analysis or x-ray or thermal-degradation. The element analysis of DMSP exhibited 99% purity. The compound has, in particular, the highest substrate specificity for the enzyme, dimethylthetine- and betaine-homocysteine methltransferase [E.C.2.1.1.3. and 2.1.1.5.], and the most active structure among closely related compounds for the attraction of feeding behaviors in fish (Nakajima, 1996).

The structures of DMSP and related compound are shown in Figure 6.1. Methylmethionine (vitamin U), glycine betaine (GB) (commercially available) and dimethylsulfonioacetate (DMSA) (prepared by refluxing DMS and acetate [halide]). DMSP is naturally and worldwide contained in the water and organisms in the sea shore and ocean: bacteria (Karsten et al., 1996), symbionts (zooxanthellae) (Van Alstyne, Schupp & Slattery, 2006; Van Alstyne, Dominiquuelll & Muller-Parker, 2009), coral (Broadbent, Jones & Jones, 2002; Raina et al., 2013; Burdett et al., 2014.), phyto- and zooplankton (Ackman, Tocher & Mclachian., 1966; Tokunaga, Iida & Nakamura, 1977; Keller & Korjeff-Bellows,1996; Daly & Tullio, 1996; Yoch, 2002), benthic algae (Broadbent, Jones & Jones, 2002), macroalgae (Challenger & Simpson, 1948; Iida, 1988; Reed, 1983), sponges (Van Alstyne, Shupp & Slattery, 2006), higher plants (Karsten et al., 1996; Otte et al., 2004), mollusks (Ackman & Hingley, 1968; Hill, Dacey & Edward, 2000; Hill & Dacey, 2007), crustaceans (Nakamura, Iida & Tokunaga, 1981) and fish (Ackman, Dale & Hingley, 1966; Ackman, Hingley & MacKay, 1972. Iida, 1988). In contrast, DMSP proves to be synthesized by marine micro- and macro-algae and coral (Karsten et al., 1996; Keller & Korjeff-Bellows, 1996; Van Alstyne, Shupp & Slattery, 2006; Tapiolas et al., 2013; Bullock, 2017). In particular, DMSP is formed and contained in large amounts in green sea macroalgae (*Enteromorpha* sp., *Ulva pertusa*, *Monostroma nitidum*) (White, 1982; Reed, 1983; Bluden et al., 1992). DMSP in other organisms are likely supplemented through diets and symbionts.

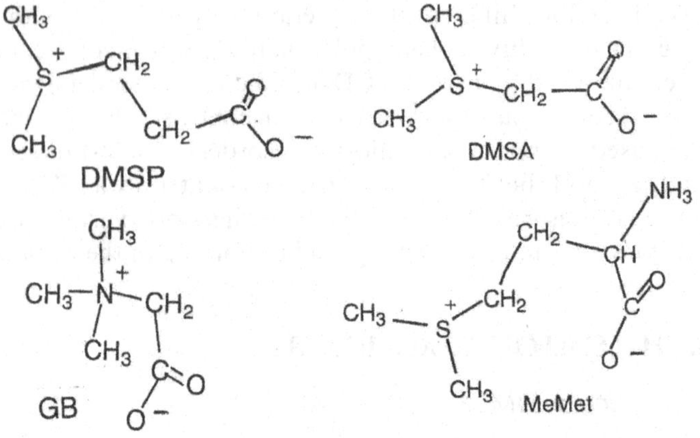

FIGURE 6.1 Dimensional structures of DMSP and related compounds. Two-dimensional structures of DMSP, DMSA, MeMet (methylmethionine, vitamin U), and GB (glycine betaine). Methyl and carboxyl carbons were arranged to be the same size as much as possible based on the three-dimensional structures. The structures were depicted using the software of CSC ChemDraw (Cambridge Scientific Computing, Inc., ver. 3.1)

6.3.2 Physiology of DMSP

The green sea macroalga (*Enteromorpha* sp.), of great interest, has been ingested by people in Japan since 701 CE for about 1300 years (Osakabe, 701) and the aquaculture of the algae has been conducted since the Edo period (1603–1868) (Hosokawa, 1999; Palmer, 2012). Since then, green sea algae has been eaten by Japanese for a long time. The green sea alga (*Monostroma nitidum*) has been generally ingested in Japan, in particular, in Okinawa Prefecture and likely supported the longevity of life in Japanese, in particular, Okinawans (Hosokawa, 1999; Health & Welfare Statistics Association, 2006; Willcox & Willcox, 2014). Accordingly, Japanese have habitually eaten green sea macroalgae (*Enteromorpha sp., Monostroma nitidum,* and *Ulva pertusa*) for a long time as food stuffs: raw (a sea lettuce), dried and roasted food stuffs in soup, and favorite processed-side foods with cooked rice (so called sushi, nori, tsukudani, furikake, etc.) (Radmer, 1996; Nagai &Yukimot, 2003). Administered DMSP infiltrates into all parts of the body and also into the brain in animals through the blood-brain barrier without side effects (Nakajima, 1996; Nakajima & Tani, 2005; Nakajima & Tani, 2006). Therefore, DMSP migrates to various damaged cells and tissues in crucial diseased animals containing cancers and disorders in central nervous system (Parkinson's, Alzheimer's, and so on) and ameliorate the diseases without toxicity and cytokine storm (Nakajima, 2015, 2020). Administration of DMSP to animals activates the delayed-type hypersensitive immune reaction (Nakajima & Minematsu, 2002; Nakajima, Yokoyama & Nakajima, 2009; Nakajima, 2015), occurring in T lymphocytes, which play a central role in cell-mediated immunity, and, of great interest, dose-dependently elevated activated macrophages (M2) (Nakajima, Tsujiwaki & Nakajima, 2014), being the first defense against foreign pathogens, cancer cells, and primers of T lymphocytes (adapted immunity) (Biedermann et al., 2000; Park et al., 2005; Bernhard et al., 2015) followed by the healing of diseases, in particular cancer, in a dose-dependent manner (Nakajima, Tsujiwaki & Nakajima, 2014; Nakajima, 2015). Moreover, DMSP proves to ameliorate the loss of learning and memory (Nakajima, 2003, 2004, 2005, 2006, 2015) and prolong the life span (Nakajima, 2002), and dried macroalgae (*M. nitidum)* in diets also ameliorated the loss of learning and memory of senescence accelerated mouse (SAM P8 strain) (Nakajima & Minematsu, 2002; Nakajima, 2015). Moreover, the healing effect of DMSP on injury to the central nervous system is strongly supported by the finding that DMSP significantly accelerates the outgrowth and elongation of neurites in a nerve model cells, Pheochromocytoma cells, compared to those of the optimally administered concentration of nerve growth factor and restore the outgrowth and elongation of neurites damaged by destruction of dopaminergic neuron in substantia nigra pars compacta which is caused by 1-methyl-4-phenyl-1,2,3,6-tetrahydroxypyrimidine (MPTP) and 1-methyl-4-phenyl-pyridinium (MPP^+) (Nakajima, 2015).

In some decades before, the role of DMSP was reported to be osmoregulant (Dickson, Wyn Jones & Davenport, 1982; Reed, 1983; Bayles & Wilkinson, 2000) and cryoprotectant (Karsten, Wiencke & Kirst, 1991; Reed, 1983.; Bluden et al., 1992; Karsten, 1996). Recently, there are many reports on the physiological roles of DMSP as an attractant for aggregation of fish (DeBose & Nevitt, 2008; Nakajima, 1996, 2020), an inducer for "foraging behavior" (mastication behaviors after attraction and aggregation) and "feeding behavior" (behaviors of mastication, swallowing, digestion, and excretion after foraging behaviors) of fish (Nakajima, Uchida &Ishida, 1989, 1990; Nakajima, 1996, 2015, 2020), an inducer for the long life of fish (Nakajima, 1996, 2020; Lee et al., 2016.); a stimulator of growth, locomotion, molt; and feeding behavior of crustaceans (Nakajima, 1991, 1996, 2020.), mollusks (top shell) (Nakajima, 2001a, 2001d, 2002, 2020), clam (Nakajima, 2001b, 2001c, 2020), and plants (vegetables and "callus") (Nakajima, 2020). Moreover, the fish (common goldfish), mollusk (topshell), and crustacean (striped prawn) showed a sharp response to concentrations of DMSP less than 10^{-10} M, at ca. 10^{-8} M in in the attraction and feeding behaviors (the previous two) and of ca. 10^{-9} M DMSP in the attraction and foraging behavior (the last), respectively (Nakajima, 2020). Furthermore, DMSP proves to afford a strong tolerance to fish and corals against rising water temperature, oxygen deficiency, ammonia stress, and environmental stresses (Nakajima, 1992, 1996, 2020; McLenon & DiTullio, 2012; Raina, 2013). The facts appear to strengthen the tolerance of freshwater and marine organisms exposed to recent abnormal climates, in particular global warming (Ripple et al., 2017), rising of water and atmosphere temperature, change of stream of oceanic water, oxygen deficiency in oceanic water, and intense rainfall, resulting in the bleaching of coral, the decrease and death of certain species of marine organisms, symbionts, other micro-, macro-algae, larval, juvenile and small fish, etc., and further leading to the inevitable decline of middle, large fish, and further cetaceans.

Therefore, the occurrence of DMSP plays likely a dramatic role for a number of aquatic and terrestrial animals.

Of great significant, the supplementation of DMSP to the eggs immediately after fertilization of fish eggs was proven to significantly promote the hatching of eggs and further the growth and locomotion (swimming power) of larval fish immediately after hatching and of larval fish treated with DMSP by sending the signal to fertilized eggs and larval fish with no abnormality and no toxicity (Nakajima, 2020).

Furthermore, DMSP proves to promote the feeding behavior, growth, and locomotion of the freshwater and marine juvenile and adult fish, crustaceans, and mollusks (Nakajima, 1996, 2020). Moreover, DMSP proves to accelerate the development, growth, and differentiation of an unorganized plant tissue, callus, but kill the undifferentiated cells and cancers and induce the germination of various plant seeds in lands (embryophytes) (Nakajima, 2020). For the animals, DMSP exerts the healing effect of various diseases (Nakajima, 2015). These facts, in particular, that DMSP accelerates the hatching and development of fertilized eggs, development, differentiation, and growth of callus of plant origins, germination of plant seeds, and exert repair effect of animal diseases, are likely comparable to the various characteristic properties of mesenchymal stem cells (MSCs) (proliferation of MSC from embryos, migration, homing of MSC to damaged cells and tissues, and repair function of damaged cells and tissues in animals).

Accordingly, DMSP is involved in its migration to and activation; renewal, regeneration, heeling of all the damaged cells and tissues in animals, the development of fertilized fish eggs (embryogenesis), callus of plants (unorganized plant tissues), and plant seeds (primordial germ cell); these effects are proved to exert the pluripotent and/or totipotent function, without tumorigenesis, cytokine storm, and side effects. Furthermore, DMSP proves to accelerate adhesion properties of a nerve model cells, Pheochromocytoma cells (PC 12 cells) along with their cultivation (unpublished data). The fact indicates possibly that DMSP well do and/or promote the implantation of fertilized eggs to endometrial membrane in animals and fix them to sea floor and algae in fish because the outer layer of the fertilized eggs consists commonly of induced pluripotent stem viscous compounds (glycoproteins) in animals (Ortholog?gene, 2018) and fish (Rizzo, 1998).

The wide and specific effect of DMSP proves likely to be attributable to its structure, tertiary sulfonium structure (Figure 6.1), which was compared with various closely related compounds to DMSP (Nakajima, 2001). These drastic outcome of DMSP comes likely from the facts that DMSP bears a sulfonum portion (dimethyl sulfide) and carboxy ethyl portion (two methylene chain and free carboxyl group) without one amino group (Nakajima, 1996) in experiments examining striking behaviors of fish (Nakajima, Uchida & Ishida, 1989; Nakajima 1996), with the highest substrate specificity of an enzyme reaction (dimethylthetine- and betaine-homocysteine methltransferase [E.C.2.1.1.3. and 2.1.1.5.] from fish, chicken, and mammals; Nakajima, 1993).

6.3.3 Preparation of DMSP

DMSP was synthesized by refluxing 3-bromopropionate and dimethyl sulfide. Crude crystals (DMSP-Br) were washed with ethylether and crystallized them from methanol, and thus purified to 99.8% as shown by element analysis, which was obtained in large amounts in laboratory experiments (Nakajima, 1996; Ishikake et al., 2003). The identification of the obtained crystals was performed by NMR, IR, elemental, and mass analyses or by x-ray or thermal degradation. The crystals are likely stable at -20°C under dryness for about one year. The degradation of the compound can be known from the smell of dimethyl sulfide (DMS) of sea algae, as the compounds have no smell.

The degradation of DMSP is caused by enzymatic (DMSP lyases (EC 4.4.1.3) (Cantoni & Anderson, 1956; Bullock, Luo & Whitman, 2017) and non-enzymatic reaction in alkaline solution (Challenger & Simpson, 1948; Ackman & Dale, 1965) to produce dimethylsulfide which forms the main source of cloud condensation nuclei in atmosphere (Yoch, 2002; Bentley & Chasteen, 2004) and has a great influence on the terrestrial animals (Nakajima, 2015) and oceanic organisms (Hay, 2009; Ferrer & Zimmer, 2012.; Nakajima, 2020).

DMSP proved to exert pluripotent and/or totipotent ameliorating effects on crucial disorders (ectodermal, endodermal, mesodermal, and neural crest diseases) in laboratory animals without toxicity and cytokine storm (Nakajima, Yokoyama & Nakajima, 2009; Nakajima, Tsujiwaki & Nakajima, 2014; Nakajima, Nakajima &Tsujiwaki, 2015; Nakajima, 2015, 2020).

6.3.4 Toxicity of DMSP

Single *i.p.* supplementation of a large amount of DMSP (7 g net weight/kg body weight in aqueous paste (about 3.2 M, 1 ml) to rats, oral and sequential administration of high concentration of 20 M DMSP solution (208 mg/rat) for up to 33weeks to young rats (Nakajima & Minematsu, 2002) or a single *ip* injection of DMSP (180 mM, 1 ml [39.6 mg/mouse]) to juvenile mice did not result in toxicity for a long time. These findings indicate that it is possible to administer orally and intraperitoneally high and more frequent doses of DMSP to diseased animals without side effects and cytokine storm in vivo.

6.4 MESENCHYMAL STEM CELLS

In contrast, so-called MSCs—human bone marrow–derived mesenchymal stem cells (hBM-MSCs), human adipose–derived mesenchymal stem cells (hAD-MSCs), human blood–derived mesenchymal stem cells (hBD-MSCs), umbilical cord blood stem cells (UCBSc), dental stem cells (DSCs), multipotent adult progenitor cells, skin-derived precursor cells, etc.—have recently been broadly studied by clinical trials (Lu, Shen & Broxmeyer, 1996; Barber & Arispe, 2006; Kitamura et al., 2008; Poll et al., 2008; Tamaki et al., 2013; Takahashi et al., 2015). Moreover, MSCs have been reported to play a significant and multifunctional role in clinical therapies for the amelioration of NCDs (Kuroda &

Dezawa, 2014). MSCs are well known to reside in various tissues (Jiang et al., 2002; Rozemuller et al., 2010) and have a variety of properties: self-renewal, non-tumorigenesis, migration and homing to injury cells and tissues, multi- or pluri-potent stem cells, and host tissue function modulator (Chamberlain et al., 2007; Ratajczak et al., 2010; Frenette et al., 2013; Dezawa, 2016; Chamberlain et al., 2017). Although the supplemented MSCs migrate, differentiate, and renew damaged cells (Bianco, 2014), these phenomena appear likely to be regenerated by paracrine factors (cytokines and chemokines) (Chen et al., 2008; Liwen et al., 2008; Kono et al., 2014) in microenvironment (Chow et al., 2014), factors (microvesicles) (Madigal, Rao & Riordan, 2014; Alessio et al., 2017), and molecules (Poll et al. 2008) secreted from AD-MSCs in complex ways. Moreover, the factors from a wound-treated BM-MSCs medium likely increase abundant macrophages, further recruit macrophages, and endothelial lineage cells into the wound leading to wound healing (Chen et al., 2008; Lolmede et al., 2009).

6.4.1 Typical Isolation and Purification Methods of BM-MSCs

Isolation and purification of MSCs are performed in brief as follows (Haas et al., 2011). The mouse at 4 to 6 weeks old is killed by cervical dislocation, the body is sterilized by ethanol, and tibias and femurs are dissected and muscles, ligaments, and tendons are removed. Moreover, the bone is moved on the sterilized culture dish containing the α-minimal essential medium (α-MEM medium). Then, two ends of bone marrow are cut, gauze needle is inserted into bone marrow cavity and the bone marrow is harvested by flushing. All bone pieces are excluded. The residual fat mass (bone marrow aspirates) is incubated on α-MEM (or Dulbecco's modified Eagle's medium [DMEM]) in carbon dioxide (5%) atmosphere. Thus, the adherent spindle-shaped cells are formed, carefully washed, and treated with trypsin. The cells are centrifuged and suspended in the α-MEM medium containing antibiotics and serum. The cells are increased to about 80%–90% confluence by further incubations and employed for MSCs therapy. The cells obtained are usually used at about 1×10^6–10^7 cells for the trial of therapeutic treatment.

The isolation methods of hAD-MSCs, BM-MSCs, and of fetal membrane (FM) (medical waste)-derived-MSCs (FM-MSCs) are also similar, following the use of human liposuctioned fat aspirates (Mizuno & Hyakusoku, 2003), mouse (human) bone marrow (Huang et al., 2015), and fetal membrane in place of the fat mass described earlier. Moreover, type II collagenase are used in the destruction of matrix for the preparation of the latter MSCs. BM-MSCs are isolated also by density gradient centrifugation of human bone marrow aspirates (Chen et al., 2008).

Embryonic mesenchymal stem cells (EB-MSCs) prepared from mouse embryo are pluripotent stem cells (Evans & Kaufman, 1981) which can develop to ectodermal- (i.e. skin epithelial and dermis cells, central and peripheral nerve cells, etc.), endodermal (digestive tract, lung, pancreas, liver, etc.), mesodermal (heart and kidney cells, skeletal muscle cells, etc.) lineage cells, and neural crest (glial cells, peripheral neuronal cells) eventually give rise to all of animal's tissues and organs through the process of organogenesis. However, the transplantation of EB-MSCs into the mouse possibly causes the occurrence of teratoma and tumor (Evans & Kaufman, 1981; Knoepfler, 2009; Stacheischeid et al., 2013). Moreover, as for the artificial preparation and clinical employment of the primordia of life, EB-MSCs have a significant ethical problem for animals, in particular humans, and the healing effect of DMSP and EB-MSCs was not compared here.

In contrast, iPS cells are well known to be made from adult fibroblast cells (skin cells) in mouse about ten years ago (Takahashi & Yamanaka, 2006; Takahashi et al., 2007). Therefore, a number of techniques preparing patient specific iPS cells also have been widely developed and have been hopefully used for screening drug candidates to heal various disorders (Yamanaka, 2012). However, the transplantation of the cells into animals proves to elicit transplant rejection (immunorejection) on the animals by the difference of human leukocyte antigen (HLA) type. Thus, it is difficult to search the blood suitable for various men for avoiding the rejection reaction. Furthermore, there is the possibility that the transplantation of iPSc gives rise to damages of the tissues and organs and causes incidence of cancer (Sel, 1993; Takahashi & Yamanaka, 2006; Knoepfler, 2009; Gutierrez-Aranda et al., 2010; Clement et al., 2017). For the reasons, the comparison of DMSP and iPSc for the amelioration of disorders was not made here.

In contrast, the employment of MSCs obtained through many passages should avoid the use of clinical therapy because rodents which were treated with MSCs undergoing long-term passage in culture sometimes elicit tumors (Uccelli, Moretta & Pistro, 2008, Li, Ezzelarab & Cooper, 2012). Moreover, there are also the possibilities that MSCs transform to a variety of peripheral cancers when MSCs enter the tissue during infection and inflammation (Houghton et al., 2007). Therefore, the prognosis of clinical therapy with MSCs transplantation needs to be examined.

Human beings immediately after birth are exposed to various crises by exogenous bacteria, viruses, and parasites and by endogenous cancer cells. Exogenous microorganisms invade into the trachea, digestive tract, and skin, but generally are in a cooperative relation to endogenous microorganisms. When their balance is destroyed by abnormal proliferation of exogenous or endogenous microorganisms and cancer cells in the body and by infiltration of pathogenic microorganisms, abnormality of immune systems functions to ameliorate and mitigate the abnormal condition of the body. The immune systems consist of recognition of self- and non-self immune reactions and exclusion of distinguished non-self immune reactions. Therefore, the immune systems are a very significant barrier for protecting the body against not only exogenous poisonous origins but also endogenous immune factors and cancers. In particular, activated macrophages (M2 macrophage) prove to play multiple and critical roles for fighting against exogenous abnormal microorganisms (host defense), participating in wound healing in animal body, and innate and adaptive immune regulation; these functions lead to homeostasis of the metabolisms in the body (Mosser & Edwards, 2008; Gordon & Martinez-Pomares,

2017). Therefore, immunological rejection is a great obstacle in the transplantation of non-self tissues and organs into the animal and human body, in particular the latter, and further also in the MSCs therapy. However, mesenchymal cells, of great significance, mediate their immunosuppressive effects via a variety of mechanisms ((Kim & Hematti, 2009; Eggenhofer & Hoogduliin, 2012).

Residential macrophages in many tissues of the body are mostly derived from the yolk sac during embryogenesis, but at a later time point, from fetal liver and hematopoietic stem cells (Spelman, Lavine & Randolph, 2014; Perdiguero et al., 2015). In the amelioration process, transference of tumor associated macrophage (M1 macrophage) to M2 macrophage is also very important because M2 macrophage is proven to be involved in wound construction, angiogenesis, reepithelization, connective tissue generation, and removal of pathogens and damaged tissues (Yacoub-Youssef et al., 2013; Das et al., 2015; Wynn & Vannella, 2016). The transference and/or increase of M1 macrophage to M2 macrophage is caused also by MSC secretions (Teixeira et al., 2013; Madigal, Rao & Riordan, 2014; Alessio et al., 2017) or by extracellular vesicles excreted from MSCs (Poll et al., 2008; Sico et al., 2017; Liew et al., 2017). The multifunctional amelioration effect of MSCs on various diseases in three germ layers (ectoderm, mesoderm, and endoderm) is very similar to that of DMSP (Nakajima, 2015).

Together, if the multipotent and/or pluripotent ameliorating effect of DMSP depend likely on similar mechanisms via macrophages (Nakajima et al., 2015) and T cells (Morgan, Lefkowitz & Everse, 1998; Nakajima & Minematsu, 2002; Nakajima, Yokoyama & Nakajima, 2009; Zamarron & Chen, 2011; Alderton & Bordon, 2012) to those of these stem cells on NCDs. DMSP is possibly replaced of or employed with MSCs in clinical therapies of NSCs. This further indicates that very convenient and oral administration of DMSP can really heal a number of NCDs with and/or in place of supplementation of various MSCs. Therefore, both effects on NSCs are briefly described next for their comparison.

6.5 GASTRIC ULCERS AND SKIN ULCERS

6.5.1 Gastric Ulcers and Structurally Similar Compounds to DMSP

A gastric ulcer is one of the NCDs which are found worldwide in recent years (Chai, 2011; FAO, 2017) and one of the mesodermal disorders. A gastric ulcer is elicited by various physiological and psychiatric stresses (Bhargava et al., 1980.; Brozowski et al., 2007.; Levenstein et al., 2015), non-steroidal anti-inflammatory drugs (NSAIDs): aspirin, ibuprofen, indomethacin, etc. (Michael et al., 2017; William et al., 2017), and by the infection of a gram-negative microaerophilic bacterium, *Helicobacter pylori* (Kelly, 1998; Kusters, Vliet & Kuipers, 2006). The amelioration of the disease is very significant because the ulcer proceeds to the recurrence after the amelioration and further gastric cancer unless the disease is rapidly healed (Aihara et al., 2016). Then, we examined the amelioration effects of DMSP of marine-algae origin, dimethylsulfonioacetate (DMSA) of animal origin (Challenger & Simpson, 1948; Maw, 1958), and methylmethionine (MeMet) (vitamin U) of plant origin (Cheney, 1950), which is well known as the ulcer healing drug (Strehler, 1955; Tanaka et al., 1969; Yoshinaka & Nakamura, 1981), on acute gastric ulcer caused by the water immersion (psychological stress) and physical restraint (physiological stress) of test rats (Takagi, Kasuy & Watanabe, 1964; Uramoto, Ohno & Ishihara, 1990; Nakajima, 1991), showing their structures in Figure 6.1. In brief, SD-Wister male rats at 6 weeks of age had free access to the commercial solid diets and either the distilled water or one of the test solutions. After 12 days, the rats were deprived of food and drink for 24 hours. Each starved rat was tightly fixed in the wire net cage with aluminum frames and was then vertically immersed to the height of the rat's xiphoid in a water bath at 23°C for 21 hours in the dark. Test rats were cut open along a greater curvature and then the area of ulcerous portions (blood clots) was calculated by perpendicularly determining diameters and totaled. Ulcer indexes were expressed as the ratio of the sum of ulcerous area of test rats against that of the control rats. The test solutions of DMSP, DMSA, and MeMet (5 mM solution each) were preliminarily and freely administered to test rats and their effect on the rats under the physical restraint and water immersion stresses was examined. Preliminary administration of 5 mM DMSP, DMSA, and MeMet solutions to test rats in this order decreased the stress-induced gastric ulcer by ca. 75% (in the previous one) and 25% (in the latter two), compared to that of the control based on their mean values with significant difference ($p < 0.05$). The effect of DMSA was almost the same as that of MetMet (Nakajima, 2004, 2015). Furthermore, the effect of DMSP on gastric ulcers needs to be examined dependently on time and dose under the same experimental conditions, because the *i.p.*-single supplementation of DMSP by 250 mM/ml to mouse proves not to provide any residual disability, showing the complete recovery from some symptoms for some hours. Moreover, ethanol abuse elicits gastric ulcer in animals and human (Hollander, Transnawski & Gergely, 1985; Razvodovsky, 2006). The gastric ulcer is caused by the *p.o.*-administration of ethanol to rats. The acute gastric ulcer caused by ethanol proves to be ameliorated to almost similar extent by the supplementation of same concentration and the same dose methods of DMSP as those of the water immersion and restrain-induced gastric ulcer in rats. Moreover, administration of two fold concentrations of DMSP to ethanol-induced ulcerous rats proves to significantly heal the gastric ulcer with significant difference ($p < 0.05$) under the same experimental conditions (Ishikake et al., 1990).

6.5.2 Skin Ulcers and DMSP

The effect of DMSP on the skin ulcers in aging-model mouse was examined. Skin ulcers are one of ectodermal diseases. The effect of DMSP on naturally occurring skin ulcer in one of senescence-prone inbred, senescence accelerated mice P1 strain (SAM P1) (Hosokawa et al., 1984; Takeda, Hosokawa & Higuti, 1991) was individually and sequentially examined (Nakajima, 2004, 2015). Oral administration (0.5 mM) of DMSP was proven to completely heal the skin ulcers which occur arbitrarily regardless of sex and age of SAM P1 (Nakajima, 2004, 2005). The finding reveals likely that DMSP plays a significant role for pressure muscus injury (Takagi, Kasuya & Watanabe, 1964) and reduced immunity with aging (Takeda, Hosokawa & Higuti, 1991; Toichi et al., 1994; Nakajima & Minematsu, 2002; Nakajima, Yokoyama & Nakajima, 2009; Nakajima, 2015). The findings reveal that he healing mechanisms due to DMSP from these ulcers are likely attributable to the recovery of impairments of immunity, the exclusion and regeneration of damaged cells and tissues (submucosa and/or muscularis externa), angiogenesis, and regeneration of contracts (Takagi, Kasuya & Watanabe, 1964; Kangwan et al., 2014; Tsuboi, Harada & Aizawa, 2016).

6.5.3 Gastric Ulcers and DMSP

The administration of DMSP to the crucial disorder, Ehrlich ascites carcinoma in mice, proves to linearly accumulate large amounts of M2 macrophage in the peritoneal cavity in a dose dependent way (Nakajima, Tsujiwaki & Nakajima, 2014). Accordingly, the accumulation of M2 macrophage and activation of delayed type-hypertension immune system (cell-mediated immune reaction) (Nakajima & Minematsu, 2002; Nakajima, Yokoyama & Nakajima, 2009; Nakajima, 2015) are considered to play the significant roles also for amelioration of stress-induced gastric ulcer and skin ulcer with rapid aging in laboratory animals (Nakajima, 2004; Nakajima, 2015). There are important links between the brain and the stomach that have significant effect on gastric function (Gerald & Nicholas, 2014). Therefore, the psychiatric stresses obligatorily reflect on the function of brain, eliciting endocrine secretion in brain. Moreover, the effect of an adrenal medulla hormone, dopamine, on the stress-induced gastric ulcer and stress response was examined by the administration of a neurotoxin into substantia nigra (Landeira-Fernadez & Grijalva, 2004), the function of which is controlled by dopaminergic neuron and by the levels of dopamine in mesoaccumbents which is closely related to the stress response (Cabib & Puglisi-Allegra, 2012). The results indicated that dopamine and its receptors are significantly involved in the restriction of the stresses.

We also examined whether or not the administration of DMSP to normal rats, mice, and fish increases the amounts of dopamine in livers and brains. The results indicated that administered DMSP incorporates into brains in rats, mice, and fish and accumulates increasing amounts of dopamine in livers and brains with the rearing time for up to 5 weeks (Nakajima, 2005; Nakajima & Tani, 2005; Nakajima & Tani, 2006). Accordingly, the facts suggest the possibility that the accumulation of dopamine in brain by supplementation of DMSP is involved in the amelioration of gastric and skin ulcers in rats and mice, accompanied by the activation of immune reaction (Nakajima & Minematsu, 2002; Nakajima, Yokoyama & Nakajima, 2009; Nakajima, Tsujiwaki & Nakajima, 2014; Nakajima, 2015).

6.5.4 Skin Ulcers, Gastric Ulcers, and MSCs

The administration of BM-MSCs is also reported to ameliorate gastric ulcers in rats (Hayashi et al., 2008). The results indicate that the subserosal injection of isolated but unselected BM-MSCs and MSCs to the ulcerous portions of rats suffered from acetic acid-induced gastric ulcer reduced by ca. 68%–78% compared to the ulcerous area (ulcer index) of non-treated ulcerous rats based on their mean values with significant difference ($p < 0.05$) during the experimental period. The amelioration rate is almost similar to that in the supplementation of DMSP to the water immersion-restraint induced ulcerous rats described earlier, although the MSCs treatment is likely to need some other factors. Secretome from AD-MSCs (Makridakis & Vlahou, 2010) is also proven to exert the effective role on acetic acid induced-gastric ulcer healing (Xia et al., 2018). Furthermore, the *i.v.*-supplementation of CD34+ cells (stem cells) isolated from umbilical cord blood to the rats with ethanol induced-gastric ulcers reduced significantly the damaged ratio (ulcer index) in parietal cells (Saleh et al., 2013).

The incident mechanisms of gastric ulcer by *H. pylori* are similar as the ones noted earlier, and the amelioration is healed by the administration of antibiotics and NSCDs (Michael et al., 2017; William et al., 2017). The skin ulcer appears to be ameliorated by the division and increase of epidermal MSCs along basal membrane, accompanied by the administration of transfer factor Tbx3 (involving in the proliferation and development of MSCs and progenitor cells) (Ichijiro et al., 2017). However, the amelioration of gastric and skin ulcers by DMSP, interestingly, does not need any factors.

6.6 DIABETES MELLITUS

6.6.1 Diabetes Mellitus and DMP

Diabetes mellitus (DM) is one of the NCDs and endodermal diseases, which principally consists of autoimmune disease (type 1 DM) and insulin resistant disease (type 2 DM). The latter (adult-onset) is much more than the former (juvenile or childhood-onset) (Salisali & Nathan, 2006; Rother, 2007; Baynest, 2015; WHO, 2017). However, type 1 DM recently increases also in adults (WHO, 2017.; Khardori, 2017). Type 1 diabetes is associated with a number of medical risks (*Web*MD, 2018). The amelioration of type 1 DM is considered to be more problematic compared to that in type 2 DM. Type 1 DM is ascribable to T-cell-mediated autoimmune response against insulin producing-beta cells in pancreas, leading to the deficiency of glucose by the destruction of beta cells, whereas type 2 DM is attributable to the impairment of beta-cells, leading to the deficiency of glucose and of insulin action (insulin resistance) and causing hyperglycemia (Salisali & Nathan, 2006; Rother, 2007; Baynest, 2015). The two DMs are similar in effectors of beta-cell failure (hyperglycemia) (Rother, 2007). In the DM model animals, DM-induced drugs, alloxan or streptozotocin, had been widely employed (Dunn & Mcletchie, 1943; Hernberg, 1952; Szkudelski, 2001; Lenzen, 2008). We have examined the effect of alloxan on rats and found the facts that daily *i.p.*-injection of alloxan (mg/10 g body wt.) to rats (Hernberg, 1952) for up to 56 days significantly destroys the beta cells in pancreas based on our microscopic observation (Nakajima, Ohgushi & Shibata, 2001) and remarkably elevates by ca. 4.8 and ca. 2300 fold (ca. 380 and ca.

2300 mg/dl) the glucose content in plasma and urine compared to those in the control at the 27 days, accompanied by the significant increase of drink and urine volumes (acute type 1 DM) (hyperglycemia). Upon the same experimental conditions, the amounts of plasma insulin without alloxan increased by ca. 185% of compared to those by administration of alloxan (from ca. 13.4 to 4.7 *pg* insulin/ml plasma) in comparison with mean values with significant difference, $p < 0.017$) (Nakajima, Ohgushi & Shibata, 1999, 2001). Then, we further investigated the effect of DMSP on the alloxan-induced DM in rats under the same experimental conditions (Nakajima, 2004). The results indicated that free supplementation of DMSP solution (2.5 mM) and *i.p.* administration of alloxan saline solution (mg/10 g body wt.) to the rats for up to 50 days decreased by 1137 mg/dl (ca. 64%) and 1205 g/dl (ca. 45%) as compared with 3120 mg/dl and 2172 g/dl in plasma and urine glucose of the alloxan-supplemented rats without DMSP based on their mean values at 50 days, followed by 1123 mg/dl and 1132 g/dl of the control and the DMSP groups, based on the total amounts of their mean values for 14 days during the experimental period, respectively. The amounts of plasma insulin were 13.4 ± 1.8 and 4.7 ± 0.9 (pg/ml) of the control and alloxan group under the same experimental conditions on 46th day (Nakajima, Ohgushi & Shibata, 2001). Therefore, the results indicate that the administration of the low concentration of DMSP to acute alloxan-diabetic rats almost completely ameliorate the acute 1 type diabetes during the experimental period, probably due to the recovery of beta-cells followed by increase of the insulin amounts (Nakajima, Ohgushi & Shibata, 1999, 2001). The amelioration due to DMSP is likely attributable to the antioxidant effect against reactive oxygen species which occurred via the function of alloxan (Heikkila, 1976; Sunda, 2002; Nakajima, 2005), temperate suppression of excessive T cells (HLA), and protection for the reaction of alloxan with SH-group on the enzyme, glucokinase, being very vulnerable to alloxan (Lenzen, Tiedge & Panten, 1987; Lenzen, Freytag & Panten, 1988; Lenzen & Munday, 1991). Moreover, the *i.p.* administration of alloxan to the untreated rats maintained the high levels of glucose in plasma and urine for up to the end of the experiments and the oral administration of low concentration (2.5 mM) of DMSP to the diabetes rats exerted almost no effect on ingestion of diets in the diabetic rats, showing the complete recovery of body weight decreased by the supplementation of alloxan. Therefore, the facts that autoimmune disorder (1 type DM) in rats caused by alloxan is effectively ameliorated by DMSP show a strong possibility that other autoimmune diseases, rheumatoid arthritis, multiple sclerosis, Graves' disease, etc. (the human leucocyte antigen [HLA] complex constitutes the most relevant susceptibility region) (Wang, Wang & Gershwin, 2015; Ganapathy et al., 2017; Navegantes et al., 2017; Andreone, Gimeno & Perone, 2018) are also healed by the moderate and suppressive control of excessive T cells due to DMSP.

6.6.2 Diabetes and MSCs

In contrast, MSCs have been widelt utilized for curing DM (Chhabra & Brayman, 2013; Gurusamy, 2017). Three time *i.p*-injection of AD-MSCs into diabetic mice (NOD (nonobese diabetic) mice) (Shoda et al., 2005) reduced by ca. 70% the blood glucose and increased by ca. 125% insulin levels compared to those in untreated diabetic mice at 35 days after treatment based on the mean values with significant differences but treated mice did not maintain optimal blood glucose amounts for a long time (Bassi et al., 2012). BM-MSCs isolated from the femurs of mice were collected by the density gradient. Furthermore, BM-MSCs were selected by their adhesion property to the culture dish. For the determination of glucose levels in blood, BM-MSCs through the passage 3 was *i.v.*-administered to streptozotocin-induced diabetic mice. The results indicated that the diabetic mice were recovered by ca. 42% in the levels of blood glucose compared to those of the control at 42 days with significant difference ($p < 0.05$) after the start of the experiments (Gao et al., 2014). Moreover, hBM-MSCs which were obtained from the Turan Center for the Preparation and Distribution of Adult Stem Cells were plated at high density in complete culture medium with FCS and incubated to ca. 70% confluent. For transplantation, the cells were suspended in Hank's medium and maintained before use. The suspension was injected into the left ventricle. Male immunodeficient NOD/scid mice were *i.p.*-injected with 35 mg/kg streptozotocin daily from 1 to 4 days. Blood glucose was assayed in tail-vein blood with a glucometer. Blood insulin was assayed on blood obtained by intracardiac puncture of anesthetized mice. Blood glucose in hMSCs-administered diabetic mice restored by ca. 40% and ca. 56% the blood glucose and insulin amounts compared to those in diabetic mice without MSCs at 32 days. However, the latter values were low by ca. 56% when compared to those in the normal rats (Lee et al., 2006).

Therefore, MSCs appears to ameliorate DM partially (Lilly et al., 2016). Moreover, the alloxan-induced diabetic rabbit model was here used to best represent the human situation (Loughlin et al., 2013). Allogenic BM-MSCs aspirates from rabbit was isolated and cultured in α-MEM with FBS and antibiotics. MSCs through three passages were employed to the alloxan-induced rabbit which was caused by i.v. injection of alloxan (150 mg/kg) to the back of the ears. The healing effect was assessed by the reduction of wound area (the ratio of wound area against the initial area for 1 week (wound closure [%]) (O'Loughlin et al., 2013) after 1 week treated with MSCs. The results indicated that the percentage in wound closure of the cutaneous ulcers 1 week after the treatment with maximal amounts (10^6 cells) of MSCs seeded on a collagen scaffold was recovered by 22% and by ca. 38% compared to that in untreated wound and in collagen alone without MSCs.

Together, more effective stem cell therapies in combination with islet transplantation, immunosuppressive-, immunomodulatory-drug, and bioengineering techniques appear to be needed in the future (Chhabra & Brayman, 2013).

Of note, supplemented MSCs themselves appear not to differentiate to beta-cells, because beta-cells are regenerated by paracrine factors in microenvironment of DM (Gao et al., 2014). Moreover, in the amelioration process of DM, transference of M1 macrophage to M2 macrophage is very important because M2 macrophage has been proven to involve in DM healing (Gordon & Taylor, 2005; Anghelina et al., 2006; Navegantes et al., 2017).

Accordingly, the supplementation of DMSP to the acute diabetic rats (type 1 DM) induced by alloxan also appears to cause the utmost healing effect by the accumulation of M2 macrophages in a dose-dependent way without side effects and cytokine storm (Nakajima, Tsujiwaki & Nakajima, 2014), accompanied by the antioxidant function of DMSP against oxidative stresses caused by alloxan (Heikkila et al., 1976; Sunda et al., 2002; Nakajima, 2005; Lee et al., 2013).

Furthermore, the transference of M1 macrophage to M2 macrophage which is caused by anti-inflammatory effects of AD-MSCs from rats promotes type 2 macrophage polarization to ameliorate the myocardial injury caused by diabetic cardiomyopathy (Jin et al., 2019). The polarization of type 2 macrophages appears to remarkably occur also by the administration of DMSP to Ehrlich Ascites Carcinoma (EAC), probably of mammary origin (Nakajima, 2015).

As the amelioration effects of MSCs therapies are, however, not complete, it needs to be used with the known medicines for DM therapies or specifically engineered MSCs in the future.

6.7 ALZHEIMER'S DISEASE

6.7.1 Alzheimer's Disease and DMSP

Senility proceeds obligatorily in all living things with aging via adult stages. The senility causes various dysfunctions of body, which elicits several critical disorders in brain of humans (WHO, 2016. Dementia). Among the dysfunctions, there is a major calamity, dementia, among aged people. Dementia is a crucial chronic or progressive disorder, in particular, confers loss of learning and memory, orientation, and every behavior to the patients. The patients suffering from dementia (neurological diseases) induced by denaturation and retardation of central nerves have to undergo heavy clinical, social and psychological impairments. Moreover, not only the patients but also their families and society have a great impact (WHO, 2016. Dementia). As people live longer in the world, people suffereing from dementia drastically increase (WHO, 2014. Ageing and life-course).

Therefore, the effective drugs and/or clinical therapies for dementia are the utmost urgent social problem to date (Jellinger, 2013).

We examined the effects of DMSP on the dementia in senescence-accelerated mice (SAM) (Takeda, Hosokawa & Higuti, 1991). The simple and naturally occurring-compound, DMSP, have been proven to exert the dramatic healing effect of gastric-, skin-ulcers, and 1 type diabetes mellitus (three germ layer-diseases) in the laboratory animals with no side effect and no cytokine storm (Nakajima, 2015). SAM have the characteristic properties of loss of learning and memory, slow growth, rapid aging, and immune impairments (Hosokawa et al., 1984; Takeda, Hosokawa & Higuti, 1991). Among them, aging of all SAM strains is initiated after birth and shows maximal difference from the R strain showing normal aging at about 64 weeks (Hosokawa et al., 1984). Moreover, all SAM strains die at about 80 weeks (Takeda et al., 1981). Among the SAM strains, inbred SAM R1 strain has the normal aging (normal growth), inbred SAM P1 and P8 strains have particularly the properties of the senile amyloidosis and immune impairments (Takeda, Hosokawa & Higuti, 1991; Hosokawa, 2004), and the loss of learning and memory (Alzheimer's disease model mice) (Miyamoto et al., 1986; Takeda, Hosokawa & Higuti, 1991; Miyamoto, 1997; Morley et al., 2004), respectively.

The loss of learning and memory of test mouse was assessed by estimating the elapse time (remember time [sec]) to avoid the movement from the light compartment without electric impact into the dark compartment beyond electric impact zone, after some trials (passive avoidance test) (Yagi et al., 1988; Nakajima, 2003). The aging was determined by counting the individual numbers of reactivity, passivity, glossiness, coarseness, loss of hair, ulcers of skin, periophthalmic lesions, corneal opacity, cataracts, and lordokyphosis, numbered from 1 (best) to 5 or 6 (worst), respectively, and by individually summing up (total grading score) (Hosokawa et al., 1984). The effect of free ingestion of DMSP solution (0.5 mM) on loss of learning and memory and growth of SAM P1 strain at the age of 20 weeks (fairly aging) was examined for up to 7 weeks (Nakajima, 2005). The results indicated that the free supplementation of DMSP solution (0.5 mM) to female SAM P1 provided no effect of growth along with the slow growth, but significantly restored the loss of learning and memory from the start of the experiment, showing about 6-fold recovery compared to that of the control at the end of the experiments. Furthermore, the effect of direct administration of 2% (93 mM) DMSP solution (1 ml) into the stomachs twice a week on the growth, aging (total grading score), and loss of learning and memory of SAM P8 at the age of 34 and 50 weeks was examined in male and female SAMP8 and R1 strains. The results demonstrated that the supplementation of DMSP on the test mice remarkably restricts the decline of the growth, significantly restores the aging and the loss of learning and memory of both aged male and female mice for up to 30 days, accompanied by ca. 3 times restoration of loss of learning and memory in male and female SAM P8 strain, in particular more aged SAMP8 male and female mice, when compared to those of the control strains at the end of experiments, respectively (Nakajima, 2003, 2004). The similar experiments with male and female SAMP8 mice indicated that the administration of DMSP solution (0.5–0.75 mM) reduces by ca. ½-fold both aging and significantly reverses the loss of learning and memory from 4- to 24-week-old mice (younger in bred SAMP8 mice) (Nakajima, 2003). Moreover, effect of free uptake of DMSP solution (0.25, 0.5 mM) by male and female SAM R1 and P8 strains on the growth, aging, and loss of learning and memory was examined for a long period (96 weeks), respectively. The results represented that the administration of 0.5 mM DMSP solution to male P8 strain fairly restores the low growth compared to that in male P8 strain without DMSP for up to 68 weeks, while the growth of male P8 without DMSP showed the lowest values during the experimental period (68 weeks). Whereas the growth of female P8 fed with 0.25 and 0.5 mM DMSP solution continued to recover till 24 weeks but thereafter female P8 groups with and without DMSP solution exhibited similar decline of the growth, showing lower values than those in the control for up to 68 weeks, respectively. Aging in male and female

R1 and P8 strains without DMSP, in particular, that in both male strains showed the highest values but aging in male and female P8 with the DMSP solution showed the similar and lowest values for up to 72 weeks, respectively. Loss of learning and memory in male P8 group with DMSP (0.5 mM) exhibited the similar behavior to that in the male R1 group without DMSP for up to 64 weeks, showing the lowest values in the male P8 group without DMSP from 40 to 64 weeks. Therefore, the administration of low concentration of DMSP restores delayed growth, suppresses rapid aging in male and female P8 group, and delayed the rapid loss of learning and memory with aging in male P8 group for the long period, respectively. During the long experimental period, loss of learning and memory in female P8 group with DMSP (0.5 mM) appeared to maintain higher values than that in female P8 group without DMSP from 24 to 64 weeks, and furthermore showed similar values to that in R1 group up to 40 weeks and thereafter showed rather higher values than those in R1 group up to 64 weeks. Especially, the administration of DMSP on loss of learning and memory in male and female P 8 strains restored by ca. two fold that in P8 strains without DMSP at 48 weeks, respectively. Survival in male and female P8 groups gradually decreased and all mice in male and female P8 groups died at 80 weeks after the experiments. However, the male and female mice in P8 group fed with 0.25 and 0.5 mM DMSP interestingly increased about 20% and 30% and about 60% (equal in both concentrations in female groups) survival rate after 80 weeks to more than 96 weeks (4 months longer). Therefore, the results of survival rate proved to significantly prolong their lives by the administration of low concentration of DMSP solutions to male and female P8 mice, in particular in female mice, without side effect, respectively. The facts were clearly confirmed also by the experiments with male and female R1 and P8 group mice which 0.75 mM DMSP solution were freely administered for 24 weeks (Nakajima, 2003). Moreover, administration effect of green sea algae in diets on the delayed growth and loss of learning and memory in male and female SAM P8 was examined. Interestingly, the results demonstrated that administration of the diets containing 5% dry green sea algae (*Monostoma nitidum)* (w/w) clearly recovers the decrease of growth and loss of learning and memory of SAM P8 at the age of 43 weeks with significant difference (p <0.05) (Nakajima & Minematsu, 2002, 2006).

Therefore, the availability of DMSP for therapeutical approach to dementia is likely of great significance up to date.

6.7.2 Alzheimer's Disease and MSCs

MSCs have been wide employed for the treatment of neurological diseases (Momin et al., 2010). Moreover, the two-third of dementia is reported to be Alzheimer's disease (WHO, 2017 Dementia.). Therefore, the physiological roles of MSCs on Alzheimer's disease and related neuronal disorders are described below.

Alzheimer's disease is the most common dementia. Many reports indicated that microglia activation and amyloid deposits appear at the early stage and then tau and Aβ fibrillary tangles were formed with aging (Wyss-Coray et al., 1997; Yan et al., 1997; Arends, 2000; Uboga & Price, 2000; Heneka et al., 2012; Boche & Nicoll, 2013). These incidences result in the disfunction of medial temporal lobe consisting of structures that are vital for declarative or long-term memory (Collie & Maruff, 2000). However, recent studies reveal that the extent of Aβ deposition and senile tangles (plaques) in the brain does not necessarily correlate with dementia because healthy elderly people can exhibit abundant plaques even in the absence of AD. Moreover, the recent findings have indicated that neurogenerative disorders are related to loss of function, aggregation, deposition and degradation of misfolding proteins in the brain (Adav & Sze, 2016). Neurofunctional studies demonstrate that age-related atrophy is incident in the hippocampal region (retrosplenial region/posterior cingulate cortex, left hippocampus, and bilateral inferior and middle frontal areas) (DE Vogelaere, 2012; Du et al., 2006). Whereas recent other reports indicated that earliest pathological diagnosis of Alzheimer's disease is possible by the determination of the mounts of several amyloid-beta-associated peptide fragments in the blood by immunoprecipitation-coupled with mass spectrometry (Nakamura et al., 2018). For the moment, the effect of MSCs on Alzheimer's disease are described in brief.

The effect of hAD-MSCs on loss of learning and memory in the 18-month-old mice was examined by passive avoidance and Morris water maze tests (the spatial memory performance). AD-MSCs were isolated from human abdominal fat tissue, selected by their adhesion, cultured in the keratinocyte-SFM (Invitrogen) medium with recombinant epidermal growth factor (EGF), and donated from the same medium (90% effluence) through three passages. hADMSCs were transplanted into the tail veins (IV) once or intracerebroventicularly (ICV) four times at 2-week intervals from a hole cut in the center of the pregma. The effect of single supplementation of IV or ICV administration four times at 2-week intervals with MSCs into 18-month-old mice on the locomotion of the test mice was examined by the passive avoidance and Morris water maize tests for 9 days. The results indicated that the IV or ICV supplementation with hMSCs restored by ca. 12-fold and ca. 17-fold compared to those of the initial values of 18-week-old mice at 9 days by first passive avoidance tests with hMSCs supplementation whereas the IV or ICV supplementation with hMSCs remained late by ca. 67%, ca. 80% and by ca. 48% and 78% compared to the initial values and those of 18-week-old mice by first Morris water maize tests at 7 days (Park et al., 2013). These test were tried two times; the second trial showed the similar values to those of the first experiments. Moreover, BM-MSCs isolated from bone marrow aspirates were transplanted intracerebroventicularly (ICV) into a hole made in the center of the pregma in 7- to 8-month-old double-transgenic mutant mice (APP/PSI), in which the cognitive impairment become apparent at the age of 6 months (Howlett et al., 2004). The results indicate that the injection of BM-MSCs suspensions to the mutant mice reversed by ca. 1.6-fold and ca. 3.5-fold in Morris water maze and single probe tests compared to those in non-injected mice at 10 days with significant difference ($p < 0.01$) (Lee et al., 2010). These reports provide the significant facts that the supplementation of MSCs improves the neuronal impairments in the brain closely related to Alzheimer's disease.

In contrast, MSCs remained equivocal to pass through into blood-brain barrier (BBB), but in vivo and in vitro experiments have recently showed the possibility that the neurovascular unit, constituted of astrocytes, neuron cells, and pericytes, in particular pericytes, significantly involve in the permeability of MSCs across BBB, and the pericytes can be replaced with MSCs, indicating that MSCs possess the ability to migrate across the BBB (Liu et al., 2013; Pombero, Garcia-Lopez & Martinez, 2016; Tian, Brookes & Battaglia, 2017).

However, it remains uncertain whether or not these recovery effect of spatial memory by intracerebroventicular administration of MSCs continues for a long time or whether the frequent and more dose of MSCs are possible for maintaining the restoring effects without side effects.

6.8 PARKINSON'S DISEASE

6.8.1 Parkinson's Disease and DMSP

Parkinson's disease is a chronic and progressive neurological disease (European Brain Council, 2011; Tieu, 2011; Blesa et al., 2012; Solari et al., 2018). The age of onset of the disease is usually over 60 (European Brain Council, 2011). The disease is the second most common neurodegenerative brain disorder after Alzheimer's disease (Tieu, 2011). The incidence of the disorder causes slow physical movements (bradykinesia), tremor at rest, muscle stiffness (rigidity), and balance difficulties (postural instability). The symptoms of PD result in the deficiency of dopamine in the substantia nigra in the striatum in the brain, and the formed Lewy bodies are mainly consisting of α-synuclein and ubiquitin (Meredith & Pademacher, 2011; Blesa et al., 2012; Solari et al., 2018). The deficiency of dopamine plays a significant role in controlling smooth movement of muscle and disappearance of 60%–80% of dopaminergic neurons appears to cease the normal movement (Meredith & Pademacher, 2011; European Brain Fact Sheet, 2011). However, the effective cure agents have been not found up to date.

L-3,4-dihydroxyphenylalanine (levodopa), a precursor of dopamine, has been the main drug for therapy of Parkinson's disease (Heikkila et al., 1984; Magner, Jarvis & Rubin, 1986; Tieu, 2011; European Brain Council, 2011). However, the long-term dose of L-dopa proves to elicit some side effects (European Brain Council, 2011; Cenci & Crossman, 2018) and its therapeutic effect reduces after 3–5 years (Goodazi et al., 2015) The more effective drug than L-dopa for Parkinson's disease has not been detected to date. Therefore, the drugs and other therapeutic techniques which heal, stop, and/or slow the PD need to be found in the near future. Several additional drugs are available in combination with levodopa but none significantly delays the proceeding of the neurodegeneration. Further problematic matter is that PD is mostly not diagnosed until overt symptoms appear. At this time, most of the dopaminergic neurons in the brain are already lost (Europian Brain Council, 2011; NIH

RePort, 2013. Parkinson's Disease; Goodazi et al., 2015). However, the administration of L-dopa has remained to be the main clinical therapy to ameliorate Parkinson's disease (Heikkila et al., 1984; Wagner et al., 1986).

Parkison's disease has been proven to be caused by the administration of 1-methyl-4-phenyl-1,2,3,6-tetrahydropyridine (MPTP) to mice and monkeys (Snyder, 1984; Snyder & D'Amato, 1985; Jenner & Marsden, 1986; Guridi et al., 1996; Tieu, 2011). Moreover, the toxicity is attributable to 1-methyl-4-phenyl-pyridinium (MPP^+) converted from MPTP via 1-methyl-4-pheny-l-2,3-dihydroxypyridinium (2,3-$MPDP^+$) formed by monoamine oxidase-B (Kalaria, Mitchell & Harik, 1987). Moreover, the supplementation of MPTP to human and animals destroys the dopaminergic neuron in substantia nigra in brain, following the depletion of dopamine (Heikkila et al., 1984; Wagner et al., 1986). Parkinson's disease has a number of non-motor symptoms; sleep disturbances, fatigue, anxiety, depression, cognitive impairment, dementia, olfactory dysfunction, pain, sweating, constipation except motor dysfunction (Barone, 2010; Blesa et al., 2012; Goodazi et al., 2015; Solari et al., 2018). Pathological aspects of neuronal degeneration in brain by MPTP function been reported in more detailed (Meredith & Pademacher, 2011; Tieu, 2011). Parkinson's model mice have been the most significant and most frequently used animal model for investigating parkinsonism in humans. In particular, the dysfunction of motor activity in Parkinson's disease is the typical, significant, and prominent symptom of the disease. Then, I focused on the effect of DMSP on idiopathic Parkinson's disease in normal mice induced by MPTP. The C57BL/6 mice (a MPTP-sensitive strain) at the age of 16 week (average body wt. 30.4 ±2.1 g) were divided into two groups which were given with the distilled water and the DMSP solution at 5×10^{-4} M for 2 wk. Then, the MPTP solution (0.5 ml) was ***i.p.***-injected daily into all the test mice at the ratio of 20 mg/kg body wt for three consecutive days from the day before the start of the experiments. From the next day (the first estimation day), motor activity (running activity) of the mice was determined at 1, 2, 3 and 5 days after the first MPTP injection using a wheel running instrument (15Φcm), which electrically turns at the constant rate of 10 rpm and expressed as the continuous turning numbers. The immobility duration was simultaneously determined with the running numbers with a polygraph. For the determination of the immobility duration, test mice **were** placed upside down and the tail of the mice was suspended by connecting it to the bottom edge of a vertical wire tied to a horizontal thin iron bar of a round sensor box 37 cm from the floor in order to record the moving stimuli (a tail suspension test). The duration of the immobility (total times (s)/6 min) after the estimation of the mobility were calculated by a computer and software. After the experiments, the amounts of catecholamines in brain without cerebellums were immediately homogenized and centrifuged 26000rpm. The obtained supernatant was subjected to the high performance liquid chromatography (LC-6A). Thus, the amounts of norepinephrine, dopamine, and serotonin (μg/g) were also determined (Nakajima & Minematsu, 2006). The administration of MPTP to the preliminarily administrated or non-treated groups with DMSP restricted by ca. 30% and ca. 70% the turning numbers and increased by ca. 20% and 90% immobility duration compared to those in the start values at 5 days, respectively. Whereas the turning numbers and the duration of immobility in the MPTP group with DMSP (DMSP/MPTP group) increased by ca. 2.3-fold (40%) and restricted by ca. 1.9-fold (70%) compared to those in the MPTP group without DMSP at 5 days with significant difference ($p < 0.05$). The results obtained at 5 days were similarly obtained at 2 days. The DMSP/MPTP group increased the amounts of epinephrine, dopamine, and serotonin, in particular, the amounts of dopamine in the administration of DMSP to the MPTP group increased by ca. 2.3-fold when compared to those in the MPTP group at 5 days with significant difference ($p < 0.05$) (Nakajima & Minematsu, 2006). Moreover, the similar experiments were performed with SAM P8 strain (Nakajima, Minematsu, & Miyamoto, 2008; Flood & Morley, 1997). The test mice (average body wt. 28.1 ± 1.2 g) were divided into two groups, one is given the distilled water and the other, DMSP (0.5 mM) solution for 2 weeks. Moreover, the two groups were divided into two groups, one further underwent the injection of MPTP (20 mg/kg), which was ***i.p.***-injected daily for 3 consecutive days before the initiation of the experiments and then at 10 mg/kg body wt on the first days of every wk from 2 to 5 weeks in the control and DMSP groups with MPTP. Therefore, the experimental groups consist of four groups; the control (distilled water), the DMSP group (preliminarily administered for two weeks), the MPTP group, and the DMSP/MPTP group. The immobility frequency and tremor durations were estimated by the tail suspension test and expressed in terms of the numbers of body movement and of resting time (sec/6 min). The amounts of norepinephrine, dopamine, and 3,4-dihydroxyphenylacetic acid (DOPAC), a metabolite of dopamine, were estimated at the indicated times for up to 5 weeks. Other experimental conditions were the same as described earlier (Nakajima & Minematsu, 2006). The supplementation of MPTP alone to the mice occurred the typical syndromes of Parkinson's disease. The turning numbers decreased by ca. 30% in the DMSP/MPTP group and by ca. 70% in the MPTP group at 5 days compared to those at the start of the experiments. The results obtained at 5 days were similarly obtained already at 2 days. The frequency of immobility duration in the MPTP group elevated by ca. 2.4-fold compared to that in the DMSP/MPTP group and by ca. 2.0-fold compared to the untreated group at 2 weeks, whereas the frequency of tremor duration in the MPTP group increased by ca. 5.3-fold compared to the those of the DMSP/MPTP group and by ca. 3.5-fold compared to that in the control (untreated) group and at 4 weeks. Furthermore, the frequency of immobility duration and tremor duration in the DMSP/MPTP group decreased to the same levels as in the control group during the experimental period 5 weeks. These results were compared based on the mean values with significant difference ($p < 0.05$). Whereas the amounts of norepinephrine, dopamine, and DOPAC in the DMSP/MPTP group increased by ca. 20.0-fold, ca. 2.3-fold, and ca. 5.4-fold compared to those in the DMSP group at 5 weeks (Minematsu & Nakajima, 2008). Whereas the administration of MPTP after the preliminary supplementation of DMSP to the SAMP 8 did not show any change of frequencies of immobility and tremor duration in spite of the significant elevation of amounts of norepinephrine, dopamine, and its derivative (DOPAC) with significant differences ($p < 0.05$) at 2 or 5 weeks. Therefore, the supplementation of MPTP after the preliminary administration of DMSP (DMSP/MPTP group) in SAM P8 seems not to cause the degeneration of dopaminergic neuron in aubatantia nigra. The facts were found in the normal mice accompanied by the addition of MPTP after preliminary administration of DMSP to a lesser extent. While the DMSP/MPTP group likely inhibited weakly the activity of monoamine oxidase causing accumulation of dopamine while strongly restricted the activity of catechol-0-methylytansferase (COMPT) leading to large accumulation of DOPAC and further activated the biosynthetic pathway of epinephrine in dopaminergic neurons and synaptic vesicles resulting in accumulation of norepinephrine in SAMP 8. The inhibition of COMT likely elicits recovery of memory processes and blocks β-amyloid fibrils (Alessandro & Paolo, 2012). Frequencies of immobility and tremor duration in the DMSP/MPTP group in SAM P8 were almost the same as the values in the DMSP group, although the administration of MPTP alone to the normal mice remarkably decreased the turning numbers and elevated the frequencies of immobility and tremor duration, probably by the deficiency of dopamine resulting from degeneration due to MPTP of dopaminergic neurons. All these results with normal mice and SAMP 8 were compared based on the mean values, accompanied by significant difference ($p <0.05$).

The results obtained with normal mice indicate that the administered MPTP causes the typical disorders of **idiopathic** Parkinson's disease by the destruction of dopaminergic neuron in nigrostriatal but the **preliminarily** administered DMSP over two weeks reverses effectively these symptoms of Parkinson's disease to almost normal conditions, resulting from the accumulation of dopamine. Furthermore, the results obtained with SAMP 8 revealed that the administration of MPTP alone significantly elevates the immobility and tremor duration, the typical hallmarks of PD, but preliminary supplementation of DMSP reverses the symptoms to the normal levels even by the addition of MPTP. The combinatory administration of DMSP and MPTP largely accumulated dopamine and further DOPAC accumulated far above the amounts of both compounds by the administration of DMSP. The facts indicate likely that Parkinson's disease in SAMP 8 likely attributes to the characteristic disorders consisted of the loss of learning and memory (cognitive impairment), impairments of immune system, delayed growth, and rapid death (rapid aging), and high oxidative stress (high content of malondialdehyde and less activity of superoxide dismutase) in a complex way (Nomura et al., 1989.; Takeda, Hosokawa & Higuti, 1991; Floor & Wetzel, 1998). In particular, antioxidant function and cell-mediated immune systems (type VI hypersensitive immune system [T cells and derivative cells] (Jones et al., 2015) and M2 macrophage (Wynn, Chawla & Pollard, 2013)) caused by DMSP for idiopathic Parkinson's disease likely contributes to ameliorate the disfunction of motor activity in **PD** to a great extent (Sunda et al., 2002; Nakajima, 2005; Nakajima & Minematsu, 2002; Nakajima, Yokoyama & Nakajima, 2009; Stone et al., 2009; Nakajima, Tsujiwaki & Nakajima, 2014; Phaniendra, Jestadi & Periyasamy, 2015).

6.8.2 Nerve Cells and DMSP

The neuronal regeneration due to DMSP was ascertained using a nerve model cell, Pheochromocytoma cell (PC 12 cell) (Tischler & Green, 1975, 1978; Green & Tischler, 1976; Green, 1978; Westerink & Ewing, 2008). The combination effect of various concentrations of DMSP with nerve growth factor (NGF) (R & D Systems Inc., USA) was examined on the 1640 incubation medium (Sigma Co., Ltd., USA) containing bovine and horse serums and antibiotics in polystyrene petri dishes (3.5 Φcm) in an atmosphere of 5% carbon dioxide for up to 5 days. The neurite-bearing cells were expressed as the cells having neurites which are the same as or longer than the diameter of the cells. The combination of 0.1 mM DMSP and NGF (5 ng/ml) which showed the maximum acceleration of neurite appearance further stimulated the outgrowth of PC 12 cells by ca. 48% compared to that in the NGF group at 3 days (Nakajima & Miyamoto, 2007). Moreover, the single and combinatory effect of DMSP, MPTP, and NGF on the outgrowth of neurites from the PC 12 cells were examined on the collagen-coated dishes under the same experimental conditions for up to 5 days (Minematsu & Nakajima, 2008). The growth of PC 12 cells with NGF (5 ng/ml) was inhibited by ca. 56% by the administration of MPTP (5 ng/ml) but the inhibition by MPTP was completely recovered by the administration of DMSP (1 mM) at 3 days. The facts were clearly confirmed by an inverted microscopic observation.

The effect of DMSP on the outgrowth and viability of nerve cells and glia was examined with the N2a (Neuro2a) cells (mouse neuroblastoma cells) and OLN-93 cells (rat oligodendroglial cells) as a model system for mammalian nerve cells and glia. The incubation of N2a cells with DMSP (1 mg/ml) for 24 hours elevated by ca. 9-fold the neurite-bearing cells compared to the control group at 24 hours. The viability of N2a and OLN-93 cells preliminarily treated with DMSP was not inhibited by the administration over 0–0.3 μ g/ml of a cytotoxic compound of marine origin, tropodithietic acid (TDA) (Wichmann et al., 2015) which inhibited the viability with increased concentrations (Wichmann et al., 2016). Moreover, administration of (TDA) to N2A and OLN-93 cells inhibited by ca. 52% and ca. 20% and by ca. 90% and ca. 72% their viability of both cells without and with pre-incubation of DMSP, while the effect of TDA (1 μg/ml) on the cell viability of DMSP-treated N2a and OLN-93 cells after preincubation with DMSP (1 mg/ml, 24 hours) was examined, which indicated that both cells restored by ca. 67% and ca. 180% compared to that without the DMSP treatment at 24 hours ($p < 0.001$).

DMSP proves to incorporate into the brain in rats and mice with the rearing times (Nakajima & Tani, 2005; Nakajima & Tani, 2006), which indicates that freely ingested DMSP easily pass through the BBB. Accordingly, the findings represent that DMSP is involved closely in the function mechanisms of the central nervous system and endocrinologically ameliorates the neurodegeneration of Alzheimer's and Parkinson's diseases and other neuronal disorders without side effects. Therefore, DMSP is likely the promising therapeutic drug for healing various disorders in the human brain.

6.8.3 Parkinson's Disease and Deep Brain Stimulation

The deep brain stimulation (DBS) for Parkinson's disease should be noted to be a therapeutical method for ameliorating the condition (Obeso et al., 2001; Schlaepfer et al., 2007; Delaloye & Holtsheimer, 2014; Fang & Tolleson, 2017). In previous reports, the stimulation conditions of pulse width: 450 and 90 μs, frequency: 185 and 145 Hz, and voltage: 10.5 and 4 V with a pulse generator were used, respectively. Moreover, the higher frequency of 130 Hz than 20 Hz exerts likely the efficacy of deep brain stimulation for the treatment-resistant depression (Delaloye & Holtsheimer, 2014). Then, the effect of DMSP on the brain of fish (carp) was examined by the electroencephalography. For the rapid determination, the bipolar electrodes were insulated into olfactory tracts of sedative carp under anesthesia immediately after the operation and the brain waves evoked by DMSP are estimated. Alternatively, the evoked brain waves of alive carp were determined under anesthesia 3 days after the operation. The operation is performed as follows. The rectangle portion of the upper skull above the olfactory tracts is cut off, bielectrodes are insulated into the tracts, and mounted the excised skull portion with the polystylene sheet. Administration of one drop of DMSP solution (0.5 μL. 1 mM) to the nasal cavity of carp (average wet. 325 g) or the supplementation of the DMSP solution (3.4 x 10^{-5} M) into the flow water to nasal cavity under external respiration activated the brain waves from olfactory tracks. These carp showed activated brain waves of 38.5 and 46.0 μV/sec (Nakajima, 1996; Nakajima, 2000 (unpul. data)). Moreover, two electrodes were insulated into the posterior zone of dorsal telencephalic area of the carp under anesthesia with respiration and fixed. The *i.p.* single administration of DMSP and methylmethionine (0.5 mL, 5 mM each) to the operated carp with respiration under anesthesia also elevated by ca. 203% and ca. 163% the activation of brain waves, and in particular activated brain waves corresponding to the β wave (13–30 Hz) from electrocorticograms compared to those in the control without DMSP 9 minutes after the operation (unpubl. data), accompanied by the large accumulation of catecholamines (Nakajima, 2005). The activation waves by administered DMSP also was confirmed at the same moment by electroolfactogram and electrocorticogram of olfactory tract, olfactory bulb, and optic lobe in carp (unpubl. data). The results reveal that the brain waves activated by DMSP were spontaneously observed from these portions of brain, which indicates that the electric stimuli induced by the activation of brain waves due to DMSP is very rapidly and widely transferred to all portions of the brain and, further, for a long time. The finding has been evident in DBS in animals and humans (Delaloye & Holtsheimer, 2014; Fang & Tolleson, 2017). Moreover, DBS elicits the increase of neurotransmitters such as dopamine, serotonin, and/or others (Delaloye & Holtsheimer, 2014; Fang & Tolleson, 2017). The facts are remarkably detected also in the SAM P8 having the dysfunctions accompanied by oxidative stress, loss of learning and memory, delayed growth, and early aging, accompanied by large accumulation of dopamine and its derivatives (DOPAC) by the addition of DMSP and further by combinatory supplementation of DMSP and MTPT (Minematsu & Nakajima, 2008). However, the sequential administration of low concentrations of DMSP during their lives significantly prolonged also the life span of SAM P8 (Nakajima, 2002). The activation of brain by the supplementation of DMSP clearly attributes to that of central nervous systems in fish and rats. Therefore, the electric stimuli for fish brains by DMSP involves likely in the healing effects of the Alzheimer's and Parkinson's diseases by similar but mild mechanisms compared to those of DBS in mammals (Mochizuki et al., 2016; Fang & Tolleson, 2017). However, DBS application for therapeutic treatments in mammals has the disadvantages resulting from dose side effects, excess current delivery, failure of hardware and surgical treatments, poor electrode positioning, etc. (Fang & Tolleson, 2017). Therefore, the long term application of DMSP which exerts the stimuli like DBS with impulses of low frequency likely is successfully tried also for mammals as seen in the case of fish mentioned earlier and in long-term administration of low concentrations of DMSP in SAM P8 (Nakajima, 2002).

6.8.4 Parkinson's Disease and MSCs

The mesenchymal stem cells have been wide employed as the therapeutic methods for Parkinson's disease so far. I would like to focus on the effect of MSCs on the typical syndrome of PD, disfunction of motor activity. The umbilical cord blood mesenchymal stem cells (UB-MSCs) (pluripotent stem cells containing neural stem cells) were obtained as follows. In brief. The tissue obtained from the cord was treated with the enzymes (hyaluronidase, trypsin, and collagenase) and crushed. The cells obtained were plated onto hyaluronic acid–coated plastic plates. The adherent cells were suspended in the defined media with low serum, and incubated at 37°C in an incubator with 5% CO_2 (70–80% confluency). The free UB-MSCs were stored in a freezing media at -135°C before use. The Parkinson rat model was prepared by injection of 6-hydroxydopamine (6-OHDA) (an inducer of animal model of Parkinson's disease) (Tieu, 2011; Blesa et al., 2012). UB-MSCs were transplanted intracerebroventicularly (ICV) into a hole made in the center of the pregma. Four weeks following the lesion, rats that exhibited at least 200 rotations/30 minutes toward the contralateral side after apomorphine (therapeutic agent) treatment were employed for the experiment. Behavioral assessment was evaluated by the number of rotations. Non-PD model animals with and without human UB-MS cell-transplantation did not rotate by the treatment of apomorphine. Whereas post transplant animals treated with 6-OHDA increased the rotations by ca. 36% and ca. 100 % at 6 and 12 weeks based on mean values without significant difference (Weiss et al., 2006). Whereas the following experiments were designed for inducing the differentiation of MSCs into the neuronal lineages (postmitotic cells) by gene transfection with a Notch (Notch gene) intracellular domain of Notch receptor and subsequent administration of basic fibroblast growth factor (bFGF), forskolin, and ciliary neurotrophic factor (CNTF) for the expression of numerous transcription factors. BM-MSCs were isolated from bone marrow aspirates in rat. Subcultures were performed four times and further treated with β-mercaptoethanol, and then a reducing agent, all-trans-retinoic acid for 3 days. The cells were cultured in the MEM medium containing forskolin, basic-FGF, PDGF, and heregulin to change to the cells resembling Schwann cells and expressing p75,S-100,GFAP and O4. MSCs were genetically engineered by transduction with retrovirus encoding green fluorescent protein (GFP) and differentiated by treatment with factors described earlier. The MSCs were transplanted into the cut ends of sciatic nerve in which the GFP-expressing MSCs regenerated myelination of nerve fibers. For preparation of PD model rats, 6-OHDA was injected into the left medial forebrain bundle in the adult Wister rats. Rat and huma-BM-MSCs obtained from commercially available sources were injected into left side of the ipsilateral striatum (a hole made around the center of the pregma). The adult rats which showed more than six turns/minutes for first 30 minutes after apomorphine supplementation were used for the test. The administration of glial cell line-derived neurotrophic factor (GDNF): "G" (glial cell-derived survival promoter) (G-rat MSC) and trophic factors-administered rat-MSCs (TF-rat MSCs) (TF: promoter of the in vitro survival of embryonic neurons from peripheral and central nervous system) to MSC transplanted rats restored by ca. 84% and by ca. 28% the rotations numbers at 10 weeks. Moreover, the supplementation of "G" to hMSCs restricted by ca. 70% the rotation numbers in untreated MSCs at 4 weeks. While the administration of G to rat-MSCs reversed by ca. 30% the values of the step test in the control (the ratio of the number of steps of the lesioned side paw to the contralateral intact side paw) and by ca. 22% the values of paw-reaching test in the control (the ratios of the number of pellets eaten by the lesioned side paw to the contralateral intact side paw) at 6 weeks in rats, respectively. Therefore, behavior test (rotation test), stepping and paw reaching tests, in particular, the former test proved to be significantly improved the dysfunction of behavior at the indicated times after grafting of G-MSCs. In this experiment, the brain slice culture demonstrated the production of dopamine based on the activation of the immune positive reaction for tyrosine hydroxylase in the G-MSCs transplanted brains (Dezawa et al., 2004). Moreover, the bone marrow aspirates were isolated from C57BL/6-Tg(UBC-GFP)30Scha/J male mice (The transgenic mice which express enhanced GFP), washed and seeded in α-MEM, Lonza (Minimum Essential Medium Eagle, 1959; Stanners, Eliceli, & Green, 1971) containing defined factors (antibiotics and serum, etc.) and cultured in humidified atmosphere with 5% CO_2. After 48 hours, the adherent cells were collected. The cells were subcultured for subsequent passages. 6-OHDA was injected unilaterally in the brain of rats. Thereafter, BM-MSCs were invasively applied to nostrils after the hyaluronidase treatment to promote migration of cells into the brain. The motor activity of the Parkinson's forepaw restored by ca. 68% of the normal values 40–110 days after the application of BM-MSCs (Daniatian et al., 2011).

So far, there have however been no successful clinical data involving MSCs for Parkinson's disease (Yasuhara et al., 2017). However, the further detailed and clinical investigations focusing on transplantational therapy of MSCs (Danielyan et al., 2014) need to be considered and to be performed in successful engineering and also in combination with medicines.

6.9 CANCERS

Cancer is a leading cause of death worldwide and the number of individuals with cancers has continued to expand over a long period. The death is expected to be 8.8 million deaths in 2015. The most common causes of cancer death are cancers of lung (1.69 million deaths), liver (788 000 deaths), colorectal (774 000 deaths), stomach (754 000 deaths), breast (571 000 deaths) (WHO, 2018). Cancer arises from the transformation of normal cells into tumor cells in a multistage process that generally progresses from a pre-cancerous lesion to a malignant tumor. These changes are the result of the interaction between a person's genetic factors and physical, chemical, and biological carcinogens. The cancers continue to grow and malignant tumors can spread other locations from the original site form (metastasis). Cancer cells don't experience aging, continue to maintain their ability to divide, and do not respond to self-termination (Provinciali et al., 2013; Fulop, 2013; Tollefsbol, 2014).

Whereas the pace of population aging around the world is increasing dramatically (WHO, 2015). Aging is the fundamental and unavoidable factor for the development of cancer. The incidence of cancer rises dramatically with age, most likely due to a build-up of risks for specific cancers that increase with age. The overall risk accumulation is combined with the tendency for cellular repair mechanisms to be less effective as a person grows older. Therefore, cancers have been the most tragic and terrible disorders for humans. In the cells of tissues and organs, the cellular senescence is exposed to nonlethal intrinsic or extrinsic stress that results in persistent growth arrest with a distinct morphological and biochemical phenotype. The engagement of senescence may represent a key component for therapeutic intervention in the eradication of cancer occurring with ageing. Normal cells in the body follow an orderly path of growth, division, and death. When this process breaks down, cancer begins to form. Unlikely, regular cells, cancer cells do not experience programmatic death and instead continue to grow and divide. This leads to a mass of abnormal cells that grow out of control. Accordingly, the cellular senescence in anti-cancer therapy may be important also in aging because of the age related-changes down-regulation at the level of both cancer suppressor genes and immune functions (Provinciali et al., 2013; Fulop et al., 2013; Tollefsbol, 2014). This indicates that people age 65 and older bear a higher risk of suffering from cancer compared to younger people (Misra, Seo & Cohen, 2004; WHO, 2018).

For several decades, a number of chemical anticancer drugs has been developed but they prove to have serious side effects for patients at present (Rates, 2001; Tsuda et al., 2004; Truong, Hindmarsh & O'Brien, 2009). In contrast, chemoprevention for cancers of products and purified compounds originating from plants, especially herbal plants, has been found in abundant numbers (Rates, 2001; Tsuda et al., 2004; Truong, Hindmarsh & O'Brien, 2009).

However, naturally occurring compound with a potent anti-cancer effect without side effects has not been detected (Rates, 2001; Tsuda et al., 2004; Truong, Hindmarsh & O'Brien, 2009). We examined the effect of DMSP in green sea algae on the most common type of cancer (carcinoma), Ehrlich ascites carcinoma (free cell cancer), and an organ cancer, liver cancer (solid cancer), with model animals.

6.9.1 Liver Cancers and 3′-Amino-4-dimethylaminoazobenzene (MeDAB)

MeDAB is one of the simplest carcinogenetic derivatives of azobenzene and a more active derivative of tumor induction of the azo dyes. MeDAB is the specific carcinogen toward the liver of rodents. The rodent liver is most common target tissue of amino azo dyes. Rat is much more susceptible to MeDAB among various animals (Arcos & Argus, 1952). Then, the effect of DMSP on the MeDAB-induced liver cancer in rats was examined with inbred Jcl-SD rats. The male rats at the age of 4 weeks had free access to the distilled water (control) and 10 and 20 mM DMSP solution and the powdery diets containing 0.06% MeDAB (w/w). The solutions and diets were given singly or in combination in the control or test solution and test diets during the experimental period, respectively. The body weights were measured at specified times for up to 28 weeks. After 30 weeks, the serum was taken by cutting the tail ends of the rats with the MeDAB-induced liver cancers with and without DMSP at specified times and γ-glutamyl-transpeptidase (γ-GTP) [E.C.2.3.2.1] (a marker enzyme of liver function) activity in serum was automatically measured with a Spotchem (SP-4410, Kyoto Daiichi Science Co., Ltd. Japan) and test paper (Kyoto Daiich Science) for determination of γ-GTP activity. The body weights in all groups rapidly increased up to 8 weeks and thereafter gradually increased up to 28 weeks. However, the body weights of all the groups with and without MeDAB diets almost unchanged for up to 28 weeks without significant difference ($p < 0.05$, Tukey-Kramer test). The activity of γ-GTP showed significantly the highest values in the cancerous rats whereas administration of 10 and 20 mM DMSP to the cancerous rats reduced the values to normal levels and provided no effect of liver weights (g/body wt.) except cancerous rats without DMSP (ca. 3-fold those in the control group), with significant difference ($p < 0.05$) (Nakajima & Minematsu, 2002). Therefore, the administration of DMSP to the cancerous rats was proven to exhibit the effective restoration of liver function. Moreover, the supplementation of MeDAB with DMSP exhibited the reddish yellow color similar to that in the normal rats liver but the administration of MeDAB without DMSP changed the liver color of cancerous rats to the yellowish white by the observation and photographs of excised livers after the operation at 33 weeks (Nakajima, 2015). The effect of DMSP on delayed type hypersensitive (DTH) immune reaction in rats with the MeDAB-induced liver cancer was examined in the ear skin of test rats with the phytohem-agglutinin (PHA) injection. The rats in the control solution group given with the control diets and the rats in the 10 and 20 mM DMSP groups with MeDAB-diets exhibited almost the same red-yellow color, showing the evident occurrence of

DTH immune reaction (Type VI hypersensitivity) ($p < 0.05$, by Tukey-Kramer test). The results reveal that the immune cells, T cells, significantly increased by the administration of DMSP to MeDAB-induced cancerous rats and thus represent likely that T cells, possibly proceeds to the immune systems: CD4+ T cells–dendritic cells–macrophage–neutrophils–B cells and further CD8+ T cells possibly occur. These immune systems had fought against the solid cancer, liver cancer, without cytokine storms for a long time. In fact, the administration of DMSP proves to accumulate macrophages and inhibit the accumulation of ascites fluid in the dose dependently, and arise high rate of neutrophils in Ehrlich ascites carcinoma-bearing mice, which further arise possibly CD8+ T cells leading to kill cancer cells.

6.9.2 Ehrlich Ascites Carcinoma

EAC is a malignant tumor originating from epithelial cells (Ehrlich & Apolant, 1905; Mayer, 1966; Ozaslan et al., 2011). EAC is also referred to a mammary adenocarcinoma with aggressive behavior and *i.p.* injected EAC cells grows in asitic form (Ehrlich & Apolant, 1905: Reddy, Nagarathna & Divvya, 2013; Fernandes et al., In press). EACs are a very terrible cancer because all the mice to which EAC cells were *i.p.* injected obligatorily die within about 2 weeks (Segura, Barbero & Marquez, 2000; Reddy, Nagarathna & Divvya, 2013). We also examined the effect of *i. p.* administration of of EAC cells on the inbred mice and found the critical scenes: large volume of mixed viscous fluid of molten tissues and organs with EAC cells from the dissected abdominal site of mice which died in a ball shape 15 days after the dissemination of EAC cells (Nakajima, 2015). The phenomena show that EAC cells proliferated on the peritoneal membrane, infiltrated, and metastasized into the surrounding viscera after the propagation. In fact, the *s.c.* injection of EAC cells into the footpad (Dagli, Guerra & Saldiva, 1992), hind quarter (Baral & Chattopadhyay, 2004), and the right thigh of the lower limb of mice (Ghoneum et al., 2008.), elicits the proliferation of solid tumors. Moreover, metastases of EAC cells occur from the footpad to lymph node (Dagli, Guerra & Saldiva, 1992) and from tail vein of mice to lung, liver, and kidney of mice in this order (Tsukada et al., 1968; Takamiya, 1993).

6.9.3 Ehrlich Ascites Carcinoma and DMSP

Here, crucial EAC, free cell cancers, were exploited to examine the effect of DMSP on the healing effect of EAC-bearing mice in various ways. The effect of preliminary administration of DMSP solutions on the body weight of EAC bearing-mice was examined with increasing rearing times up to 300 days. DMSP solutions (5, 10 and 20 mM, 1 ml) were *i.p.* administered every other days prior to the injection of EAC cells for 2 weeks. Then, EAC cells (5×10^5 cells/ml) were *i.p.* supplemented to the test mice and the body weight in these group mice were estimated at the indicated times for up to 300 days. The body weights appeared to rapidly increase in the order of the mice in the Carcinoma-,

5, 10 and 20 mM DMSP-Carcinoma group, especially in the former two groups, and in this order reached maxima on the 40th, 44th, 44th, and 48th day, respectively. Thereafter, the body weight of other group except for the Control and 5 mM DMSP-Carcinoma group gradually increased. However, the increase of body weights in the Carcinoma and 5 mM DMSP-Carcinoma group ceased and died on the 47th day and that in the 10 and 20 mM DMSP-Carcinoma group simultaneously stayed on the 54th day. Thereafter, the 10 and 20 mM DMSP-Carcinoma group exhibited almost the same growth curve as that in untreated control during the experimental period, continuing to survive for more than 300 days (63% survival rate). Of great interest is that the preliminary administration of DMSP prolongs by 63% the survival of EAC-bearing mice to more than 300 days, although all of the EAC bearing-mice have died at much earlier days when DMSP is not administered.

The effect of 5 and 10 mM on the DTH immune reaction in EAC bearing-mice were examined by the footpad test in which bovine serum albumin and Freund incomplete adjuvant were injected the into the dorsal portion of the mice and, three week later, albumin was injected again into rear food arch. After 24 hours, the results were obtained by estimating the vertical height from the foot instep to the foot arch. The preliminary administration of saline and 5 and 10 mM DMSP solutions, described earlier, especially the last solutions, proved to activate the DTH immune system of the diseased mice. The facts indicate likely that the supplementation of DMSP has preliminarily elicited the activation of T lymphocytes in the normal mice before the injection of EAC cells, which elicits the long term survival (more than 300 days) of more than half the EAC bearing mice likely due to increased T cells (Morgan, Lefkowitz & Everse, 1998). Moreover, the effect of DMSP at the concentrations of 5 to 30 mM on EAC cells was examined on Ham's medium (Ham, 1965) for 5 h. The results indicated that the dead cells occur much more slightly than those in the control group up to 5 hour and the proliferation of EAC cells reduced in higher concentrations of DMSP among these groups at the indicated times with significant difference ($p < 0.05$, Tukey-Kramer test) (Nakajima, Yokoyama & Nakajima, 2009).

Moreover, the effect of DMSP and closely related compounds, DMSA, and methylmethionine (MeMet), a precursor of DMSP in the biosynthetic pathway of flowering plants (McNeil, Nuccio & Hanson, 1999; Stefels, 2000; Otte et al., 2004; Bullock, Luo & Whitman, 2017), on EAC bearing-mice was examined as follows. An aliquot (0.5 ml) of the saline solution and the 10 mM MeMet, DMSA, DMSP solution was *i.p.* supplemented singly or in combination with EAC cells (5×10^5 cells/ ml) into two group inbred mice each (two series of four test-solution groups). The body weights of all the mice increased with the increasing rearing-times up to 10 days to a varied extent. Especially, the body weights of the Carcinoma-MeMet group very rapidly increased after 6 days and thereafter maintained a plateau region at highest levels from 8 to 10 days. The Carcinoma-Control group tended to show the lowest values up to 6 days but then rapidly continued to increase up to 10 days. The body weights of the Carcinoma-DMSA and -DMSP group gave almost the same values as those in the Control, MeMet, DMSA and DMSP group among them without significant differences ($p < 0.05$, Tukey-Kramer test). Especially, the body weights of the Carcinoma-DMSA group showed the lower values, significantly distinguishing from those in the Carcinoma-Control and Carcinoma-MeMet group with significant difference ($p <0.05$). The effects of MeMet, DMSA and DMSP on the accumulation of ascites fluid, the amounts of RBC and WBC and the ratio (%) of lymphocyte, neutrophil and monocyte of EAC bearing-mice were examined on the 10th day. The volume of the ascitic fluid in the Control and MeMet group was negligible. However, the volume of the fluid in the Carcinoma-Control and Carcinoma-MeMet group were largest among all the groups. The volume of the fluid in the Carcinoma-DMSA and -DMSP group nearly approached to one-third of the volume in the Carcinoma-Control and -MeMet group on the 10th day. Moreover, all other hematological values in the Carcinoma-DMSA and Carcinoma-DMSP group showed almost the same values as those in the Control group. The amounts of red blood cells (RBC) in the Carcinoma-Control and Carcinoma-MeMet group were lowest among all the groups. The amounts of white blood cells (WBCs) in the Carcinoma-Control group and next the Carcinoma-MeMet group were much higher and fairly higher than those in the Control group, respectively. The values of lymphocyte in the Carcinoma-Control and Carcinoma-MeMet group were lowest among all carcinoma groups. And further lower when compared to those in the Control group, respectively. The values of neutrophil in the Carcinoma-Control group were much higher than those of the Control group, which were followed by the Carcinoma-MeMet group. The values of monocyte in the Carcinoma-Control and-MeMet group were lowest among these groups. Accordingly, the results represent likely that WBC, especially, neutrophils in the Carcinoma and Carcinoma-MeMet group significantly fight against EAC cells at 10 days. The body weights, the volume of ascites fluid and hematological values in the Control, MeMet, DMSA and DMSP group exhibited the same values during the experimental period, respectively. Furthermore, the EAC cells obtained in the same way as mentioned earlier were suspended in Ham's medium containing 8% fetal bovine serum (2.5×10^4 cells/ml) and incubated with the DMSA, DMSP, and MeMet solutions at a concentration of 5 mM (final) under a humidified atmosphere of 5% CO_2 in air at 37°C for 5 hours (in vitro). Then, dead cells were counted in a Neubauer haematocytometer using trypan blue dye and expressed in terms of the ratios (%) (mean ± SD, $n = 5$) of the dead cells versus all the EAC cells in each group. The ratios of dead cells in all the groups increased with increasing incubation times up to 5 hours but the ratios at the specified times decreased in the order of the Control, MeMet and DMSA, and DMSP group, respectively (Nakajima & Nakajima, 2011).

Accordingly, the supplementation of DMSP and closely related compounds proved to provide no death of EAC cells and rather prevented the death. The results demonstrate that DMSP show no direct relation to EAC cells. Furthermore, the effect of higher concentrations of DMSP on body weights of EAC-bearing mice was examined for up to 10 days. The suspension (0.5 ml) of EAC cells (2×10^{-6} cells) and 10, 20, 30, 50, and 70 mM DMSP solution (final) were *i.p.* administered to the inbred ICR/Jcl male mice. Initial *i.p.*

supplementation of the control, 10, and 20 mM DMSP solutions in this order linearly increased dose-dependently the body weight in the form of ascites fluid (R^2 = 0.944), whereas administration of 30, 50, and 70 mM DMSP solutions maintained similar and normal body weight with the increasing rearing times for up to 10 days. Moreover effect of various concentrations of DMSP on accumulation of ascites fluid were examined on the 10th day. Administration of 10–70 mM DMSP solutions almost linearly restricted accumulation of ascites fluid in a dose dependent manner (R^2 = 0.964). The effect on accumulation of macrophages and the inflammation by administration of DMSP and thioglycol acid was examined for 3 days. Thioglycolic acid (TG) proved to cause the accumulation of macrophages with inflammation (Rabinovitch et al., 1977; Melnicoff, Horan & Morahan, 1989). For the isolation of purified macrophages, 1 ml of 0, 20 40 mM DMSP solution and thioglycollate (3.7 mg/mouse) was *i.p.*-injected into normal 4-week-old mice (n = 5, each), the peritoneal cells were collected from the peritoneal cavity 3 days after the injection. The cells were then allowed to adhere in the petri dishes for 2 hours at 37°C; non-adhering cells were removed by aspiration and washing. The adherent cells (macrophages) were carefully collected. Interestingly, the administration of the DMSP solutions linearly accumulated macrophages in the peritoneal cavity in a dose dependent manner (R^2 = 0.999). Furthermore, DMSP and TG (inflammatory agent) (Rabinovitch & De Stefano, 1973) administrations (40 mM each) accumulated large numbers of macrophages at the same levels (without significant difference) 3 days after the injection of DMSP or TG. Moreover, to examine the occurrence of inflammation and cell death following administration of DMSP and TG, 1 ml aliquots of DMSP and TG solutions (40 mM each) were *i.p.*-injected into normal 4-week-old mice (n = 3). The abdominal site of the test mice was widely exposed aseptically 3 days after the injection and a trypan blue dye solution was equally sprayed on the exposed viscera. Photographs of viscera were immediately taken. The area which dyed blue was carefully calculated. Results of the area showed that the control and the DMSP group had the same values (without significant difference) whereas the TG group increased by ca. 3-fold compared to those in the Control and DMSP group. The results were clearly confirmed also by photographs. Moreover, the effect of 30, 50, and 70 mM (final) DMSP solutions on EAC cells and macrophages (5 × 104 cells/each) were examined in vitro (in Harm's medium) up to 5 hours. EAC cells tended to multiply with increasing incubation times and concentrations of DMSP, whereas macrophages almost remained unchangeable under both conditions. However, proliferation of macrophages and EAC cells was similarly restricted only at 5 hours at 70 mM solution. In addition, proliferation of EAC cells and macrophages were examined singly or in combination in vitro for up to 5 hours. EAC cells. Macrophages proliferated rapidly with increasing incubation times, although the proliferation rate of macrophages was fairly low compared to that in EAC cells for up to 5 hours. In contrast, EAC-cells in the mixture containing macrophages were highest at 1 h and then continued to decrease up to 5 hours, showing very low values. Macrophages in the mixture containing EAC cells gradually increased up to 3 hours, thereafter reached maximal values at 3 hours and then slightly decreased up to 5 hours, showing smaller numbers than those in macrophage cells alone (Nakajima, Tsujiwaki & Nakajima, 2014). Therefore, macrophages likely gave rise to phagocytosis against EAC cells in this experiment.

6.9.4 Ehrlich Ascites Carcinoma and Betaines

Furthermore, the effect of DMSP (sulfobetaine) and glycine betaine (GB); natural osmolytes (Bentley & Chasteen, 2004; Lever & Slow, 2010: Hoffmann et al., 2013), methyl donors (Nakajima, 1993; Lever & Slow, 2010; Collinsova et al., 2006; Strakova et al., 2011), and twitterion (Yoo et al., 2011), at 5 mM each on the EAC-bearing mice and accumulation of ascitic fluid was examined for up to 10 days. The body weights of the Carcinoma-GB and Carcinoma-Control groups in this order rapidly increased over the 10-day period. The body weights in the Carcinoma-DMSP group increased more slowly than those in the Carcinoma-GB group and tended to decrease less than those of the Carcinoma-control group for up to 10 days. The body weights of the Control and DMSP group increased slightly and similarly up to 10 days, whereas those of the GB group remained unchangeable, showing almost no growth. Whereas ascitic fluid was not found in the Control, DMSP and GB group on the 10th day. However, the volumes of the fluids in the Carcinoma-control and Carcinoma-GB groups were larger than that of the Carcinoma-DMSP group (p <0.05).

Moreover, 5, 10, and 20 mM DMSP solutions and TG medium on the proliferation were *i.p.* administered to normal 4-week-old male mice, and the number of accumulated macrophages was examined 3 days after injection. DMSP administration linearly increased the number of macrophages in a dose-dependent manner (R^2 = 0.980) in the peritoneal cavity, without apparent side-effects. Whereas administration of TG (inflammatory agent) also increased macrophage accumulation similarly to 20 mM DMSP solution.

Incubation of EAC cells with 5 and 20 mM DMSP or GB solutions was performed in in vitro for up to 5 hours. The proliferation of EAC cells in the control, 5 and 20 mM DMSP solutions and the 5 mM GB solution remained unchanged for the duration of the experiment, whereas proliferation of EAC cells in the 20 mM GB group was restricted only at 5 hours (p <0.05). Similarly, proliferation of macrophages in the control, 5 and 20 mM DMSP and the 5 mM GB groups also remained unchanged during the experimental period, whereas macrophages in the 20 mM GB-treated group were linearly restricted in a time-dependent manner (R^2 = 0.998) (p <0.05).

The effect of 5 mM DMSP and GB solutions on the amounts and types of WBCs in EAC-bearing mice were estimated at 10 days. The number of WBCs in the Carcinoma control group corresponded by ca. 8-fold increase compared to the Control group and by ca. 3.6-fold increase compared to those in the EAC-DMSP and EAC-GB group. In contrast, the number of WBCs in the Carcinoma DMSP and Carcinoma GB group corresponded to ca. 2-fold increase compared to that of the control group. The rate of lymphocytes in the Carcinoma control group was the highest among all Carcinoma groups, and decreased by ca. ½-fold compared to those of the control, DMSP and GB group. All the groups showed the similar rates except the rate of lymphocytes in the Carcinoma Control. Additionally, the rate of neutrophils in the Carcinoma control group corresponded to ca. 3-fold that of the control group, whereas the rate of neutrophils in the Carcinoma Control group increased by ca. 3.0-fold and ca. 2.4-fold compared to those in the Carcinoma-DMSP and -GB group, respectively, when compared to those in the control group (Nakajima, Nakajima & Tsujiwaki, 2015). Hematological parameters demonstrate that WBCs and the rates of neutrophils in the Carcinoma Control group increases by ca. 4–7-fold and 3-fold compared to those in the Control group but lymphocytes decreased by ca. ½-fold of those in the Control group at the later stage (10 days) of the carcinoma, whereas both values in the Carcinoma-DMSP and -DMSA group reversed to those in the Control group at the later sage. The amounts of WBCs, rates of lymphocytes and neutrophils in the Carcinoma-MeMet and -GB group were similar to those in Carcinoma Control group to a lesser extent. As tumor associated neutrophils (TANs) show heterogeneity in their capacity to promote or inhibit T cell immunity. Therefore, TANs promote likely the proliferation of EAC cells without DMSP (Simon & Kim, 2010; Singel & Segel, 2016; Pcana et al., 2017) while neutrophils appear to prime a long-lived M2 macrophage by DMSP (Chen et al., 2014). The dose-dependent accumulation of M2 macrophage was confirmed by the administration of high concentrations of DMSP to the EAC-bearing mice (Nakajima, 2014, 2015). These results reveal that M2 macrophage plays a significant role in amelioration of crucial Ehrlich ascites carcinoma at the later stage.

Together, these results demonstrated that the administration of 3′-methyl-4-dimethylamino-azobenzene into the normal rats for a long time (28 weeks) caused liver cancer without almost any side effect in body and liver weights but with ca. 3-fold liver weights in the MeDAB-treated group, whereas preliminarily and sequentially administered DMSP solutions (10 and 20 mM) to the cancerous rats reversed the function of liver in the MeDAB-treated rats without elevation of liver weights and also activates the delayed type hypersensitive immune reaction at 30 weeks.

6.9.5 Cancers and MSCs

The healing effect of MSCs on various cancers have been widely investigated. BM-MSCs were exploited to examine its effect on breast cancer in female mice. MSCs were obtained from bone marrow (tibia and femur) aspirates of BALB/c mice and cultured in defined MEM medium with serum and antibiotics. Murine 4T1 cells (breast cancer cells) were cultured in RPMI-1640 medium with serum and antibiotics. Mouse Sitrt 1 cDNA and GFP gene were employed with adenoviral vector. In this experiment, MSCs were utilized for tumor formation. The tumor cells were subcutaneously injected into armpit of 6- to 8-week-old female BALB/c mice and used as the control. In this experiment, it was examined whether breast cancers in female mice are mitigated by the combination of 4T1 breast cancer cells and Sirt 1. All tumor-bearing mice were sacrificed on the 18th day. Then, the length, width, and weights of tumors were measured. Tumor volume and tumor weight by administration of Sirtl to MSC reduced by ca. 67% and ca. 60 compared to those in MSCs alone at 18 days, based on the mean values (Yu et al., 2016). Moreover, the effect of MSCs on multiple lung tumors of 10 weeks old of BALB/c mice was examined with BM-MSCs. BM-MSCs were isolated from bone marrow aspirates and cultured in Dulbecco's MEM with serum and antibiotics. Seventy-two hours after the culture the cells reached confluence, plastic adherent cells were treated with trypsin and semi-adherent cells were subcultured and expanded. Large and polygonal cells were passaged 3 to 4 times for the experimental use. BM-MSCs to which NK4 was transduced by NK4 gene, an antagonist of hepatocyte growth factor (HGF) (by the aid of hematocyte growth factor), were exploited with an adenoviral vector with an RGD motif (Arg-Gly-Asp, specific attachment site) were employed. MSCs which express NK4 were referred to NK4-MSCs, which were injected to a tail vein of tumor-bearing mice. After the *i.v.* injection (0 day) of C 26 cells to form the Colon-26 lung metstasis model mice and further *i.v.* injection of NK4-MSCs at 5, 7, and 9 days, the NK4-MSCs mice group survived by ca. 30% up at 55 days with fewer adverse effects whereas the Control (Null-MSCs) group mice completely died at 33 days (Kanehiro et al., 2007). Next, the effect of engineered hBM-MSC-IFN-β (MSCs as carrier) on the survival of the 6- to 8-week-old femals BALB/c mice with the U87 intracranial glioma was examined. BM-MSCs were isolated by gradient centrifugation and suspended in α-MEM with serum, glutamine, and antibiotics followed by plating. Then, the adherent cells were cultured until they reached confluence and treated with trypsin. For the experiments, the cells through passages 3 to 4 were used. Human glioma cells (tumor cells) were injected into the right frontal lobe of nude mice under anesthesia. hBM-MSCs were injected into the internal carotid artery. To track hBM-MSCs, the cells were incubated with SP-Dil fluorescent dye, washed, and used for experiments. U87 glioma cells were implanted (intercranially) into the frontal lobe of nude mice. After 10 days, tumors were injected with hBM-MSCs-β IFN. The survival after the intratumoral injection of hBM-MSCs-β IFN into the U87 glioma of test mice were examined for up to 60 days. The intratumoral injection (2.5×10^5 cells) of hBM-MSCs-β IFN prolonged the survival of the mice up to 60 days whereas the corresponding *s.c.* injection of hBM-MSCs-β IFN died at 30 days. Morover, the survival of the mice with U87 glioma was examined after injection of hBM-MSCs-β IFN into the carotid artery. U 87 cells were implanted into the frontal lobe of nude mice. After 10 days, hBM-MSCs-β IFN were injected into the carotid artery of the mice. The mice treated with hBM-MSCs-β IFN survived up to 50 days while the mice treated with phosphate buffer, IFN-β *i.v.* injected, and the mice with the flank to which hBM-MSCs-β IFN was injected similarly died at about 33 days.

Therefore, the treatment with MSCs engineered to hMSCs-β IFN proved to prolong the life of the mice with glioma ($p < 0.05$) (Nakamizo et al., 2005).

These healing effect was likely exerted by the effect of NK4 or β IFN rather than MSCs on the carotid artery and glioma involving in brain. Therefore, MSCs have been exploited as engineered MSCs but not as direct curing agents in these experiments.

MSCs are one of key players within the tumor microenvironment and can either inhibit or promote tumor cell growth through complex cellular interaction (Yagi & Kitagawa, 2013; Melzer, Yang & Haas, 2016). If MSCs undergo transformation (possibly by silencing of tumor suppressor), transformed MSCs show characteristics of cancer cells with loss of contact inhibition, mesenchymal lineages, and adherent growth. MSCs conditioned by tumor cells also promote metastasis of tumors (Lazennec & Lam, 2016; Ridge, Sullivan & Glynn, 2017). Accordingly, it is a great problematic matter that MSCs have the ability to act on all steps of carcinogenesis. Moreover, it is very difficult but important to consider adaptive amounts and sites of MSCs, donor variability (different tissue sources), possibility of transformation of MSCs, injection timing of MSCs, usage of viral vector, and engineered conditions in detail for clinical application of MSCs in future therapy. Therefore, the clinical application of MSCs is in the course of the study at present although the clinical therapy with MSCs is the promising approach for various diseases.

The clinical test with Muse cells will be initiated for patients with acute myocardial infarction at Gifu University Hospital by a research group of Prof. M. Dezawa in Tohoku University and Prof. S. Hamaguchi of Gihu University and other institutions. Human Bone marrow-derived Muse cell allografts and xenografts are proven to ameliorate the damaged tissue and function of acute heart infarction in rabbits by intravenous injection. SIP-S1PR2 axis mediates homing of Muse cells into damaged heart for long-lasting tissue repair and functional recovery after acute myocardial infarction (Uchida et al., 2015; Yamada et al., 2018).

The clinical trial with DMSP have been not performed although DMSP proves to be safe and useful for humans. Okinawans who have eaten a lot of green sea algae had maintained a long life in Japan until their lives became Westernized in recent years. Moreover, the diets fed with 5% dried green sea algae (*Monosodium nitidun*) proved to ameliorate the loss of learning and memory of SAM P8 strain (Nakajima & Minematsu, 2002). The clinical test should be singly tried with raw and dried green algae and in combination with MSCs in various ways. The clinical trial with iPSc was performed for patients suffereing from Parkinson's disease in Kyoto University Hospital in November 9, 2018, in which 2.4 million allogeneic dopaminergic progenitors formed from iPSc were introduced into bilateral corpus striatum through a kind of syringe needle from a hole (1.2 Φcm.) on the brain skull by Prof. Takahashi, J. et al. by the same methods as those in DBS operation. Human iPS cell-derived dopaminergic neurons proves to function in Parkinson's disease of monkey (Kikuchi et al., 2017). However, there remain two big problems, tumorigenesis and immune rejection, in clinical trials with iPSc. Recently, iPSc which is easier to undergo immune rejection seems to have been developed using gene editing method with CRIPS-Cas9 (Xu et al., 2019). The clinical trials in heart infarct, Parkinson's disease, spinal cord injury, and others are expected to succeed in obtaining various detailed conditions by the safety and availability of these treatments for these diseases in humans.

6.10 CONCLUDING REMARKS

The obtained results indicate that the injection of 10 mM DMSP-Br solution (1 ml) every other day for 2 weeks ameliorates critical Ehrlich ascites carcinoma (free cell cancers) at a high rate in mice, and that the free ingestion of 10 mM solution for 33 weeks reinstates the 3′-MeDAB-inducecd liver cancers (solid cancers) in rats. Our experiments further revealed that the intraperitoneal administration (0.5 ml) of 70 mM DMSP-Br solution, of great significance, does not provide the effect on the growth of the inbred mice at the age of 4 weeks for 15 days. Of great interest, these results and findings may support that the corresponding amounts (2.1 g DMSP-Br) per person (1 ml of 10 mM DMSP solution (1.7 mg DMSP-Br) per mouse) hold no toxins also for humans. The green sea macro algae contained large amounts of DMSP and have been habitually ingested by people in Japan since ancient times. Furthermore, DMSP proved to activate the mobility and/or growth in normal chicken, mice, rats, and fish along with the accumulation of catecholamines by the incorporation of DMSP into brain and liver of mice, rats, and fish. Moreover, DMSP ameliorates and/or reinstates skin and stomach ulcers in mice, 1 type diabetes mellitus in rats, and to recover the delayed growth, loss of learning and memory, aging in mice, and the shorter life in aged mice, hyperhomocysteinemia in mice (Nakajima, 2006), Alzheimer's disease in mice, MPTP-induced Parkinson's disease in lower concentrations, and MeDAB-induced liver cancer in rats, and crucial Ehrlich ascites carcinoma in mice in high concentrations, respectively. Whereas the amounts of DMSP needed for healing of cancers are equivalent to 10–100 garlic cloves (Bianchini & Vainio, 2001) and 10 cups of green tea per day per person (Fujiki, 2005). These favorable response of DMSP to various diseases attribute likely to antioxidant function, immune activation (in particular, well balanced amounts of T cells, neutrophils, and in particular increase of activated (M2) macrophages, hormonal function (accumulation of catecholamines), and activated methylation reaction with DMSP (as methyl donor for the methylation enzymes through S-adenosylmethionine accumulated by the addition of DMSP), in various and well-balanced ways (Nakajima, 2015, 2020).

Injection of 1 ml of 10 mM solution (1.7 mg DMSP-Br) is equivalent to the ingestion of ca. 0.28 g of wet algae (*Monosodium nitidum*) (6.1 mg DMSP-Br/g (Reed, 1983)) or ca. 0.04 g the dry algae (41.2 mg DMSP/g) per 40 g body wt. (mouse) per day, which may require the ingestion of ca. 350 g wet algae (2.1 g DMSP-Br) or ca. 50 g dry algae (2.1 g DMSP) per 50 kg body wt. (person), every other day for 2 weeks. In contrast, the amounts of DMSP in the commercially available wet and dry algae appear to be able to be elevated by about 5-fold to 10-fold by a newly developed dry procedure, in which the biosynthesis of DMSP probably proceed in the course of drying the commercial wet and dry algae under reduced pressure and moderate heating (unpubl. data). The methods may reduce by 1/5–1/10 the amounts of the algae needed for humans every second day for 2 weeks. Therefore, these facts indicate that ca. 350 g wet algae (2.1 g DMSP-Br) or ca. 50 g dry algae (2.1 g DMSP) per 50 kg body wt. (person) are decreased to ca 35–70 g wet algae or ca. 5–10 g dry algae every other day for 2 weeks. The ingestion of –~10 g dry algae every other day for 2 weeks is quite available for people. The amounts can be replaced by ingestion of less amounts of dry algae every days. Accordingly, the ingestion of the dry algae is likely to reinstate almost all the diseases described earlier. Moreover, DMSP is effectively available as an ingredient for aquaculture of fresh water and marine fish, crustaceans, marine mollusks, and amphibians (juvenile frog and salamander (Nakajima, 1995, 1996, 1998). These are possible by using the green sea algae themselves. In facts, the diets containing the dry algae (*Monosodium nitidum*) healed the delayed growth and the loss of learning and memory in the aged mouse (SAM P8 strain) and the green sea algae soaked in DMSP solution accelerated the growth and metamorphosis of juvenile Brown frog (Nakajima, 1998). Therefore, DMSP proves to be favorably involved in normal and diseased aquatic and terrestrial animals and terrestrial plants (Nakajima, 2020) and likely in not only the hatching of fertilized fish eggs (origin of embryonic stem cell) ("embryogenesis") but also the growth of larval and juvenile fish and young adult carp. All the findings suggest likely that a sulfur containing-simple natural compound, DMSP, is likely to have been synthesized in more or less amounts by the simple reaction of dimethylsulfidfe and propionic acid in primitive times, possibly occurring under the troposphere or the conditions in the volcanic or sea floor hydrothermal vents over the high temperature or the large glacial period and thus DMSP had an intimate involvement in the origin of life. The consideration is supported by the facts that sulfur, carbon, hydrogen, nitrogen, and oxygen atoms (in the form of H_2S, H_2S_2, CH_4, H_2, NH_3, H_2O, FeS_2, etc.) and further methyl alc*o*hol, some amino acids, organic acids, and pre-proteins had arisen under the similar conditions in earth's primitive environments or the troposphere in a laboratory (Miller & Urei, 1959; Ishigami, 1969; Schwartz, 1981; Martin et al., 2008; Baross, Kelley & Russell, 2008). Furthermore, DMSP likely is closely related to the birth, development, differentiation, and growth of life (evolution), considering the acceleration of the occurrence, development and differentiation of unorganized plant tissues (callus), the promotion of the hatching of fish fertilized eggs (embryogenesis), germination of terrestrial plant seeds (embryogenesis), the acceleration of the mobility and growth of larval and juvenile fish, terrestrial plants and animals, and the repairing effect of diseased animals due to DMSP. Moreover, the acceleration of hatching of fertilized fish eggs, germination plant seeds and the growth and mobility (swimming) of larval fish is considered to have a strong potential also for resident mesenchymal stem cells. However, the experimental and clinical trial of MSCs need likely to prepare the well thought-engineered MSCs although MSCs have a high potential for the therapeutic effect for crucial diseases. In contrast, DMSP, a most simple, primitive, and natural compound, is believed to have to further play a pivotal role for aquatic and terrestrial plants and normal and diseased animals in varied ways in the future. Furthermore, all the results and findings represent that DMSP has always been a wide, miraculous, primordial, and natural compound over various ways in the world of life.

COMPETING INTERESTS

The authors declare that they have no competing interests

ACKNOWLEDGMENTS

I heartily thank Asst. Prof. Y. Nakajima, M. Sakamoto, Mr. Y. Miyamoto, Miss. M. Tsujiwaki, Prof. M. Minematsu, and other colleagues in Laboratory of Biochemistry, the graduate School of Koshien.

REFERENCES

Ackman, R. G., & Dale, J. (1965) Reactor for determination of dimetinhyl-β-propiothetin in tissue of marine origin by gas-liquid chromatography. *J. Fish. Res. BD. Canada*. 22(4): 875–883.

Ackman, R. G., & Hingley, J. (1968) The occurrence and retention of dimethyl-β-propiothetin in some folter-feeding organisms. *J. Fis. Res. Bd. Canada*. 25(2): 267–284.

Ackman, R. G., Dale, J., & Hingley, J. (1966) Deposion of dimetinhyl-β-propiothetin in Atlantic cod during feeding experiments. *J. Fish. Res. BD. Canada*. 23(4): 487–497.

Ackman, R. G., Hingley, J., & MacKay, K. T. (1972) Dimethyl sulfide as an odor component in Nova Scotia fall mackerel. *J. Fish. Res. BD. Canada*. 29(7): 1085–1088.

Ackman, R. G., Tocher, C. S., & Mclachian, J. (1966) Occurrence of dimethyl-β-propiothetin in marine phytoplankton. *J. Fish. Res. BD. Canada*. 23(3): 357–364.

Adav, S. S., & Sze, S. K. (2016) Insight of bain degenerative protein modification and dementia by proteomic profiling. *Mol. Brain*. 9: 92–114.

Aihara, E., Matthis, A. L., Kams, R. A., Engevik, K. A., Jiang, P., Wang, J., Yacyshyn, B. R., Marshall, H., & Montrose, M. H. (2016) Epithelial regeneration after gastric ulceration causes prolonged cell-type alteration. *Cell Mol. Gastroentheerol. Hepatol*. 2(5): 624–647.

Alderton, G., & Bordon, Y. (2012) Tumor immunotherapy—leucocytes take up the fight. *Nat. Rev*. 12: 237.

Alessandro, S., & Paolo, O. (2012) Catechol-O-methyltransferase and Alzheimer's disease: A review of biological and genetic findings. *Current Drug Target—CNS & Neurol. Disorder*. 11(3): 299–305.

Alessio, N., Ozacan, S., Tatsumi, K., Murat, A., Peliso, G., & Dezawa, M. (2017) The secretome of MUSE cells contains factors that may play a role in regulation of stemness, apoptosis and immunomodulation. *Cell Cycle*. 16(1): 33–44.

Andreone, L., Gimeno, M. L., & Perone, M. J. (2018) Interactions between the neuroendocrine system and T lymphocytes in diabetes. *Front. Endocrinol*. 9: 229–248.

Anghelina, M., Krishnan, P., Moldova, L., & Moldovan, N. I. (2006) Monocytes/Macrophages cooperate with progenitor cells during neovasscularrization and tissue repair. Conversion of cell columns into into fibrovascular bundles. *Am. J. Pathol*. 168(2): 529–541.

Arcos, J. C., & Argus, M. C. (1952) Structural bases and biochemical mechanisms. *Chemical Induction of Cancer* (Vol. II B). New York and London: Academic Press Inc., p. 144.

Arends, Y. M., Duyckaerts, C., Rozemuller, J. M., Eikelenboom, P., & Hauw, J. J. (2000) Microglia, amyloid and dementia in Alzheimer desease. A correlative study. *Neurobiol. Aging*. 21: 39–47.

Baral, R., & Chattopadhyay, U. (2004) Neem (Azadirachta indica) leaf mediated immune activation causes prophylactic growth inhibition of murine Ehrlich carcinoma and B16 melanoma. *Inter. Immunopharmacol.* 4: 355–366.

Barber, C., & Arispe, M. L. I. (2006) The ever-elisive endothelial progenitor cell: Identities, functions and clinical implication. *Pedriatr. Res.* 59(4): 26R–32R.

Barone, P. (2010) Neurotransmission in Parkinson's disease: Beyond dopamine. *Eur. J. Neurol.* 17(3): 364–376.

Baross, J., Kelley, D., & Russell, M. J. (2008) Hydrothermal vents and the origin of life. *Nat. Rev* 6: 805–814.

Bassi, E., Moraes-Vieira, P. M. M., Moreira-Sa, C. S. R., Almeida, D. C., Vieira, L. M., Cunha, C. S., Hiyane, M. I., Basso, A. S., Pachenco-Silva, A., & Camara, N. O. (2012) Immune regulatory properties of allogeneic adipose-derived mesenchymal stem cells in the treatment of experimental autoimmune diabetes. *Diabetes*. 61: 2534–2545.

Bayles, D. O., & Wilkinson, B. J. (2000) Osmoprotectants and cryoprotectants for Listeria monocytogenes. *Lett. Appl. Microbiol.* 30: 23–27.

Baynest, H. W. (2015) Classification, pathophysiology, diagnosis, and management of diabetes. *J Diabetes Metab.* 6(5): 541–550.

Bentley, R., & Chasteen, T. G. (2004) Environmental VOSCS-formation and degradation of dimethylsulfide, methanthiol and related materials. *Chemosphere*. 55: 291297.

Bernhard, C. A., Ried, C., Kochanek, S., & Brocker, T. (2015) CD169+ macrophages are sufficient for priming of CTLs with specificities left out by cross-priming dendrictic cells. 2015. *PNAS*. 112(17): 5461–5466.

Bhargava, K., Daas, K., Gupta, G. P., & Gupta, M. (1980) Study of central neurotransmetters in stress-induced gastric ulceration inalbino rats. *Br. J. Pharmaciol.* 68(4): 765–772.

Bianchini, F., & Vainio, H. (2001) Allium vegetables and organosulfur compounds: Do they help prevent cancer? *Environ. Health Perspect.* 109: 893–902.

Bianco, P. (2014) Mesenchymal stemcells. *Ann. Rev. Cell Dev. Biol.* 30: 677–704.

Biedermann, T., Knelling, M., Mailhammer, R., Maler, K., Sander, C., Kollis, G., Kunkel, S., Hultner, L., & Rocken, M. (2000) Mast cells control neutrophil recrument during T cell-mediated delayed-type hypesensitivity reaction through tumor necrosis factor and macrophage inflammatory protein2. *J. Exp. Med.* 192(10): 1441–1452.

Blesa, J., Phani, S., Jackson-Lewis, V., & Predborski, S. (2012) Classic and New animal models of Parkinson's disease. *J. Biomed. Biotechnol.* Article ID 845618 (10 pages).

Bluden, G., Smith, B. E., Iron, M. W., Yang, M., Roch, O. G., & Patel, A. V. (1992) Betaines and reverses the effects of MPP^{+} toxicity. *Psychoharmacology*. 88(3): 401–402.

Boche, D., & Nicoll, J. A. R. (2013) Neuroinflammation in aging and in neurodegenerative disease. *Neuropathol. Appl. Neurobiol.* 39(1): 1–2.

Broadbent, A. D., Jones, G. B., & Jones, R. J. (2002) DMSP in corals and benthic algae rom the great barrier reef. *Estur. Coast. Shelf Sci.* 55(4): 547–555.

Brozowski, T., Zwirska-Korczala, K., Konturec, P., C., Silwowski, Z., Pawlik, M., Kwiecien, S., Drozdowicz, D., Mazurkiewicz-Jank, M., Bielanski, W., & Pawlik, W. W. (2007) Role of circadian rhythm and endogenous melatonin in pathogenesis of acute gastric bleeding rsions induced by stress. *J. Physiol. Pharmacol.* 58 Suppl 6: 53–64.

Bullock, H. A., Luo, H., & Whitman, W. B. (2017) Evolution of dimethylsulfoniopropionate metabolism in marine phytoplankton and bacteria. *Front Microbiol.*, 8: 637–655.

Burdett, H. L., Carruthers, M., Donohue, P. J. C., Wicks, L. C., Hennige, S. J., Roberts, J. M., & Kamenos, N. A. (2014) Effects of high temperature and CO_2 in the cold-water coral *Lophelia pertusa*. *Mar. Biol.* 161(7): 1499–1506.

Cabib, S., & Puglisi-Allegra, S. (2012) The mesoaccumbens dopamine in coping with stress. *Neur. Biobehav. Rev.* 36: 799–789.

Cantoni, G. L., & Andersonm, D. G. (1956) Enzymatic cleavage of dimethylpropiothetin by Polysiphonia lanosa. *J. Biol. Chem.* 222: 171–177.

Cenci, M. A., & Crossman, A. R. (2018) Animal models of L-dopa-induced dyskinesia in Parkinson's disease. *Mov. Disord.* 2018, February 28.

Chai, J. (2011) Gastric ulcer healing-role of serum response factor. In: Chai, J. (ed.) *Gastric Ulcer Disease*. New Delhi: In Tech. Chapter 8. pp. 143–149.

Challenger, F., & Simpson, M. I. (1948) Studies on biological methylation. Part XII. A precursor of the dimethyl sulfide evolved by *Plysiphonia fastigiata*. Dimethyl-2-carboxyethylsulfonium hydroxide and its salts. *J. Chem. Soc.* 1591–1597.

Chamberlain, C., Saether, E. E., Aktas, E., & Vanderby, R. (2017) Mesenchymal stem cell therapy on tendon/ligament healing. *J. Cytokine Biol.* 2(1): 112–119.

Chamberlain, G., Fox, J., Ashton, B., & Middleton, J. (2007) Concise review: Mesenchymal stem cell: Their phenotype, differentiation capacity, immunological, and potential for homing. *Stem Cells.* 25: 2739–2749.

Chen, F., Wu, W., Milliman, A., Craft, J. F., Chen, E., Patel, N., Boucher, J. L., Urban Jr, J. F., Kim, C. C., & Gause, W. (2014) Neutrophils prome a long-lived effector macrophage phenotype that mediates accelerated helminth expulsion. *Nat. Immunol.* 15(10): 938–946.

Chen, L., Tredget, E. E., Wu, Y. G., & Wu, Y. (2008) Parachrine factors of mesenchymal stem cells recruit mcrophages and endothelial lineage cells and enhance wound healing. *PlosOne*. 3(4): e1886, pp. 1–12.

Cheney, G. (1950) The nature of the antipeptic-ulcer dietary factor. *Stanford Med. Bull.* 8: 144–161.

Chhabra, P., & Brayman, K. (2013) Stem cell therapy to cure type 1 diabetes: From hype to hope. *Stem Cells Translat. Med.* 2: 328–336.

Chow, A., Lucas, D., Hidalgo, A., Merdez-Ferrer, S., Hashimoto, D., Schelermann, C., Prophere, C., Van Rooijen, N., Merad, M., & Frenette P., S. (2014) Bone marrow $CD169^+$ macrophages promote the retention of hematopoietic stem and progenitor cells in the mesenchymal stem cell niche. *J. Exp. Med.* 208(2): 261–271.

Clement, F., Grockkwiak, E., Zyberrsztein, F., Fossard, G., Gobert, S., & Maquer-Satta, V. (2017) Stem cell manipulation, gene therapy and the risk of cancer stem cell emergence. *Stem Cell Investig.* 4: 67–83.

Collie, A., & Maruff, P. (2000) The neuropsychology of preclinical Alzhaimer's disease and mild cognitive impairment. *Neur. Biobehav. Rev.* 2: 365–374.

Collinsova, M., Strakova, J., Jiracek, J., & Garrow, T. A. (2006) Inhibition of betaine-homocysteine S-methyltransferase causes hyperhomocysteinemia in mice. *J Nutr.* 136: 1493–1497.

Dagli, M. L. Z., Guerra, J. L., & Saldiva, P. H. M. (1992) An experimental stydy on the lymphatic dissemination of the solid Ehrlich tumor in mice. *Braz. J. vet. Res. Anim. Sci.* 29(1): 97–103.

Daly, K. L., & Tullio, G. R. (1996) Particulate dimethylsulfonioproponate removal and dimethylsulfid produvtion by zooplankton in the southern ocean. In: Kien, R. P., Visscher, P. T., Keller, M. D., & Kirst, G. O. (eds.) *Biological and Environmental Chemistry of DMSP and Related Sulfonium Compounds*. Heiderberg, New York and London: Springer International Publishing, pp. 230–232.

Daniatian, L., Schafer, R., Ameln-Mayerhofer, A., Bernhard, F., Verleysdonk, S., Buadze, M., Lourhmati, A., Klopfer, T., Schaumann, F., Schmid, B., Koehle, C., Proksch, B., Weissert, R., Reichardt, H. M., Brandt, J., Buniatian, G. H., Schwab, M., Christoph, H., Gleiter, C. H., William, H., & Frey II, W. H. (2011) Therapeutic efficacy of intranesally delivered mesenchymal stem cells in a rat model of Parkinson's disease. *Rejuvenation Res*. 14(1): 3–16.

Danielyan, L., Beer-Hammer, B., Stolzing, S., Schfer, R., Siegel, G., Fabian, C., Lourhmati, A., Buadze, M., Novakovic, A., Proksch, B. C. H., Frey II, W. H., & Schwab, M. (2014) Intranasal delivery of bone marrow-mesenchymal stem cells, macrophages, and microglia to the brain in mouse models of Alzheimer's and Parkinson's disease. *Cell Transplantat*, 23. Suppl 1: S123–S139.

Das, A., Sinha, M., Datta, S., Abas, M., Chaffee, S., Sen, C. K., & Roy, S. (2015) Monocyte and macrophage plasticity in tissue repair and generation. *Am. J. Pathol.* 185(10): 2596–2606.

De Vogelaere, F., Santens, P., Achten, E., Boon, P., & Vingerhoets, G. (2012) Altered default-mode network activation in mild cognitive impairment compared with healthy aging. *Neuroradiol.* 54(11): 1195–1206.

DeBose, J. L., & Nevitt, G. A. (2008) Dimethylsulfoniopropionate (DMSP) is linked to coral spawning, fish abundance and squid aggregations over a coral reef. Proc. 11th Internatinal coral reef symposium, Ft. Laudaerdale, FL, 7–11.

Delaloye, S., & Holtsheimer, P. E. (2014) Deep brain stimulation in the treatment of depression. *Dialogues Clin. Neuro.* 16(1): 83–91.

Dezawa, M. (2016) Muse Cells Provide the Pluripotency of Mesenchymal Stem Cells: Direct Contribution of Muse Cells to Tissue Regeneration. *Cell Transplant.* 25(5): 849–861.

Dezawa, M., Kanno, H., Hoshino, M., Cho, H., Matsumoto, N., Itokazu, Y., Tajima, N., Yamada, H., Sawada, H., Ishikawa, H., Mimura, T., Kitada, M., Suzuki, Y., & Ide, C. (2004) Specific induction of neuronal cells from bone marrow stromal cells and application for autologous transplantation. *J. Clin. Invest.* 11: 1701–1710.

Dickson, D. M., Wyn, Jones, R. G., & Davenport, J. (1982) Osmotic adaptation in Ulva lactuca under fluctuating salinity regimes. *Plant*. 155(5): 409–415.

Du, A-T., Schuff, N., Chao, L. L., Kornak, J., Jagust, W. J., Kramer, J. H., Reed, B R., Miller, B. L., Norman, D., Helena C. Chui, H. C., & Michael W. Weiner M. W. (2006) Age effects on atrophy rates of entorhinal cortex and hippocampus. *Neurobiol. Aging*. 27(5): 733–740.

Dunn, J. S., & Mcletchie, N. G. B. (1943) Experimental alloxan diabetes in the rat. *Lancet*. 245: 384–387.

Eggenhofer, E., & Hoogduliin, M. (2012) Mesenchymal stem cell-educated macrophages. *Translant Res*. 1(1): 12–17.

Ehrlich, P., & Apolant, H. (1905) Beobachtungen uber maligne mausen-tumoren. *Berlin Klin Wochenschr*. 42: 871–874.

European Brain Council. Parkinson's disease Fact Sheet 2011.

Evans, M. J., & Kaufman, M. H. (1981) Establishment in culture of pluripotential cells from mouse embryos. *Nature*. 292(5819): 154–156.

Fang, J. Y., & Tolleson, C. (2017) The role of deep brain stimulation in Parkinson's disease: An overview and update on new developments. *Neuropscychiatr. Dis. Treat*. 13: 723–732.

Fernandes, P. D., Guerra, F., Sala, N. M., Ssardella, T. T. B., Jancar, S., & Nerves, J. S. (In press) Characterization of the inflamamatory response during Ehrlich ascitic tumor development. *J. Pharmaccol. Toxicol. Methods* (Content lists available at ScienceDirect).

Ferrer, R. P., & Zimmer, R. K. (2012) Community ecology and the evolution of molecules of keystone significant. *Biol. Bull*. 223: 167–177.

Flood, J. F., & Morley, J. E. (1997) Learnning and memory in the SAMP 8 mouse. *Neuro. Biobehav. Rev.* 22(1): 1–20.

Floor, E., & Wetzel, M. G. (1998) Increased protein oxidation in human substantia nigra pars compacta in comparison with basal ganglia and prefrontal cortex measured with an improved dinitrophenylhydrazine assay. *J. Neurochem.* 70(1): 268–275.

Frenette, P. S., Pinho, S., Lucas, D., & Scheiermann, C. (2013) Mesenchymal stem cell: Keystone of the hematopoietic stem cell niche and a stepping-stone for regenerative medicine. *Annual Rev. Immunol.* 31: 285–316.

Fujiki, H. (2005) Green tea: Health benefits as cancer preventive for humans. *Chem. Res.* 5: 119–132.

Fulop, T., Labi, A., Kotb, R., & Pawelec, G. (2013) Cancer and Immunology of aging and cancer development. *Cancer Aging*. 38: 38–48.

Ganapathy, S., Vedam, V., Rajee, V., & Arunachalam, R. (2017) Autoimmune disorders-immunopathogenesis and potential therapies. *J. Young Pharm.* 9(1): 14–22.

Gao, X., Song, L., Shen, K., Wang, H., Qian, M., Niu, W., & Qin, X. (2014) Bone marrow mesenchymal stem cells promote the repair of islets from diabetic mice through paracrine actions. *Mol Cell Endocrinol.*, 388(1–2): 41–50.

Gerald, H., & Nicholas, J. (2014) The stomach-brain axis. *Best Pract. Res. Clin. Gastroenterol.* 28(6): 967–979.

Ghoneum, M., El-Din, N. K. B., Noaman, E., & Tolentiino, L. (2008) Saccharomyces cerevisiae, the Baker's Yeast, suppresses the growth of Ehrlich carcinoma-bearing mice. *Cancer Immunol. Immunother.* 57: 581–592.

Goodazi, P., Aghayan, H. Z., Larijiani, B., Soleimani, M., Dehpour, A. R., Sahebjam, M., Ghaden, F., & Arjimand, B. (2015) Stem Cell-based approach for the treatment of Parkinson's disease. *Med. J. Islam Repub. Iran*. 29: 168–178.

Gordon, S., & Martinez-Pomares, L. (2017) Physiological roles of macrophages. *Eur. J. Physiol.* 469: 365–374.

Gordon, S., & Taylor, P. R. (2005) Monocyte and macrophage heterogeneity. *Nat. Rev. Immunol.* 5: 953–964.

Green, L. A. (1978) NGF-responsive clonal PC12 pheochromocytoma cells as tools for neuropharmacological investigation. In: Adopolphe, M. (ed.) *Advances in Pharmacology and Therapeutics*. Oxford and New York: Pergamon Press, pp. 197–203.

Green, L., & Tischler, A. S. (1976) Establishment of a noradrenaergic clonal line of rat aderenal pheochromocytoma cells which respond to nerve growth factor. *Proc. Natl. Acad. Sci., USA*. 73(7): 2424–2428.

Guridi, J., Herrero, T., Luquin, M. R., Guillen, J., Ruberg, M., Laguna, J., Vila, M., Javoy-Agid, F., Agid, Y., Hirsch, E., & Obeso, J. A. (1996) Subthalamotomy in Parkinsonian monkeys. *Behav. Biochem. Analy. Brain*. 119: 1717–1727.

Gurusamy, N. (2017) Mini review: The potential of mesenchymal stem cells in diabetes mellitus. *Diabesity*. 3(1): 1–4.

Gutierrez-Aranda, I., Ramos-Meijia, V., Bueno, C., Munoz-Lopez, M., Real, P. J. A., Sanchez, L., Ligero, G., Garcia-Parez, J. L., & Menendez, P. (2010) Human induced pluripotent stem cells develop tetratoma mor efficiently and faster than human embryonic stem cells regardless the site of injection. *Stem Cells*. 28(9): 1568–1570.

Ham, R. G. (1965) Clonal growth of mammalian cells in a chemically defined, synthetic medium. *Microbiology*. 53: 288–293.

Hass, R., Cornelia Kasper, C., Stefanie Böhm & Jacobs, R. (2011) Different populations and sources of human mesenchymal stem cells (MSC): A comparison of adult and neonatal tissue-derived MSC. *Cell Communication and Signaling*. 9: 12.

Hay, M. E. (2009). Marine chemical ecology: Chemical signals and cues structure marine populations, cimunities, and ecosystem. *Ann Rev. Mar. Sci.* 1: 193–212.

Hayashi, Y., Tsujii, S., Tsujii, M., Nishida, T., Ishii, S., Iijima, H., Nakamura, T., Eguchi, H., Miyoshi, E., Hayashi N., Kawano, S., & Takeda, T. (2008) Topical transplantation of mesenchymal stem cells accelerates gastric ulcer healingin rats. *Am. J. Physiol. Gastrointest. Liver Physiol.* 294: G778–G786.

Heikkila, R. E., Manzino, L., Cabbat, F. S., & Duvoisin, R. C. (1984) Protection against the domaminergic neurotoxicity of 1-methyl-4-phenyl-1,2,5,6-tetrahtdropyridine by mnoamine oxidase inhibitors. *Nature.* 311: 467–469.

Heikkila, R. E., Winston, B., Cohen, G., & Barden, H. (1976) Alloxan induced diabetes, evidence for hydroxyl radicals as a cytotoxic intermediate. *Biochem Pharmacol.* 25: 1085–1092.

Heneka, M. T., Kummer, M. P., Stutz, A., Delekate, A., Schwartz, S., Vieira-Saecker, A., Korte, M., Latz, E., & Golenbock, D. T. (2012) NLRP3 is activated in Alzheimer's disease and contruibutes to pathology in AOO/PSI mice. *Nature.* 493(7434).

Hernberg, C. A. (1952) The bone structure in alloxan-induced diabetes mellitus in rats. *J. Int. Med.* 142(4): 274–283.

Hill, R. W., & Dacey, J. W. H. (2007) Processing of ingested dimethylsulfoniopropionate by mussels Mytilus edulis and scallops Argopecten irradians. *Mar. Ecol. Prog. Ser.* 343: 131–140.

Hill, R. W., Dacey, J. W., & Edward, A. (2000) Dimethlsulfoniopropionate in gaiant clams (Tridacnidae). *Biol. Bull.* 1999(2): 108–115.

Hoffmann, L., Brauers, G., Gehrmann, T., Haussinger, D., Mayatepek., E., Schliess, F., & Schwahn, B. C. (2013) Osmotic regulation of hepatic betaine metabolism. *Am J Physiol Gastrointest Liver Physiol.* 304: G835–G846.

Hollander, D., Transnawski, A., & Gergely, H. (1985) Protective effect of sucralfate against alcohol-induced gastric mucosal injury in the rat. *Gastroenterolog.* 88: 366–374.

Hosokawa, K., Kasai, R., Higuchi, K., Takeshita, S., Shimizu, K., Hamamoto, T., Honda, A., Honma A., Irino, M., Toda, K., Matsumura, A., Matsushita, M., & Takeda T. (1984). Grading scoresystem: A method of evaluation of the degree of senescence acceleratedmouse (SAM). *Mech. Aging Dev.* 26(1): 91–102.

Hosokawa, T. (2004) Immune system deficiencies in SAM. *Int. Cong. Ser.* 1260: 41–46.

Hosokawa, Y. (1999) Food ingestin in long span of life. In: Shou, H., & Yamamoto, M. (eds.) *Okinawa Prefecture Business Center for Academic Societies Japan (Japanese).* Tokyo, pp. 43–44.

Houghton, J., Morozov, A., Smirnova, I., & Wang, T. C. (2007) Stem cells and cancer. *Seminars in Cancer Biology*, 191–203.

Howlett, D. R., Richardson, J. C., Austin, A., Parsons, A. A., Bate, S. T., Davies, D. C., & Gonzalez, M. I. (2004) Cognitive correlations of Abeta deposition in male and female mice bearing amyloid precursor protein and presenilin-1 mutant transgenes. *Brain Res.* 1017(1–2): 130–136.

Huang, S., Xu, L., Sun, Y., Wu, T., Wang, K., & Li, G. (2015) An improved protocol for isolation and cultutre of mesenchymal stem cells from mouse bone marrow. *J. Orthopaedic Transl.* 3(1): 26–33.

Ichijiro, R., Kobayashi, H., Yoneda, S., Lizuka, Y., Matsumura, S., Honnda, T., & Toyoshima, F. (2017) Tbx3-dependent proliferation of transit-amplifying cells drives interfollicular epidermal expansion during pregnancy and regeneration. *J. Dermatol. Sci.* 86(2): e53–e54.

Iida, H. (1988) Studies on the accumulation of dimethyl-β-propiothetin and the formation of dimethylsulfide in aquatic organisms. *Bull. Tokai. Reg. Fish. Res. Lab* (Japanese). 124, 46–47, 52–53.

Ishigami, M. (1969) 87. Synthesis of organic compounds by high frequency discharge. *Proc. Japan. Acad.* 45(51): 399–405.

Ishikake, S. O., Yamahara, K., Sada, M., Harada, K., Mishima, K., Iwasaki, K., Fujiwara, M., Mizuno, H., & Hyakusoku, H. (2003) Mesengenic potential and future clinical perspective of human proceeded lipoaspirate cells. *J. Nippon Med. Sch.* 70(4): 300–306.

Ishikake, S., Ohnishi, S., Yamahara, K., Sada, M., Hrada, K., Mishima, K., Iwasaki, K., Fujiwara, M., Ishida, Y., Ogihara, Y., & Okabe, S. (1990) Effects of a crude extract of a marine dinoflagellate, containing dimethyl-β-propiothetin, on HCL-ethanol induced gastric erosions and gastric secreation in rats. *Japan Pharmacol.* 54: 333–338.

Jellinger, K. A. (2013) Neurological approaches to cerebral aging and neuroplasticity. *Dialogues Clin. Neurosci.* 15(1): 29–43.

Jenner, P., & Marsden, C. D. (1986) The actions of 1-methyl-4-phenyl-1,2,3,6-tetrahydroxy-pyridine in animals as a model of Parkinson's sdisease. *J Neural Transm.* Supple. 20: 11–39.

Jiang, Y., Vaessen, B., Lenvik, T., Blackstad, M., Reyes, M., & Verfailli, C. M. (2002) Multipotent progenitor cels can be isolated from postnatal murine marrow, muscle, and brain. *Experimenal Hematol.* 30(8): 896–904.

Jin, L., Deng, Z., Zhang, J., Yang, C., Liu, J., Han, W., Ye, P., Si, Y., & Chen, G. J. (2019) Mesentimal stem cells promote type 2 machrophage polarization to ameliorate the myocardial injury caused by diabetic cardiomyopathy. *Transl. Med.* 17: 251–264.

Jones, J., Lovett-Racke, A. E., Walker, C. L., & Sanders, V. M. (2015) CD_4+T cells and neuroprotection: Relevance to motoneuron injury and disese. *Neuroimmune Pharmacol.* 10: 587–594.

Kalaria, R., Mitchell, M. J., & Harik, S. I. (1987) Correlation of 1-methyl-4-Phenyl-1,2,3,6-tetrapyridine neurotoxin with blood-brain barrier monoamine oxidase activity. *Proc. Natl. Acad. Asi., USA.* 84: 3521–3525.

Kanehiro, M., Xin, H., Hoshino, K., Maemondo, M., Mizugichi, H., Hayakawa, T., Matsumoto, K., Nakamuura, T., Nukiwa, T., & Saijo, Y. (2007) Targeted delivery of NK4 to multi lung tumors by bone marrow-derived mesenchymal stem cells. *Cancer Gene Ther.* 14: 894–903.

Kangwan, N., Park, J. M., Kim, E. H., & Hahm, K. B. (2014) Quality of healing of gastric ulcers: Natural products beyond acid suppression. *World Gastrointest. Pathophysiol.* 5(1): 40–47.

Karsten, U., Wiencke, C., & Kirst, G. O. (1991) The effect of salinity changes upon the physiology of eulittoral green macroalgae from Antarctica and southern Chile: II intracellular inorganic ions and organic compounds. *Journal of Experimental Botany.* 42(245).

Karsten, U., Kuck, K., Vogt, K. C., & Kirst, G. O. (1996) Dimethylsulfoniopropionate production in phototrophic organisms and its physiological function as a cryoprotectant. In: Kien, R. P., Visscher, P. T., Keller, M. D., & Kirst, G. O. (eds.) *Biological and Environmental Chemistry of DMSP and Related Sulfonium Compounds.* Heiderberg, New York and London: Springer International Publisher, pp. 136–138, 143–153.

Keller, M.D., & Korieff-Bellows, W. (1996) Physiolgical aspect of the production of dimethylsulfoniopropionate (DMSP) by marine phytoplankton. In: Kien, R. P., Visscher P. T., Keller, M. D., & Kirst, G. O. (eds.) *Biological and environmental chemistry of DMSP and related sulfonium compound.* Heiderberg, New York and London: Springer International Publisher, pp. 131–142.

Kelly, D. J. (1998) The physiology and metabolism of the human gastric pathogen *Helicobacter pylori. Adv. Microb. Physiol.* 40: 137–189.

Khardori, R. (2017) Type 1 Diabetes mellitus. *Medscape.* (http://emedicine.medscape.com/article/117739-overview).

Kikuchi, T., Morizane, A., Doi, D., Magotani, H., Onoe, H., T. Hayashi., H., Mizuma, S., Takara, R., Takahashi, H., Inoue, S., Morita, M., Yamamoto, K., Okita, M., Nakagawa, M., Parmar., & Takahashi, J. (2017) Human iPS cell-derived dopaminergic neurons function in a primate Parkinson's disease model. *Nature*. 548(3): 592–596.

Kim, J., & Hematti, P. (2009) Mesenchymal stem cell-educated macrophages: A novel type of alternatively activated macrophages. *Exp. Hematol*. 7(12): 1445–1453.

Kitamura, S., Nagaya, N., & Ikeda, T. (2008) Allogeneic injection of fetal membrane-derived mesenchymal stem cells induces therapeutic angiogenesis in a rat model of hind ischemia. *Stem Cell*. 26(10): 2625–2633.

Knoepfler, P. S. (2009) Deconstructing stem cell tumorigenicity: A roadmap to safe regenerative medicine. *Stem Cells*. 27: 1050–1056.

Kono, T. M., Sims, E. K., Moss, D. R., Yamamoto, W., Ahn, G., Diamond, J. M., Tong, X., Day, K. H. A., Territo, P. R., Hanenberg, H., Traktuev, D. O., March, K. L., & Evans-Molina, C. (2014) Human adipose-derived stromal/stem cells protect against STZ-induced hyperglycemia: Analysis of hASC-derived paracrine effectors. *Stem Cells*. 32: 1831–1841.

Kuroda, Y., & Dezawa, M. (2014) Mesenchymal stem cells and their subpopulation, plurpotent muse cells, in basic research and regenerative medicine. *Anat. Rec*. (Hoboken), 297(1): 98–110.

Kusters, J. G., Vliet, A. H. M., & Kuipers, E. J. (2006) Pathogenesis of *Helicobacter Pylori* infection. *Clin. Microbiol. Rev*, 449–490.

Landeira-Fernadez, J., & Grijalva, C. V. (2004) Participation of the substatia nigra dopaminergic neurons in the occurrence of gastric mucosal erosion. *Physiol. Behav*. 81: 91–99.

Lazennec, G., & Lam, P. Y. (2016) Recent discoveries concerning the tumor-mesenchymal stem cell interactions. *Biochimica Biophysica Acta*. 1866(2016): 290–299.

Lee, J. K., Jin, H. K., Endo, S., Edward, H., Schuchman, E. H., Carter, J. E., & Bae, J. (2010) Intracerebral transplantation of Bone marrow-derived mesenchymal stem cells reduces amyloid-beta deposition and reduces memory deficits in Alzheimer's disease mice by modulation of immune responses. *Stem Cells*. 28: 329–343.

Lee, J. S. F., Poretsky, R. S., Cook, M. A., Reyes-Tomassini, J. J., Berejikian, B. A., & Goet, F. W. (2016) Dimethylsulfoniopropionate (DMSP) increase survival of larval Sablefish. *J. Chem Ecol*. 42: 533–536.

Lee, J-C., Hou, M-F., Huang, H-W., Chang, F-R., Yeh, C-C., Tang, J-Y., & Chang, H-W. (2013.) Marine algal natural products with anti-oxidative, anti-inflammatory, and anticancer properties. *Cancer cell International*. 13: 55–61.

Lee, R. H., Seo, M. J., Reger, R. L., Spees, J. L., Pulin, A. A., & Olson, S. D. (2006) Multipotent stromal cells from human marrow home to and promote repair of pancreatic islets and renal glomeruli in diabetic NOD/*scid* mice. *Pros. National. Acad Sci*. 103(46): 17348–17443.

Lenzen, S. (2008) The machanisms of alloxan- and streptozotocin-induced diabetes. *Diabetologia*. 51(2): 216–226.

Lenzen, S., & Munday, R. (1991) Thiol-group reactivity, hydrophilicity and stability of alloxan, its reduction products and its N-methyl derivatives and a comparison with ninhydrin. *Biochem Pharmacol*. 42: 1385–1391.

Lenzen, S., Freytag, S., & Panten, U. (1988) Inhibition of glucokinase by alloxan through interaction with sugar-binding site of the enzyme. *Mol Pharmacol*. 34: 395–400.

Lenzen, S., Tiedge, M., & Panten, U. (1987) Glucokinase in pancreatic B-cells and its inhibition by alloxan. *Acta Endocrinol* (Copenh). 115: 21–29.

Levenstein, S., Rosenstein, S., Jacobsen, K., & Jorgensen, T. (2015) Psychological stress increases risk for peptic uncer, regardless of Helicobacter pylori or use of nonsteroidal anti-inflammatory drugs. *Clin. Gastroenterol. Hepathol*. 13(3): 498–506.

Lever, M., & Slow, S. (2010) The clinical significance of betaine, an osmolyte with a key role in methyl group metabolism. *Clin Biochem*. 43: 732–744.

Li, J., Ezzelarab, M. B., & Cooper, D. K. C. (2012) Do mesenchymal stem cells function across species marries? Relevance for xenotrransplantation. *Xenotransplantation*. 19(5): 273–285.

Liew, L. C., Katsuda, T., Gailhouste, L., Nakagami, H., & Ochiya, T. (2017) Mesenchymal stem cell-derived excellular vesicles: A glimmer of hope in treating Alzheimer's diseases. *Int Immunol*. 29(1): 11–19.

Lilly, M. A., Fabie, J. E., Terhune, E. B., & Gallicano, G. I. (2016) Current stem cell based therapies in diabetes. *Am J Stem Cells*. 2016; 5(3): 87–98.

Liu, L., Eckert, M. A., Rizifar, H., Kang, D., Agalliu, D., & Zhao, W. (2013) From blood to the brain: Can systemically transplanted mesenchymal stem cells cross the blood brain marrier? *Stem Cell Internatinal*. 2013: 435093 (7 pages).

Liwen, C., Edwawd, E. T., Philip, Y. G. W., & Yaojiong, W. (2008) Paracrine factors of mesenchymal stem cells recruit macrophages and endotherial lineage cells and enhanve wound heeling. *PloS One*. 3(4): e1886, 1–12.

Lolmede, K., Campana, L., Vezzoli, M., & Rovere-Querini, L. P. (2009) Inflamatory and alternatively activated human macrophages attract vessel-associated stem cells, relying on separate HMGB1- and MMP-9-dependent pathways. *J. Leukoc. Biol*. 85(5): 779–787.

Loughlin, A., Kulkarni, M., Creane, M., Vaigham, E. E., Mooney, E., Shaw, G., Murphy, M., Dockery, P., Pandit, A., & O'Brien, T. (2013) Topical administration of allogeneic mesenchymal stromal cells and increases angiogenesis in the diabetic rabbit ulcer. *Diabetes*. 62: 2588–2594.

Lu, L., Shen, R. N., & Broxmeyer, H. E. (1996) Stem cells from bone marrow, umbilical cord blood and peripheral blood for clinical application: Current status and future application. *Crit. Rev. Oncol./Hematol*. 22: 61–78.

Madigal, M., Rao, K. S., & Riordan, N. H. (2014) A review of therapeutic effects of mesenchymal stem cell secreations and induction of secretory modification by different culture methods. *J Transl Med*. 12(1): 260–274.

Magner, G. C., Jarvis, M. F., & Rubin, J. G. (1986) L-dopa reverses the effects of MPP^+ toxicity. *Psychopharmacology*. 88(3): 401–402.

Makridakis, M., & Vlahou, A. (2010) Secretome proteomics for discovery of cancer biomarkers. *J Proteomics*. 73: 2291–2305.

Martin, W., Baross, J., Kelley, D., & Michael J Russell, M. J. (2008) Hydrothermal vents and the origin of life. *Nat Rev Microbiol*., 6(11): 805–814.

Maw, G. A. (1958) Thetin-homocysteine transmethylase. The distribution of the enzyme, studied with the aid of trimethyl-sulfonium chloride as substrate. *Biochem J*. 72: 602–608.

Mayer, K. D. (1966) The pathogenicity of the Ehrlich ascites tumour. *Br J Exp Pathol*., 47(5): 537–544.

McLenon, A. L., & DiTullio, G. R. (2012) Effects of increased temperature on dimethylsulfoniopropionate (D<SP) concentration and methionine synthetase activity in *Symbiodinium microadriaticum*. *Biogeochem*. 110: 17–29.

McNeil, S. D., Nuccio, M. L., & Hanson, A. D. (1999) Betaine and relate osmoprotectants, targets for metabolic engeneering of stress reasistance. *Plant Physiol*. 120: 945–949.

Melnicoff, M., Horan, P. K., & Morahan, P. S. (1989) Kinetics of changes in peritoneal cell populations following acute inflammation. *Cellular Imunolgy*, 118: 178–191.

Melzer, C., Yang, Y., & Haas, R. (2016) Interaction of MSC with tumor cells. *Cell Commun. Signaling*. 14(1): 20–32.

Meredith, G. E., & Pademacher, D. J. (2011) MPTP mouse models of Parkinson's disease: An update. *J Parkinsons Dis*. 1(1): 19–33.

Michael, J., Hongtao, W., Brigitta, P., Ellis, L, Rabiha, I., JamesWilliam, S. I., & Andrej, S. (2017) Inhibition of angiogenesis by nonsteroidal anti-inflammatory drugs: Insight into mechanisms and implications for cancer growth and ulcer healing. *Nat. Med.* 5(12): 1418–1424.

Miler, S. L., & Urei, H. C. (1959) Organic compound synthesis on the primitive earth. *Science*. 130(3370): 245–251.

Minematsu, H., & Nakajima, K. (2008) Significant effect of dimethylsulfoniopropionate on Parkinson's disease of senescence-accelerated mice induced by 1-methyl-4-phenyl-1,2,3,6-tetrahydropyridine. *J Nutr. Sci. Vitaminol.*, 54: 335–338.

Ministry Statics of health, Labour and Welfare, Japan. Average life. In: Health and Welfare Statistics Association (ed.). *Health and Statistics*. (2006). Tokyo (in Japanese). pp. 414–415.

Misra, D., Seo, P. H., & Cohen, H. J. (2004) Aging and cancer. *Clin Adv Hematol Oncol* 2:457–465.

Miyamoto, M. (1997) Characteristics of age-related behavioral changes in senescence-accelerated mouse SAMP8 and SAMP10. *Exp Gerontrol.* 32(1–2): 139–148.

Miyamoto, M., Kiyota, Y., Yamazaki, A., Nagaola, A., Matsui, T., Nagawa, Y., & Takeda, T. (1986) Age-related changes in learning and memory in the senescence-accelerated mouse. *Ohysiol. Behav*. 38(3): 399–406.

Mizuno, H., & Hyakusoku, H. (2003) Mesengenic potential and future clinical perspective of human processed lipoaspirate cells. *J Nippon Med Sch.*, 70(4): 3000–3006.

Mochizuki, Y., Onaga, T., Shimazaki, H., Shimikawa, T., Tsubo, Y., Kimura, R., Saiki, A., Sakai, Y., Isomura, Y., Fujisawa, S., Shibata, K., Hirai, D., Furuta, T., Kaneko, T., Takahashi, S., Nakazono, T., Ishino, S., Sakurai, Y., Kitsukawa, T., Lee, J., W., Lee, H., Jung, M., W., Babul, C., Maldonado, P., E., Takahashi, K., Arce-McShane, F., I., Ross, C., F., Sessle, B., J., Hatsopoulos, N., G., Brochier, T., Riehle, A., Chorley, P., Grün, S., Nishijo, H., Ichihara-Takeda, S., Funahashi, S., Shima, K., Mushiake, H., Yamane, Y., Tamura, H., Fujita, I., Naba, N., Kawano, K., Kurkin, S., Fukushima, K., Kurata, K., Taira, M., Tsutsui, K., Ogawa, T., Komatsu, H., Koida, K., Toyama, K., Richmond, B. J., & Shinomoto, S. (2016) Similarity in neuronal firing regimes across mammalian species. *J. Neuro*. 36(21): 5736–5747.

Momin, E., Mohyeldin, A., Zaidi, H. A., Vela, G., & Quifiones-Hinojosa, A. (2010) Mesenchymal stem cells: New approaches for the treatment of neurological diseases. *Current Stem Cell Res. Ther*. 5: 326–344.

Morley, J. E., Banks, W. A., Kumar, V. B., & Farr, S. A. (2004) The SAMP8 mouse as a model for Alzheaimer disease: Studies from Saint Louis University. *Int. Cong, Ser.* 1260: 23–28.

Mosser, D. M., & Edwards, J. P. (2008) Exploring the full spectrum of macrophage activation. *Nat. Rev. Immunol*. 8: 958–969.

Morgan, C., Lefkowitz, S. S., & Everse, J. (1998) Synergism of dimethoxybenzosemiquinone free radicals and CD4+ T-lymphocytes to suppress Ehtlich ascites tumor (44209). *Soc. Exp. Biol. Med*. 217: 89–96.

Nagai, T., & Yukimoto, T. (2003) Preparation and functional properties of beverages made from sea algae. *Food Chem*. 81: 327–332.

Nakajima, K. (1991) Dimethyl-β-propiothetin, a potent growth and molt stimulant for striped prawn. *Nippon Suisan Gakkaishi*. 57(9): 1717–1722.

Nakajima, K. (1991) Dimethyl-β-propiothetin, new potent resistive-agent against stress-induced gastric ulcers in rats. *J. Nutr. Sci. Vitaminol*. 37: 229–238.

Nakajima, K. (1992) Activation effect of a short term of dimethyl-β-propiothetin supplementation on goldfish and rainbow trout. *Nippon Suisan Gakkaishi*. 58(8): 1453–1458.

Nakajima, K. (1993) Dimethylthetin-and. Betaine-homocysteine. Methyltransferase. Activities from Livers of Fish, Chicken, and Mammals. *Nippon Suisan Gakkaishi.* 59(8): 1389–1393.

Nakajima, K. (1993) Effect of a sulfonium compound, dimethyl-β-propiothetin, on growth, molt, and survival of Kuruma prawn (in English). *Bull Koshien Univ.* 21(A): 11–15.

Nakajima, K. (1995) Stimulation effect of dimethylsulfoniopropionate on the growth of Kasumi salamander and sawagani. *Bull. Koshien Univ.*, 23(a): 7–10.

Nakajima, K. (1996) Biological and enviromental of DMSP and related sulfonium compounds. In: Kien, R. P., Visscher P. T., Keller, M. D., & Kirst, G.O. (eds.). Heiderberg, New York, Dordrecht and London, Springer International Publisher, Chapter 15, pp. 165–176.

Nakajima, K. (1998) Effects of dimethylsulfoniopropionate-added spinach leaves on growth, metamorphosis, and survival of juvenile Brown frog. *Bull. Koshien Univ.* 26A: 5–11.

Nakajima, K., (2001) Presentative effects of dimethylthetine, dimethyl-β-propiothetin and vitamin U on stress-induced gastric ulcers in rats. *Bull. Kosien Univ.*, 18(A): 15–22.

Nakajima, K. (2001a) A new determination method of topshell feeding activators (in English). *Bull. Koshien Univ.* 2(2): 2–8.

Nakajima, K. (2001b) Effects of sulfoniums, betaines, their components and amino acids on feeding behaviors of Top shells. *Bull Koshien Univ.* 29(A): 3–15.

Nakajima, K. (2001c) Activation of body movements of topshell and clams by dimethyl-sulfoniopropionate. *Fisheries Sci.* 67(4): 767–769.

Nakajima, K. (2001d) A method to determine the activated muscle strength by dimethyl-sulfoniopropionate in the clam, bivalve. *ITE Lett* 2(3): 424–426.

Nakajima, K. (2002) Activation of body movements of striped prawns by dimethylsulfonio-propionate. *ITE Lett.* 3: 104–107.

Nakajima, K. (2002) The long-term effect of dimethylsulfoniopropionate on the senility of senescence accelerated mouse-R/1 and P/8. *ITE Lett.* 3(3): 616–622.

Nakajima, K. (2003) Direct effects of high concentrations of dimethylsulfoniopropionate, vitamin E, and ferulic acid to the adult male and female senescence accelerated mouse (SAM-P/8). *ITE Lett.* 4(3): 357–361.

Nakajima, K. (2003) Simultaneous estimation of the loss of memory and lerning of senescence-accelerated-R/1 and P/8 mice. *ITE Lett.* 4(1): 87–91.

Nakajima, K. (2004) Ameliorating effects of dimethylsulfoniopropionate on the ulcers in senescence accelerated mouse (SAM P1) (in English). *Bull. Koshien Univ.* 32(A): 1–6.

Nakajima, K. (2004) Direct effects of high concentrations of dimethylsulfoniopropionate, vitamin E, and ferulic acid on the senility of aged senescence-accelerated mouse (SAMP8). *J. Nutr. Sci. Vitaminol.* 50: 231–237.

Nakajima, K. (2004) Effects of a sulfur-containing compound, dimethylsulfoniopropionate, on acute alloxan-diabetic rats. *ITE Lett.*5(4): 394–398.

Nakajima, K. (2005) Remarkable effects of dimethylsulfoniopropionate on the loss of learning and memory of senescence-accelerated mouse (SAMP1). *ITE Lett.* 6(1): 68–72.

Nakajima, K. (2005). Effects of dimethylsulfoniopropionate on growth and moving ability and on catecholamines in brain of carp. *ITE Lett.* 16: 65–69.

Nakajima, K. (2006) Hyperhomocysteinemia can be ameliorated by dimethylsilfoniopropionate in place of folic acid in mice. *J. Nutr. Sci. Vitaminol.* 5261–65.

Nakajima, K. (2015) Amelioration effect of a tertiary sulfonium compound, dimethyl-sulfonipropionate, in green sea algae on Ehrlich ascitic-tumor, solid tumor and related diseases. In: Se-Kwon, K. (ed.) *Hundbook of Anticancer Drugs of Marine Origin.* Heidelberg, New York, Dordrecht and London: Springer International Publisher. Chapter 11, pp. 205–238.

Nakajima, K. (2020). Epigenetic activation of animals, fish, crustaceans, mollusks, and plants by dimethlsulfoniopropionate in green sea algae. In: Se-Kwon, K. (ed.), *Encyclopedia of Marine Biothechnology*. Bern, London, Paris and Toronto: John Wiley & Sons Ltd. Publishing. Chapter 11, pp. 359–406.

Nakajima, K., & Minematsu, M. (2002) Effect of the diets containing green sea alga, Monostroma nitidum, on the senile phenomenon of scenescence-acceleraed mouse. *ITE Lett*. 3(3): 367–370.

Nakajima, K., & Minematsu, M. (2002) Suppressive effect of dimethylsulfoniopropionate on 3′-methyl-4-dimethylaminoazobenzene-induced liver cancers in rats. *ITE Lett*. 3(3): 371–74.

Nakajima, K., & Minematsu, M. (2006) Amelioration of dimethylsulfoniopropionate on the 1-methyl-4-Phenyl-1,2,3,6-tetrapyridine-induced Parkinson's disease of mice. *J. Nutr. Sci. Vitaminol*. 52: 70–74.

Nakajima, K., & Miyamoto, Y. (2007) Effects of Nerve growth factor and dimethylsulfonio-propionate in green sea algae on the out growth of neurites from Pheochromocytoma cells. *J Nutr. Sci Vitaminol*. 53: 441–445.

Nakajima, K., & Nakajima, Y. (2011) Carcinoma infection and Immune systems of Ehrlich ascitescarcinoma-bearing mice treated with structurally similar sulfonium compounds. *Biosci. Biotechnol. Biochem*. 75: 808–811.

Nakajima, K., & Tani, J. (2005) Relation of moving ability to catecholamines accumulated by dimethylsulfoniopropionate in rats. *ITE Lett*. 6: 160–165.

Nakajima, K., & Tani, J. (2006) Relation of stimulation of moving ability to accumulation of catecholamines by dimethylsulfoniopropionate in mice. *ITE Lett*. 7: 92–98.

Nakajima, K., Minematsu, M., & Miyamoto, Y. (2008) Inhibition of the outgrowth and elongation of neurites deom Pheochromocytoma cells by 1-methyl-4-phenyl-1,2,5,6-tetrahydroxypyridine and preventive effects of dimethylsulfoniopropionate in the presence of nerve growth factor. *J. Nutr. Sci. Vitaminol*. 54: 176–180.

Nakajima, K., Nakajima, Y., & Tsujiwaki, S. (2015) Sulfobetaine (dimethylsulfoniopropionate) and glycine betaine show imcompatible involvement in crucial Ehrlich ascites carcinoma in mice. *Ant Res*. 35: 1474–1480.

Nakajima, K., Ohgushi, T., & Shibata, K. (1999) Effects of rice bran fluid treated with enzymes and *B. Bifidum* on alloxandiabetes and stress-induced gastric ulcers in rats. *Ant. Res*. 19: 3798.

Nakajima, K., Ohgushi, T., & Shibata, K. (2001) Long-term effects of rice bran treated with enzymes and *Bifidobacterium bifidum* on acute alloxan-daibetic rats. *ITE Lett*. 2(5): 63–69.

Nakajima, K., Tsujiwaki, S., & Nakajima, K. (2014) A tertiary sulfonium compound, dimethylsulfoniopropionate in green sea algae, completely syppress cruial Ehrlich ascites carcinoma in mice. *Anticancer Res*. 34(8): 4045–4050.

Nakajima, K., Uchida, A., & Ishida, Y. (1989) A new feeding attractant, dimethyl-β-propiothetin, for freshwater fish. *Nippon Suisan Gakkaishi*. 55(4): 689–695.

Nakajima, K., Uchida, A., & Ishida, Y. (1990) Effect of a feeding attractant, dimethyl-β-propiothetin, on growth of marine fish. *Nippon Suisan Gakkaishi*. 56(7): 1151–1154.

Nakajima, K., Yokoyama, A., & Nakajima, Y. (2009) Anticancer effect of a tertiary sulfonium compound, dimethylsulfoniopropionate, in green sea algae on Ehrlich asites carcinoma-bearing mice. *J. Nutr. Sc. Vitaminol*. 55: 434–438.

Nakamizo, A., Marini, F., Amano, T., Khan, A., Studeny, M., Gumin, J., Chen, J., Hentschel, S., Vecil, G., Dembinski, J., Andreeff, M., & Lang, F. F. (2005) Human bone marrow-derived mesenchymal stem cells in the treatment of gliomas. *Cancer Res*. 65(8): 3307–3318.

Nakamura, A., Kaneko, N., Villemagne, L., Kato, J., & Doecke, J. (2018) High performance plasma amyloid-β biomarkers for Alzheimer's disease. *Nature*. 554: 249–254.

Nakamura, K., Iida, H., &Tokunaga, T. (1981) Offensive odor of crab *Palalomis multistima. Bull. Japn. Soc. Sci. Fish.* 47(9): 1241.
Navegantes, K.C., de Souza Gomes, R., Pereira, P. A. T., Czaikoski, P. G., Azevedo, C. H. M., & Monteiro, M. C. (2017) Immune modulation of some autoimmune diseases: the critical role of macrophages and neutrophils in the innate and adaptive immunity. *J Transl Med* 15, 36.
Navegantes, K. C., Gomes, R. S., Priscilla P. A. T., Czaikoski, P. C. G., Azevedo, C. H. M., & Monteirol, M. C. (2017) Immune modulation of some autoimmune diseases: The critical role of macrophages and neutrophils in the innate and adaptive immunity. *J Transl. Med.* 15: 36–57.
NIH RePort. (2013) Accelerated Medicines Partnership (AMP). *Parkinson's Disease.*
Nomura, Y., Wang, B. X., Namba, T., & Kaneko, S. (1989) Biochemical changes related to aging in the senescence-accelerated mouse. *Exp. Gerontol.* 24(1): 49–55.
Obeso, J. A., Olanow, C. W., Rodriquez-Oroz, M. C., Krack, P., Kumar, R., & Lang, A. (2001) Deep-Brain stimulation of the subthalamic nucleus or the pars interna of the globus pallidus in Parkinson's disease. *New Eng. J. Med.* 345(13): 956–963.
Ortholog?gene?3956[group] (2018).
Osakabe Shinnnou, Hujiwara H. Awata M., Shimotuke M. Taihou Rituryou10th (701).
Otte, M., Wilson, G., Morris, J. T., & Moran, B. M. (2004) Dimethylsulfonio-propionate (DMSP) and related compounds in higher plants. *J. Exp. Botany*. 55(404): 1919–1925.
Ozaslan, M., Karagoz, I. D., Kilic, I. H., & Guldur, M. E. (2011) Ehrlich ascites carcinoma. *African Biotechnol.* 10: 2375–2378.
Palmer, R. D. (2012) *Seafood in Japan*. Responsible seafood advocate, p. 2.
Park, D., Yang, G., Bae, D., K., Lee, S., H., Yang, Y-H, Kyung J, Kim, D, Choi, K-K., Choi, K., Kim, S. U., Kang, S., K., Ra, J., C., & Kim, Y-B. (2013) Human adipose tissue-derived mesenchymal stem cells improve cognitive function and physical activity in ageing mice. *J. Neurosci. Res.* 91(5): 660–670.
Park, H., Li, Z., Yang, X. O., Chang, S. H., Nurieva, R., Wang, Y-H., Hood, L., Zhu, Z., Tian, Q., & Chen Dong, C. (2005) A distinct lineage of CD4 T cells regulates tissue inflammation by producing interleukin 17. *Nat. Immunol.* 16(11): 1133–1141.
Pcana, A., Niet-Jimenez, C., Pandiella, A., & Templeton, A. J. (2017) Neutrophils in cancer: Prognostic role andtherapeutic strategies. *Molecular Cancer*. 16: 137–143.
Perdiguero, E. G., Klapproth, K., Schulz, C., Busch, K., Bruijin, M., Rodewald, H. R., & Geissman, F. (2015) The origin of tissue-resident macrophages: When an erythro-myeroid progenitor is an embryo-myeroid progenitor. *Immunity*. 1023–1024.
Phaniendra, A., Jestadi, D., B., & Periyasamy, L. (2015) Free radicals: Properties, sources, targets, and theor implication in various diseases. *Ind. Clin. Biochem.* 30(1): 11–26.
Poll, D., Parekkadan, B., Cho, C. H., Berthiaume, F., Nahmias, Y., Tilles, A., & Yarmamush, M. (2008) Mesenchymal stem cell-drived molecules directly modulate hepatocellular death and regeneration in vivtro and in vivo. *Hematology*. 47(5): 1634–1643.
Pombero, A., Garcia-Lopez, R., & Martinez, S. (2016) Brain mesenchymal stem cells: Physiology and pathological implications. *Develop. Growth Deffer*. 58: 469–480.
Provinciali, M., Cardell, I. M., Marchegiani, F., & Pierpaoli, F. E. (2013) Impact of cellular scenescence in aging and cancer. *Curr Pharm Des*. 19: 1699–1709.
Rabinovitch, M., & De Stefano, M. J. (1973) Particle recognition by cultivated macrophages. *J. Immun*. 110: 695–701.
Rabinovitch, M., Manejias, E., Russo, M., & Abbey, E. E. (1977) Increased spreading of macrophages from mice treated with interferon inducer. *Cellular Immunol.* 29: 86–95.
Radmer, R. J. (1996) Algal diversity and commercial algal products. *BioSci.* 46(4): 263–270.

Raina, J. P., Tapiolas, D. M., Foret, A., Lutz, A., Abrego, D., Ceh, J., Seneca, F. O., Clode, P. L., Bourne, D. G., Willis, B. T., & Motti, C. A. (2013) DMSP biosynthesis by an animal and its role in coral thermal stress response. *Nature Letter*. 502(7473): 677–680. terrestrial sulfonium compounds from 63 specie of marine algae. *Biochem. Syst*. 20: 373–388.

Rao, M. S., Velagaleti, G., & Troyer, D. (2006) Human umbilical cord matrix stem cells: Preliminary characterization and effect of transplantation in a rodent model of Parkinson's disease. *Stem Cells*. 24(3): 781–792.

Ratajczak, N. Z., Kim, C. H., Vojanowski, W., Janowska-Wieczorek, A., Kucia, M., & Ratajczak, J. (2010) Innate immune as orchestrator of stem cell mobilization. *Leukemia*. 24: 1667–1675.

Rates, S. M. K. (2001) Plants as source of drugs. *Toxicon*. 39: 603–613.

Razvodovsky, Y. (2006) Aggegate level association between alcohol and the peptic ulcer mortality rate. *Alcoholism*. 42(2): 61–69.

Reddy, V. N., Nagarathna, P. K. M., & Divvya, M. (2013) A study on the survival time, sex, weight and blood content of the Ehrlich ascites carcinoma involved tumor. *Int. J. Phaem. Res*. 3: 27–34.

Reed, R. H. (1983) Measurement and osmotic significance of dimethylsulfonioproponate in marine macroalgae. *Mar. Biol. Lett*. 4(3): 173–191.

Ridge, S. M., Sullivan, F. J., & Glynn, S. A. (2017) Mesenchymal stem cells: Key players in cancer progression. *Molecular Cancer*. 16: 31–41.

Ripple, W. J., Wolf, C., Newsome, T. M., Galetti, M., Alamgir, M, Crist, E., Mahmoud, M. I., Laurance, W. F., Kapel, T. Y. S. et al. (15364 scientists) (2017) World scientists' warning to humanity: A second notice. *BioScience Bix*. 125(12).

Rizzo, E., Moura T. F.C., Yoshimi, Y., & Bazzoli.S. (1998) Oocyte surface in four teleost fish species postspawning and fertilization. *Braz. Arch. Biol. Technol*. 41(1).

Rother, K. I. (2007) Diabetes treatment—bridging the divide. *N. Engl. J. Med*. 356(15): 1499–1501.

Rozemuller, H., Prins, H. J., Naaijkens, B., Staal, J., Buhring, H. J., & Martens, A. C. (2010) Prospective isolation of mesenchymal stem cells from multiple mammalian species using cross-reacting anti-human monochlonal antibodies. *Stem Cells Dev*. 19(12): 1911–1921.

Saleh, S., El-Ridi, M., Atia, F., El-Kotb, S., & El-Gizawy, E. (2013) Treatment of experimentally-induced peptic ulcer in rats by hematopoietic stem cells. *Med. J. Cairo Univ*. 81(2): 229–236.

Salisali, A., & Nathan, M. (2006) A review of types 1 and 2 diabetes mellitus and their treatment with insulin. *Am. J. Ther*. 13(4): 349–361.

Schlaepfer, T. E., Cohen, M. X., Frick, C., Kosel, M., Brodesser, D., Axmacher, N., Joe, A. Y., Kreft, M., Lenartz, D., & Sturm, V. (2007) Deep brain stimulation to reward circuitry alleviates anhedonia in refractory major depression. *Neuropsychopharmacology*. 1–10.

Schwartz, A. W. (1981) Chapter 2. Chemical evolution-The genesis of the first orgnic compounds. *Elsevier Oceanogr. Ser*. 31: 7–30.

Segura, J. A., Barbero, L. G., & Marquez, J. (2000) Ehrlich ascites tumour unbalances splenic cell populations and reduces responsiveness of T cells to Staphylococcus aureus enterotoxin B stimulation. *Immunol. Lett*. 74: 111–115.

Sell, S. (1993) Cellular origin of cancer: Dedifferentiation or stem cell maturation arrest? *Environ Health Perspect*. 101(Suppl 5): 15–26.

Shoda, L. K. M., Young, D., Ramanujian, S., Bluestone, J. A., Eisenbarth, G. S., Mathis, D., Rossini, A. A., Campbell, S. E., Kahn, R., & Kreuwel, H. T. C. (2005) A comprehensive Review of interventions in the NOD mouse and implications of translation. *Immunity*. 23: 115–126.

Sico, C., Reverber, D., Balbi, C., Ulivi, V., Principi, E., Pascucci, L., Becherini, P., Bosco, M. C., Varesio, L., Franzin, C., Pozzobon, M., Cancedda, R., & Tasso, R. (2017) Mesenchymal stem cell-derived extracellular vesicles as mediators of anti-inflammatory effect: Endorsement of macrophage polarization. *Atem Cell Transl. Med.* 6: 1018–1028.

Simon, S. I., & Kim, M. H. (2010) A day (or 5) in a neutrophil's life. *Blood.* 116(4): 511–512.

Singel, K. L., & Segel, B. H. (2016) Neutrophils in the tumpr microevent: Truing to heal the wound that cannot heal. *Immunolo. Rev.* 273(1): 329–343.

Snyder, S. H. (1984) Cluestoaetiology from a toxin. *Nature.* 311: 314.

Snyder, S. H., & D'Amato, R. J. (1985) Predicting Parkinson's disese. *Nature.* 317: 198–199.

Solari, N., Bonito-Oliva, A., Fisone, G., & Brambilla, R. (2018) Understanding cognitive deficits in Parkinson's desease: Lessons from preclinical animal models. *CSHL Press.* 20: 592–600.

Spelman, S., Lavine, K. J., & Randolph, G. J. (2014) Origin and function of tissue macrophages. *Immuniy.* 41: 21–35.

Stacheischeid, H., Wulf-Goldenberg, A., Eckert, K., Jensen, J., Edsbagge, J., Biorquist, M., Strehl, R., Jozefczuk, J., Prigione, A., Adjaye, J., Urbaniak, T., Bussmann, P., Zeilinger K., & Gerlach J. C. (2013) Teratoma formation of human embryonic stem cells in three-dimensional perfusion culture bioreactors. *J. Tissue Eng. Regerat. Med.* 7(9): 729–741.

Stanners, C. P., Eliceli, G. L., & Green, H. (1971) Two types of ribosome in mouse-hamster hybrid cells. *Nature New Biol.* 230: 52–54.

Stefels, J. (2000) Physiological aspects of the production and conversion of DMSP in marine algae and higher plants. *J. Sea Res.* 43: 183–197.

Stone, D. K., Reynolds, A. D., Mosley, R. L., & Gendelman, H. E. (2009) Innate and adaptive immunity for the pathobiology of Parkinson's disease. *Antioxid. Redox Signal.* 11: 2151–2166.

Strakova, J., Gupta, S., Kruger, W. D., Diliger, R. N., Tryon, K., Li, L., & Garroul, T. A. (2011) Inhibition of betaine-himocysteine Smethyltransferase in rats causes hyperhomocysteinemia and reduces cystanionine-β synthase activity methylation capacity. *Nutr Res.* 31: 563–571.

Strehler, E. (1955) Gagenwartigen Stand der seg. "Vitamin U"-Therapie der Magenderm-Ulcera. *Gastroenterlogia*, 84: 119–131.

Sunda, W., Klieber, D. J., Kien, R. P., & Huntsman, S. (2002) An antioxidant function for DMSP and DMS in marine algae. *Nature.* 418(6895): 317–320.

Szkudelski, T. (2001) The mechanism of alloxan and streptozotocin action in B cells of the rat pancreas. *Physiol. Res.* 5: 536–546.

Takagi, K., Kasuya, Y., & Watanabe, K. (1964) Studies on the drugsfor peptic ulcers. A rliable method for producing stress ulcer in rats. *Chem. Pharm. Bull.* 12(4): 465–472.

Takahashi, K., & Yamanaka, S. (2006) Induction of pluripotent stem cells from mouse embryonic and adult fibroblast cultures by defined factors. *Cell.* 126(4): 663–676.

Takahashi, K., Kakuda, Y., Munemoto, S., Yamazaki, H., Nozaki, I., & Yamada, M. (2015) Differentiation of donor-derived cells into microglia after umbilical cord blood stem cell transplantation. *J. Neuropathol. Exp. Neurol.* 74(9): 862–866.

Takahashi, K., Tanabe, K., Ohnuki, M., Narita, M., Ichisaka, T., Tomoda, K., & Yamanaka, S. (2007) Induction of pluripotent stem cells from adult human fibroblasts by defined factors. *Cell.* 131(5): 861–872.

Takamiya, K. (1993) Malignancy of Ehrlich ascites carcinoma and effect of vitamin C on EAC, *Vitamins.* 67(10): 525–530 (Japanese).

Takeda, T., Hosokawa, M., & Higuti, K. (1991) Senescence-accelerated mouse (SAM): A novel murine model of accelerated senescence. *Am. Geom. Soc.* 39: 911–919.

Takeda, T., Hosokawa, M., Takeshita, S., Irino, M., Higuchi, K., et al. (1981) A new murine model of accelerated senescence. *Machnisms of Age. Devel.* 17: 183–194.

Tamaki, Y., Nakahara, T., Ishikawa, H., & Sato, S. (2013) In vitro analysis of mesenchymal stem cells derived from human teeth and bone marrow. *Odontology*. 101(2): 121–132.

Tanaka, T., Ohara, J., Takezoe, K., Ando, T., & Hokari, N. (1969) Healing effects of methylmethionine chloride on gastric ulcers and chronic dyspepsia. *New Remedies Ther*. 8: 20–22.

Tapiolas, D. M., Raina, J-B., Lutz, A., Willis, B. L., & Motti, C. A. (2013). Direct measurement of dimethylsulfoniopropionate (DMSP) in reef-building corals using quantitative nuclear magnetic resonance (NMR) spectroscopy. *Journal of Experimental Marine Biology and Ecology*, 443: 85–89.

Teixeira, F. G., Carvalho, M. M., Sousa, N., & Salgado, A. J. (2013) Mesenchymal stem cells secretome: A new paradigm for central nervous system regeneration? *Cell Mol Life Sci.*, 70(20): 3871–3882.

The current US Census Bureau world population (2019) The U. S. and World Population Clocks. https://www.census.gov/popclock/.

Tian, X., Brookes, O., & Battaglia, G. (2017) Pricytes from mesenchymal stem cells as a model for the blood-brain barrier. *Scientif. Rep.* 7: 39676–39683.

Tieu K. (2011) A guid to neurotoxin animal models of Parkinson's disease. Cold Spring Harbor Perspectives in medicine. *Cold Spring Harb. Perspect. Med.* 1: a009316.

Tieu, K. (2011) A guide to neurotoxic animal models of Parkinson's disease. *Cold Spring Harb. Perspect. Med.* 11: a009316. (CSHL Press) pp. 1–20.

Tischler, A. R., & Green, L. (1975) Nerve growth factor-induced process formation by cultured rat pheochromocytoma cells. *Nature*. 258: 341–342.

Tischler, A. R., & Green, L. (1978) Morphologic and cytochemical properties of a clonal line of rat adrenal Pheochromocytoma cells which respond to nerve growth factor. *Lab. Invest*. 39: 77–89.

Toichi, E., Katoh, H., Hosokawa, T., & Hosonoda, M. (1994) Immune activities in SAM mice: cellular and genetic basis for the impared responsiveness of helper T cells in humoral immunity. In: Takeda, T. (ed.) *The SAM Model of Senescence*. Amsterdam and New York: Excepta Medica, pp. 41–46.

Tokunaga, T., Iida, H., & Nakamura, K. (1977) Formation of dimethylsulfide in Antarktic krill, *Euphausia superuba*. *Bull. Jap. Soc. Sci. Fish.* 43(9): 1209–1217.

Tollefsbol, T. O. (2014) Dietary epigenetics in cancer and aging. *Cancer Treat Res*. 159: 257–267.

Truong, D., Hindmarsh, W., & O'Brien, P. J. (2009) The molecular mechanisms of diallyl disulfideand diallyl sulfide induced hepathocyte cytotoxicity. *Chemo-Biol Interact*. 180: 79–88.

Tsuboi, I., Harada, T., & Aizawa, S. (2016) Age-related functional changes in hematopoietic microenvironment. *J. Phy. Fitness Sports Med.* 5(2): 167–175.

Tsuda, H., Ohshima, Y., Nomoto, H., Fujita, K., Matsuda, E., Iigo, M., Takahashi, N., & Moore, M. (2004) Cancer preventionby natural compounds. *Drug Metab Pharmacokimet*. 19: 345–326.

Tsukada, H., Fujiwara, S., Ezoe, M., & Fujiwara, F. (1968) Appraisal of several biological properties of Ehrlich ascites tumor to malignancy. *Gann*, 54: 311–321.

Uboga, N. V., & Price, J. L. (2000) Formation of diffuse and fibrillar tangles in aging and early Alzheimer's disease. *Neurobiol. Aging*. 21: 1–10.

Uccelli, A., Moretta, L., & Pistro, V. (2008) Mesenchymal stem cells in health and disease. *Nature*. 8: 726–736.

Uchida, H., Morita, T., Niizuma, K., Kushida, Y., Kuroda, Y., Wakao, S., Satake, H., Matsuzaka, Y., Mushikake, H., Tominaga, T., Borlongan, C. V., & Dezawa, M. (2015) Transplantation of unique subpopulation of fibroblasts, Muse cells, ameliorates experimental stroke possibly via robust neuronal differentiation. *Stem Cells*. 34(1): 160–173.

Uramoto, T., Ohno, T., & Ishihara, T. (1990) Gasric mucosal protection induced by restraint and water immersion stress in rats. *J. Pharm*. 54(3): 287–298.

Van Alstyne, K. L., Dominiquuelll, V. J., & Muller-Parker, G. (2009) Is dimethyl-sulfoniopropionate (DMSP) produced by the synbionts or the host in an anemone-zooxanthella symbiosis? *Coral Reefs*. 28(1): 167–176.

Van Alstyne, K. L., Schupp, P., & Slattery, M. (2006) The distribution of dimethylsulfoniopropionate in tropical pacific coral reef invertebrates. *Coral Reefs*. 25(3): 321–327.

Van Alstyne, K. L., Shupp, P., & Slattery, M. (2006) The distribution of dimethylsulfoniopropionate in tropical Pacific coral reaf invertebrate. *Coral Reefs*. 25(3): 321–327.

Wagner, G. C., Jarvis, M. F., & Rubin, J. G. (1986) L-dopa reverses the effects of MPP+ toxicity. *Psychopharmacology* (Berl). 88(3): 401–402.

Wang, L., Wang, F. S., & Gershwin, M. E. (2015) Humanautoimmune diseases: A comprehensive update. *J. Int. Med*. 278: 369–395.

*Web*MD. (2018) *Diabetes Health Center*. Types of diabetes mellitus.

Weiss, M. L., Medicetty, S., Bledsoe, A. R., Rachakatla, R. S., Choi, M., Merchav, S., Luo, Y., Rao, M. S., Velagaleti, G., & Troyer, D. (2006) Human umbilical cord matrix stem cells: Preliminary characterization and effect of transplantation in a rodent model of Parkinson's disease. *Stem Cells*. 24(3): 781–792.

Westerink, R. H. S., & Ewing, A. G. (2008) The PC 12 cell as model for neurosecretion. *Acta Physiol*. 192(2): 273–285.

White, R. H. (1982) Analysis of dimethyl sulfonium compounds in marine algae. *J. Mar. Res*. 40: 529–537.

WHO. (2014) *Ageing and Life-Cource*. Facts About Ageing.

WHO. (2015) *Ageing and Health*.

WHO. (2015–01) *Diabetes*. Fact Sheet (Report).

WHO. (2016) *Dementia*. Assessed April.

WHO. (2017) *Department of Information*. Geneva: Evidence and Research WHO, p. 55.

WHO. (2018) *Cancer*. Fact Sheet.

WHO methods and data sources for country level causes of death. (2000–2015). *Department of Information*. Geneva: Evidence and Researh. WHO, January 2017, *Ageing and Health*, p. 7.

Wichmann, H., Brinkhoff, T., Simon, M., & Richter-Landsberg, C. (2016) Dimethylsulfonio-propionate promote process outgrouth in neural cells and exerts protective effects against tropodithietic acid. *Marine Drugs*. 14: 89–102.

Wichmann, H., Vocke, F., Brinkhoff, T., Simon, M., & Richter-Landsberg, C. (2015) Cytotoxic effects of tropoditheetic acid on mammaloan clonal cell lines of neuronal and glia origin. *Marine Drugs*. 13: 7113–7123.

Willcox, B. J., & Willcox, D. C. (2014) Caroric restriction, caloric restriction mimetics, and healthy aging in Okinawa: Controversies and clinic implication. *Cur. Opinion Clin. Nutr. Metab. Care*. 17: 51–58.

William, D. C., Leontiadis, G. I., Howden, C. W., & Moss, S. F. (2017) Treatment of Helicobacter pylori infection. *Am. J. Gastroenterol*. 112: 212–238.

World population sections. Current world population.

World population sections. Current world population. p. 6.

Wynn, T. A., & Vannella, K. M. (2016) Macrophages in tissue repair, regeneration, and fibrobrosis. *Immunity*. 44(3): 450–462.

Wynn, T. A., Chawla, A., & Pollard, J. W. (2013) Macrophage biology in development, homeostasis and disease. *Nature*. 496: 445–455.

Wyss-Coray, T., Masliah, E. M., McConlogue, L., Johnson-Wood, K., Lin, L. & Mucke, C. L. (1997) Amyloidogenic role of cytokine TGF-β1 in transgenic mice and in Alzheimer's disease. *Nature,* 389(9): 603–606.

Xia, X., Chlu, P. W. Y., Lam, P. K., & Lau, J. Y. W. (2018) Secretome from hypoxia-conditioned adipose-derived mesenchymal stem cells promotes the healing of gastric mucosal injury in a rodent model. *Biochimia Biophysica Acta*. 1864(1): 178–188.

Xu, H., Wang, B., Ono, M., Kagita, A., Fujii, K., Sasakawa, N., Ueda, T., Gee, P., Nishikawa, S., Nomura, M., Kitaoka, F., Takahashi, T., Okita, K., Yoshida, Y., Kaneko, S., & Hotta, A. (2019) Targeted disruption of HLA genes via CRISPR-Cas9 generates iPSCs with enhanced immune compatiilty. *Cell Stem Cell*. 24(4): P566–578.E7.

Yacoub-Youssef, M., Saclier, H., Macky, A. L., Arnold, L., Ardjoune, H., Magnan, M., Saihan, S., Chelly, J., Pavlath, G. K., Mounier, R., Kjaer, M., & Chazaud, B. (2013) Differentially activated macrophages orchestrate myogenic precursor cell fate during human skeletal muscle regeneration. *Stem Cells*. 31: 38964–38973.

Yagi, H., & Kitagawa, Y. (2013) The role of mesenchymal stem cells in cancer development. *Front. Genet*. 4(261): 1–6.

Yagi, H., Katoh, S., Akiguchi, I., & Takeda, T. (1988) Age-ralated deterioration of ability of acquisition in memory and leaning in senescenced mouse: SAM-P8 as as animal model of disturbances in recent memory. *Brain Res*. 474: 86–93.

Yamada, Y., Wakao, S., Kushida, Y., Minatoguchi, S., Mikami, A., Higashi, K., Baba, S., Shigemoto, T., Kuroda, Y., Kanamori, H., Amin, M., Muramatsu, C., Dezawa, M., & Nishiguchi, S. (2018) SIP-S1PR2 axis mediates homing of Muse cells into damaged heart for long-lasting tissue repair and functional recovery after acute myocardial infarction. *Circulat. Res*. 122: 1069–1083.

Yamanaka, S. (2012) Induced pluripotent stem cells: Past, present, & future. *Cell Stem Cell*. 10(6): 678–684.

Yan, S. D., Fu, J., Soto, C., Chen, X., Zhu, H., Al-Mohanna, F., Collison, K., Zhu, A., Stem, E., Saido, T., Tohyama, M., Ogawa, S., Roher, A., & Stem, D. (1997) An intracellular protein that binds amyloid-βpeptide and mediates neurotpxicity in Alzheimer's disease. *Nature*. 359(16): 689–695.

Yasuhara, T., Kameda, M., Sasaki, T., Tajiri, N., & Date, I. (2017) Cell therapy for Parkinson's disease. *Cell Transpl*. 26(9): 1551–1559.

Yoch, D. C. (2002) Dimethylsulfoniopropionate: Its souces, role in the marine food web, and biological degradation to dimetylsulfide. *Appl. Environ. Microbiol*. 68(12): 5804–5815.

Yoo, E. J-H., Feketeová, L., Khairallah, G.N., White, J. M., & O'Hair, R. A. J. (2011) Structure and unimolecular chemistry of protonated sulfur betaines, $(CH_3)_2S^+(CH_2)_nCO_2H$ ($n = 1$ and 2). *Organic & Biomolecular Chemistry*, 8: 2557–3056.

Yoshinaka, Y., & Nakamura, M. (1981) Conbined effects of methylmethionione sulfonium chloride and antacids on various experimental gastric ulcers in rats. *Pharmacometrics*. 21: 921–925.

Yu, Y., Liu, Y., Zong, C., Yu, Q., Yang, X., Liang, L., Ye, F., Nong, L., Jia, Y., Lu, Y., & Han, Z. (2016) Mesenchymal stem cells with Sirt1 overexpression suppress breast tumor growth via chemokine-dependent natural killer cells recruitment. *Scientific Reprt*. 16: 35998–36009.

Zamarron, B. Z., & Chen, W. (2011) Dual roles of immune cells and their factors in cancer development and progression. *Int. J. Biol. Sci*. 7(5): 651–658.

7 Seaweeds as Sustainable Sources for Food Packaging

Y. S. M. Senarathna, I. Wickramasinghe, and S. B. Navaratne

CONTENTS

7.1 INTRODUCTION

The use of natural materials in food packaging has been practiced for centuries. With the industrial revolution, synthetic polymers were invented and were applied in the packaging industry too. However, the aftermath of this invention was the accumulation of synthetic polymers which have become a huge environmental issue even for unborn future generations (Mohamed et al. 2020). Hence, it will be vital, if we can go for natural food packaging materials like in the past as it directly and positively affects the hostile environment formed

DOI: 10.1201/9781003303916-7

by synthetic polymers, and they are highly biodegradable and environmentally friendly. So also, their user-friendly nature and abundance in the natural habitat they can make more advantageously in applying to food packaging (Galus et al. 2020). Hence, the development of biodegradable and edible packaging materials from biopolymers with innovative techniques has gained attention in the recent past for food packaging and extension of shelf-life, which is the real barrier still associated with the bio-polymers against synthetic packaging (Yerramathi et al. 2021). Besides, developing a packaging material with comparable properties to synthetic packaging material is a challenge (Sid et al. 2021). Further, when selecting a biopolymer for the development of packaging films and coatings, it is very important to consider the physical, mechanical and barrier properties that can be achieved in the packaging film or the coating. Polysaccharides, proteins and lipids are the common biopolymers that are considerably investigated in the application as food packaging materials (Qin et al. 2019; Sharmin et al. 2021). Among them, polysaccharides-based materials are attracted more attention owing to their better compatibility, outstanding film formation capability and high abundancy at low cost (Carissimi et al. 2018). Cellulose derivatives, pectin, starches, chitosan, pullulan and several seaweed-based polysaccharides are the popular polysaccharides that are noticeably studied for developing natural packaging materials for foods (Falguera et al. 2011).

Seaweeds, marine macroalgae, are a group of lower plants that exhibit a variety of biological properties desirable for making biofilms. Under this context, seaweeds can be grouped into three main varieties considering their pigmentation and the main polysaccharide type as shown in Figure 7.1. Green algae which are identified as Chlorophyta contains chlorophyll a, b and ulvan as the main polysaccharide meanwhile brown algae (Phaeophyta) which contains carotenoid fucoxanthin and the polysaccharides namely alginate, fucoidan

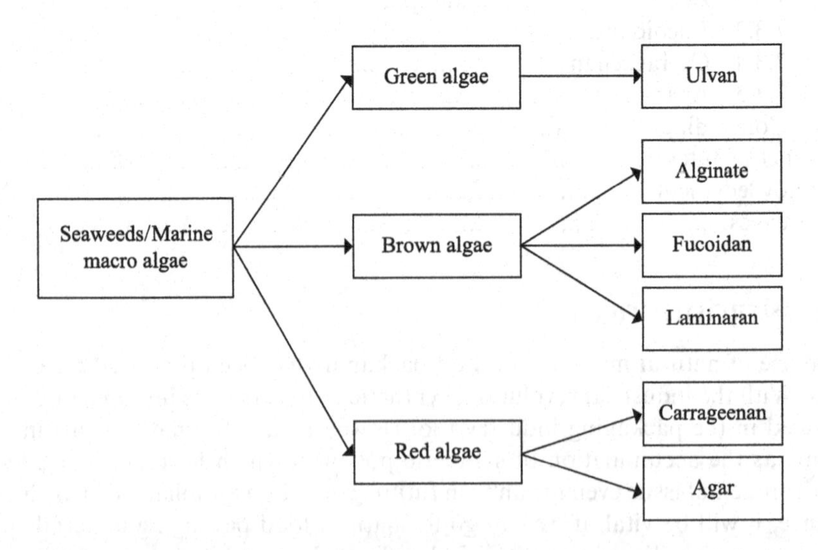

FIGURE 7.1 Classification of seaweed-based polysaccharides.

and Laminarans. The third group is, Rhodophyta or red algae which consist of phycoerythrin and phycocyanin as the main pigments with carrageenan and agar as primary polysaccharides (Anil et al. 2017; Fauziee et al. 2021). The polysaccharides in seaweed that accounts for around 4–76% of the dry weight contributes to the nutritional aspect of seaweeds as they act as dietary fibres in diets (Kadam et al. 2015). Owing to the colloidal nature, good film formation ability, high abundance and low cost of marine macroalgal polysaccharides, they have gained wide attention recently in the application of the packaging industry (Ili Balqis et al. 2017; Guidara et al. 2020). Moreover, the better properties of the films and coatings made out of seaweed-based polysaccharides such as gas barrier property, good mechanical strength, transparency, solubility and non-toxicity make them ideal for use in packaging appliances (Ganesan et al. 2018; Guidara et al. 2020). Besides, the availability of bioactive properties in seaweed-based polysaccharides such as antioxidants and antimicrobial can fulfil the requirements of active packaging. Thus, seaweed-based polysaccharides are superior and sustainable in developing natural packaging materials for the food industry (Ganesan et al. 2018; Carina et al. 2021).

7.2 STRUCTURAL APPLICABILITY OF MARINE MACROALGAL POLYSACCHARIDES IN FOOD PACKAGING

A comprehensive understanding of the structures of marine macroalgal polysaccharides is vitally important to use widely in the food packaging industry as the structure is the main responsible factor that determines their final properties. Further, understanding of the structure facilitates conducting modification experiments for the structure to achieve target properties to be possessed with a packaging film and also for coatings. Hence, this section summarizes the structural properties of the main polysaccharides, derived from each algal type and their applicability in making packaging films and coatings for the food industry.

7.2.1 Structural Applicability of Green Algal Polysaccharides in Food Packaging

Ulvan is the major sulfated polysaccharide isolated from green algae that accounts for 9–36% of alga mass (Kidgell et al. 2019). It is identified as an anionic, water-soluble and stable polysaccharide that contains glucose, xylose, rhamnose and glucuronic acid monomers (Madany et al. 2021). As reported by the previous studies, a common ulvan structure (Figure 7.2) composed of sulfated rhamnose is linked to glucuronic acid in ulvanobiuronic acid 3-sulfate or iduronic acid in ulvanobiuronic acid 3-sulfate (Ganesan et al. 2018: Madany et al. 2021). Apart from this common structure, two other structures of ulvan can be identified in several green algal species as branching glucuronic acid to O-2 rhamnose3-sulfate and O-2 partially sulfated xylose (Madany et al. 2021). Although ulvan has been subjected to research for developing films and coatings, no scientific evaluation has been reported on the applicability of ulvan structure in making packaging

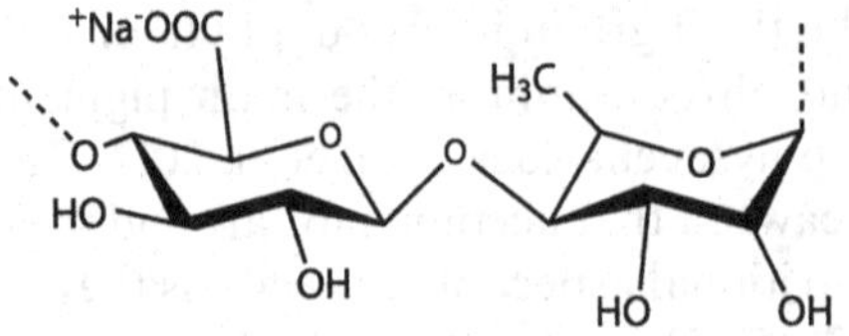

FIGURE 7.2 Chemical structure of ulvan.

materials. However, ulvan is a potential source of active packaging as it contains a broad range of bioactive properties such as antioxidant, anticancer, antifungal, anticoagulant and antitumor activities (Kidgell et al. 2019).

7.2.2 Structural Applicability of Brown Algal Polysaccharides in Food Packaging

Alginate, fucoidan and laminaran are the main polysaccharides contained in brown algae and alginate is widely used in commercial-scale food as well as non-food industries such as textile, paper, medicinal and pharmaceutical as stabilizing, thickening, gel-forming agents (Tavassoli-Kafrani et al. 2016). The commercial sources of alginate are *Sargassum* sp., *Laminaria* sp., *Eclonia* sp., *Ascophyllum nodosum*, *Lessonia nigrescens*, *Durvillae antarctica* and *Macroscystis pyrifera* (Draget 2009).

Alginate is considered a filmogenic material owing to its linear structure that can form fibrous structures and strong film matrices (Blanco-Pascual et al. 2014). Moreover, alginate is water-soluble and anionic (Hambleton et al. 2011; Vu and Won 2013) polysaccharide (Figure 7.3) with monomeric units of 1–4-linked α-D mannuronate (M) and ß-L-guluronate (G) (Tavassoli-Kafrani et al. 2016). So also, the structure consists of two homopolymeric blocks named M, G and one heteropolymeric block that is identified as MG block. The G block is derived from L-guluronic acid and it is responsible for the gel strength of alginate. Meanwhile, the M block is derived from D-mannuronic acid and the MG blocks contain both units that account for the acid solubility of alginates (Tavassoli-Kafrani et al. 2016). Nevertheless, the fraction of each unit is dependent on the species, age, seasonal variations and the part of the seaweed from which the alginate is extracted. The composition of these units in alginate determines the final gel properties leading to film properties also. If the M/G ratio is less than 1, this means the presence of a high amount of guluronic acid leads to forming strong junctions in networks. Further, the lower guluronic acid content that is indicated by M/G ratio higher than 1, leading to forming more elastic and softer film matrices (Tavassoli-Kafrani et al. 2016). Further, alginate is considered a pH-sensitive polymer due to the presence of a high number of carboxyl (-COO^-) groups in the structure (Sharmin et al. 2021). Therefore, the gel formation from alginate is induced by acidic crosslinking (Hua et al. 2010; Sharmin et al. 2021). In acidic conditions, owing to the improved hydrogen bond interaction, -COOH groups are formed and the molecular chain of

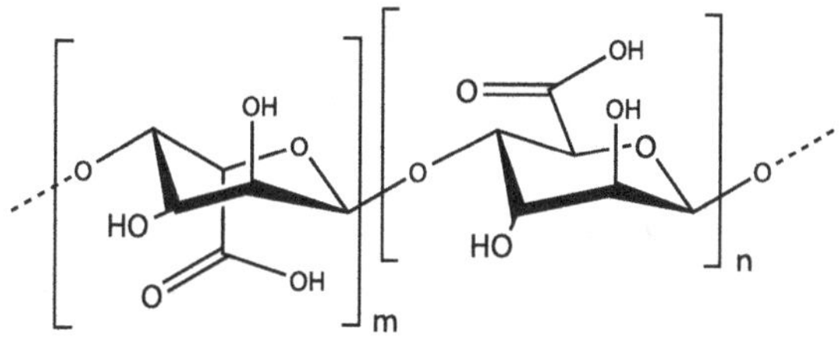

FIGURE 7.3 Chemical structure of alginate.

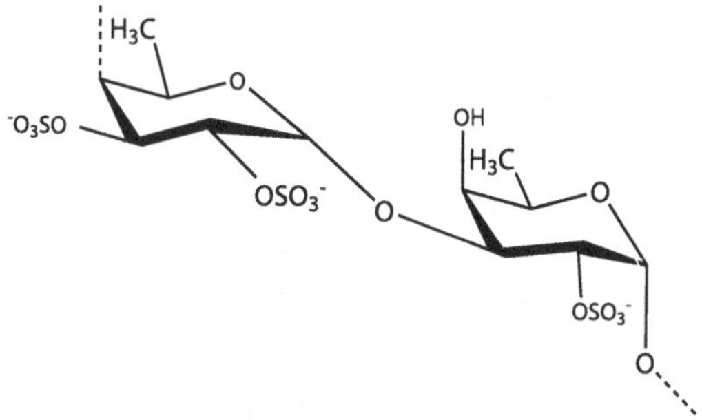

FIGURE 7.4 Chemical structure of fucoidan.

alginate is shrunk which can considerably change the rheological, mechanical and barrier properties (Zou et al. 2020). On contrary, with the increase of pH value, -COOH is ionized to -COO- and the molecular chain is extended which leads to an increase in the hydrophilic nature of alginate (Zou et al. 2020).

Although alginates and fucoidans are derived from brown seaweeds, their structures are much different (Fernando et al. 2019) because unlike in alginates, fucoidans contain a high number of L-fucose monosaccharides and some other monosaccharides such as glucose, mannose, xylose and galactose. These monosaccharide units are randomly connected within its polymeric network with randomly substituted sulfate groups (Fernando et al. 2017) that make the fucoidan structure more complex. Moreover, broad structural variation of heteropolysaccharides and repeating monomer sequence can be observed among the brown algal species used in fucoidan isolation. Despite that, a specific structure can be described as in Figure 7.4 that is commonly seen in fucoidans. Thus, most of the fucoidan types contain a backbone of (1→3) and (1→4) linked-α-l-fucopyranosyl residues along with sulfate substitution at the positions of C-2 and/or C-4 (Adhikari et al. 2006; Bilan et al. 2006). Further, in some fucoidans, it has been reported that 1→2 linkages between α-l-fucopyranosyl and the random and occasional acetyl substitution at C-4 in 3-linked residues and C-3 in 4-linked residues can be observed in some fucoidans (Bilan et al. 2006; Fernando et al. 2019). Even though fucoidan contains prominent bioactive properties such as antioxidant, antivirus, anticoagulant, anticancer, anti-inflammatory, antimicrobial and

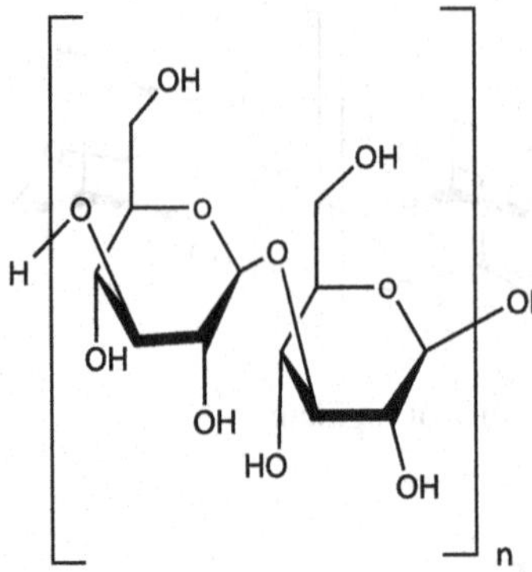

FIGURE 7.5 Chemical structure of laminaran.

anti-diabetic properties (Hifney et al. 2016; Fernando et al. 2019), it is lack of capability in making hydrogels like alginate which causes to limit its applications in making packaging materials; films and coatings (Fernando et al. 2019; Gomaa et al. 2018). As fucoidan cannot form gels itself, the combination of it with other polymers has been studied to get more advantageous packaging materials, as it leads to active packaging too (Venkatesan et al. 2014).

Laminaran as a sulfated polysaccharide derived from brown algae also exhibits a range of bioactive properties including antioxidant, antitumor, anti-coagulant, antiapoptotic and anti-inflammatory activities (Kadam et al. 2015) that can be applied in making bioactive packaging materials. Although laminaran (Figure 7.5) has been identified as a water-soluble, non-toxic polysaccharide that contains (1,3)-b-D-glucan with b-(1,6)-intrachain links (Kadam et al. 2015), its applicability in making packaging materials has not been studied yet to the best of our knowledge.

7.2.3 Structural Applicability of Red Algal Polysaccharides in Food Packaging

Carrageenan and agar are the main polysaccharides derived from red algae and *Kappaphycus alvarezii* and *Eucheuma denticulatum* are the most common green seaweeds that used in carrageenan extraction (Tavassoli-Kafrani et al. 2016). However, commercially popular sources in producing carrageenan are *Gelidium* sp. and *Gracilaria* sp. (Mostafavi and Zaeim 2020).

Carrageenan is a hydrophilic polysaccharide that is applied both in food and non-food industries and it is used as a stabilizer in dairy products, meat products, nutritional supplements, infant and pet food as well as in pharmaceutical, textile, cosmetic and printing industries too. However, its applicability in making packaging materials is also researched for years (Tavassoli-Kafrani et al. 2016). Based on the chemical structure and the properties, carrageenan (Figure 7.6) has been classified into three categories as kappa (κ), iota (ι) and lambda (λ). κ-Carrageenan contains a 3-linked, 4-sulfated galactose

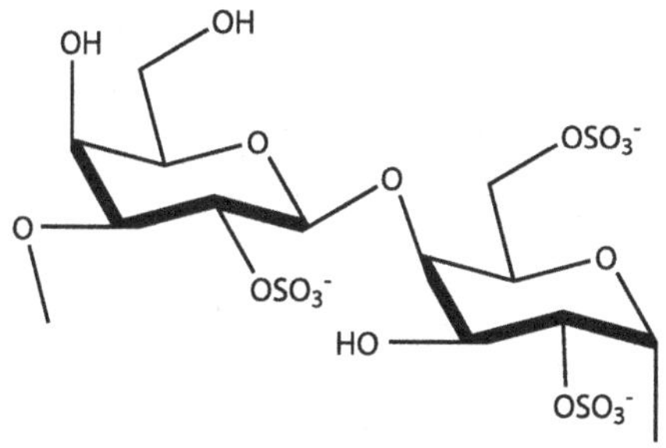

FIGURE 7.6 Chemical structure of carrageenan.

and a 4-linked 3,6-anhydrogalactose and ι-carrageenan contains a 3-linked, 4-sulfated galactose and a 4-linked 3,6-anhydrogalactose with additional sulfate ester group at the position of C-2 of the 3,6-anhydrogalactose residue. Meanwhile, λ-carrageenan consists of a 2-sulfated, 3-linked galactose unit, and a 2,6-disulfated 4-linked galactose unit (Cosenza et al. 2014). The water solubility of carrageenan is dependent on the number of ester sulfate groups attached to the structure and the availability of associated cations. Also, the linear chain of partially sulfated galactans of carrageenan increases its potential in film formation. Nevertheless, the hydrophilicity of κ- and ι-carrageenan is decreased owing to the presence of anhydro-bridges within the structure (Tavassoli-Kafrani et al. 2016). Moreover, only the κ- and ι-forms can form gels that lead to being applied in making packaging films and coatings. Therein, κ-carrageenan can form hard, brittle and strong gels, while ι-carrageenan forms soft, elastic and weak gels that lead to widening the applications of κ-carrageenan in the packaging industry (Farhan and Hani 2017).

Agar is another type of sulfated polysaccharide (Figure 7.7) derived from red algae that consists of α (1–4)-3, 6-anhydro-L-galactose and β 9(1–3)-D-galactose residues (Carina et al. 2021). Two main parts can be identified in agar structure: agaropectin, which is the non-gelling fraction, and agarose, the gelling fraction. In producing agar, the agaropectin part is removed and only the agarose part is used to get higher gel strength (Mostafavi and Zaeim 2020). Hence, the applicability of agar is improved in food and medical industries as a thickening, emulsifying, gelling and texture modification agent (Carina et al. 2021). The gel formation capacity of agar leads to make consistent coatings and film matrices (Mostafavi and Zaeim 2020), thus widening its applicability in making packaging materials.

7.3 CURRENT USE OF MARINE MACROALGAL POLYSACCHARIDES IN NATURAL FOOD PACKAGING

7.3.1 Ulvan

Ulvan, which is the main sulfated polysaccharide from green algae, is considered a good film-forming agent. Nevertheless, the extraction conditions

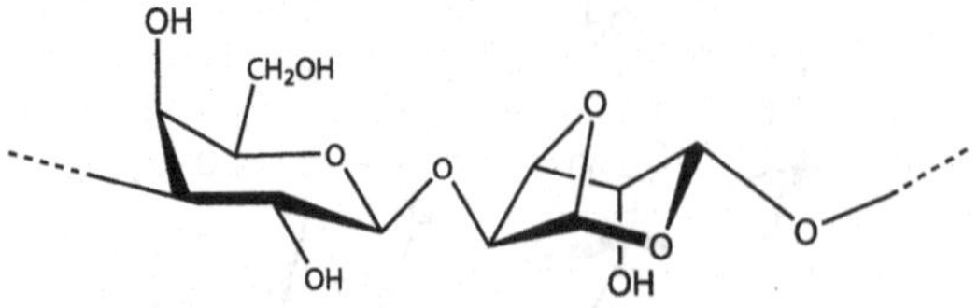

FIGURE 7.7 Chemical structure of agar.

of ulvan, the plasticizer and its concentration used in film formation directly affect properties of the resulting film such as optical, thermal, structural and bioactive properties (Guidara et al. 2019, 2020). As an example, the ulvan films developed using glycerol as the plasticizing agent exhibits better film structure when compared to plasticizing with sorbitol. Meanwhile, Amin (2021) proposed a novel film formula from ulvan with silver nanoparticles as a potential alternative for food packaging with better physical and bioactive properties that helps in food preservation during processing, storage and transportation. However, in-depth attention on property enhancement of ulvan-based films and coatings is important and essential because *Ulva* is the most abundant alga source in the world; thus, it would be a cost-effective biopolymer source.

7.3.2 Alginate

Alginate-based films exhibit good flexibility, transparency, gel formation capability with low water vapor and oxygen permeability (Shahabi-Ghahfarrokhi et al. 2020; Santos et al. 2021). Along with better organoleptic properties of films formulated with alginates such as relatively good taste and odorless properties, it makes more applicable in edible packaging (Puscaselu et al. 2019). However, relatively lower mechanical strength, thermal stability, high water sensitivity and non-antibacterial properties of alginate films make its packaging applications limited (Hou et al. 2019; Chen et al. 2021).

With the aim of improving the properties of alginate films and coatings and to add more value by embedding antioxidant and antimicrobial properties, several research studies have been carried out in incorporating other sources of biopolymers, nanoparticles, bioactive compounds rather than using only the alginate as the film-forming material (Aziz et al. 2018; Chen et al. 2021; Fan et al. 2021). Therein, Santos et al. (2021) developed alginate-based film by incorporating purple onion peel extracts which contain a high number of phenolic compounds. According to the findings, the use of purple onion peel extract in alginate film has resulted in low water-soluble films consist of better polymeric network with active packaging properties owing to high antioxidant and antimicrobial activities that lead this film to be applied for extending shelf-lives of the food products that inclined to lipid oxidation over the high water activity. Moreover, Chen et al. (2021) investigated developing films from alginate-thymol composite and suggested it as an alternative to the polyethylene cling wrap that can be used in the packaging of fresh-cut fruits. The study further revealed that the combination of alginate with thymol which has prominent antimicrobial, antioxidant led to increasing the film performance such as tensile strength, elongation at break, light transmittance, swelling index, water-solubility, water vapor permeability along with considerable scavenging activity with 1,1-diphenyl-2-picrylhydrazyl (DPPH) radicals and outstanding inhibition of *Staphylococcus aureus* and *Escherichia coli*. Fan et al. (2021) have provided basic data on novel alginate-based film blended with pectin and xanthan gum using glycerol as the plasticizer and $CaCl_2$ as a cross-linking agent for the preservation of fresh-cut products with better mechanical and barrier properties. According to a study carried out by Aziz et al. (2018), the mechanical strength of alginate-based films could be significantly increased by combining alginate with castor oil in film making. Further, the composite film can be proposed as an alternative and viable candidate for the packaging industry owing to its lower water vapor permeability, and the antibacterial inhibition for gram-positive bacteria namely *S. aureus* and *B. subtilis*. With the aim of increasing the bioactivity of alginate-based films, tea polyphenols were incorporated into the film by Biao et al. (2019). Therein, the antioxidant activity against DPPH and 2,2′-azino-bis(3-ethylbenzthiozoline-6)-sulphonic acid (ABTS) radical has considerably increased. Also, other film properties such as tensile strength and elongation were increased however, the optical transmittance of the film was decreased.

Park et al. (2020) have used alginate dialdehyde which is an environmentally friendly crosslinking agent in film formation, to overcome the low mechanical properties and stability of fish-derived gelatin that limit its applicability in packaging. The findings further revealed that the crosslinking with alginate dialdehyde could increase the mechanical strength and Young's modulus of the developed films by, respectively, 600% and 800%. Moreover, the developed composite film exhibited higher stability with moisture and improved the antioxidant activity suggesting an application in edible packaging. Active edible alginate films were prepared by Aziz and Salama (2021) by incorporating aloe vera and garlic oil and evaluated their applicability in extending the shelf-life of tomatoes. The findings suggest the potentiality of using the developed formulation in food packaging as it could preserve the tomatoes with prominent antimicrobial properties. Further, the incorporation of aloe vera and garlic oil could significantly increase the mechanical, thermal, and UV shielding properties of the film. The incorporation of nanoparticles in making alginate films has been identified as a successful method to increase their mechanical strength and thermal stability (Hou et al. 2019; Yang et al. 2018). Junior et al. (2021) were able to increase the thermal stability of sodium alginate films by blending sodium alginate with a lower concentration of nano-SiO_2 significantly. Moreover, according to the findings of the investigation, the nano incorporation has not significantly affected the positive film properties of the film such as light, water vapor barrier and mechanical properties, which also suggested using lower concentrations of nano-SiO_2 for making alginate composite films as an economic method applicable in future food packaging. Further, the development of bioactive sodium alginate films blending with mesoporous silica nanoparticles embedded in oregano essential oil was studied by Lu et al. (2021). The results indicated that the dosage of 0.8% nano embedded essential oil incorporated alginate films exhibited the optimum mechanical and physical properties. So also, the barrier property of films against UV light has been improved by the addition of the nano embedded essential oil. Meanwhile, the strongest scavenging activity (75.31%) against DPPH and the prominent antimicrobial activity for *Curvularia lunata* was shown by the film contains 1.0% nano embedded essential oil. Further, the strong hydrogen bonds in the film matrix caused to increase the mechanical properties and the resistance for the water of films. However, the market applicability of the developed alginate-based film is proposed to be conducted.

7.3.3 Fucoidan

Fucoidan is a minimally researched polysaccharide used in making packaging films and coatings, and its applicability in packaging is limited owing to its inability to form the hydrogels, although it has excellent bioactive properties. Therefore, making composite films with fucoidan by incorporating other potential biopolymers is important to derive the benefit of its bioactive compounds. Hence, Gomaa et al. (2018) investigated developing edible films with natural antioxidant properties using fucoidan. The results revealed that the incorporation of fucoidan into alginate chitosan films could increase the UV shielding property while decreasing the water solubility of the films. Moreover, these developed films exhibited good antioxidant properties after they were analyzed in terms of ferric reducing antioxidant power, hydroxyl radical scavenging activity and total antioxidant assay. Nevertheless, Xu and Wu (2021) were able to develop a fucoidan based coating for the shelf-life extension of mango fruits. The findings of the study indicate that the developed coating could reduce the weight and nutrition loss from mangoes while protecting from physical and biological damage, suggesting its applicability for other fruits too.

7.3.4 Carrageenan

Carrageenan is a widely researched algal polysaccharide used in making packaging films and coatings. Mahajan et al. (2021) suggested a carrageenan-based edible film that was combined with aloe vera for the preservation of kulfi, one of the popular ice cream types in India. The findings showed that aloe vera could increase the film thickness and decrease water vapor permeability. Moreover, the antioxidant and antimicrobial potentials of the developed film were improved owing to the addition of aloe vera extract to the film formulation. Since the outcome of this study directly affected to increase the stability of kulfi leading to apply commercially in frozen dairy products. In order to increase the properties of the pearl millet starch-based films, a composite was developed by Sandhu et al. (2020). Based on the results of the study, carrageenan could increase the film transparency and tensile strength while decreasing the solubility and water vapor permeability, suggesting a better alternative for food packaging or coating. Further, Farhan and Hani (2020) have used the water extract of germinated fenugreek seeds with semi-refined κ-carrageenan in film making with the aim of revealing another potential way of active packaging from carrageenan. The findings of the study revealed that the combination of semi-refined κ-carrageenan and the water extract of germinated fenugreek seeds could result, films with better color attributes, enhanced antioxidant activity and antimicrobial properties while inhibiting the microbial growth in chicken breasts too. Further, positive effects for extending the shelf-life of chicken meat were observed by incorporating camellia oil as an active and emulsification agent into κ-carrageenan and konjac glucomannan based coating. According to the results of this study, the developed coating could reduce the weight loss, thiobarbituric acid reactive substance, total volatile nitrogen, pH and microbial counts of the chicken meat samples. Meanwhile, antioxidant and antimicrobial properties have been impregnated into the carrageenan-based coating by adding camellia oil; it causes to extend the shelf-life of meat samples by retardation of lipid, protein oxidation and microbial growth (Zhou et al. 2021). Thakur et al. (2016) developed a composite film from rice starch and ι-carrageenan, the results of which revealed that the incorporation of ι-carrageenan could improve the tensile strength and elongation at break of films suggesting the developed film formulation for applying in fruits and vegetable packaging.

7.3.5 Agar

Agar is a red algal polysaccharide that is researched for applying in making packaging materials owing to its good film formation capability, thermoplasticity, moderate resistance for water and better transparency. However, the commercialization of agar-based films and coatings has become limited due to the lower mechanical properties, barrier properties, and thermal stability. Therefore, recent studies focus on improving its packaging properties by combining hydrophobic compounds, biopolymers and nanoparticles (Wang et al. 2018; Mostafavi and Zaeim 2020). Wongphan and Harnkarnsujarit (2020) suggested a composite formula by combining agar, starch and maltodextrin to control the water solubility of films. Based on the findings, incorporating agar has led to forming film matrices with enhanced solidity and gelation at lower temperatures that result in increased film-forming capability and hydrophobic nature. Moreover, agar has contributed to producing consistent film matrices in the starch network. Conversely, for the improvement of agar-based films, the incorporation of nano bacterial cellulose has been studied by Wang et al. (2018). According to the findings, the addition of nano bacterial cellulose could improve the film crystallinity, thermal stability and mechanical strength while decreasing the water solubility and water vapor permeability, revealing the applicability of developed film as a good material for food packaging. However, more scientific investigations should be conducted on using agar in making films and coatings to improve their mechanical, physical and barrier properties.

7.4 CONCLUDING REMARKS

The use of natural packaging materials for food packaging has been identified as a successful solution for the global threat as a result of the environmental accumulation of synthetic polymers. Recent research findings prove that seaweed-based polysaccharides are a sustainable source for natural food packaging. However, further scientific studies are required in improving the properties of films and coatings developed from seaweed-based polysaccharides with both aspects of conducting structural modifications and also developing composite films.

CONFLICT OF INTEREST

The authors declare no conflict of interest, financial or otherwise.

ACKNOWLEDGMENT

Authors would like to acknowledge University Research Grants (Grant No. ASP/01/RE/SCI/2019/11), University of Sri Jayewardenepura, Nugegoda, Sri Lanka.

REFERENCES

Adhikari, U., Mateu, C. G., Chattopadhyay, K., Pujol, C. A., Damonte, E., B., and B. Ray. 2006. Structure and antiviral activity of sulfated fucans from *Stoechospermum marginatum*. *Phytochemistry* 67:2474–2482.

Amin, H. H. 2021. Safe Ulvan silver nanoparticles composite films for active food packaging. *American Journal of Biochemistry and Biotechnology* 17:28–39.

Anil, S., Venkatesan, J., Chalisserry, E. P., Nam, S. Y., and Kim, S. 2017. Applications of seaweed polysaccharides in dentistry. In *Seaweed Polysaccharides*, ed. J. Venkatesan, S. Anil, and S. Kim, 331–340. Elsevier.

Aziz, M. S. A., and H. E. Salama. 2021. Developing multifunctional edible coatings based on alginate for active food packaging. *International Journal of Biological Macromolecules* 190:837–844.

Aziz, M. S. A., Salama, H. E., and M. W. Sabaa. 2018. Biobased alginate/castor oil edible films for active food packaging. *LWT* 96:455–460.

Biao, Y., Yuxuan, C., Qi, T., et al. 2019. Enhanced performance and functionality of active edible films by incorporating tea polyphenols into thin calcium alginate hydrogels. *Food Hydrocolloids* 97:105197.

Bilan, M. I., Grachev, A. A., Shashkov, A. S., Nifantiev, N. E., A. I. Usov. 2006. Structure of a fucoidan from the brown seaweed *Fucus serratus* L. *Carbohydrate Research* 341:238–245.

Blanco-Pascual, N., Montero, M. P., and M. C. Gomez-Guillen. 2014. Antioxidant film development from unrefined extracts of brown seaweeds *Laminaria digitata* and *Ascophyllum nodosum*. *Food Hydrocolloids* 37:100–110.

Carina, D., Sharma, S., Jaiswal, A. K., and S. Jaiswal. 2021. Seaweeds polysaccharides in active food packaging: A review of recent progress. *Trends in Food Science & Technology* 110:559–572.

Carissimi, M., Flores, S. H., and R. Rech. 2018. Effect of microalgae addition on active biodegradable starch film. *Algal Research* 32:201–209.

Chen, J., Wu, A., Yang, M., et al. 2021. Characterization of sodium alginate-based films incorporated with thymol for fresh-cut apple packaging. *Food Control* 126:108063.

Cosenza, V. A., Navarro, D. A., Fissore, E. N., Rojas, A. M., and C. A. Stortz. 2014. Chemical and rheological characterization of the carrageenans from *Hypnea musciformis* (Wulfen) Lamoroux. *Carbohydrate Polymers* 102:780–789.

Draget, K. I. 2009. Alginates. In *Handbook of Hydrocolloids*, ed. G. O. Phillips and P. A. Williams, 807–828. Woodhead Publishing.

Falguera, V., Quintero, J. P., Jimenez, A., Munoz, J. A., and A. Ibarz. 2011. Edible films and coatings: Structures, active functions and trends in their use. *Trends in Food Science & Technology* 22:292–303.

Fan, Y., Yang, J., Duan, A., and X. Li. 2021. Pectin/sodium alginate/xanthan gum edible composite films as the fresh-cut package. *International Journal of Biological Macromolecules* 181:1003–1009.

Farhan, A., and N. M. Hani. 2017. Characterization of edible packaging films based on semi-refined kappa-carrageenan plasticized with glycerol and sorbitol. *Food Hydrocolloids* 64:48–58.

Farhan, A., and N. M. Hani. 2020. Active edible films based on semi-refined κ-carrageenan: Antioxidant and color properties and application in chicken breast packaging. *Food Packaging and Shelf Life* 24:100476.

Fauziee, N. A. M., Chang, L. S., Mustapha, W. A. W., Nor, A. R. M., and S. J. Lim. 2021. Functional polysaccharides of fucoidan, laminaran and alginate from Malaysian brown seaweeds (*Sargassum polycystum*, *Turbinaria ornata* and *Padina boryana*). *International Journal of Biological Macromolecules* 167:1135–1145.

Fernando, I. P. S., Kim, D., Nah, J., and Y. Jeon. 2019. Advances in functionalizing fucoidans and alginates (bio)polymers by structural modifications: A review. *Chemical Engineering Journal* 355:33–48.

Fernando, I. P. S., Sanjeewa, K. K. A., Samarakoon, K. W., et al. 2017. A fucoidan fraction purified from *Chnoospora minima*; a potential inhibitor of LPS-induced inflammatory responses. *International Journal of Biological Macromolecules* 104:1185–1193.

Galus, S., Kibar, E. A. A., Gniewosz, M., and K. Krasniewska. 2020. Novel materials in the preparation of edible films and coatings—a review. *Coatings* 10:674.

Ganesan, A. R., Munisamy, S., and R. Bhat. 2018. Producing novel edible films from semi refined carrageenan (SRC) and Ulvan polysaccharides for potential food applications. *International Journal of Biological Macromolecules* 112:1164–1170.

Gomaa, M., Hifney, A. F., Fawzy, M. A., and K. M. Abdel-Gawad. 2018. Use of seaweed and filamentous fungus derived polysaccharides in the development of alginate-chitosan edible films containing fucoidan: Study of moisture sorption, polyphenol release and antioxidant properties. *Food Hydrocolloids* 82:239–247.

Guidara, M., Yaich, H., Benelhadj, S., et al. 2020. Smart Ulvan films responsive to stimuli of plasticizer and extraction condition in physico-chemical, optical, barrier and mechanical properties. *International Journal of Biological Macromolecules* 150:714–726.

Guidara, M., Yaich, H., Richel, A., et al. 2019. Effects of extraction procedures and plasticizer concentration on the optical, thermal, structural and antioxidant properties of novel ulvan films. *International Journal of Biological Macromolecules* 135:647–658.

Hambleton, A., Voilley, A., and F. Debeaufort. 2011. Transport parameters for aroma compounds through ι-carrageenan and sodium alginate-based edible films. *Food Hydrocolloids* 25:1128–1133.

Hifney, A. F., Fawzy, M. A., Abdel-Gawad, K. M., and M. Gomaa. 2016. Industrial optimization of fucoidan extraction from *Sargassum* sp. and its potential antioxidant and emulsifying activities. *Food Hydrocolloids* 54:77–88.

Hou, X., Xue, Z., Xia, Y., et al. 2019. Effect of SiO_2 nanoparticle on the physical and chemical properties of eco-friendly agar/sodium alginate nanocomposite film. *International Journal of Biological Macromolecules* 125:1289–1298.

Hua, S., Ma, H., Li, X., Yang, H., and A. Wang. 2010. pH-sensitive sodium alginate/poly(vinyl alcohol) hydrogel beads prepared by combined Ca^{2+} crosslinking and freeze-thawing cycles for controlled release of diclofenac sodium. *International Journal of Biological Macromolecules* 46:517–523.

Ili Balqis, A. M., Nor Khaizura, M. A. R., Russly, A. R., and Z. A. Nur Hanani. 2017. Effects of plasticizers on the physicochemical properties of kappa-carrageenan films extracted from *Eucheuma cottonii*. *International Journal of Biological Macromolecules* 103:721–732.

Junior, L. M., da Silva, R. G., Anjos, C. A. R., Vieira, R. P., and R. M. V. Alves. 2021. Effect of low concentrations of SiO_2 nanoparticles on the physical and chemical properties of sodium alginate-based films. *Carbohydrate Polymers* 269:118286.

Kadam, S. U., Tiwari, B. K., and C. P. O'Donnell. 2015. Extraction, structure and biofunctional activities of laminarin from brown algae. *International Journal of Food Science & Technology* 50:24–31.

Kidgell, J. T., Magnusson, M., de Nys, R., and C. R. K. Glasson. 2019. Ulvan: A systematic review of extraction, composition and function. *Algal Research* 39:101422.

Lu, W., Chen, M., Cheng, M., et al. 2021. Development of antioxidant and antimicrobial bioactive films based on *Oregano* essential oil/mesoporous nano-silica/sodium alginate. *Food Packaging and Shelf Life* 29:100691.

Madany, M. A., Abdel-Kareem, M. S., Al-Oufy, A. K., Haroun, M., and S. A. Sheweita. 2021. The biopolymer ulvan from *Ulva fasciata*: Extraction towards nanofibers fabrication. *International Journal of Biological Macromolecules* 177:401–412.

Mahajan, K., Kumar, S., Bhat, Z. F., Naqvi, Z., Mungure, T. E., and A. E. A. Bekhit. 2021. Functionalization of carrageenan based edible film using *Aloe vera* for improved lipid oxidative and microbial stability of frozen dairy products. *Food Bioscience* 43:101336.

Mohamed, S. A. A., El-Sakhawy, M., and M. A. El-Sakhawy. 2020. Polysaccharides, protein and lipid -based natural edible films in food packaging: A review. *Carbohydrate Polymers* 238:116178.

Mostafavi, F. S., and D. Zaeim. 2020. Agar-based edible films for food packaging applications—A review. *International Journal of Biological Macromolecules* 159:1165–1176.

Park, J., Nam, J., Yun, H., Jin, H., and H. W. Kwak. 2020. Aquatic polymer-based edible films of fish gelatin crosslinked with alginate dialdehyde having enhanced physicochemical properties. *Carbohydrate Polymers* 254:117317.

Puscaselu, R., Gutt, G., and S. Amariei. 2019. Rethinking the future of food packaging: Biobased edible films for powdered food and drinks. *Molecules* 24:3136.

Qin, Z., Mo, L., Liao, M., He, H., and J. Sun. 2019. Preparation and characterization of soy protein isolate-based nanocomposite films with cellulose nanofibers and nano-silica via silane grafting. *Polymers* 11:1835.

Sandhu, K. S., Sharma, L., Kaur, M., and R. Kaur. 2020. Physical, structural and thermal properties of composite edible films prepared from pearl millet starch and carrageenan gum: Process optimization using response surface methodology. *International Journal of Biological Macromolecules* 143:704–713.

Santos, L. G., Silva, G. F. A., Gomes, B. M., and V. G. Martins. 2021. A novel sodium alginate active films functionalized with purple onion peel extract (*Allium cepa*). *Biocatalysis and Agricultural Biotechnology* 35:102096.

Shahabi-Ghahfarrokhi, I., Almasi, H., and A. Babaei-Ghazvini. 2020. Characteristics of biopolymers from natural resources. In *Processing and Development of Polysaccharide-Based Biopolymers for Packaging Applications*, ed. Y. Zhang, 49–95. Elsevier.

Sharmin, N., Sone, I., Walsh, J. L., Sivertsvik, M., and E. N. Fernandez. 2021. Effect of citric acid and plasma activated water on the functional properties of sodium alginate for potential food packaging applications. *Food Packaging and Shelf Life* 29:100733.

Sid, S., Mor, R. S., Kishore, A., and V. S. Sharanagat. 2021. Bio-sourced polymers as alternatives to conventional food packaging materials: A review. *Trends in Food Science & Technology* 115:87–104.

Tavassoli-Kafrani, E., Shekarchizadeh, H., and M. Masoudpour-Behabadi. 2016. Development of edible films and coatings from alginates and carrageenans. *Carbohydrate Polymers* 137:360–374.

Thakur, R., Saberi, B., Pristijono, P., et al. 2016. Characterization of rice starch-ι-carrageenan biodegradable edible film. Effect of stearic acid on the film properties. *International Journal of Biological Macromolecules* 93:952–960.

Venkatesan, J., Bhatnagar, I., and S. Kim. 2014. Chitosan-alginate biocomposite containing fucoidan for bone tissue engineering. *Marine Drugs* 12:300–316.

Vu, C. H. T., and K. Won. 2013. Novel water-resistant UV-activated oxygen indicator for intelligent food packaging. *Food Chemistry* 140:52–56.

Wang, X., Guo, C., Hao, W., et al. 2018. Development and characterization of agar-based edible films reinforced with nano-bacterial cellulose. *International Journal of Biological Macromolecules* 118:722–730.

Wongphan, P., and N. Harnkarnsujarit. 2020. Characterization of starch, agar and maltodextrin blends for controlled dissolution of edible films. *International Journal of Biological Macromolecules* 156:80–93.

Xu, B., and S. Wu. 2021. Preservation of mango fruit quality using fucoidan coatings. *LWT* 143:111150.

Yang, M., Shi, J., and Y. Xia. 2018. Effect of SiO_2, PVA and glycerol concentrations on chemical and mechanical properties of alginate-based films. *International Journal of Biological Macromolecules* 107:2686–2694.

Yerramathi, B. B., Kola, M., Muniraj, B. A., Aluru, R., Thirumanyam, M., and G. V. Zyryanov. 2021. Structural studies and bioactivity of sodium alginate edible films fabricated through ferulic acid crosslinking mechanism. *Journal of Food Engineering* 301:110566.

Zhou, X., Zong, X., Zhang, M., et al. 2021. Effect of konjac glucomannan/carrageenan-based edible emulsion coatings with camellia oil on quality and shelf-life of chicken meat. *International Journal of Biological Macromolecules* 183:331–339.

Zou, Z., Zhang, B., Nie, X., et al. 2020. A sodium alginate-based sustained-release IPN hydrogel and its applications. *RSC Advances* 65.

8 Seaweed Fucoidans and Their Marine Invertebrate Animal Counterparts

Mauro Sérgio Gonçalves Pavão and Fernanda de Souza Cardoso

CONTENTS

8.1 SEAWEED FUCOIDANS

Fucoidans are sulfated polysaccharides from brown algae, composed mainly of α-L-fucose and sulfate. Fucoidans, alginate, and cellulose are the main components of the cell wall and the extracellular matrix (Mabate et al. 2021).

The glycosidic linkage between the α-L-fucose residues and the position of sulfation also varies in fucoidans: 1,3 linked fucose prevails in type 1 fucoidan, while 1–3 and 1–4 linked fucose in type 2 fucoidan. Thus, fucose residues can be sulfated at carbons 2, 3, and 4 (Van Weelden et al. 2019). Various amounts of other monosaccharides can also be present.

The chemical composition of fucoidan varies among species and locations. Depending on the species, variation in the molecular weight and the content of sulfate, fucose, and monosaccharide has been reported. Table 8.1 presents partially characterized fucoidans from different seaweed species, showing their

DOI: 10.1201/9781003303916-8

molecular weight, sulfate and fucose content, and monosaccharide composition (adapted from Mabate et al. 2021).

According to the extraction method used, fucoidans might also vary in yield, molecular weight, sulfate, and monosaccharide components. The fucoidan extracted by different methods might also present distinct in vitro antioxidant activities evaluated by DPPH and hydroxyl radical scavenging assays (Liu et al. 2020). Different from other marine polysaccharides, a universal protocol for fucoidan extraction has not been established. Hence, scientific papers and patents alike present additional pretreatment, extraction, and even purification methods for this sulfated polysaccharide from marine sources (Zayed and Ulber 2020).

The biomass is usually washed with tap water, eliminating sand particles and epiphytes, which could otherwise damage the extraction process. The biomass is also traditionally dried and milled to increase the area-to-mass ratio (Zayed and Ulber 2020). Some also apply a bleaching step (Zhang et al. 2021) or a solvent defatting or degreasing step, where the biomass is subjected to ethanol or acetone washing to remove lipids and pigments (e.g., chlorophyll, flavins, and carotenoids) (Torabi et al. 2020; Wang et al. 2021; Fawzy and Gomaa 2021; G. Y. Li et al. 2017).

The extraction processes might vary from acidic (Sugiono and Ferdiansyah 2019; Lorbeer et al. 2017), hot water (L. Wang et al. 2020; Huang et al. 2017), subcritical (G. Y. Li et al. 2017; Zhang et al. 2021), microwave (S. Wang et al. 2021), ultrasonic (Alboofetileh et al. 2018), and enzymatic (Baba et al. 2018; Dörschmann 2020). A combination of the extractions mentioned earlier might also be applied to obtain a fucoidan with specific characteristics or with a higher yield (Torabi et al. 2020; Sopelkina et al. 2020; Hanjabam 2019). An observed trend is the addition of calcium chloride to precipitate the alginate in the form of calcium alginate and separate it from the fucoidan fraction (C. H. Wang and Chen 2016; Baba et al. 2018; Saravana et al. 2016; Baba et al. 2018).

Similarly, the purification methods are diverse and depend on the degree of purity of the desired fucoidan extract. Deproteination might be carried out by protein precipitation (Zou et al. 2020; C. H. Wang and Chen 2016). The application of anion exchange or dye affinity chromatography using a salt gradient is usually followed by a chromatographic gel permeation (Zou et al. 2020) or a dialysis step against water using small molecular weight cut-off (MWCO) membranes (Zou et al. 2020; Sichert et al. 2021; Jayawardena et al. 2019) to remove salts, but this combination usually increases the production costs (Zayed and Ulber 2020). Dialysis might also be applied with different MWCO membranes to separate fucoidans according to their molecular weight, making it possible to separate the low-molecular-weight fucoidans (LMWF) from the high-molecular-weight fucoidans (HMWF) (Zayed and Ulber 2020). Another option is the application of ultrafiltration, which also allows the separation of different fucoidan sizes and the removal of low-molecular-weight impurities, such as laminarin (Zhang et al. 2021).

The most recurrent purification step is the precipitation of fucoidan with concentrated ethanol (Fawzy and Gomaa 2021; Liu et al. 2020; Nguyen et al. 2020), which might be followed by its dissolution in purified water and further

TABLE 8.1
Partially Characterized Fucoidan from Different Seaweed Species

Source	Mw (KDa)	Sulfate Content (W/W)	Fucose Content (W/W)	Monosaccharide Composition (W/W)
Fucus vesiculosus	98	15.5%	94.8%	2.3% xyl, 1.9% gal
Ascophyllum nodosum	420	20.6%	80.1%	14.3% xyl, 5.6% gal
Sargassum wightii	637	36%	53%	ND
Sargassum honeri	ND	ND	32%	23.2% man, 27.6%gal, 4.2% xyl
Ecklonia maxima	470	6.01%	4.45%	12.7% fru, 1.44%gal, 26.55% glc, 4.3% man, 0.78%xyl
Turbinaria ornata	ND	33%	59%	ND
Undaria pinnatifida	378	15.02%	39.24%	28.85% xyl, 26.4% gal, 5.04% man, 0.95% glc

Addapted from: Mar. Drugs 2021, 19, 30

TABLE 8.2
Extraction Methods and Fucoidan Properties

Browm Seaweed	Extraction Method	Fucoidan Yield (w/w)	Mol Weight	Sulfate Content
U. pinnatifida	HCl	3.2–16%	30–80	14–29%
U. pinnatifida	HCl/$CaCl_2$	3.9%	2100	7.4%
U. pinnatifida	Distilled water/$CaCl_2$	8.8%	500–23,600	45%
U. pinnatifida	H_2SO_4	4.5%	150–710	16–48%
L. japonica	Hot water/$MgCl_2$	2.3%	nd	33.1%
F. evanescens	HCL/distilled water/ H_2SO_4	11–18%	nd	nd
F. vesiculosus	Microwave-assisted	18%	nd	16–35%
F. vesiculosus	Autohydrolysis	16%	nd	20%

ethanol precipitation in order to obtain a purer product (Zou et al. 2020; Jayawardena et al. 2019). Finally, the received product might be dried by lyophilization, spray-drying, or even oven-drying (Zou et al. 2020; Fawzy and Gomaa 2021; Liu et al. 2020; C. H. Wang and Chen 2016).

Table 8.2 shows the differences in yield, sulfate content, and monosaccharide composition of fucoidans obtained by different methods. For example, fucoidan yield from *U. pinnatifida* varied from 3.9% when extracted with HCl/CaCl2 to 45% when extracted with distilled water/$CaCl_2$. The sulfate content of the fucoidan increased by 37.8% when HCl was replaced by distilled water (Wacker 2020).

TABLE 8.3
Analytical Methods

Method	Fucoidan Characterization, Determination of Purity, and Physicochemical and Chemical Properties
FT-IR	Preliminary identification of fucoidan functional groups by FT-IR between 400 and 4000 cm^{-1}.
1D NMR 1H- and 13C-NMR	O–H group of monomeric monosaccharides, C–H, asymmetric stretching of S=O and C–O–S of sulfate ester groups, and O–C–O and C–O–C of glycosidic and intramolecular linkages In the 1H-NMR spectrum: singlet peaks at around 1.2 ppm of –CH3 groups (H-6) of the fucose monomer; - Peaks slightly shifted between 3.8 and 4.5 ppm of H-2, H-3, H-4, and H-5; Anomeric proton H-1 deshielded at 5.2 ppm (α-linked sugar monomers); The presence of other sugars such as galactose, mannose, and xylose can be deduced from signals in regions lower than 3.7 ppm.
2D NMR	COSY, TOCSY, NOESY, ROESY and HSCQ – 2d NMR techniques have been used to further reveal the potential structural secrets of numerous fucoidans.
Mass Spectrometry	Mass spectrometry (MS) – glycosidic linkages and/or sulfation patterns
Fucoidan-Degrading Enzymes	Fucoidanases and sulfatases in combination with spectrometric methods. Such enzymes are mainly isolated from symbionts (e.g., Proteobacteria and Bacteroidetes) associated with brown algae or marine invertebrates.

Mar. Drugs 2020, 18, 571

Several analytical methods might be employed for the characterization of fucoidans. Fourier transform infrared spectroscopy (FT-IR) is commonly used for the preliminary identification of functional groups. This can be followed by complex analysis by 1D and 2D nuclear magnetic resonance (NMR) to identify fucose anomeric protons, methyl groups, alpha linked sugars, etc. Mass spectrometry is usually employed for linkage and sulfation patterns. Enzymatic treatment using specific fucoidases and sulfatases complements the spectrometric methods (Table 8.3) (Shikov et al. 2020).

Several pharmacological activities have been linked to fucoidans. Inhibition of coagulation involving activation of plasma serpins, antithrombin (AT), and heparin cofactor II (HCII) (Ustyuzhanina et al. 2013); inhibition of inflammation by blockage of inflammatory signaling pathway (Park et al. 2011); viral infection by blocking virus-host cell interaction (Park et al. 2011); cancer involving the induction of cell cycle arrest and apoptosis (Atashrazm et al. 2015); and immune stimulation by the potentiation of natural killer (NK) cells (Yoo et al. 2019).

8.2 FUCOIDAN COUNTERPARTS IN MARINE INVERTEBRATE ANIMALS

Seaweed fucoidans counterparts are frequently found in marine invertebrate animals. The sulfated fucans from sea urchins and the fucosylated chondroitin sulfate from sea cucumbers are interesting examples (Figure 8.1). The sulfated fucose glycans can be extracted from the sea urchin egg jelly by papain digestion and purified by ethanol precipitation followed by ion-exchange chromatography. Analysis by 1D and 2D NMR (Vilela-Silva et al. 1999) indicated two types of fucans (a 6-deoxy-L-galactose) with the same sulfation position, differing in the glycosidic linkage and molecular mass. The glycan from *Strongylocentrotus franciscanus* is a linear polymer of α-L-fucopyranose units linked through 1–3 glycosidic bound and sulfated at carbon 2. A linear polymer of α-L-fucopyranose units also sulfated at carbon 2, but linked through 1–4 glycosidic bond is found in *Strongylocentrotus droebachiensis* (Teixeira et al. 2018). Different from the two species of *Strongylocentrotus*, the sulfated glycan found in the sea urchin *Echinometra lucunter* is a polymer of α-L-galactopyranose (a fucose containing a hydroxyl group at C6) units linked through 1–3 glycosidic bound and sulfated at carbon 2 (Figure 8.1) (Teixeira et al. 2018).

An additional level of complexity of sulfated glycans is found in other echinoderms. A non-linear type of chondroitin sulfate glycosaminoglycan, containing sulfated fucose branches occurs in the holothurian *Ludwigothurea grisea* (Figure 8.1) (Lubor Borsig et al. 2007). This glycan is isolated in significant amounts from the body wall of the sea cucumber. It is composed of a backbone similar to that of mammalian chondroitin sulfate: [4-β-D-GlcA-1→3-β-D-GalNAc-1]n, but substituted at the 3-position of the β-D-glucuronic acid residues with sulfated α-L-fucopyranosyl branches (Vieira et al. 1991). Fucose 4-O-monosulfate residues abound in the glycan chain, but 2,4- and 3,4-di-O-sulfated residues are also present. The sulfated α-L-fucose residues are concentrated toward the non-reducing end of the polysaccharide chains.

8.3 BIOLOGICAL ACTIVITIES OF SULFATED GLYCANS

8.3.1 Sulfated Fucans and Galactans from Sea Urchins

Anticoagulant activity is among the most widely studied effect of sulfated glycans. Among them, heparin is the most used anticoagulant in the prevention and treatment of thrombosis (Kakkar, V. V. and Hedges, A. R. 1989); dermatan sulfate is also an anticoagulant glycan, although of lower potency than heparin (Bourin and Lindahl 1993; Maggi et al. 1987). In addition, sulfated fucans from brown seaweed have anticoagulant activity due to the ability to potentiate inhibition of thrombin by AT or HCII (Colliec et al. 1991).

The analysis of the anticoagulant activity provided exciting information about the relationship between the structure and biological activity of the sulfated fucans and galactans isolated from sea urchins (Figure 8.2). For the sulfated fucans, variations in the molecular weight and glycosidic bound do not affect anticoagulant activity. However, the introduction of a hydroxyl group on the C6 of fucose drastically reduces the anticoagulant effect of this type of glycan (Vilela-Silva et al. 1999; Teixeira et al. 2018).

One of the hallmarks of carcinoma cells is the presence of exacerbated amounts of highly branched or sialylated oligosaccharides, mainly fucosylated glycans

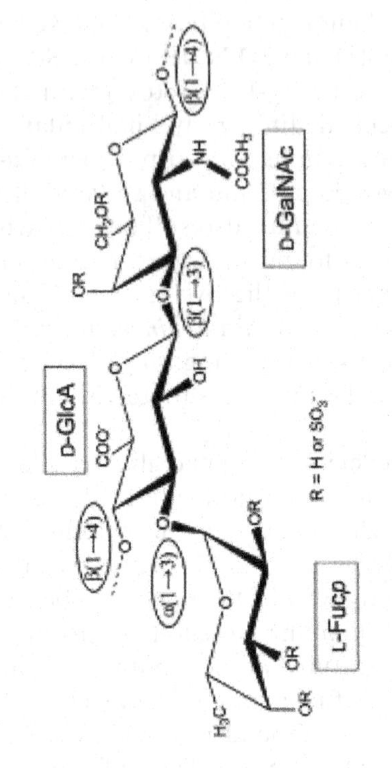

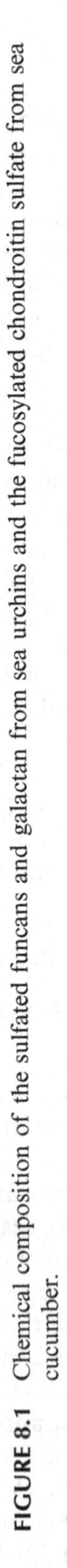

FIGURE 8.1 Chemical composition of the sulfated funcans and galactan from sea urchins and the fucosylated chondroitin sulfate from sea cucumber.

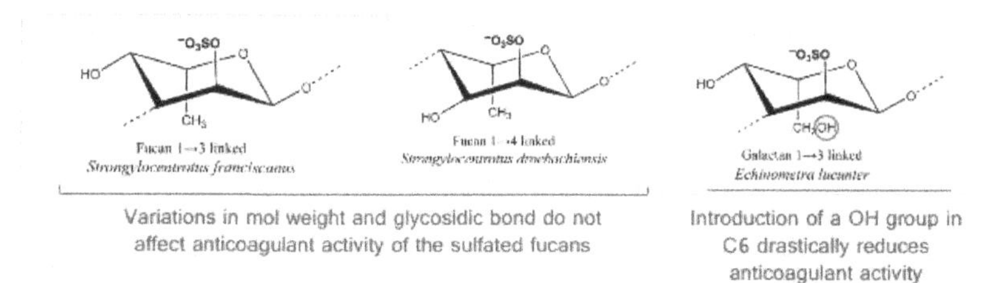

Variations in mol weight and glycosidic bond do not affect anticoagulant activity of the sulfated fucans
Introduction of a OH group in C6 drastically reduces anticoagulant activity

Structure, average molecular mass and anticoagulant activity of sulfated glycans from sea urchins

Species	Structure	Average Molecular Mass (Kda)	aPTT (IU/mg)
Echinometra lucunter	[→3)-α-L-Gal*p*-2(OSO$_3^-$)-(1→]$_n$	100	20
Strongylocentrotus droebachiensis	[→4)-α-L-Fuc*p*-2(OSO$_3^-$)-(1→]$_n$	80	<1
Strongylocentrotus franciscanus	[→3)-α-L-Fuc*p*-2(OSO$_3^-$)-(1→]$_n$	100	~2

FIGURE 8.2 Structure versus anticoagulant activity relationship of the sulfated glycans from sea urchins.

such as sialyl-Lewis$_x$ (Siaα2–3Galβ1–4(Fucα1–3)GlcNAc) and sialyl-Lewis$_a$ (Siaα2–3Galβ1–3(Fucα1–4)GlcNAc) (Hakomori 1996). The presence of these oligosaccharides in carcinoma cells is associated with a poor prognosis for cancer patients due to tumor progression and metastatic spread (Dennis and Laferte 1987; Hakomori 1996; Y. S. Kim et al. 1996; Y. J. Kim et al. 1999).

During the dynamic cellular events of hematogeneous metastasis, some of the tumor cells at the primary tumor site suffer epithelial-to-mesenchymal transition and acquire the ability to migrate and enter the blood vessel (Banyard and Bielenberg 2015). Inside the blood vessel, tumor cells are rapidly surrounded by activated platelets, which provide protection. The platelets prevent the attack of circulating monocytes, allowing tumor cells to migrate to distant sites, forming metastasis (Figure 8.3A).

One of the most fascinating biological activities of sulfated glycans is their ability to inhibit the interaction of sialyl Lewis$_{x,a}$-rich oligosaccharides on tumor cells with P-selectin on activated platelets (L. Borsig et al. 2001; Fuster et al. 2003) (Figure 8.3B). Thus, in the presence of exogenous sulfated glycans, circulating tumor cells lose the protection conferred by the platelets, allowing the cytotoxic action of immune effector cells, resulting in a significant reduction of metastasis (Figure 8.3C).

The anti-P-selectin activity of the sulfated fucans and galactans from sea urchins was accessed by their ability to inhibit the binding of sialyl Lewis$_{X,A}$-rich LS180 tumor cells to immobilized P-selectin (Figure 8.4) (Teixeira et al. 2018). Differences in glycosidic bond and the presence of OH at C6 do not affect the anti-P-selectin activity of the sulfated fucans and galactan from the sea urchins. However, it seems that a slight increase in the molecular weight of 2-sulfated fucose glycans decreases the anti-P-selectin activity (Figure 8.4).

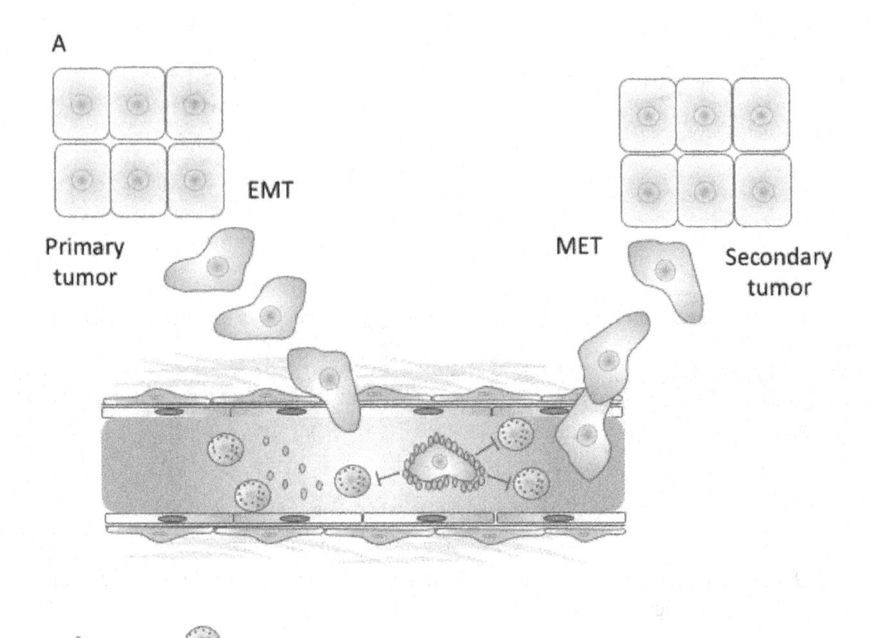

FIGURE 8.3A Cellular events involved in tumor cell dissemination. A, Metastasis: stromal migration and invasion, vessel intravasation, vascular dissemination, vascular extravasation, stromal migration and invasion, and tissue colonization. EMT, epithelial-mesenchymal transition. MET, mesenchymal-epithelial transition. B, Tumor cell–platelet complex formation involving the binding of platelet P-selectin to tumor cell sialyl Lewis oligosaccharides and inhibition by sulfated glycans. C, Metastasis in the presence of exogenous sulfated glycans in the vessel.

This differs from the relationship between the structure of the sulfated glycans from the echinoderms and anticoagulant activity, where the presence of a hydroxyl group on the C6 of fucose drastically reduces the anticoagulant effect of this type of glycans (Figures 8.2 and 8.4).

8.3.2 Fucosylated Chondroitin Sulfate from Sea Cucumber

The fucosylated chondroitin sulfate from the holothurian *L. grisea* has an anticoagulant activity in vitro of 40 IU/mg, compared to that of low-molecular-weight heparin (Mourão et al. 1996). Ex vivo studies also revealed that 10 minutes after intravenous injection of the *L. grisea* fucosylated chondroitin sulfate (200 and 500 μg/mouse) produced 1.4-fold and 2.5-fold increments in the plasma activated partial thromboplastin time (aPTT), respectively (Lubor Borsig et al. 2007). Removal of the sulfated α-L-fucose branches abolishes the anticoagulant activity of the polysaccharide. The specific anticoagulant activity estimated

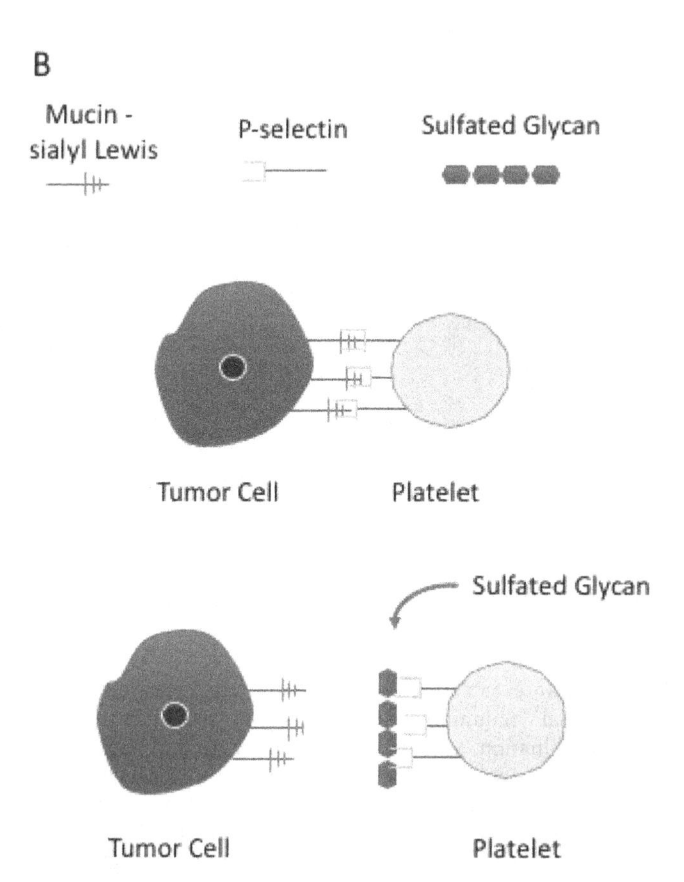

FIGURE 8.3B Cellular events involved in tumor cell dissemination. A, Metastasis: stromal migration and invasion, vessel intravasation, vascular dissemination, vascular extravasation, stromal migration and invasion, and tissue colonization. EMT, epithelial-mesenchymal transition. MET, mesenchymal-epithelial transition. B, Tumor cell-platelet complex formation involving the binding of platelet P-selectin to tumor cell sialyl Lewis oligosaccharides and inhibition by sulfated glycans. C, Metastasis in the presence of exogenous sulfated glycans in the vessel.

by the activated partial thromboplastin time assay of a linear homopolymeric α-L-fucan with about the same degree of sulfation is lower than that of the fucosylated chondroitin sulfate. The anticoagulant mechanism of the holothurian glycan involves the potentiation of the thrombin inhibition activity of both AT and HCII (Mourão et al. 1996).

The interaction of the *L. grisea* fucosylated chondroitin sulfate with different selectins was evaluated by measuring the ability of the glycan to block binding of P-, L- and E-selectin to immobilized sialyl $Lewis_x$ (Lubor Borsig et al. 2007) (Figure 8.5). The holothurian glycan inhibits the binding of P- and L-selectin to sialyl $Lewis_x$, but does not interfere with the binding to E-selectin.

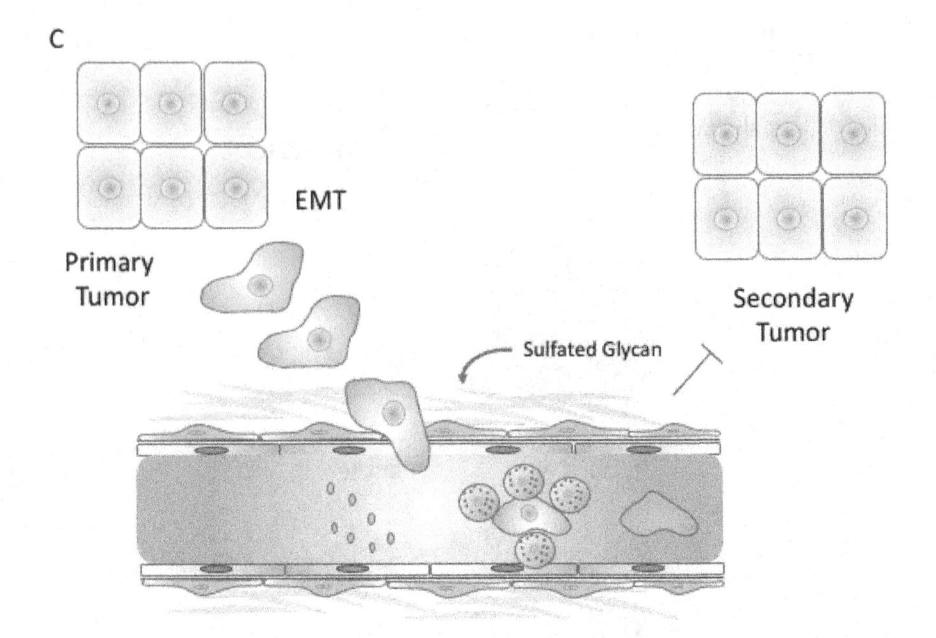

FIGURE 8.3C Cellular events involved in tumor cell dissemination. A, metastasis: stromal migration and invasion, vessel intravasation, vascular dissemination, vascular extravasation, stromal migration and invasion, and tissue colonization. EMT, epithelial-mesenchymal transition. MET, mesenchymal-epithelial transition. B, Tumor cell-platelet complex formation involving the binding of platelet P-selectin to tumor cell sialyl Lewis oligosaccharides and inhibition by sulfated glycans. C, Metastasis in the presence of exogenous sulfated glycans in the vessel.

The inhibition is dose-dependent for P- and L-selectin and requires a lower concentration than that of mammalian heparin. IC_{50} P-selectin blocking values from dose-response curves yielded 0.3 and 2.0 μg/ml for the *L. grisea* glycan and heparin, respectively. Values for L-selectin blocking were 0.25 and 0.5 μg/ml. Removal of the sulfated fucose branches by mild acid hydrolysis abolishes the anti-P- and L-selectin effect (Lubor Borsig et al. 2007) (Figure 8.5).

8.4 PHARMACOLOGICAL EFFECTS OF THE SULFATED FUCANS

8.4.1 Inhibition of Tumor Cell–Platelet Complex

One of the essential steps for the successful dissemination of cancer or metastasis is the P-selectin-mediated binding of platelets to the surface of tumor cells inside blood vessels. Disruption of the formation of tumor cell–platelet complex exposes tumor cells to the physical stress and the immune cells in the bloodstream, resulting in a drastic attenuation of metastasis (Figure 8.3).

Experiments to investigate how the anti-P-selectin activity of the sea urchins sulfated fucans and galactan and the holothurian fucosylated chondroitin sulfate

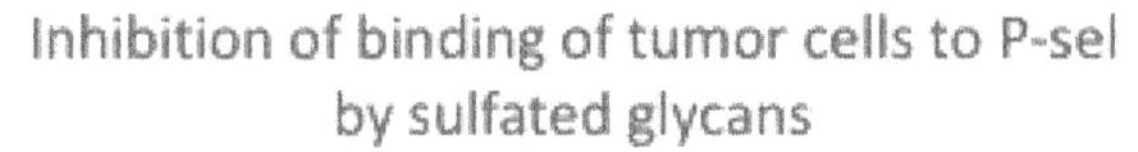

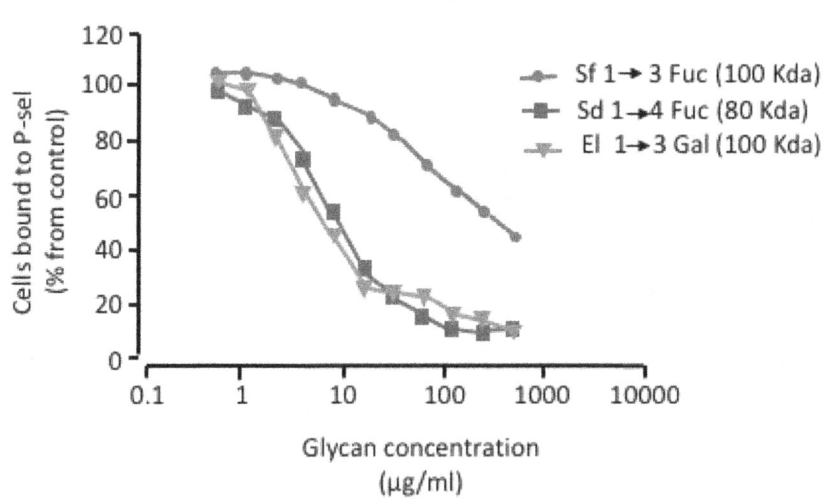

FIGURE 8.4 Inhibition of tumor cells (LLC) binding to immobilized P-selectin by sea urchins sulfated glycans.

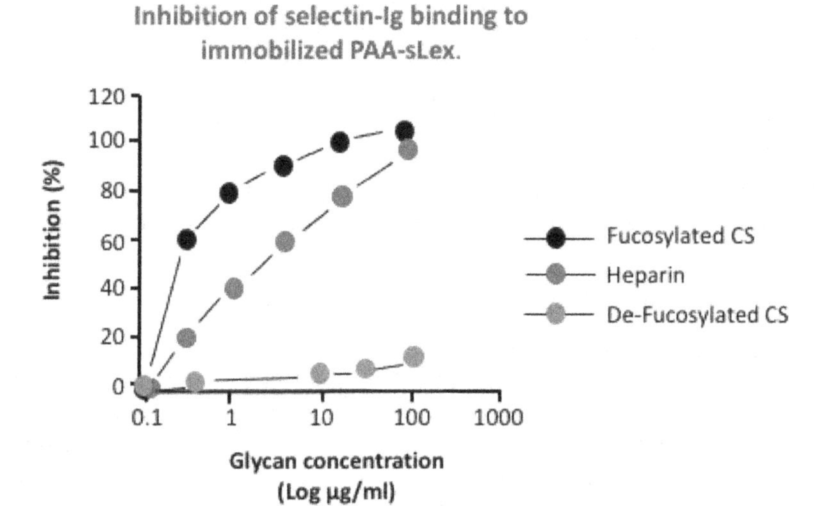

FIGURE 8.5 Inhibition of selectin binding to immobilized PAA-sLex by sea cucumber fucosylated chondroitin sulfate.

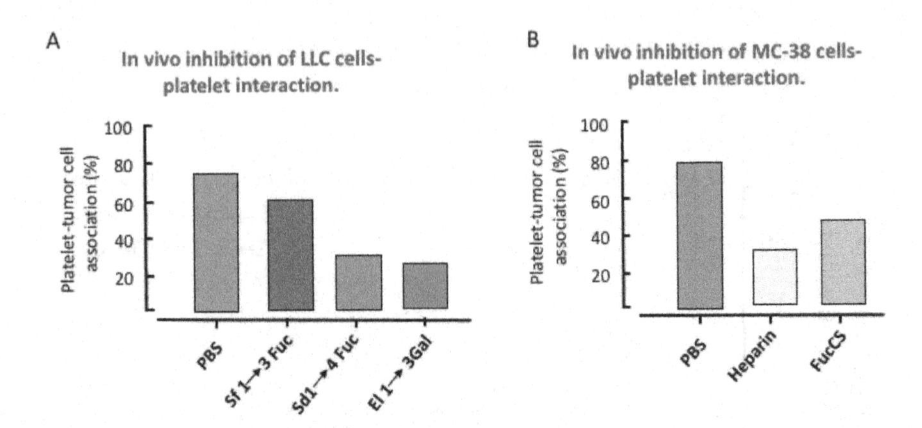

FIGURE 8.6 Inhibition of tumor cell platelet interaction in vivo by sea urchin sulfated fucans and galactan (A), and sea cucumber fucosylated chondroitin sulfate (B). LLC, Lung Lewis Carcinoma. MC-38, murine colon adenocarcinoma cells.

(Figure 8.4) correlates to their ability to inhibit the formation of platelet-tumor cell complex in vivo showed a direct association between the anti-P-selectin activity of the sulfated glycans and the inhibition of platelet–tumor cell complex formation in blood vessels. Therefore, the *E. lucunter* galactan and *S. droebachiensis* fucan reduced tumor cell-platelet complex in pulmonary capillaries by 24% and 27%, respectively. On the other hand, as expected from the in vitro studies, the *S. franciscanus* fucan did not significantly reduce this association (Figure 8.6A) (Teixeira et al. 2018).

Regarding the ability of the holothurian fucosylated chondroitin sulfate to inhibit tumor cell–platelet complex formation, similar results to those obtained for the sea urchin sulfated fucans and galactan were observed. In addition, the fucosylated chondroitin sulfate drastically reduced tumor cell–platelet complex formation in the vasculature (Figure 8.6B) (Lubor Borsig et al. 2007).

8.4.2 Attenuation of Metastasis

Since the sulfated glycans from sea urchins and sea-cucumber demonstrated potential to inhibit P-selectin and platelet–tumor cell aggregation and given the central importance of these two factors in metastasis, the anti-metastatic potential of the sulfated polysaccharides from sea urchins and sea cucumber were investigated. The experiments consisted of injecting 50 μg of the different sulfated glycans into mice 15 minutes before the injection of MC-38 or LLC tumor cells. Then, after 21 to 29 days, the mice were sacrificed, and the dissected lungs were analyzed for the presence of metastatic foci by measuring fluorescent GFP-tagged tumor cells and by counting tumor foci on the lung surface (Teixeira et al. 2018).

The treatment with the *E. lucunter* galactan or the *S. droebachiensis* fucan completely prevented metastasis. As expected, the *S. franciscanus fucan* had

TABLE 8.4
Antimetastatic Effect of the Sulfated Glycans *In Vivo*

Glycan	Number of Mice with 5 or More Metastatic Focci (LLC Cells) in the Lungs (N=6 Animals)
Sea urchins1	
Control (PBS)	5
Sf 1→3 Fuc	3
Sd 1→4 Fuc	0
El 1→3 Gal	0
Sea cucumber2	
Glycan	**Number of mice with 75 or more metastatic focci (MC-38 cells) in the lungs (n=8 animals)**
Control (PBS)	7
L. grisea	0

no significant antimetastatic activity, despite slightly decreasing the formation of metastasis (Table 8.4) (Teixeira et al. 2018). As for the *L. grisea* fucosylated chondroitin sulfate, a drastic reduction of the number of metastatic lung foci was observed in the glycan treated mice (Table 8.4) (Lubor Borsig et al. 2007).

Overall, these results strongly support the assumption that the anti-selectin effect of the sulfated glycans from sea urchins and sea cucumber correlates with their antimetastatic outcome.

8.4.3 Anti-Heparanase Activity

LMWF and its fractions, extracted from *Laminaria japonica*, have renal protective effects. LMFW was found to have an inhibitory effect on the heparanase protein expression in the human renal proximal tubular cell line (HK-2) (X. Li et al. 2017).

8.4.4 Antioxidant and Inhibition of Cancer Cell Lines

The crude fucoidan extracted from *Sargassum muticum* via ultrasound-assisted aqueous extraction contained 25 mg phlorotannins/g and presented an ABTS [2,2'-azinobis-(3-ethylbenzothiazoline-6-sulfonate)] radical scavenging activity equivalent to 40 mg Trolox/g extract measured with the Trolox equivalent antioxidant capacity (TEAC) assay. In addition, the cell viability of selected tumoral human cell lines was also reduced up to 30% in the presence of crude extracts at 500 mg/L (Flórez-Fernández et al. 2017).

Another fucoidan extracted via ultrasound-assisted extraction, but from *Nizamuddinia zanardinii*, presented interesting DPPH radical scavenging and reducing power. It also exhibited cell growth inhibition against HeLa (62.36%) and HepG2 (56.83%) cancer cell lines and induced RAW264.7 murine macrophage cells to release a considerable amount of nitric oxide (41 μmol)

(Alboofetileh et al. 2018). Meanwhile, the purified fucoidan extracted from *Turbinaria ornata* through an enzyme-assisted extraction presents potential anti-inflammatory activity, increasing the macrophage cellular and zebrafish embryo resistance against LPS-induced inflammation (Jayawardena et al. 2019).

Fucoidan extracts obtained by hydrothermal extraction method from *Himantalia elongata* present cytotoxic effects against lung, ovarian, and breast carcinoma cells and achieve the half-maximal inhibitory concentration to 1.31 ± 0.98 mg/ml of autohydrolysis extract at 220°C for the ovarian cell line tested (Cernadas et al. 2019).

8.5 FUTURE PERSPECTIVES

A critical aspect of anti-P-selectin sulfated glycans from sea urchins and sea cucumbers is the shallow risk of contamination with pathogens, considering their evolutionary distance from mammals. Also, regarding the therapeutic use of an animal-derived drug is the technical and economic possibility to obtain unlimited quantities in an ecologically friendly manner. Overall, the sulfated glycans from the marine invertebrate animals are isolated at reasonable yields (about 0.5%–2% of the dry weight compared to 0.022% from pig intestinal mucosa (Linhardt et al. 1992) by procedures similar to those already employed in the preparation of pharmaceutical heparin). Several marine invertebrates, including those containing high quantities of sulfated glycans, have been successfully cultivated in different parts of the world. The current aquaculture technologies are capable of producing ton-quantities of starting material. In 2001, the world's production of sea cucumber reached about 21,000 tons ('Advances in Sea Cucumber Aquaculture and Management' n.d.). Possibly, significant limitations for medical application of polysaccharides from marine organisms are a more profound analysis of their effects on mammalian systems and their mechanisms of action. Probably, these analyses could be performed in a shorter time than that required for the artificial synthesis of these glycans. Another essential aspect of fucoidan obtention is the lack of standard extraction and purification methods, which produce fucoidan extracts from the same marine source with highly different yields, molecular weight, purity, composition, and even antioxidant activity. The identification of a fucoidan production method that combines cost-effective techniques, high fucoidan yields, and conditions that preserve its native chemical backbone and physicochemical characteristics, which are associated with its desired application (Zayed and Ulber 2020).

REFERENCES

'Advances in Sea Cucumber Aquaculture and Management'. n.d. Accessed 18 October 2021. www.fao.org/3/y5501e/y5501e07.htm.

Alboofetileh, M., M. Rezaei, M. Tabarsa, and S.G. You. 2018. 'Ultrasound-Assisted Extraction of Sulfated Polysaccharide from Nizamuddinia Zanardinii: Process Optimization, Structural Characterization, and Biological Properties'. https://doi.org/10.1111/jfpe.12979.

Atashrazm, Farzaneh, Ray M. Lowenthal, Gregory M. Woods, Adele F. Holloway, and Joanne L. Dickinson. 2015. 'Fucoidan and Cancer: A Multifunctional Molecule with

Anti-Tumor Potential'. *Marine Drugs* 13 (4): 2327–2346. https://doi.org/10.3390/md13042327.

Baba, B.M., W. Mustapha, and L.S. Joe. 2018. 'Effect of Extraction Methods on the Yield, Fucose Content and Purity of Fucoidan from Sargassum Sp. Obtained from Pulau Langkawi, Malaysia'. *Malaysian Journal of Analytical Sciences* 22 (1): 87–94. https://doi.org/10.17576/mjas-2018-2201-11.

Banyard, Jacqueline, and Diane R. Bielenberg. 2015. 'The Role of EMT and MET in Cancer Dissemination'. *Connective Tissue Research* 56 (5): 403–413. https://doi.org/10.3109/03008207.2015.1060970.

Borsig, L., R. Wong, J. Feramisco, D. R. Nadeau, N. M. Varki, and A. Varki. 2001. 'Heparin and Cancer Revisited: Mechanistic Connections Involving Platelets, P-Selectin, Carcinoma Mucins, and Tumor Metastasis'. *Proceedings of the National Academy of Sciences of the United States of America* 98 (6): 3352–3357. https://doi.org/10.1073/pnas.061615598.

Borsig, Lubor, Lianchun Wang, Moises C. M. Cavalcante, Larissa Cardilo-Reis, Paola L. Ferreira, Paulo A. S. Mourąo, Jeffrey D. Esko, and Mauro S. G. Pavąo. 2007. 'Selectin Blocking Activity of a Fucosylated Chondroitin Sulfate Glycosaminoglycan from Sea Cucumber: Effect on Tumor Metastasis and Neutrophil Recruitment'. *Journal of Biological Chemistry* 282 (20): 14984–14991. https://doi.org/10.1074/jbc.M610560200.

Bourin, M.C., and U. Lindahl. 1993. 'Glycosaminoglycans and the Regulation of Blood Coagulation.' *Biochemical Journal* 289 (Pt 2): 313–330.

Cernadas, H., N. Flórez-Fernández, M.J. González-Muñoz, et al. 2019. 'Retrieving of High-Value Biomolecules from Edible Himanthalia Elongata Brown Seaweed Using Hydrothermal Processing'. *Food and Bioproducts Processing* 117. https://doi.org/10.1016/j.fbp.2019.07.015.

Colliec, S., A.M. Fischer, J. Tapon-Bretaudiere, C. Boisson, P. Durand, and J. Jozefonvicz. 1991. 'Anticoagulant Properties of a Fucoïdan Fraction'. *Thrombosis Research* 64 (2): 143–154. https://doi.org/10.1016/0049-3848(91)90114-c.

Dennis, J. W., and S. Laferte. 1987. 'Tumor Cell Surface Carbohydrate and the Metastatic Phenotype'. *Cancer Metastasis Reviews* 5 (3): 185–204. https://doi.org/10.1007/BF00046998.

Dörschmann, Philipp. 2020. 'Effects of a Newly Developed Enzyme-Assisted Extraction Method on the Biological Activities of Fucoidans in Ocular Cells'. *Marine Drugs* 18 (6): 282.

Fawzy, M.A., and M. Gomaa. 2021. 'Optimization of Citric Acid Treatment for the Sequential Extraction of Fucoidan and Alginate from Sargassum Latifolium and Their Potential Antioxidant and Fe(III) Chelation Properties'. *Journal of Applied Phycology* 33: 2523–2535.

Flórez-Fernández, N., M. López-García, M.J. González-Muñoz, et al. 2017. 'Ultrasound-Assisted Extraction of Fucoidan from Sargassum Muticum'. *Journal of Applied Phycology* 29: 1553–1561. https://doi.org/10.1007/s10811-016-1043-9.

Fuster, Mark M., Jillian R. Brown, Lianchun Wang, and Jeffrey D. Esko. 2003. 'A Disaccharide Precursor of Sialyl Lewis X Inhibits Metastatic Potential of Tumor Cells'. *Cancer Research* 63 (11): 2775–2781.

Hakomori, S. 1996. 'Tumor Malignancy Defined by Aberrant Glycosylation and Sphingo(Glyco)Lipid Metabolism'. *Cancer Research* 56 (23): 5309–5318.

Hanjabam, M.D. 2019. 'Isolation of Crude Fucoidan from Sargassum Wightii Using Conventional and Ultra-Sonication Extraction Methods'. *Bioactive Carbohydrates and Dietary Fibre* 20: 100200.

Huang, C.Y., S.W. Wu, W.N. Yang, A.W. Kuan, and C.Y. Chen. 2017. 'Antioxidant Activities of Crude Extracts of Fucoidan Extracted from Sargassum Glaucescens by a Compressional-Puffing-Hydrothermal Extraction Process'. *Food Chem* 197 Part B: 1121–1129. https://doi.org/10.1016/j.foodchem.2015.11.100.

Jayawardena, T.U., et al. 2019. 'Isolation and Purification of Fucoidan Fraction in Turbinaria Ornata from the Maldives; Inflammation Inhibitory Potential under LPS Stimulated Conditions in in-Vitro and in-Vivo Models'. *International Journal of Biological Macromolecules* 131: 614–623.

Kakkar, V.V., and A.R. Hedges. 1989. *Heparin*. Lane: Lindahl U. Edward Arnold.

Kim, Y.J., L. Borsig, H.L. Han, N.M. Varki, and A. Varki. 1999. 'Distinct Selectin Ligands on Colon Carcinoma Mucins Can Mediate Pathological Interactions Among Platelets, Leukocytes, and Endothelium'. *The American Journal of Pathology* 155 (2): 461–472. https://doi.org/10.1016/S0002-9440(10)65142-5.

Kim, Y.S., J. Gum, and I. Brockhausen. 1996. 'Mucin Glycoproteins in Neoplasia'. *Glycoconjugate Journal* 13 (5): 693–707. https://doi.org/10.1007/BF00702333.

Li, G.Y., Z.C. Luo, F. Yuan, and X.B. Yu. 2017. 'Combined Process of High-Pressure Homogenization and Hydrothermal Extraction for the Extraction of Fucoidan with Good Antioxidant Properties from Nemacystus Decipients.' *Food and Bioproducts Processing* 106: 35–42. https://doi.org//10.1016/j.fbp.2017.08.002.

Li, X., X. Li, Q. Zhang, and T. Zhao. 2017. 'Low Molecular Weight Fucoidan and Its Fractions Inhibit Renal Epithelial Mesenchymal Transition Induced by TGF-B1 or FGF-2'. *International Journal of Biological Macromolecules* 105 part 2: 1482–1490. https://doi.org/10.1016/j.ijbiomac.2017.06.058.

Linhardt, R.J., S.A. Ampofo, J. Fareed, D. Hoppensteadt, J.B. Mulliken, and J. Folkman. 1992. 'Isolation and Characterization of Human Heparin'. *Biochemistry* 31 (49): 12441–12445. https://doi.org/10.1021/bi00164a020.

Liu, J., S.I. Wu, Q.J. Li, et al. 2020. 'Different Extraction Methods Bring About Distinct Physicochemical Properties and Antioxidant Activities of Sargassum Fusiforme Fucoidans.' *International Journal of Biological Macromolecules* 155: 1385–1392. https://doi.org/10.1016/j.ijbiomac.2019.11.113.

Lorbeer, A.J., S. Charoensiddhi, and J. Lahnstein. 2017. 'Sequential Extraction and Characterization of Fucoidans and Alginates from Ecklonia Radiata, Macrocystis Pyrifera, Durvillaea Potatorum, and Seirococcus Axillaris'. *Journal of Applied Phycology* 29: 1515–1526. https://doi.org/10.1007/s10811-016-0990-5.

Mabate, Blessing, Chantal Désirée Daub, Samkelo Malgas, Adrienne Lesley Edkins, and Brett Ivan Pletschke. 2021. 'Fucoidan Structure and Its Impact on Glucose Metabolism: Implications for Diabetes and Cancer Therapy'. *Marine Drugs* 19 (1): 30. https://doi.org/10.3390/md19010030.

Maggi, A., M. Abbadini, and P.J. Pagella. 1987. 'Antithrombotic Properties of Dermatan Sulphate in a Rat Venous Thrombosis Model'. *Haemostasis* 17 (6): 329–335. https://doi.org/10.1159/000215765.

Mourão, P.A., M.S. Pereira, M.S. Pavão, B. Mulloy, D.M. Tollefsen, M.C. Mowinckel, and U. Abildgaard. 1996. 'Structure and Anticoagulant Activity of a Fucosylated Chondroitin Sulfate from Echinoderm. Sulfated Fucose Branches on the Polysaccharide Account for Its High Anticoagulant Action'. *The Journal of Biological Chemistry* 271 (39): 23973–23984. https://doi.org/10.1074/jbc.271.39.23973.

Nguyen, Thuan Thi, et al. 2020. 'Enzyme-Assisted Fucoidan Extraction from Brown Macroalgae Fucus Distichus Subsp. Evanescens and Saccharina Latissima'. *Marine Drugs* 18 (6): 296. https://doi.org/10.3390/md18060296.

Park, Hye Young, Min Ho Han, Cheol Park, Cheng-Yun Jin, Gi-Young Kim, Il-Whan Choi, Nam Deuk Kim, Taek-Jeong Nam, Taeg Kyu Kwon, and Yung Hyun Choi. 2011. 'Anti-Inflammatory Effects of Fucoidan Through Inhibition of NF-KB, MAPK and Akt Activation in Lipopolysaccharide-Induced BV2 Microglia Cells'. *Food and Chemical Toxicology* 49 (8): 1745–1752. https://doi.org/10.1016/j.fct.2011.04.020.

Saravana, P.S., Y.J. Cho, Y.B. Park, et al. 2016. 'Structural, Antioxidant, and Emulsifying Activities of Fucoidan from Saccharina Japonica Using Pressurized Liquid Extraction'. *Carbohydr Polym* 153: 518–525. https://doi.org/10.1016/j.carbpol.2016.08.014.

Shikov, Alexander N., Elena V. Flisyuk, Ekaterina D. Obluchinskaya, and Olga N. Pozharitskaya. 2020. 'Pharmacokinetics of Marine-Derived Drugs'. *Marine Drugs* 18 (11): 557. https://doi.org/10.3390/md18110557.

Sichert, A., S. Le Gall, L.J. Klau, et al. 2021. 'Ion-Exchange Purification and Structural Characterization of Five Sulfated Fucoidans from Brown Algae'. *Glycobiology* 31 (4): 352–357. https://doi.org/10.1093/glycob/cwaa064.

Sopelkina, K.I., I.V. Geide, and I.S. Selezneva. 2020. 'Search for Ways to Obtain Fucoidan from Brown Algae Fucus Vesiculosus and Laminariae Thalli'. 2313. https://doi.org/10.1063/5.0032834.

Sugiono, S., and D. Ferdiansyah. 2019. 'Biorefinery Sequential Extraction of Alginate by Conventional and Hydrothermal Fucoidan from the Brown Alga, Sargassum Cristaefolium'. *Bioscience Biotechnology Research Communications* 12 (4): 894–903.

Teixeira, F.C.O.B., E.O. Kozlowski, K.V.A. Micheli, et al. 2018. 'Sulfated Fucans and a Sulfated Galactan from Sea Urchins as Potent Inhibitors of Selectin-Dependent Hematogenous Metastasis'. *Glycobiology* 28 (6): 427–434. https://doi.org/10.1093/glycob/cwy020.

Torabi, P., N. Hamdami, and J. Keramat. 2020. 'Microwave-Assisted Extraction of Fucoidan from Brown Seaweeds of Nizimuddinia Zanardini and Assessment of the Chemical and Antioxidant Properties of the Extracted Compound'. https://doi.org/10.52547/nsft.16.1.61.

Ustyuzhanina, Nadezhda E., Natalia A. Ushakova, Ksenia A. Zyuzina, Maria I. Bilan, Anna L. Elizarova, Oksana V. Somonova, Albina V. Madzhuga, et al. 2013. 'Influence of Fucoidans on Hemostatic System'. *Marine Drugs* 11 (7): 2444–2458. https://doi.org/10.3390/md11072444.

Van Weelden, Geert, Marcin Bobiński, Karolina Okła, Willem Jan Van Weelden, Andrea Romano, and Johanna M.A. Pijnenborg. 2019. 'Fucoidan Structure and Activity in Relation to Anti-Cancer Mechanisms'. *Marine Drugs* 17 (1): 32. https://doi.org/10.3390/md17010032.

Vieira, R.P., B. Mulloy, and P.A. Mourão. 1991. 'Structure of a Fucose-Branched Chondroitin Sulfate from Sea Cucumber. Evidence for the Presence of 3-O-Sulfo-Beta-D-Glucuronosyl Residues'. *The Journal of Biological Chemistry* 266 (21): 13530–13536.

Vilela-Silva, Ana-Cristina E.S., Ana-Paula Alves, Ana-Paula Valente, Victor D. Vacquier, and Paulo A.S. Mourão. 1999. 'Structure of the Sulfated α-L-Fucan from the Egg Jelly Coat of the Sea Urchin Strongylocentrotus Franciscanus: Patterns of Preferential 2-O- and 4-O-Sulfation Determine Sperm Cell Recognition'. *Glycobiology* 9 (9): 927–933. https://doi.org/10.1093/glycob/9.9.927.

Wacker. 2020. 'Sustainability Report 2019/2020'. https://reports.wacker.com/2020/sustainability-report/.

Wang, C.H., and Y.C. Chen. 2016. 'Extraction and Characterization of Fucoidan from Six Brown Macroalgae'. *Journal of Marine Science and Technology* 24 (2): 319–328. https://doi.org/10.6119/JMST-015-0521-3.

Wang, L., J. Fan, S. Guo, et al. 2020. 'Extraction Process Optimization of Fucoidan from Dealginated Kelp Waste'. *Journal of Physics: Conference Series* 1622. https://doi.org/10.1088/1742-6596/1622/1/012039.

Wang, S., et al. 2021. 'Isolation and Purification of Brown Algae Fucoidan from Sargassum Siliquosum and the Analysis of Anti-Lipogenesis Activity'. *Biochemical Engineering Journal* 165: 107798. https://doi.org/10.1016/j.bej.2020.107798.

Yoo, H.J., D.J. You, and K.W. Lee. 2019. 'Characterization and Immunomodulatory Effects of High Molecular Weight Fucoidan Fraction from the Sporophyll of Undaria Pinnatifida in Cyclophosphamide-Induced Immunosuppressed Mice'. *Marine Drugs* 17 (8): 447. https://doi.org/10.3390/md17080447.

Zayed, Ahmed, and Roland Ulber. 2020. 'Fucoidans: Downstream Processes and Recent Applications'. *Marine Drugs* 18 (3): 170. https://doi.org/10.3390/md18030170.

Zhang, Xueqian, et al. 2021. 'Environmental Life Cycle Assessment of Cascade Valorisation Strategies of South African Macroalga Ecklonia Maxima Using Green Extraction Technologies'. *Algal Research* 58.

Zou, Ping, et al. 2020. 'Purification and Characterization of a Fucoidan from the Brown Algae Macrocystis Pyrifera and the Activity of Enhancing Salt-Stress Tolerance of Wheat Seedlings'. *International Journal of Biological Macromolecules* 180: 547–558.

9 Anti-Inflammatory Compounds Derived from Marine Macroalgae

Snezana Agatonovic-Kustrin and David W. Morton

CONTENTS

9.1 INTRODUCTION

Inflammation is a part of the body's immune response as a first line of defense to injury or invasion by pathogenic bacteria, viruses, or cancer cells in the host (Calder 2006). The purpose of inflammation is to control damage and to identify and destroy invading pathogens. Therefore, inflammation is vital for host health in acute disease states. However, if inflammation continues long-term, uncontrolled, at a subclinical level, the activated immune system can start to damage host tissues and enhance chronic disease states such as cardiovascular disease (CVD) (Hansson 2005), inflammatory bowel disease (Fiocchi 1998; Okamoto and Watanabe 2015), cancer (Crusz and Balkwill 2015), diabetes (Pickup 2004), asthma (Murdoch and Lloyd 2010), and Alzheimer's disease (Akiyama et al. 2000). Nonsteroidal anti-inflammatory drugs (NSAIDs) and steroidal anti-inflammatory drugs (SAIDs) that are commonly used in the treatment of these diseases have serious side effects (Marcum and Hanlon 2010). Thus, there is great interest in the natural anti-inflammatory compounds in products such as dietary supplements and herbal remedies due to the lower incidence of side effects.

Many of these natural compounds also work by inhibiting the inflammatory pathways in a similar manner to NSAID drugs. Current interest in natural products as anti-inflammatory drug leads has resulted in an increased focus on the potential of marine organisms as a resource for these types of compounds. The biological and chemical diversity of marine environments means there is an enormous number of potential compounds in marine organisms that may have anti-inflammatory activity. Extensive studies on the health benefits

DOI: 10.1201/9781003303916-9

of brown, red, and green macroalgae, their uses as food or as drug carriers, and on the bioactive natural products they produce, have been published in recent years. Marine algae are a large heterogeneous group of ancient plants, with different evolutionary roots. Diverse metabolic pathways have led to an abundance of novel chemical compounds with unique chemical structures and a broad range of biological activities. Due to the current advancement in chemical structure characterization, organic synthesis, and bioassays, natural products from marine organisms or their derivatives with specific biological activities can be easily identified, synthesized, and evaluated (Simmons et al. 2005). Several diterpenes, sesquiterpenoids (tri isoprene units), steroids, polysaccharides, and other chemical compounds isolated from marine organisms, have shown anti-inflammatory activity. Diterpenes are often present in soft corals as secondary metabolites. Several sesquiterpenoids isolated from the soft coral *Rumphella antipathies* (family Gorgoniidae) were found to exhibit in vitro anti-inflammatory activity (Chang et al. 2020). Several steroids, with potent inhibitory effects on the production of all three pro-inflammatory cytokines (interleukin [IL]-1, IL-6, and tumor necrosis factor [TNF]-α), have been isolated from the starfish *Astropecten polyacanthus* (Thao et al. 2013). Marine sponges are also particularly a rich source of steroids with potent anti-inflammatory activity (Anjum et al. 2016).

9.2 PHLOROTANNINS

Phlorotannins are a group of polyphenolic tannin derivatives with diverse molecular structures, that are produced by the polymerization of phloroglucinol (benzene-1,3,5-triol). Among all marine algae, the members of the Laminariaceae (brown algae) family are reported to be the richest source of phlorotannins (Thomas and Kim 2011). The structures of many of these phlorotannin compounds could drive the development of many new drugs, since they exhibit antioxidant, anti-inflammatory, antidiabetic, antitumor, antihypertensive, and anti-allergic activities (Wijesekara et al. 2010).

Increasing attention has been paid to the anti-inflammatory activity of phlorotannins from seaweeds, especially from the *Eisenia* and *Ecklonia* genera due to their wide distribution and ecological importance (Bolton 2010). These compounds have the ability to down-regulate inducible nitric oxide synthase (iNOS) expression in cells exposed to lipopolysaccharide (LPS), thus exerting an anti-inflammatory action by the reduction of nitric oxide (NO) production. Nitric oxide is an important biological mediator, a signaling molecule that plays a key role in the pathogenesis of inflammation. It gives an anti-inflammatory effect under normal physiological conditions (Giles 2006). However, excessive production of NO, catalyzed by iNOS, is pathogenic for host tissues (Aktan 2004). Hence, inhibition of NO accumulation is a beneficial therapeutic strategy for the treatment of NO-mediated conditions. It is becoming evident that NO is also a major signaling molecule in plants and is involved in multiple plant physiological functions (Palavan-Unsal and Arisan 2009).

Jung et al. (2013) evaluated the anti-inflammatory activity of edible brown alga *Eisenia bicyclis* by studying the inhibition of LPS-induced NO and tert-butylhydroperoxide (t-BHP)-induced reactive oxygen species (ROS), along with suppression against expression of iNOS, and cyclooxygenase-2 (COX-2). The six phlorotannins that they isolated from *Eisenia bicyclis*, phloroglucinol, eckol, dieckol, 7-phloroeckol, dioxinodehydroeckol, and phlorofucofuroeckol A (Figure 9.1a–f), showed a dose-dependent inhibition of LPS-induced NO production at non-toxic concentrations in LPS-induced RAW 264.7 murine macrophage cells. Sugiura and co-workers showed that eckol, phlorofucofuro-eckol A and B, and 8,8′-bieckol (Figure 9.1c, f–h) from the same seaweed species had good anti-inflammatory activity in an in vivo model (mouse ear edema) (Sugiura et al. 2013). Kim and colleagues evaluated both the antioxidant and anti-inflammatory activity of phlorotannins isolated from *Ecklonia* species and confirmed that phlorofucofuroeckol A, significantly inhibited the LPS-induced production of NO and PGE2, through the downregulation of iNOS and COX-2 protein expression in LPS-induced RAW 264.7 murine macrophage cells (Kim et al. 2009).

The anti-inflammatory activity of many other brown seaweed extracts has also been attributed to these polyphenols. Organic fractions (chloroform, ethyl acetate and methanol) from the Mediterranean brown seaweeds, *Cystoseira sedoides* and *Cystoseira compressa*, were evaluated for both in vivo anti-inflammatory activity, using a carrageenan-induced rat paw edema model, and in vitro antiproliferative effects using different cell lines. The chloroform and ethyl acetate fractions were found to exhibit dose-dependent, anti-inflammatory activity (Mhadhebi et al. 2011, 2012; Mhadhebi et al. 2012).

9.3 MONOGALACTOSYLDIACYLGLYCEROLS AND DIGALACTOSYLDIACYLGLYCEROLS

Galacto-glycerolipids are widely found in plants and photosynthetic bacteria. They are a class of compounds, in which galactose is bound at the glycerol sn-3 position in O-glycosidic linkage to diacylglycerol. While phosphoglycerolipids make the primary building blocks of eukaryotic and prokaryotic cell membranes, non-phosphorous galactoglycerolipids are the main building blocks of plant cells (Benning and Ohta 2005). Galacto-glycerolipids are an important part of plant cell membranes where they constitute the bulk of the polar lipids in photosynthetic membranes. Moreover, galactolipids are the most widespread group of non-phosphorous lipids, being the major constituents of the photosynthetic membranes of higher plants, algae, and bacteria (Dörmann and Benning 2002). They account for 80% of the membrane lipids found in green plant tissues. In contrast, to membranes of animals and yeasts, where phospholipids are the main lipid group, galactolipids are major constituents of the photosynthetic membranes of higher plants, algae and bacteria (Dörmann and Benning 2002). Lipids in plants consists mainly of monogalactosyldiacylglycerols and digalactosyldiacylglycerols (MGDG and DGDG) containing one or two saturated or unsaturated fatty acids linked to the glycerol part of the molecule (Figure 9.2).

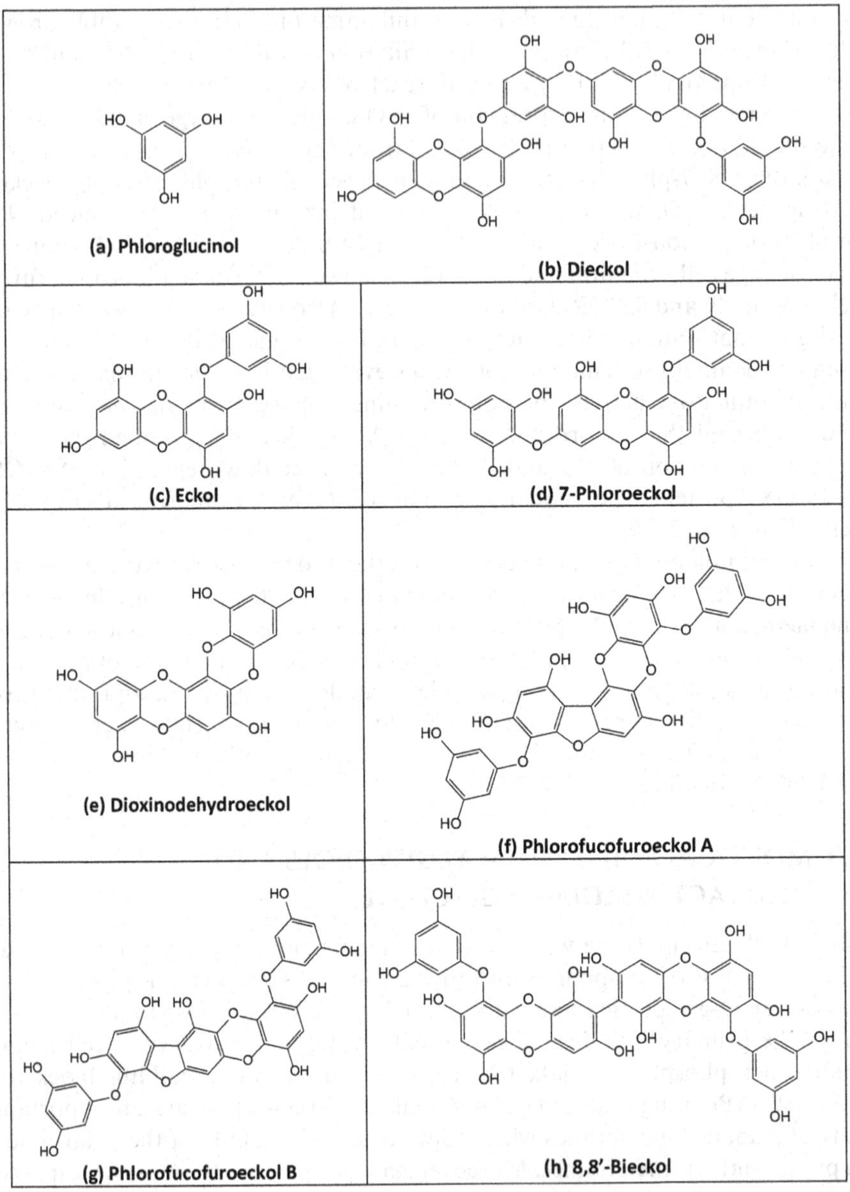

(a) Phloroglucinol

(b) Dieckol

(c) Eckol

(d) 7-Phloroeckol

(e) Dioxinodehydroeckol

(f) Phlorofucofuroeckol A

(g) Phlorofucofuroeckol B

(h) 8,8′-Bieckol

FIGURE 9.1 The molecular structures of anti-inflammatory phlorotannins, (a) phloroglucinol, (b) dieckol, (c) eckol, (d) 7-phloroeckol, (e) dioxinodehydroeckol, (f) phlorofucofuroeckol A, (g) phlorofucofuroeckol B, and (h) 8,8′-bieckol, isolated from *Eisenia bicyclis*.

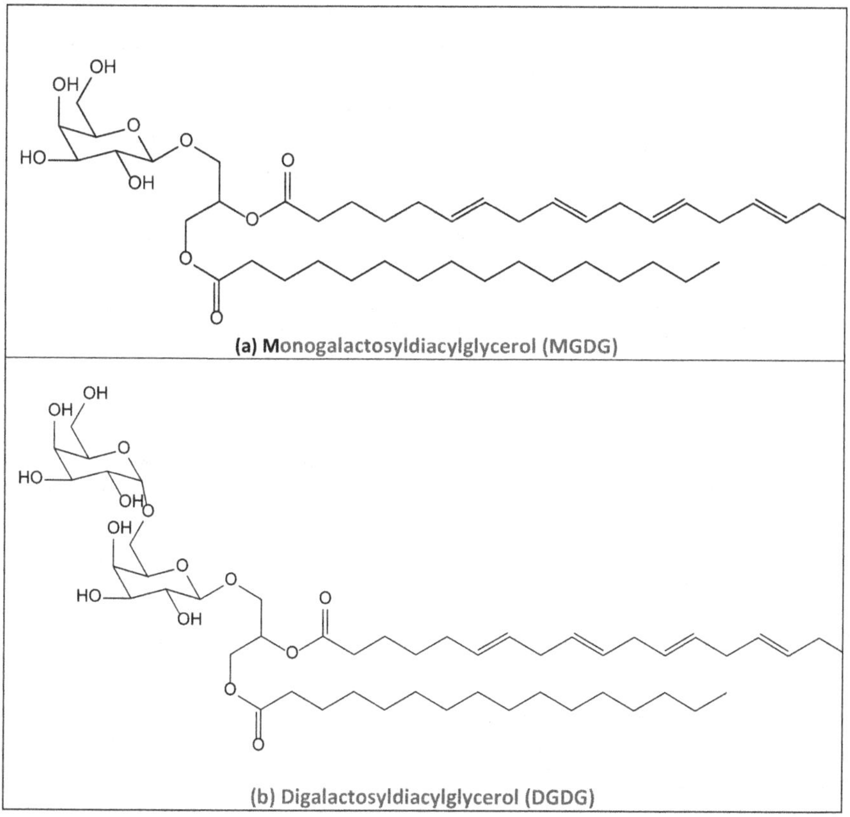

(a) Monogalactosyldiacylglycerol (MGDG)

(b) Digalactosyldiacylglycerol (DGDG)

FIGURE 9.2 The molecular structures of (a) monogalactosyldiacylglycerol (MGDG) and (b) digalactosyldiacylglycerol (DGDG).

As in higher plants, the proportions of MGDG are higher in the lipids from marine algae exposed to white light (Radwan et al. 1988).

Potent in vitro anti-inflammatory activity of MGDG in cultured human articular chondrocytes through the p38 and transcription nuclear factor-kappa B (NF-κB) pathways inhibition has been reported (Ulivi et al. 2011a). The p38 MAPK (mitogen-activated protein kinase) is an integral component of proinflammatory signaling cascades in various cell types (Kumar et al. 2003), while the NF-κB serves as a key mediator of inflammatory responses.

The freshwater *Phormidium* sp. ETS-05 thermophile deep blue-green alga, is a typical species of cyanobacteria found in the muds of the therapeutic thermal hot springs of Abano and Montegrotto, Italy. Montegrotto has been famous for its mud and spa treatments since Roman times. The physical and chemical features of the waters in the thermal hot springs severely limit the survival of photosynthetic organisms. The Cyanidiales, a group of asexual unicellular red algae (Ciniglia et al. 2004), are limited to acidic (pH <3) high temperature conditions, while in hot springs with an alkaline/neutral pH, cyanobacteria dominate. Since most species cannot tolerate higher temperatures (50–60 °C), bacterial diversity is limited to blue-green cyanobacteria. MGDG and DGDG galactolipids with a high content of polyunsaturated fatty acids, have been purified from *Phormidium* sp. that colonize the thermal mud (Lenti et al. 2009). A potent in vitro anti-inflammatory activity of isolated MGDG through the p38 and NF-κB pathways inhibition has been reported (Ulivi et al. 2011b). Cyanobacterium also transforms the surface part of the thermal mud, with its characteristic green color, and during this process releases exopolysaccharide (EPS) molecules in large quantities (Gris et al. 2020).

When algae and plants are exposed to abiotic or biotic stress, polyunsaturated fatty acids can be released from galacto-glycerolipids (Dhondt et al. 2000). Phosphoglycerolipids can also provide the polyunsaturated fatty acids required for the generation of oxylipins, and in the red alga *Gracilaria*, both galactoglycerolipids and phosphoglycerolipids are utilized (Lion et al. 2006). Although the mechanism of their anti-inflammatory action is not completely understood, it is known that glycerolipids with higher levels of unsaturation are more effective in inhibiting iNOS. In the same way, this action is also dependent on the position of the unsaturated carbon atoms in a glycerolipid.

9.4 OXYLIPINS AND POLYUNSATURATED FATTY ACIDS

Oxylipins are bioactive lipid metabolites, oxidized derivatives of polyunsaturated fatty acids (PUFAs) that function as tissue hormones in mammals but also as central hormones in plants. The oxylipin pathway is initiated by the formation of fatty acid hydroperoxides, either by oxidation catalyzed by enzymes, such as COX, lipoxygenase, and cytochrome P450, or by chemical (auto) oxidation induced by free radicals and ROS (Harwood 2019). The PUFA precursors for oxylipins synthesis are derived from the diet, or through the elongation and desaturation of essential fatty acids, released from membranes by lipase activity.

Besides common oxygenated fatty acid derivatives, macroalgae also contain several unique and complex oxylipins, such as bicyclic cymathere ethers, cyclopropyl hydroxyeicosanoids, cymatherelactones, cymatherols, egregiachlorides, ecklonialactones, and hybridalactones (Gabbs et al. 2015; Harwood 2019; Jiang and Gerwick 1997; Choi et al. 2012) (Figure 9.3). The great diversity of oxylipins in macroalgae can be partially explained by the variation in the oxygenation position, due mainly to the way in which lipoxygenase catalyzes oxylipin formation.

Oxylipins are important signaling molecules that regulate a variety of events associated with physiological and pathological processes. In mammals, eicosanoids that are derived from unsaturated C_{20} fatty acids, regulate the initiation and resolution of inflammation (Jagusch et al. 2020). In plants, oxylipins mainly serve as signal molecules regulating developmental processes, plant stress response, and innate immunity (Blée 2002). Marine algae also produce structurally diverse bioactive oxylipins as a defense to physical or abiotic stress (e.g. freeze-thawing), biotic stress (attack by pathogens) and to provide innate immunity (Weinberger et al. 2011). Plant oxylipins are mostly derived from

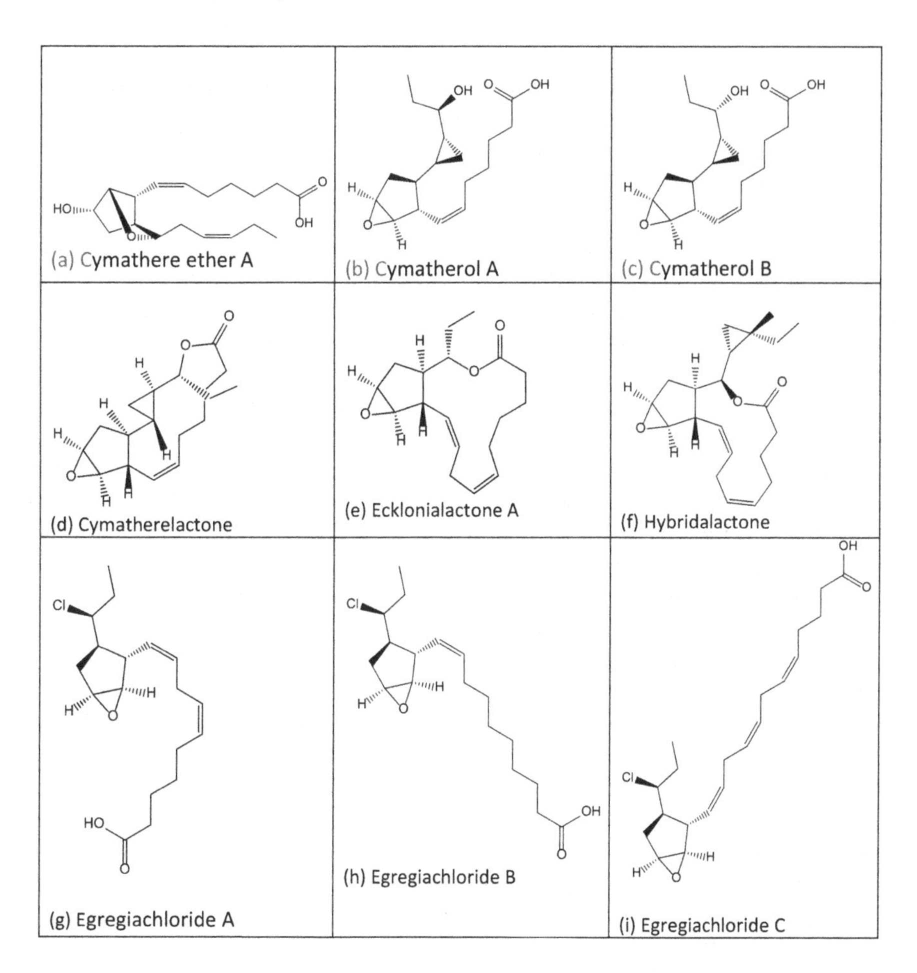

FIGURE 9.3 The molecular structures of some oxylipins unique to algae: (a) cymathere ether A, (b) cymatherol A, (c) cymatherol B, (d) cymatherelactone, (e) ecklonialactone A, (f) hybridalactone, (g) egregiachloride A, (h) egregiachloride B, and (i) egregiachloride C.

linoleic and more importantly α-linolenic acids that are released from their lipid associations by various acyl hydrolases (lipases). Kaye and co-workers found that some algae species required more α-linoleic acid when exposed to lower temperature, higher salinity or nitrogen starvation (Kaye et al. 2015). Algal oxylipins may help in interactions with the environment and with other organisms, helping algae survive in extreme conditions, being continuously challenged by a number of potentially pathogenic organisms and exposed to ecological changes.

Although the proportions of the polar membrane lipids vary considerably depending on the species (Li-Beisson et al. 2019), all algae have a high concentration of PUFAs. PUFAs are represented by two classes, either omega-3 (ω-3) or omega-6 (ω-6), depending on the position of the terminal double bond. They are vital for the formation and functioning of cell membranes, as well as the immune system. While ω-3 PUFAs are mostly found in fish oil and animal sources, the ω-6 PUFAs are present in vegetable oils.

The anti-inflammatory and hypotriglyceridemic effects of ω-3 PUFAs, including α-linolenic acid, docosapentaenoic acid, and eicosapentaenoic acid are well known, whereas pro-inflammatory properties have been recognized in the ω-6 PUFAs (D'Angelo et al. 2020). Oxylipins derived from the ω-6 fatty acids (arachidonic acid, γ-linoleic acid) generally increase inflammation, hypertension, and platelet aggregation and generally have more adverse than beneficial cardiovascular effects. Most oxylipins derived from ω-3 fatty acids (α-linolenic acid), have anti-inflammatory, antiaggregatory, and vasodilatory effects that help explain the cardioprotective effects of these fatty acids. For instance, a study by Caligiuri and co-workers found that the addition of a daily dose of 30 g of milled flaxseed (rich in α-linolenic acid) into the diet of patients with peripheral artery disease, resulted in a significant reduction in central blood pressure after 6 months (Caligiuri et al. 2016).

Both ω-3 and ω-6 fatty acids are important for health, however there is an important distinction. ω-3 fatty acids regulate cellular metabolic functions and gene expression in a manner that reduces inflammation (Deckelbaum et al. 2006), while ω-6 -fatty acids, particularly arachidonic acid, can promote inflammation when consumed in excessive amounts (Patterson et al. 2012). Although both classes of fatty acids promote health, there should be an equilibrium between ω-3 and ω-6 in our diet due to this distinction. Omega-3 fatty acids regulate cellular metabolic functions and gene expression in a manner that reduces inflammation, while omega 6-fatty acids, particularly arachidonic acid, can promote inflammation when consumed in excessive amounts (Deckelbaum and Torrejon 2012; Deckelbaum et al. 2012). Increased amounts of ω-6 over ω-3 PUFAs seems to be directly proportional to the increased the risk of acute diseases like cardiovascular disease and the major cause of chronic diseases as cancer, cardiovascular diseases, and diabetes.

Marine macroalgae are an excellent source of PUFAs with a ω-6/ω-3 fatty acid ratio lower than 10/1. This ratio is recommended by the World Health Organization (WHO) to prevent inflammatory, cardiovascular and neurological chronic diseases (Molendi-Coste et al. 2011). The Western human diet is characterized by an increased consumption of fat and vegetable oils which are rich in ω-6 PUFAs and a decreased consumption of ω-3 PUFA-rich foods. This leads to an increase in ω-6/ω-3 ratio of 10–20/1 compared to a ratio of around 1/1 in the diet of our ancestors (Molendi-Coste et al. 2011).

The type and composition of fatty acids in marine algae greatly depends on species and even more on habitat conditions. Brown algae can have significant amount of arachidonic acid (ω-6), while red algae which are most likely to be found in deep water, are often enriched in eicosapentaenoic acid (ω-3) (Harwood 2019). Eicosapentaenoic acid (ω-3) can be found together with arachidonic acid in algae such as *Chondrus crispus*, or without arachidonic acid as found in *Palmaria palmata* (Fleurence et al. 1994). In the red alga *Palmaria palmata*, eicosapentaenoic acid (ω-3) is a major fatty acid and represents half of the total fatty acids present, while arachidonic and linoleic acids (ω-6) are in negligible concentration. Out of seventeen macroalgae investigated as a potential dietary source of PUFAs, species from the red and brown phyla showed higher concentrations of the ω-3 PUFA family. However, the *Ulva sp.* was the only green alga investigated that had a high concentration of α-linolenic acid (ω-3) (Pereira et al. 2012).

Marine algae in their natural habitants occur at various depths and are consequently exposed to light of different qualities through the day. Thus, algal habitat conditions can quantitatively affect the characteristics of the fatty acids (Khotimchenko et al. 2002). It has been found that it is possible to enrich certain marine algae with arachidonic acid by keeping them under continuous illumination with light of particular wavelengths. An optimum yield of arachidonic acid in the total algal lipids present in *Ulva (Enteromorpha) intestinalis* (green grass kelp) was 45% when it was grown under white light. For *Sargassum salicifolium* (brown algae) an optimum yield of 25% was observed when it was grown under red light (Radwan et al. 1988). The genus *Ulva* has become a model for investigating complex metabolic pathways, due to its high growth rate and natural ability to grow under a wide range of environmental conditions. It was observed that intertidal alga *Ulva lactuca* is able to alter its lipid metabolism and biosynthesis, including a change in lipid classes, fatty acids, and oxylipins, when exposed to nitrate and phosphate nutritional stress (Kumari et al. 2014).

9.5 POLYSACCHARIDES

Polysaccharides of marine origin (especially from macroalgae), such as fucans (fucoidans), carrageenans, galactans, agarans, and ulvans, have shown to exert a variety of biological effects. They are components of the cell wall (extracellular matrix), with important roles in mechanical, osmotic and ionic regulation. Although they are mainly used as thickening agents in food, cosmetics and pharmaceutical products, due to their unique physicochemical and rheological properties, there is an increasing interest in their bioactive properties (Laurienzo 2010).

In contrast to land polysaccharides, most marine polysaccharides are highly sulfated. Their biodegradation requires glycoside hydrolase enzymes to cleave the glycosidic bonds of the carbohydrate backbone, and polysaccharide sulfatases to cleave the sulfate ester groups. The introduction of sulfated groups improves the bioactivity of polysaccharides, especially their antioxidant and anti-inflammatory activities (Chen et al. 2015).

Note that the type of sugar units, glycosidic linkaging and branching, and degree and position of sulfation in these negatively charged polymers varies between algal species. Some of these polysaccharides are linear, and some are multi-branched. The fucans, often called fucoidans, are polysaccharides commonly extracted from brown (Phaeophyta) algae (Figure 9.4). Galactans found in red algae (Rhodophyta) consist entirely of galactose or modified galactose units, such as agar and carrageenans. The major polysaccharides in green

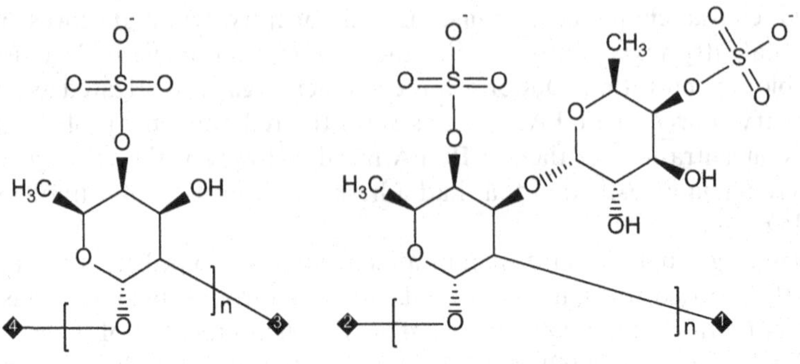

FIGURE 9.4 The backbone structure of fucoidan polysaccharides.

algae (Chlorophyta) are a group of polydisperse heteropolysaccharides known as ulvans.

Fucoidans are complex fucose enriched sulfated anionic polymers found in cell walls of brown algae. They are primarily composed of repeating units of disaccharides with α–1,3-linked and α–1,4-linked fucose (Berteau and Mulloy 2003). Besides fucose, other monosaccharides such as galactose, xylose, glucose, and mannose are also present, together with minor amounts of uronic acids and glucosamines (Cumashi et al. 2007). Sulphation of fucoidans is found at 2-O, 4-O, or both positions, and in rare cases at the 3-O position of the fucopyranose residue. Their polysaccharide chains are not only linear but can also be highly branched. The molar mass of fucoidans differs from one species to the other and can range from 13 kDa to 1300 kDa (Holtkamp 2009).

Their mechanism of anti-inflammatory activity is believed to be similar to the action of heparins (Colliec-Jouault et al. 2012). Fucoidans extracted from different sources such as *Fucus vesiculosus* and *Laminaria japonica*, down-regulate the NF-κB signaling pathway and reduce the levels of several pro-inflammatory cytokines such as IL-6 and TNF-α and matrix metalloproteinases (MMPs) (Fernando et al. 2017). The adhesion of leukocytes to vascular endothelium is a hallmark of the inflammatory process. Fucoidans interfere with P- (platelet) and L- (leukocyte) selectins, cell adhesion molecules essential in the recruitment process. Selectins play important roles in leukocyte trafficking to the sites of inflammation. The rolling of leukocytes along the endothelium is mediated by selectins. Thus, fucoidans prevent leukocyte rolling on the endothelium before their adhesion and extravasation from circulation into the inflamed site (Carvalho et al. 2014). The anti-inflammatory effect of fucoidans is also attributed to the suppression of inducible iNOS expression.

The sulfated polysaccharides that are present in cell walls of red algae (Rhodophyta) are known as galactans. This large family of hydrocolloids, well known for their gelling properties, is made up of linear chains of galactose, with alternating α(1,3) and β(1,4) linkages. The enantiomeric configuration of galactose, D- or L-, in 1,3-linked residues, distinguish agaroids from carrageenans (Stortz and Cerezo 2000). In agarose, the α-linked galactose units are

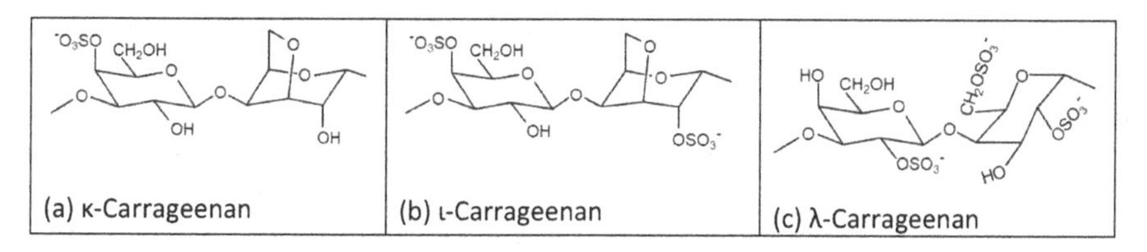

FIGURE 9.5 Monomeric units for the three most commercially used carrageenans: (a) κ-carrageenan, (b) ι-carrageenan, and (c) λ-carrageenan.

in the L configuration (L unit), whereas they are in the D configuration (D unit) in carrageenans (Rees 1969).

Carrageenan is formed by a disaccharide repeating unit which consists of alternating 3-linked β-d-galactopyranose or 4-linked α-d-galactopyranose or 4-linked 3,6-anhydro-α-d-galactopyranose. Carrageenans are classified as iota-carrageenan (ι-carrageenan), kappa-carrageenan (κ-carrageenan), and lambda-carrageenan (λ-carrageenan) according to the number and the position of sulfated ester groups and by the occurrence of 3,6 anhydro-bridges in the α-linked residues (DA unit) found in gelling carrageenans (Knutsen et al. 1994; Rees 1969). The three most commercially used carrageenans, kappa (κ), iota (ι) and lambda (λ) (Figure 9.5), are distinguished by the presence of one, two, and three ester-sulfate groups per repeating disaccharide unit, respectively. They are said to be nontoxic, biodegradable, and biocompatible, and so are of great interest to researchers in the pharmaceutical, cosmetic, and food industries (Prajapati et al. 2014). Carrageenans are approved for use as food additive by the U.S. Food and Drug Administration (USFDA) (FDA 2020) and in most other countries. They are widely used in the food industry as thickeners, food stabilizers, and gelling agents. Due to their strong negative charge, controlled drug release and gel forming abilities, carrageenans are also widely used in drug formulations (Li et al. 2014).

However, there is still substantial controversy about their safety and potential health risks associated with human consumption. Some studies have provided evidence to show that carrageenan is highly inflammatory and toxic to the digestive tract. However, the specific mechanism by which carrageenan induces inflammation in experimental animal models is not clearly defined. Carrageenan has been demonstrated to decrease the amount of epithelial glycoproteins in the colon (Al-Suhail et al. 1984). Rabbits fed with a 1% aqueous solution of degraded carrageenan developed a progressive colitis, characterized by severe inflammation and mucosal ulceration after 5 days. Further experiments revealed that carrageenan (especially kappa-carrageenan), is capable of inhibiting the interaction between macrophages and lymphocytes (Chong and Parish 1985).

The study of kappa-carrageenans from the red alga *Solieria filiformis* on in vivo animal models of nociception and inflammation, showed the involvement of prostaglandins, NO, and primary cytokines (de Araújo et al. 2011). The results suggested that this sulfated polysaccharide may be used as a tool to study the inflammatory processes associated with nociception (de Araújo et al. 2011). It was also reported that a sulfated polysaccharide from the red marine algae *Champia feldmannii* Díaz-Piferrer (Champiaceae), shows edematogenic activity, paralleled with increase in vascular permeability and leukocyte migration (Assreuy et al. 2008).

The cell wall matrix of green seaweed contains highly sulfated heteropolysaccharides named ulvans. Ulvan molecules are mainly composed of sulfated L-rhamnose, uronic acids (D-glucuronic acid and its C5-epimer L-iduronic acid) and of a minor fraction of D-xylose (Brading et al. 1954; Lahaye and Robic 2007; McKinnell and Percival 1962) in a linear arrangement (Jaulneau et al. 2010). The information on the structures and applications of the polysaccharides from green algae is scarce. Several bioactivities were reported such as antioxidant activity (Qi et al. 2005) and potential antihyperlipidemic activity (Pengzhan et al. 2003).

REFERENCES

Akiyama, H., S. Barger, S. Barnum, B. Bradt, J. Bauer, G.M. Cole, N.R. Cooper, P. Eikelenboom, M. Emmerling, B.L. Fiebich, C.E. Finch, S. Frautschy, W.S. Griffin, H. Hampel, M. Hull, G. Landreth, L. Lue, R. Mrak, I.R. Mackenzie, P.L. McGeer, M.K. O'Banion, J. Pachter, G. Pasinetti, C. Plata-Salaman, J. Rogers, R. Rydel, Y. Shen, W. Streit, R. Strohmeyer, I. Tooyoma, F.L. Van Muiswinkel, R. Veerhuis, D. Walker, S. Webster, B. Wegrzyniak, G. Wenk, and T. Wyss-Coray. 2000. "Inflammation and Alzheimer's disease." *Neurobiol Aging*. 21 (3): 383–421.

Aktan, F. 2004. "iNOS-mediated nitric oxide production and its regulation." *Life Sci.* 75 (6): 639–653.

Al-Suhail, A.A., P.E. Reid, C.F.A. Culling, W.L. Dunn, and M.G. Clay. 1984. "Studies of the degraded carrageenan-induced colitis of rabbits. II. Changes in the epithelial glycoprotein O-acylated sialic acids associated with the induction and healing phases." *Histochem J.* 16 (5): 555–564.

Anjum, K., S.Q. Abbas, S.A.A. Shah, N. Akhter, S. Batool, and S.S. ul Hassan. 2016. "Marine sponges as a drug treasure." *Biomol Ther*. 24 (4): 347–362.

Assreuy, A.M.S., D.M. Gomes, M.S.J. da Silva, V.M. Torres, R.C.L. Siqueira, A. de Freitas Pires, D.N. Criddle, N.M.N. de Alencar, B.S. Cavada, and A.H. Sampaio. 2008. "Biological effects of a sulfated-polysaccharide isolated from the marine red algae *Champia feldmannii*." *Biol Pharm Bull*. 31 (4): 691–695.

Benning, C., and H. Ohta. 2005. "Three enzyme systems for galactoglycerolipid biosynthesis are coordinately regulated in plants." *J Biol Chem*. 280 (4): 2397–2400.

Berteau, O., and B. Mulloy. 2003. "Sulfated fucans, fresh perspectives: Structures, functions, and biological properties of sulfated fucans and an overview of enzymes active toward this class of polysaccharide." *Glycobiology*. 13 (6): 29R–40R.

Blée, E. 2002. "Impact of phyto-oxylipins in plant defense." *Trends Plant Sci.* 7 (7): 315–322.

Bolton, J.J. 2010. "The biogeography of kelps (Laminariales, Phaeophyceae): A global analysis with new insights from recent advances in molecular phylogenetics." *Helgol. Mar. Res*. 64: 263–279.

Brading, J.W.E., M.M.T. Georg-Plant, and D.M. Hardy. 1954. "The polysaccharide from the alga *Ulva lactuca*. Purification, hydrolysis, and methylation of the polysaccharide." *J Chem Soc*. 319–324.

Calder, P.C. 2006. "n−3 Polyunsaturated fatty acids, inflammation, and inflammatory diseases." *Am J Clin Nutr*. 83 (6): 1505S–1519S.

Caligiuri, S.P.B., D. Rodriguez-Leyva, H.M. Aukema, A. Ravandi, W. Weighell, R. Guzman, and G.N. Pierce. 2016. "Dietary flaxseed reduces central aortic blood pressure without cardiac involvement but through changes in plasma oxylipins." *Hypertension*. 68 (4): 1031–1038.

Carvalho, A.C., R.B. Sousa, Á.X. Franco, J.V. Costa, L.M. Neves, R.A. Ribeiro, R. Sutton, D.N. Criddle, P.M. Soares, and M.H. de Souza. 2014. "Protective effects of fucoidan, a P- and L-selectin inhibitor, in murine acute pancreatitis." *Pancreas*. 43 (1): 82–87. https://doi.org/10.1097/MPA.0b013e3182a63b9d.

Chang, Y.-C., C.-C. Chiang, Y.-S. Chang, J.-J. Chen, W.-H. Wang, L.-S. Fang, H.-M. Chung, T.-L. Hwang, and P.-J. Sung. 2020. "Novel Caryophyllane-Related Sesquiterpenoids with Anti-Inflammatory Activity from *Rumphella antipathes* (Linnaeus, 1758)." *Mar Drugs*. 18 (11): 554. https://doi.org/10.3390/md18110554.

Chen, Y., H. Zhang, Y. Wang, S. Nie, C. Li, and M. Xie. 2015. "Sulfated modification of the polysaccharides from *Ganoderma atrum* and their antioxidant and immunomodulating activities." *Food Chem* 186: 231–238. https://doi.org/10.1016/j.foodchem.2014.10.032.

Choi, H., P.J. Proteau, T. Byrum, A.R. Pereira, and W.H. Gerwick. 2012. "Cymatherelactone and cymatherols A-C, polycyclic oxylipins from the marine brown alga *Cymathere triplicata*." *Phytochemistry*. 73 (1): 134–141. https://doi.org/10.1016/j.phytochem.2011.09.014.

Chong, A.S., and C.R. Parish. 1985. "Nonimmune lymphocyte-macrophage interaction: I. Quantification by an automated colorimetric assay." *Cell Immunol*. 92 (2): 265–276.

Ciniglia, C., H.S. Yoon, A. Pollio, G. Pinto, and D. Bhattacharya. 2004. "Hidden biodiversity of the extremophilic Cyanidiales red algae." *Mol Ecol*. 13 (7): 1827–1838.

Colliec-Jouault, S., C. Bavington, and C. Delbarre-Ladrat. 2012. "Heparin-like entities from marine organisms." In *Heparin—A Century of Progress. Handbook of Experimental Pharmacology*, edited by R. Lever, B. Mulloy and C. Page, 423–449. Berlin and Heidelberg: Springer.

Crusz, S.M., and F.R. Balkwill. 2015. "Inflammation and cancer: Advances and new agents." *Nat Rev Clin Oncol*. 12 (10): 584–596. https://doi.org/10.1038/nrclinonc.2015.105.

Cumashi, A., N.A. Ushakova, M.E. Preobrazhenskaya, A. D'Incecco, A. Piccoli, L. Totani, N. Tinari, G.E. Morozevich, A.E. Berman, and M.I. Bilan. 2007. "A comparative study of the anti-inflammatory, anticoagulant, antiangiogenic, and antiadhesive activities of nine different fucoidans from brown seaweeds." *Glycobiology*. 17 (5): 541–552.

D'Angelo, S., M. L. Motti, and R. Meccariello. 2020. "ω-3 and ω-6 Polyunsaturated fatty acids, obesity and cancer." *Nutrients*. 12 (9): 2751. www.mdpi.com/2072-6643/12/9/2751.

de Araújo, I.W.F., E. de S. O. Vanderlei, J.A.G. Rodrigues, C.O. Coura, A.L.G. Quinderé, B.P. Fontes, I.N.L. de Queiroz, R.J.B. Jorge, M.M. Bezerra, A.A.R. e Silva, H.V. Chavese, H.S.A. Monteiro, R.C.M. de Paula, and N.M.B. Benevides. 2011. "Effects of a sulfated polysaccharide isolated from the red seaweed Solieria filiformis on models of nociception and inflammation." *Carbohydr Polym*. 86 (3): 1207–1215.

Deckelbaum, R.J., P.C. Calder, W.S. Harris, C.C. Akoh, K.C. Maki, J. Whelan, W.J. Banz, and E. Kennedy. 2012. "Conclusions and recommendations from the symposium, heart healthy omega-3s for food: Stearidonic acid (SDA) as a sustainable choice." *J Nutr*. 142 (3): 641S–643S. https://doi.org/10.3945/jn.111.149831.

Deckelbaum, R.J., and C. Torrejon. 2012. "The omega-3 fatty acid nutritional landscape: Health benefits and sources." *J Nutr*. 142 (3): 587S–591S. https://doi.org/10.3945/jn.111.148080.

Deckelbaum, R.J., T.S. Worgall, and T. Seo. 2006. "n−3 fatty acids and gene expression." *Am J Clin Nutr*. 83 (6): 1520S–1525S.

Dhondt, S., P. Geoffroy, B.A. Stelmach, M. Legrand, and T. Heitz. 2000. "Soluble phospholipase A2 activity is induced before oxylipin accumulation in tobacco mosaic virus-infected tobacco leaves and is contributed by patatin-like enzymes." *Plant J*. 23: 431–440.

Dörmann, P., and C. Benning. 2002. "Galactolipids rule in seed plants." *Trends Plant Sci*. 7 (3): 112–118. https://doi.org/10.1016/s1360-1385(01)02216-6.

FDA. 2020. *Sec. 172.620 Carrageenan*. U.S. Food and Drug Administration (Silver Spring, MD, USA: U.S. Food and Drug Administration, 10903 New Hampshire Avenue and MD 20993 Silver Spring).

Fernando, I.P.S., K.K.A. Sanjeewa, K.W. Samarakoon, W.W. Lee, H.-S. Kim, N. Kang, P. Ranasinghe, H.-S. Lee, and Y.-J. Jeon. 2017. "A fucoidan fraction purified from *Chnoospora minima*: A potential inhibitor of LPS-induced inflammatory responses." *Int J Biol Macromol*. 104: 1185–1193.

Fiocchi, C. 1998. "Inflammatory bowel disease: Etiology and pathogenesis." *Gastroenterology*. 115 (1): 182–205.

Fleurence, J., G. Gutbier, S. Mabeau, and C. Leray. 1994. "Fatty acids from 11 marine macroalgae of the French Brittany coast." *J Appl Phycol*. 6 (5–6): 527–532. https://doi.org/10.1007/BF02182406.

Gabbs, M., S. Leng, J.G. Devassy, M. Monirujjaman, and H.M. Aukema. 2015. "Advances in our understanding of oxylipins derived from dietary PUFAs." *Adv Nutr*. 6 (5): 513–540. https://doi.org/10.3945/an.114.007732.

Giles, T.D. 2006. "Aspects of nitric oxide in health and disease: A focus on hypertension and cardiovascular disease." *J Clin Hypertens (Greenwich)*. 8: 2–16.

Gris, B., L. Treu, R.M. Zampieri, F. Caldara, C. Romualdi, S. Campanaro, and N. La Rocca. 2020. "Microbiota of the therapeutic euganean thermal muds with a focus on the main cyanobacteria species." *Microorganisms*. 8 (10): 1590.

Hansson, G.K. 2005. "Inflammation, atherosclerosis, and coronary artery disease." *N Engl J Med*. 352 (16): 1685–1695. https://doi.org/10.1056/NEJMra043430.

Harwood, J.L. 2019. "Algae: Critical sources of very long-chain polyunsaturated fatty acids." *Biomolecules*. 9 (11): 708. https://doi.org/10.3390/biom9110708.

Holtkamp, A. 2009. "Isolation, characterisation, modification and application of fucoidan from fucus vesiculosus." PhD, Faculty of Life Sciences, Technical University of Braunschweig.

Jagusch, H., T.U.H. Baumeister, and G. Pohnert. 2020. "Mammalian-like inflammatory and pro-resolving oxylipins in marine algae." *ChemBioChem*. 21 (17): 2419–2424.

Jaulneau, V., C. Lafitte, C. Jacquet, S. Fournier, S. Salamagne, X. Briand, M.T. Esquerré-Tugayé, and B. Dumas. 2010. "Ulvan, a sulfated polysaccharide from green algae, activates plant immunity through the jasmonic acid signaling pathway." *J Biomed Biotechnol*. 2010: 525291.

Jiang, Z.D., and W.H. Gerwick. 1997. "Novel oxylipins from the temperate red alga *Polyneura latissima*: Evidence for an arachidonate 9(S)-lipoxygenase." *Lipids* 32 (3): 231–235. https://doi.org/10.1007/s11745-997-0029-9.

Jung, H.A., S.E. Jin, B.R. Ahn, C.M. Lee, and J.S. Choi. 2013. "Anti-inflammatory activity of edible brown alga *Eisenia bicyclis* and its constituents fucosterol and phlorotannins in LPS-stimulated RAW264.7 macrophages." *Food Chem Toxicol*. 59: 199–206.

Kaye, Y., O. Grundman, S. Leu, A. Zarka, B. Zorin, S. Didi-Cohen, I. Khozin-Goldberg, and S. Boussiba. 2015. "Metabolic engineering toward enhanced LC-PUFA biosynthesis in Nannochloropsis oceanica: Overexpression of endogenous Δ12 desaturase driven by stress-inducible promoter leads to enhanced deposition of polyunsaturated fatty acids in TAG." *Algal Res.* 11: 387–398.

Khotimchenko, S.V., V.E. Vaskovsky, and T.V. Titlyanova. 2002. "Fatty acids of marine algae from the Pacific coast of North California." *Botanica Marina*. 42: 17–22.

Kim, A.R., T.S. Shin, M.S. Lee, J.Y. Park, K.E. Park, N.Y. Yoon, J.S. Kim, J.S. Choi, B.C. Jang, D.S. Byun, N.K. Park, and H.R. Kim. 2009. "Isolation and identification of phlorotannins from Ecklonia stolonifera with antioxidant and anti-inflammatory properties." *J Agric Food Chem.* 57 (9): 3483–3489. https://doi.org/10.1021/jf900820x.

Knutsen, S.H., D.E. Myslabodski, B. Larsen, and A.I. Usov. 1994. "A modified system of nomenclature for red algal galactans." *Bot Mar*. 37: 163–169.

Kumar, S., J. Boehm, and J.C. Lee. 2003. "p38 MAP kinases: Key signalling molecules as therapeutic targets for inflammatory diseases." *Nat Rev Drug Discov*. 2 (9): 717–726. https://doi.org/10.1038/nrd1177.

Kumari, P., M. Kumar, C.R.K. Reddy, and B. Jha. 2014. "Nitrate and phosphate regimes induced lipidomic and biochemical changes in the intertidal macroalga *Ulva lactuca* (Ulvophyceae, Chlorophyta)." *Plant Cell Physiol.* 55 (1): 52–63.

Lahaye, M., and A. Robic. 2007. "Structure and functional properties of ulvan, a polysaccharide from green seaweeds." *Biomacromolecules*. 8 (6): 1765–1774.

Laurienzo, P. 2010. "Marine polysaccharides in pharmaceutical applications: An overview." *Mar Drugs*. 8 (9): 2435–2465.

Lenti, M., C. Gentili, A. Pianezzi, G. Marcolongo, A. Lalli, R. Cancedda, and F.D. Cancedda. 2009. "Monogalactosyldiacylglycerol anti-inflammatory activity on adult articular cartilage." *Nat Prod Res*. 23 (8): 754–762. https://doi.org/10.1080/14786410802456956.

Li-Beisson, Y., J.J. Thelen, E. Fedosejevs, and J.L. Harwood. 2019. "The lipid biochemistry of eukaryotic algae." *Prog Lipid Res*. 74: 31–68.

Li, L., R. Ni, Y. Shao, and S. Mao. 2014. "Carrageenan and its applications in drug delivery." *Carbohydr Polym*. 103: 1–11. https://doi.org/10.1016/j.carbpol.2013.12.008.

Lion, U., T. Wiesemeier, F. Weinberger, J. Beltran, V. Flores, S. Faugeron, J. Correa, and G. Pohnert. 2006. "Phospholipases and galactolipases trigger oxylipin-mediated wound-activated defence in the red alga *Gracilaria chilensis* against epiphytes." *ChemBioChem*. 7: 457–462.

Marcum, Z.A., and J.T. Hanlon. 2010. "Recognizing the risks of chronic nonsteroidal anti-inflammatory drug use in older adults." *Ann Longterm Care*. 18 (9): 24–27.

McKinnell, J., and E. Percival. 1962. "606. Structural investigations on the water-soluble polysaccharide of the green seaweed *Enteromorpha compressa*." *J Chem Soc (resumed)*. 3141–3148.

Mhadhebi, L., A. Dellai, A. Clary-Laroche, R.B. Said, J. Robert, and A. Bouraoui. 2012. "Anti-inflammatory and antiproliferative activities of organic fractions from the Mediterranean brown seaweed, *Cystoseira Compressa*." *Drug Dev Res*. 73 (2): 82–89.

Mhadhebi, L., A. Laroche-Clary, J. Robert, and A. Bouraoui. 2011. "Antioxidant, anti-inflammatory, and antiproliferative activities of organic fractions from the Mediterranean brown seaweed *Cystoseira sedoides*." *Can J Physiol Pharmacol*. 89 (12): 911–921.

Molendi-Coste, O., V. Legry, and I.A. Leclercq. 2011. "Why and how meet n-3 PUFA dietary recommendations?" *Gastroenterol Res Pract*. 2011: 364040. https://doi.org/10.1155/2011/364040.

Murdoch, J.R., and C.M. Lloyd. 2010. "Chronic inflammation and asthma." *Mutat Res*. 690 (1–2): 24–39.

Okamoto, R., and M. Watanabe. 2015. "Perspectives for regenerative medicine in the treatment of inflammatory bowel diseases." *Digestion*. 92 (2): 73–77. https://doi.org/10.1159/000438663.

Palavan-Unsal, N., and D. Arisan. 2009. "Nitric oxide signalling in plants." *Botanical Review*. 75 (2): 203–229. www.jstor.org/stable/40389387.

Patterson, E., R. Wall, G.F. Fitzgerald, R.P. Ross, and C. Stanton. 2012. "Health implications of high dietary omega-6 polyunsaturated fatty acids." *J Nutr Metab*. 2012: 539426. https://doi.org/10.1155/2012/539426.

Pengzhan, Y., Z. Quanbin, L. Ning, X. Zuhong, W. Yanmei, and L. Zhi'en. 2003. "Polysaccharides from *Ulva pertusa* (Chlorophyta) and preliminary studies on their antihyperlipidemia activity." *J Appl Phycol*. 15: 21–27.

Pereira, H., L. Barreira, F. Figueiredo, L. Custódio, C. Vizetto-Duarte, C. Polo, E. Rešek, A. Engelen, and J. Varela. 2012. "Polyunsaturated fatty acids of marine macroalgae: Potential for nutritional and pharmaceutical applications." *Mar Drugs*. 10 (9): 1920–1935.

Pickup, J.C. 2004. "Inflammation and activated innate immunity in the pathogenesis of type 2 diabetes." *Diabetes Care*. 27 (3): 813–823. https://doi.org/10.2337/diacare.27.3.813.

Prajapati, V.D., P.M. Maheriya, G.K. Jani, and H.K. Solanki. 2014. "Carrageenan: A natural seaweed polysaccharide and its applications." *Carbohydr Polym*. 105: 97–112. https://doi.org/10.1016/j.carbpol.2014.01.067.

Qi, H., T. Zhao, Q. Zhang, Z. Li, Z. Zhao, and R. Xing. 2005. "Antioxidant activity of different molecular weight sulfated polysaccharides from *Ulva pertusa* Kjellm (Chlorophyta)." *J Appl Phycol*. 17: 527–534.

Radwan, S.S., A.-S. Shaaban, and H.M. Gebreel. 1988. "Arachidonic acid in the lipids of marine algae maintained under blue, white and red light." *Z Naturforsch C*. 43: 15–18.

Rees, D.A. 1969. "Structure, conformation, and mechanism in the formation of polysaccharide gels and networks." *Adv Carbohydr Chem Biochem*. 24: 267–332.

Simmons, T.L., E. Andrianasolo, K. McPhail, P. Flatt, and W.H. Gerwick. 2005. "Marine natural products as anticancer drugs." *Mol Cancer Ther*. 4 (2): 333–342.

Stortz, C.A., and A.S. Cerezo. 2000. "Novel findings in carrageenans, agaroids and 'hybrid' red seaweed galactans." *Curr Top Phytochem*. 4: 121–134.

Sugiura, Y., R. Tanaka, H. Katsuzaki, K. Imai, and T. Matsushita. 2013. "The anti-inflammatory effects of phlorotannins from *Eisenia arborea* on mouse ear edema by inflammatory inducers." *J Funct Foods*. 5 (4): 2019–2023.

Thao, N.P., N.X. Cuong, B.T.T. Luyen, T.H. Quang, T.T.H. Hanh, S. Kim, Y.-S. Koh, N.H. Nam, P. Van Kiem, and C. Van Minh. 2013. "Anti-inflammatory components of the starfish *Astropecten polyacanthus*." *Mar Drugs*. 11 (8): 2917–2926.

Thomas, N.V., and S.-K. Kim. 2011. "Potential pharmacological applications of polyphenolic derivatives from marine brown algae." *Environ Toxicol Pharmacol*. 32 (3): 325–335.

Ulivi, V., M. Lenti, C. Gentili, G. Marcolongo, R. Cancedda, and F. Descalzi Cancedda. 2011a. "Anti-inflammatory activity of monogalactosyldiacylglycerol in human articular cartilage in vitro: Activation of an anti-inflammatory cyclooxygenase-2 (COX-2) pathway." *Arthritis Res Ther*. 13 (3): R92. https://doi.org/10.1186/ar3367.

Ulivi, V., M. Lenti, C. Gentili, G. Marcolongo, R. Cancedda, and F. Descalzi Cancedda. 2011b. "Anti-inflammatory activity of monogalactosyldiacylglycerol in human articular cartilage in vitro: Activation of an anti-inflammatory cyclooxygenase-2 (COX-2) pathway." *Arthritis Res Ther*. 13 (3): 1–12.

Weinberger, F., U. Lion, L. Delage, B. Kloareg, P. Potin, J. Beltrán, V. Flores, S. Faugeron, J. Correa, and G. Pohnert. 2011. "Up-regulation of lipoxygenase, phospholipase, and oxylipin-production in the induced chemical defense of the red alga *Gracilaria chilensis* against epiphytes." *J Chem Ecol.* 37 (7): 677–686.
Wijesekara, I., N.Y. Yoon, and S.-K. Kim. 2010. "Phlorotannins from *Ecklonia cava* (Phaeophyceae): Biological activities and potential health benefits." *BioFactors.* 36 (6): 408–414.

10 Marine Chondroitin Sulfate and Its Potential Applications

Hari Eko Irianto and Giyatmi

CONTENTS

10.1 INTRODUCTION

Over the past few decades, various types of diseases have emerged in the world, including degenerative diseases, infectious diseases and diseases caused by improper behavior. At the beginning of 2020, the world community was shocked by the COVID-19 pandemic that hit almost all countries regardless of developed, developing and poor status. In addition, degenerative diseases are still a problem, such as diabetes mellitus, stroke, coronary heart disease, cardiovascular disease, obesity, dyslipidemia, osteoporosis, prostatitis and osteoarthritis (Suiraoka, 2012). Pharmacists and herbalists as well as scientists from other relevant fields including medical experts and marine biotechnologists are working hard to explore bioactive compounds from terrestrial and marine

DOI: 10.1201/9781003303916-10

resources and then develop them into affordable, effective and efficient drugs to treat those diseases.

One of the bioactive compounds, i.e. marine chondroitin sulfate, has attracted the attention of scientists from various backgrounds to study individually or collaboratively. Chondroitin sulfate is a naturally occurring biomolecule that can be found widely in almost all invertebrates and vertebrates, including humans, and the many biological processes that involve it (Volpi, 2009). Chondroitin sulfate is a supplement that can help delay the course of osteoarthritis while also reducing inflammation and discomfort. Joint function improves as a result of this. Chondroitin sulfate is frequently combined with glucosamine. The prevalence of osteoarthritis in various areas, growing awareness towards joint health, development of innovative chondroitin sulfate combination products, etc., has a favorable influence on the growth of the global chondroitin sulfate market. Religious and cultural barriers to chondroitin sulfate usage, particularly in Middle Eastern nations, are some of the reasons restricting the worldwide chondroitin sulfate market's growth (Transparency Market Research, 2017), especially chondroitin obtained from non-halal raw materials. Therefore, chondroitin sulfate processed from marine resources which can be classified as halal according to Islam will not be disputed by any religions.

In 2020, the global chondroitin sulfate market was valued at US$1.17 billion and is expected to grow at a compound annual growth rate (CAGR) of 3.0% from 2020 to 2028. The expected increase in demand for nutritional products and prevalence of osteoarthritis are driving the market growth. Nutraceuticals dominated the market in 2020, accounting for more than 36.0% of worldwide sales. In the nutraceutical sector, sodium chondroitin sulfate is commonly used as a dietary supplement for the treatment of osteoarthritis and joint discomfort. Sodium chondroitin sulfate is also utilized as thickeners, additives, forming agents and preservatives in nutritional supplements and food items, as well as in health food and animal feed (Grand View Research, 2021). The rising acceptance of nutraceutical products that meet important nutritional needs, as well as the availability of new chondroitin sulfate compositions, can be linked to the rise of the chondroitin sulfate market. Furthermore, growing incidence and prevalence of osteoarthritis are expected to add to the worldwide chondroitin sulfate market's growth. By age 85, an estimated 40% of Americans would have symptoms of osteoarthritis in at least one hand. Furthermore, among men and women aged 45 years and over in the United States, the incidence of symptomatic knee osteoarthritis was 5.9% and 13.5%, as well as 7.2% and 18.7%, respectively (Medgadget, 2021).

Exploratory studies on the use of chondroitin sulfate have been carried out quite intensively so that future application opportunities may be wider, efficient, effective, efficacious and targeted. This chapter discusses potential applications of chondroitin sulfate for drug and nutraceutical purposes.

10.2 CHONDROITIN SULFATE

Chondroitin sulfate is a natural polymer that belongs to the glycosaminoglycans family of macromolecules with a high molecular weight (10,000–50,000

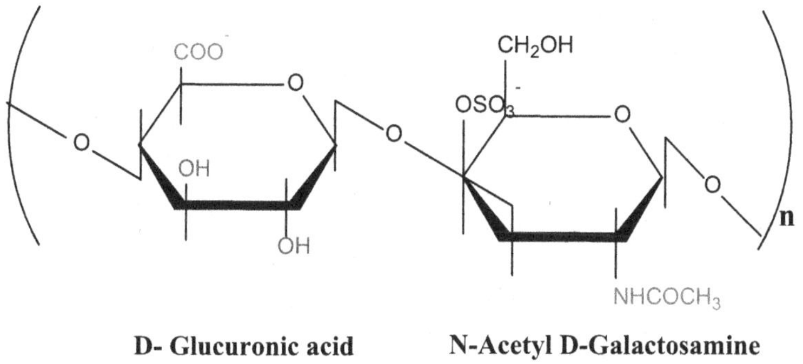

D- Glucuronic acid N-Acetyl D-Galactosamine

FIGURE 10.1 Chemical structure of chondroitin sulfate.

Da) that is a component of cartilage and connective tissue (Maccari et al., 2010; Konovalova et al., 2019; EC Huskisson, 2008). Glycosaminoglycans are polysaccharide molecules that are polymers of disaccharide units made up of various monosaccharides. The structure of glycosaminoglycan compounds, which is frequently dominated by disaccharide compounds, is more stressed in their categorization (Wikanta et al., 2002). There are several kinds of glycosaminoglycans, which are typically classified into four categories: (1) hyaluronic acid or hyaluronan, (2) keratan sulfate, (3) heparan sulfate/heparin and (4) chondroitin sulfate/dermatan sulfate (Krichen et al., 2018).

Chondroitin sulfate is a linear, complex, sulfated, polydispersity polysaccharide (Maccari et al., 2010). It is made up of glucuronic acid (GlcA) and N-acetylgalactosamine (GalNAc) repeating disaccharide units (Figure 10.1), with a sulfate group at position 4 or 6 of the GalNAc residue. Various amounts of repeated 4-sulfate and 6-sulfate disaccharides can be found in chondroitin sulfate chains having various sulfate group configurations at certain hydroxyl groups (Lee et al., 1998, Tamura et al., 2009)). In addition, chondroitin sulfate as a polysaccharide with a linking tetrasaccharide at the reducing terminal and a major moiety, called a repeating disaccharide region, belongs to the glycosaminoglycans (GAG) family. The fundamental component of the repeating disaccharide region is N-acetyl galactosaminyl glucuronic acid (βGalNAc-βGlcA), which is complemented by a distinct pattern of sulfate groups in certain hydroxyl groups (Tamura et al., 2009).

There are four major kinds of chondroitin sulfate polysaccharides: (1) chondroitin sulfate A (CS-A), (2) chondroitin sulfate C (CS-C), (3) chondroitin sulfate D (CSD) and (4) chondroitin sulfate E (CS-E). Each subtype is distinguished by a unique sulfation pattern. CS-A has a GalNAc 4-O(oxygen)-sulfate residue, CS-C has a 6-Osulfate residue, CS-D has a 2-O sulfated GlcA residue, and CS-E has a GlcNAc 4,6-O-disulfate residue (Li et al., 2020).

10.3 MARINE SOURCES OF CHONDROITIN SULFATE

Chondroitin sulfate is produced in vertebrates and invertebrates as part of proteoglycan molecules (Lamari and Karamanos, 2006). According to the source, the global market of chondroitin sulfate, also known as sulfated glycosaminoglycan, is divided into bovine, porcine, poultry, shark and synthetic. In 2020, bovine dominated the market, accounting for more than 35.0% of worldwide sales. Sharks have been designated as an endangered species, posing a threat to the industry's future growth. Sodium chondroitin sulfate from sharks, on the other hand, is preferable. A two-stage fermentation-based technique is used to manufacture synthetic sodium chondroitin sulfate (Grand View Research, 2021). However, due to concerns over bovine spongiform encephalopathy (BSE) and other reasons, exploration of microorganisms and marine species as potential sources of chondroitin sulfate has been carried out. Potential producers of chondroitin sulfate include sponges, sea cucumbers, squids, mollusks, invertebrates and mostly cartilaginous debris from fish (shark, salmon, ray, etc.) (Vázquez et al., 2013). Marine sources of chondroitin sulfate have been explored by scientists around the world, as shown in Table 10.1.

Chondroitin sulfate has been extracted from shark cartilage and stingray bone by 2.37% and 1.47%, respectively (Hanindika et al., 2014). Another observation reported that chondroitin sulfate in shark fin and ray cartilage were 15.05 and 7.49% accordingly. Fourier transform infrared spectroscopy (FTIR) of the potassium bromide pellet technique revealed that the spectrum of dry chondroitin sulfate samples extracted from various cartilage sources including shark fin and ray cartilage showed peaks at 857 and 826 cm^{-1} wave numbers, which were used to identify chondroitin-4-sulfate and chondroitin-6-sulfate each. Thus, the spectra of all extracts showed that the two cartilage samples consisted of chondroitin-4-sulfate and chondroitin-6-sulfate (Garnjanagoonchorn et al., 2007). Research on spotted dogfish (*Scyliorhinus canicula*) exhibited that head waste is the best source of chondroitin sulfate production compared to skeleton and fins (Blanco et al., 2015). Chondroitin sulfate extracted from skate (*Raja clavata*) (rays also known as skates) was 15% (w/w) yield (Murado et al., 2010).

Fucosylated chondroitin sulfates were identified in sea cucumbers, particularly *Stichopus tremulus* (Western Indian Ocean), *Pearsonothuria graeffei* (Indo-Pacific), *Isostichopus badionotus* (Western Atlantic), and *Holothuria vagabunda* (Norwegian coast). Fucosylated chondroitin sulfate is a structurally unique glycosaminoglycan discovered in the body wall of sea cucumbers, with a chondroitin sulfate backbone and connected fucose sulfate or non-sulfate side chains. This molecule has a broad range of biological action and plays a vital function in keeping the integrity of the body wall. The polysaccharides isolated from *S. tremulus, P. graeffei, I. badionotus*, and *H. vagabunda* were 7.0%, 11.0%, 9.9% and 6.3% by weight, respectively. *S. tremulus, P. graeffei* and *H. vagabunda* had higher levels of fucosylated chondroitin sulfates, whereas *I. badionotus* had more abundant fucan (Chen et al., 2011; Myron et al., 2014). Fucosylated chondroitin sulfates have also been isolated from *Acaudina molpadioides* (Hu et al., 2014), *Holothuria forskali* (Panagos et al., 2014), *Paracaudina chilensis* and *Holothuria hilla* (Ustyuzhanina et al., 2020).

Preparation of chondroitin sulfate from salmon nasal cartilage was also carried out by Takeda et al. (1998), Han et al. (2000), Kobayashia et al. (2017), and

TABLE 10.1
Exploration of Marine Chondroitin Sulfate Sources

Marine Biota		Organ Parts	References
Shark	Silky shark (*Carcharinus falciformes*)	Fin bone	Hanindika et al. (2014)
	Black-shark (Galeus melastomus)	Cartilage	Konovalova et al.)
	Spotted dogfish (*Scyliorhinus canicula*)	Head, Skeleton and Fins	Blanco et al. (2015)
	Shark	Cartilage, by-products from shark fin soup restaurant	Garnjanagoonchorn et al. (2007)
	Blacktip shark (*Carcharhinus limbatus*)	Cartilage	Wikanta et al. (2000)
Ray	Stingray (*Raja* sp.)	Cartilage	Hanindika et al.)
	Paleedged/Sharpnose Stingray (*Dasyatis zugei*)	Cartilage	Garnjanagoonchorn et al. (2007)
	Northern stingray (*Raja hyperborean*)	Cartilage	Konovalova et al.)
	Skate (*Raja clavata*)	Cartilage	Murado et al.)
Sea cucumber	*Stichopus tremulus*	Body wall	Chen et al. (2011)
	Pearsonothuria graeffei	Body wall	Chen et al. (2011)
	Isostichopus badionotus	Body wall	Chen et al. (2011)
	Holothuria vagabunda	Body wall	Chen et al. (2011)
	Acaudina molpadioides	Body wall	Hu et al. (2014)
	Holothuria forskali	Body wall	Panagos et al. (2014)
	Paracaudina chilensis	Body wall	Ustyuzhanina et al. (2020)
	Holothuria hilla	Body wall	Ustyuzhanina et al. (2020),
Cuttlefish	Pharaoh cuttlefish (*Sepia pharaonis*)	Cuttlebone	Hanindika et al. (2014)
Squid	Diamond squid (*Thysanoteuthis rhombus*)	fins, arms, skin, head, eyes, and mantle	Tamura et al.)
Salmon	Atlantic salmon (*Salmo salar*)	Cartilaginous tissue	Konovalova et al.)
	Salmon (no scientific names)	Nasal cartilage	Han et al. (2000); Kobayashia et al. (2017); Tatara et al. (2015); Goto et al. (2011); Takeda et al. (1998)
Tuna	Bluefin tuna (*Thunnus thynnus*)	Skin	Krichen et al. (2018)
Sturgeon	Sturgeon (no scientific name)	Bone	Maccari et al. (2010)
Mollusca	*Pomacea* sp.	Tissues	Nader et al. (1984)
	Tagelus gibbus,	Tissues	Nader et al. (1984).
	Anomalocardia brasiliana	Tissues	Nader et al. (1984).

(Continued)

TABLE 10.1 ***(Continued)***
Exploration of Marine Chondroitin Sulfate Sources

Marine Biota		Organ Parts	References
	Mud snail (*Cipangopaludina chinensis*)	Tissues	Lee et al. (1998)
	Clams (*Anodonta anodonta*)	Body tissues	Volpi and Maccari (2005)
Sponges	*Polymastia janeirensis,*	Sponge tissues	Maia et al. (2016)
	Echinodictyum dendroides,	Sponge tissues	Maia et al. (2016)
	Dragmacidon reticulatum	Sponge tissues	Maia et al.)
Crocodile	no scientific name	Cartilage (trachea, hyoid, sternum and rib)	Garnjanagoonchorn et al. (2007)

Tatara et al. (2015). Cartilage has the main extracellular component in the form of the proteoglycan chondroitin sulfate (Kobayashia et al., 2017). Chondroitin sulfate is reported to be commercially produced by enzymatication of salmon nasal cartilage (Goto et al., 2011).

Chondroitin sulfate was also isolated from sturgeon bone by extracting a single polysaccharide from the bone at a concentration of 0.28–0.34% for dry tissue and identifying it as chondroitin sulfate. These polymers were found to include 55% monosulfate disaccharides at position 6 GalNAc, 38% monosulfate disaccharides at position 4 GalNAc, and 7% unsulfate disaccharides after specific chondroitinase and high performance liquid chromatography (HPLC) separation of the resulting repeat unsaturated disaccharides. As a bony fish, sturgeon can be a source of chondroitin sulfate, even though it is usually discarded following ovary retrieval (Maccari et al., 2010), while the skin of bluefin tuna (*Thunnus thynnus*) was used as raw material to produce chondroitin sulfate (Krichen et al., 2018).

Chondroitin sulfate has been extracted from various tissues of diamond squid (*Thysanoteuthis rhombus*). The non-edible skin, head and eyes of diamond squid can be used as an alternative source of chondroitin sulfate with the appropriate sulfate concentration. Approximately 14 kg of diamond squid having 380 g skin, 780 g head and 290 g eyes (wet weight) obtained 238, 386 and 47 mg of pure chondroitin sulfate from these tissues, respectively (Tamura et al., 2009).

The use of mollusk species as source of chondroitin sulfate is possible. Chondroitin sulfate has already been isolated from three mollusk species, including *Pomacea* sp., *Tagelus gibbus* and *Anomalocardia brasilianu*. Bivalves (*Tagelus* and *Anomalocardia*) and gastropods (*Pomacea* sp.) have different chondroitin sulfate structures. Chondroitin sulfate from *Pomaces*, for example, has no disaccharide 6-sulfate or disulfate (Nader et al., 1984). Chondroitin sulfate extracted from mud snails (*Cipangopaludina chinensis*) contain chondroitin 4-, 6- and O- sulfates

(Lee et al., 1998). Glycosaminoglycans from the bodies of giant freshwater clams *Anodonta anodonta* are detected in around 0.6 mg/g dry tissue and are made up of chondroitin sulfate (about 38%), non-sulfate chondroitin (about 21%), and heparin (about 41%) (Volpi and Maccari, 2005).

Chondroitin sulfates are also isolated from marine sponges, such as *Polymastia janeirensis, Echinodictyum dendroides, and Dragmacidon reticulatum* (Maia et al., 2016), while chondroitin sulfate can be extracted from a nematode of *Caenorhabditis elegans* as well (Dierker et al., 2016).

10.4 EXTRACTION OF MARINE CHONDROITIN SULFATE

Basically, the extraction techniques developed for chondroitin sulfate rely on chemical hydrolysis of the tissue to break the proteoglycan core, followed by protein removal to recover the glycosaminoglycans (Abdallaha et al., 2020). Modern chondroitin sulfate production methods are multi-stage extraction procedures. Defatting of raw materials, alkaline and enzymatic hydrolysis, sedimentation of chondroitin sulfate from solution, further purification of the preparation and drying are the key steps of chondroitin sulfate production (Konovalova et al., 2019). Currently, there are three techniques for extracting chondroitin sulfate, i.e. alkali, enzyme and ultrasonic. The alkali technique was easy; however, the high concentration of alkali caused chondroitin sulfate breakdown, reducing its biological activity. The enzyme technique considerably increased chondroitin sulfate purity, but the yield was about the same as the alkali method, with the drawbacks of high enzyme cost and long extraction time. When compared to the alkali and enzyme methods, the ultrasonic technique considerably extended the extraction time, but the yield and purity of chondroitin sulfate were not significantly improved (He et al., 2014; Syed et al., 2017).

Improvements in extraction methods through a combination of enzymatic and chemical hydrolysis, selective precipitation, and membrane technology resulted in the development of a fast and highly efficient method (15% w/w) with minimal use of reagents and high purity for the resulting chondroitin sulfate (I = 99%) (Murado et al., 2010). The following describes the methods that have been applied experimentally to the extraction of chondroitin sulfate from mollusks and cartilage.

The extraction method employed to extract chondroitin sulfate from mud snail was first by removing the shell and then three days defatting with acetone. Defatted snail powder (4 g) was dissolved in 40 ml of 0.05 M Na_2CO_3 (pH 9.2) buffer, added 2 ml subtilisin and stirred 200 rpm at 60°C for 2 days. The mixture was cooled to 4°C, added 5% trichloroacetic acid and centrifuged at 8000xg for 20 minutes. The supernatant was added with three times volume of 5% potassium acetate with ethanol, kept overnight at 4°C and then centrifuge at 10000xg for 30 minutes. The precipitate was washed with alcohol, dissolved in 40 ml of 0.2 M NaCl and centrifuged at 8000xg for 30 minutes. The supernatant was mixed with 0.5 ml of 5% cetylpyridinium chloride and centrifuged at 8000xg for 15 minutes. Finally, the precipitate was dissolved in 10 ml of 2.5 M NaCl, added with five volume ethanol and centrifuged at 10000xg for 30 minutes (Lee et al., 1998).

In addition, the method used to isolate chondroitin sulfate from fish cartilages, particularly cuttlebone, ray and shark cartilages, were by soaking the raw material in papain solution for 24 hours to free the remaining muscle tissue and then dried. The cartilage is ground and added with distilled water and acetic acid to maintain acidity at pH 4.5. After warming the mixture in an oven at 37°C for 7 hours, it was filtered using filter paper and the obtained solution was centrifuged. The resulting supernatant was added with 3% w/v cetylpyridinium chloride in 0.8 M NaCl, then put in the freezer for 10 minutes and centrifuged at 5,000 rpm for 30 minutes. Subsequently, 2 M NaCl solution was added and centrifuged again at 5,000 rpm for 30 minutes. The supernatant obtained was added with methanol, and centrifuged at 5,000 rpm for 15 minutes at 4°C. The precipitate was added with 95% ethanol and centrifuged at 3,000 rpm for 15 minutes. The precipitate was removed and then dried at room temperature (Hanindika et al., 2014). A similar method was employed with chondroitin drying using freeze drier and the yield was 6.06% (Sulityowati et al., 2015).

Optimal chondroitin sulfate recovery and purification from cartilage waste of *S. canicula* was obtained at 58°C and pH 8.5 for enzymatic hydrolysis, and 0.53–0.64 M NaOH and 1.14–1.20 volumes of ethanol for chemical treatment. Furthermore, head wastes were discovered to be the most prospective source of chondroitin sulfate synthesis from *S. canicula*. The ultrafiltration and diafiltration procedure was used to prepare extracts from the alkaline hydroalcoholic treatment at 30 kDa molecular weight cut-off for differential retention of chondroitin sulfate and concurrent rejection of protein components (Blanco et al., 2015).

The process incorporating biological enzymolysis, mixed microbial fermentation and contemporary biological separation has been used to extract chondroitin sulfate from shark cartilage. The biological enzymolysis technology employed the enzyme compound system consisting of one or more combinations of papain, trypsin, neutral protease, pepsin, and flavorzyme. Fermented strains for mixed microbial fermentation include mostly *Aspergillus oryzae, Bacillus psychrosaccharolyticus* and *Bacillus subtilis* (Xuan et al., 2020).

An available review informs that enzymes that have been employed for chondroitin sulfate, including alcalase, papain, neutrase, bromelain, acid protease, actinase and pepsin (Abdallaha et al., 2020). Alcalase is a well-known endoprotease that has a high capacity to hydrolyze a variety of marine substrates (Vázquez et al., 2018).

Chondroitin sulfate quality can be improved by employing protein isoelectric point. After washing and steam boiling the cartilage for extraction, NaOH and NaCl are added to the mixture. For the enzymatic procedure, pancreatin was added, followed by deproteination and concentration. Primary sedimentation and pH value adjustment were used to eliminate impurity protein, and secondary sedimentation was used to get chondroitin sulfate crude product. After initial sedimentation, precise pH correction is required, followed by removing impurity proteins mixed in chondroitin sulfate. The chondroitin sulfate obtained has a purity of above 97% (Nanjing University of Science and Technology, 2008).

In recent years, high-intensity pulsed electric fields (PEF) technology has emerged as a growing focus of worldwide, where it has the potential to be applied to the production of chondroitin sulfate. PEF technology is a very effective method for preserving and processing a wide range of food products without significantly compromising their quality features (Syed et al., 2017). As the electrolyte, the material is placed in a chamber. To operate on the material, instantaneous pressure was produced between the two electrodes (Yin et al., 2006). PEF is widely used to extract active chemicals from natural products because of its non-thermal performance, speed, efficiency, low power consumption and low pollution. The PEF extraction system has been designed with a triangular pulse power waveform, an oscilloscope used to directly view the output voltage, a 40–3000 Hz adjustable frequency and a 70 mL processing pipe volume (He et al., 2014).

10.5 QUALITY AND SAFETY

The quality of chondroitin sulfate is influenced by the processing techniques applied to produce it. Different extraction and purification processes may bring about different consequences on the structural characteristics and properties of chondroitin sulfate, resulting in extracts with varying degrees of purity, limited biological effects, contaminants causing safety and reproducibility issues and unknown origin (Volpi, 2019). Of course, those aspects can have serious implications for consumers of end products, including pharmaceuticals, dietary supplements and nutraceuticals related to the traceability of chondroitin sulfate and the statement of the true origin of the active compounds and its content.

Symptomatic medicines for osteoarthritis such as acetaminophen and non-steroidal anti-inflammatory drugs (NSAIDs) are only effective in alleviating symptoms. Unfortunately, these drugs can have serious adverse effects in certain people and are sometimes contraindicated (Utami et al., 2012). Chondroitin sulfate, alone or in combination with glucosamine, has been employed for the treatment of osteoarthritis that is better tolerated at the gastrointestinal, cardiovascular, and renal levels than NSAIDs and cyclooxygenase 2 inhibitors (COXIBs) (Rubio-Terrés et al., 2020). One of its primary advantages for the treatment of aging patients with comorbidities is its safety profile. If the financial implications are ignored, there is no limit to its use in osteoarthritis patients. However, caution should be applied while selecting the kind and formulation of chondroitin sulfate (Henrotin et al., 2010).

The LD-50 of chondroitin sulfate extracted from shark cartilage using mice intraperitoneally is 60.8 mg/20 g body weight and after extrapolating to mice orally, the LD-50 value is 21.3 g/kg body weight which can be classified as a "Practically Non-Toxic" compound, according to the table for classification of toxic properties (Lestari et al., 2000). In addition, Chondroitin along with glucosamine for osteoarthritis drugs is reported to have a high safety profile (Narvy and Vangsness, 2010).

However, the administration level of chondroitin sulfate for therapeutic purposes should be of concern. The frequency of side effect incidences in arthritis patients reported in the consumption range of 800–1,200 milligrams per day (mg/day) was in most cases equivalent to that of the placebo group, which was not administered the drug. The most common adverse effects in the chondroitin sulfate group were gastrointestinal adverse effects, which occurred at a lower incidence than in the placebo group (BfR, 2018).

Regarding commercial products, only 5 of the 16 chondroitin sulfate samples from throughout the world had more than 90% chondroitin sulfate and matched the label. Maltodextrin or lactose were the predominant impurities in the remaining 11 samples with poor chondroitin sulfate. Those chondroitin sulfate preparations originate from various sources, and they all obviously fail to fulfill label requirements. As a result, more efficient and strict standards for both licensing and quality control of raw ingredients, as well as the final pharmacological formulations derived from them, should be implemented (Cunhaa et al., 2015). Because some nutraceuticals have poor chondroitin sulfate quality, stricter quality control regulations should be implemented to ensure the manufacture of high-quality products for nutraceutical use and to protect customers from low-quality, ineffective and potentially dangerous products (Volpi, 2009).

For evaluating the quality of chondroitin sulfate raw materials used in nutritional supplements, the capillary isotachophoretic (cITP) technique is recommended as an alternative to HPLC or CZE (capillary zone electrophoresis). Estimation of chondroitin sulfate origin, estimation of sulfatation degree of chondroitin sulfate and its homogeneity, estimation of chondroitin sulfate molecular mass and molecular weight dispersion, determination of chondroitin sulfate purity and detection of chondroitin sulfate degradation products are all possible with the cITP method (Václavíková and Kvasnička, 2015).

10.6 POTENTIAL APPLICATIONS OF MARINE CHONDROITIN SULFATE

Chondroitin sulfates can have varying chain lengths (Mr: 5,000–50,000) (Peter, 2012) or sulfation degrees depending on their origin, which has a significant impact on their biological characteristics. Dermatan sulfate is a chondroitin sulfate isomer. It also includes -L-iduronic acid, which is produced via epimerization, and N acetyl-D-galactosamine, which is mainly sulfated at position 4. It also includes -L-iduronic acid, which is produced via epimerization, and N acetyl–d-galactosamine, which is mainly sulfated at position 4 (Volpi, 2009). Chondroitin is hydrophilic and water soluble, forming a viscous fluid similar to sodium hyaluronate. Chondroitin sulfate is important for the structural and functional integrity of joints, as it is the major constituent of GAGs in articular cartilage. Chondroitin is known to help maintain joint viscosity, stimulate cartilage repair and inhibit enzymes that degrade cartilage (Zaki, 2013).

Chondroitin sulfate has important properties that influence its application, including biocompatible, anti-inflammatory, biodegradable, non-immunogenic and non-toxic (Abdallaha et al., 2020). Due to these properties, chondroitin sulfate has broad prospective applications in various fields.

10.6.1 Potential Applications for Therapeutic Agents

Chondroitin sulfate is a substance that is already present in the human body. This substance is believed to have a function to draw water and nutrients into human cartilage so that human cartilage remains healthy and supple (Joseph, 2021). The use of this substance is usually aimed at overcoming health problems as therapeutic agents such as pain in the joints and is also commonly used as an antithrombotic, an ischemic heart disease treatment, and an extravasation therapy agent along with hyperlipidemia (Archiando, 2020). The diversity of species and tissues causes chondroitin sulfate to have a heterogeneous structure and physico-chemical profile, which is responsible for the various and specific activities of this macromolecule (Volpi, 2019). The Indonesian Agency for Drug and Food Control through the decree of HK.00.05.23.3644 of 2004 stipulates that the maximum limit of chondroitin sulfate that can be consumed from dietary supplements is 1200 mg per day.

10.6.1.1 Osteoarthritis

Osteoarthritis is a degenerative illness that affects the elderly and is the most prevalent kind of arthritis. Affected people may have severe pain and functional impairment as a result of the condition. Treatments for osteoarthritis are easy to obtain, but the outcomes are uncertain and even controversial as for chondroitin sulfate and glucosamine (Utami et al., 2012; van Blitterswijk et al., 2003; Agiba, 2017; Fernández-Martín et al., 2021). The use of chondroitin sulfate to treat osteoarthritis has been around for a long time. Glucosamine and chondroitin sulfate should be useful in the treatment of osteoarthritis, according to an abundant fundamental scientific data, i.e. biochemical, tissue culture, cell culture and animal models of arthritis. They are helpful in decreasing the symptoms of arthritis in both animals and humans, according to veterinary and human clinical data (Hungerford and Jones, 2003; Singh et al., 2015; Rubio-Terrés et al., 2020).

Most treatment algorithms do not account for the use of glucosamine and chondroitin sulfate, the long-standing outpatient adjuvant being acetaminophen, followed by NSAIDs (including the newer COX-2 inhibitors), steroid or hyaluronic injections, physical therapy and joint replacement (Schenck, 2000). Several studies suggested that nutraceutical supplementation of individuals with knee/hip osteoarthritis may improve pain severity and physical function (Aghamohammadi et al., 2020; Šimánek et al., 2005). Because osteoarthritis is a slow-progressing illness, evidence of disease change following an intervention may take years to appear (Agiba, 2017).

Chondroitin sulfate is reported as a promising substance for the treatment of osteoarthritis of the knee. Depending on the assessment technique (model, dose and duration) and the source of chondroitin sulfate (origin and quality), various findings have been reported. In mild knee osteoarthritis, clinical data suggests a slow-acting effect on symptoms (Henrotin et al., 2010). In the total group of individuals with osteoarthritis of the knee, however, glucosamine and chondroitin sulfate, alone or in combination, did not significantly decrease pain. According to preliminary research, glucosamine and chondroitin sulfate together may be helpful in the subgroup of individuals with moderate-to-severe knee discomfort (Clegg et al., 2006). Classified as slow-acting medicines, chondroitin sulfate and glucosamine sulfate relieve pain and partially restore joint function in osteoarthritis patients (Bottegoni et al., 2014).

Even in meta-analyses, glucosamine and chondroitin have shown mixed efficacy in reducing knee pain and increasing joint function associated with osteoarthritis, however there is some evidence to suggest that these treatments may help delay disease progression radiographically (Narvy and Vangsness, 2010). Chondroitin sulfate lowers the levels of pro-inflammatory cytokines and transcription factors that play a role in inflammation. By blocking hydrolytic enzymes and limiting the oxidation of lipids and proteins, glucosamine sulfate improves cartilage specific matrix components and inhibits collagen degradation in chondrocytes (Bottegoni et al., 2014).

In comparison to their individual effects, the combination of glucosamine and chondroitin sulfate may be less effective. The validity and mechanism of this new finding are unknown, although it might be linked to changes in glucosamine absorption. Satisfied radiographic criteria Grade 2 knees may represent a more potentially responsive population in future osteoarthritis trials evaluating structural modification; however, a larger sample size, longer study duration and/or improved methods of measurement will be required, as the rate of joint space width loss seen on plain radiographs is much slower than previously appreciated (Sawitzke et al., 2008). When compared to glucosamine-chondroitin sulfate and placebo, the combination of glucosamine–chondroitin sulfate–methylsufonylmethane (GCM) supplements was more effective in lowering pain and improving function in individuals with Kellgren Lawrence I-II knee osteoarthritis, while the glucosamine-chondroitin sulfate supplement was shown to be no better than a placebo in terms of clinical effectiveness (Siagian, 2014). In the symptomatic treatment of knee osteoarthritis, the novel formulation of a fixed-dose combination of glucosamine sulfate (1,500 mg) and chondroitin sulfate (1,200 mg) was non-inferior to the reference product, with a high responder rate and favorable tolerability profile (Lomonte et al., 2021).

Although glucosamine and chondroitin as osteoarthritis therapy show results vary from patient to patient; simple fixes can be used in clinical settings. These medications can be used safely as an initial treatment for some osteoarthritis patients before beginning treatment with NSAIDs, acetaminophen, and other conventional medications (Narvy and Vangsness, 2010). Under mixed-mode lubrication conditions, chondroitin sulfate demonstrated to be an excellent cartilage lubricant. Supplemental chondroitin sulfate that diffused into the specimens, on the other hand, had no effect on cartilage fluid load support (Katta et al., 2009). Oral chondroitin in the prescribed dosage is more effective than placebo for pain relief and physical function. Glucosamine had a substantial influence on the result of stiffness when compared to placebo (Zhu et al., 2018).

Chondroitin sulfate patches are developed as an alternative to oral chondroitin sulfate which is more difficult to administer. With a partition value of

2.22, 150 mg of chondroitin sulfate in a transdermal patch provides continuous penetration in 19 hours (Sopyan et al., 2017). A topical cream combining glucosamine sulfate, chondroitin sulfate, and camphor is more effective than a placebo in reducing joint discomfort in individuals with knee osteoarthritis (Cohen et al., 2003).

It can be concluded that clinically, chondroitin sulfate can result in reduced pain and increased joint mobility in osteoarthritis patients and slows down joint damage (Zaki, 2013).

10.6.1.2 Gastric Ulcer

Gastric ulcer is one type of inflammatory disease that is often found both in adults and adolescents as a result of a decrease or damage to the resistance of the gastric mucosa. Treatment of gastric ulcers has shifted to cytoprotective drugs, namely drugs that can increase protection and resistance of the gastrointestinal mucosa, especially the stomach. Oral chondroitin sulfate showed a protective effect on the gastrointestinal mucosa of rats. The higher the chondroitin sulfate dosage, the greater the protective effect. A dosage of 4 g/kg body weight of chondroitin sulfate had the same excellent protective effect as sucralfate as a positive control at a level of 500 mg/kg body weight. Because chondroitin sulfate had a pH of 7.4 in the stomach of rats, it was an antacid or neutralized gastric acid (Wikanta et al., 2000).

10.6.1.3 Atherosclerosis

Atherosclerosis is one of the most prevalent illnesses in the elderly, generating fatty liver, renal failure, myocardial infarction, coronary heart disease and other cardiovascular disorders, and excessive cholesterol is one of the primary causes. A research employing a hypercholesterolemic mice model demonstrated that chondroitin sulfate has the ability to prevent atherosclerosis and coronary heart disease. Meanwhile, chondroitin sulfate was shown to be beneficial in lowering cholesterol and triglycerides in both serum and liver, as well as decreasing low-density lipoprotein cholesterol levels and raising high-density lipoprotein cholesterol levels in serum. Chondroitin sulfate was found to enhance the activities of hepatic lipoprotein lipase, hepatic lipase, superoxide dismutase and glutathione peroxidase (Wang et al., 2017).

10.6.1.4 Cancer

Muscle-invasive bladder cancer (MIBC) was studied for treatment using chondroitin sulfate. When cisplatin-resistant MIBC overexpresses certain sugar chains compared to chemotherapy-naive MIBC, enhanced cellular expression of chondroitin sulfate was discovered in a study using a mouse model. This fact shows that targeting chondroitin sulfate with VDC886 drug conjugate is a promising strategy, especially in cisplatin-resistant MIBC. This discovery might pave the way for a new therapy paradigm for human MIBC patients who aren't responding to cisplatin (Seiler et al., 2017).

Chondroitin sulfate and/or glucosamine has been explored as a colorectal cancer (CRC) chemoprevention agent, but clear evidence of an independent preventive effect of chondroitin sulfate and/or glucosamine on CRC has not been obtained, as the observed effect could be associated with concomitant use of non-steroidal anti-inflammatory drugs (NSAIDs). Their low toxicity, however, warrants additional investigation into their impact and prospective function as chemopreventive drugs (Ibáñez-Sanz et al., 2018).

10.6.1.5 Inhibition of Mast Cells

Chondroitin sulfate appears to be a powerful mast cell inhibitor of both allergic and non-immune activation, with therapeutic consequences. Mast cells are responsible for the development of allergies and potential inflammatory responses and originate in the bone marrow. Chondroitin sulfate inhibited mast cell secretagogue compound 48/80 (48/80)–induced histamine production in rat peritoneal mast cells in a dose-dependent manner. Inhibition by chondroitin sulfate increased with pre-incubation time and remained after the medication was washed off, but cromolyn's impact was restricted by rapid tachyphylaxis. Histamine production from rat connective tissue mast cells (CTMCs) was also suppressed immunologically. Ultrastructural autoradiography reveals that chondroitin sulfate is mostly linked with the plasma and perigranular membrane (Theoharides et al., 2000).

10.6.1.6 Reconstructed Cornea

A porous collagen/glycosaminoglycan-based scaffold seeded with stromal keratocytes and then epithelial and endothelial cells was used to imitate the corneal extracellular matrix. The collagen-chondroitin sulfate scaffold is an excellent substrate for the creation of artificial corneas due to its capacity to sustain long-term culture, lifespan and handling qualities, which make it appropriate for pharmacotoxicological and drug safety testing. This technology may be developed into a full-thickness man-made cornea by integrating non-transforming human endothelial cells, which is a step toward a substitute for corneal transplantation (Vrana et al., 2008).

10.6.1.7 Coating Composite Dental Implant

The chewing function of dental implants is based on osseointegration. Nerve growth factor (NGF) can stimulate bone healing as a neurotrophic agent. The impact of an NGF-chondroitin sulfate/hydroxyapatite (CS/HA)–coated composite implant on osseointegration and innervation is an essential factor to consider when using it. NGF-CS/HA coating has been demonstrated to significantly accelerate implant osseointegration and enhance peri-implant nerve regeneration in the mandible of beagle dogs. This might provide a scientific basis for using NGF-CS/HA–coated implants in oral implants (Ye et al., 2021).

10.6.2 Potential Application for Food Supplements

The European Medicines Agency regulates chondroitin sulfate as an anti-osteoarthritis prescription medication, whereas the U.S. Food and Drug

Administration (FDA) classifies it as a dietary supplement, allowing U.S. customers more freedom to purchase and use chondroitin. Chondroitin sulfate is one of the most popular joint health supplements in the United States (You, 2021). A review based on available data prefers to classify glucosamine and chondroitin sulfate as food supplements rather than drugs (Šimánek et al., 2005).

In cases of symptomatic spinal disc degeneration, daily intake of glucosamine and chondroitin sulfate-based capsules as a food supplement for two years for disc recovery, where long-term intake of this supplement can counteract the symptoms of spinal disc degeneration, especially in the early stages. The disc degeneration/regeneration process was objectively observed using magnetic resonance imaging (MRI) in clinical studies of these food supplements. Why these medicines may have a cartilage structure and symptom modifying effect has been biochemically explained, demonstrating their therapeutic effectiveness against osteoarthritis in general (van Blitterswijk et al., 2003).

There were no histopathological alterations in intestinal tissue in rabbit tests, indicating that the liposomal formulation employed was safe. As a result, liposomes can be regarded as a viable oral permeation enhancer system for glucosamine sulfate and chondroitin sulfate, even though bioavailability studies are still warranted (Agiba et al., 2018). In comparison to placebo, glucosamine and chondroitin sulfate offered statistically significant pain alleviation (Barrow, 2010). A study revealed that a single dosage of up to four capsules containing 500 mg chondroitin sulfate and 400 mg glucosamine sulfate was well tolerated, with a profile that was consistent with 12-hour treatment (Toffoletto et al., 2005). Preventive therapy, larger dosages and multimodality methods with certain combination treatments were all linked to beneficial effects of glucosamine and chondroitin sulfate treatments (Fernández-Martín et al., 2021). In long-term usage, a combination of oral chondroprotective glucosamine and chondroitin sulfate has shown efficacy in modifying osteoarthritis while also having a favorable safety profile (Agiba, 2017).

The dosage form quality control carried out for products containing glucosamine and chondroitin sulfate does not all reflect the claims stated on the label on the tablet or capsule. Weight variation tests for nutritional supplements proved insufficient to evaluate dosage forms because they did not truly reflect the amount of active ingredient present. Therefore, there needs to be regulation to protect consumers of products containing glucosamine or chondroitin sulfate (Adebowale et al., 2000). Using a supplement based on results from another product does not guarantee effectiveness. The quality and content of nutritional supplements varies. It is in the best interests of patients to carefully examine items and provide informed recommendations (Schenck, 2000).

10.6.3 Potential Application for Nutraceutical Products

Nutraceuticals dominated the market in 2020, accounting for more than 36.0% of worldwide sales. In the nutraceutical sector, sodium chondroitin sulfate is commonly used as a dietary supplement for the treatment of osteoarthritis and joint discomfort. They are consumed to relieve arthritic pain and to strengthen joints, cartilages and bones. Chondroitin sulfate is employed in a variety of foods, including drinks, chewing gums and dairy products (Grand View Research. 2021).

New probiotic drink product development was carried out by combining the healthy properties of milk-based products with the properties of non-dairy probiotic drinks, namely the presence of chondroitin sulfate as a prebiotic and supplement that modulates the immune system and the number and diversity of probiotic strains that have health benefits (Pacini and Ruggiero, 2017). Joint-protecting beverages containing one or more active substances that enhance cartilage and synovial fluid, or a combination of these chemicals with bone-building calcium and phosphorus additions are invented. The active substances in these drinks are consumed in liquid form, which allows the human body to absorb and utilize them more effectively. Glucosamine sulfate, glucosamine hydrochloride, chondroitin sulfate, methyl sulfonylmethane, and hyaluronic acid may be included in the active component combination. High fructose corn syrup, citric acid, ascorbic acid, arabic gel, sodium benzoate and potassium sorbate may all be used to make various concentration drinks (Li and Xu, 2004).

A nutritional supplement in the form of a beverage containing cartilage added with cetyl myristolate was developed. The beverage is a combination of juice drinks manufactured from water-based fruit-flavored juice pasteurized at a relatively high temperature and a cartilage supplement solution created from cartilage supplements prepared at a relatively low temperature. Carbonated drinks, non-carbonated drinks, and concentrated drinks are all options. Chondroitin, glucosamine, and hyaluronic acid are examples of cartilage supplements (Stone, 2010).

Deep processing in the blueberry beverage industry requires stable blueberry anthocyanins through the development of a nanocomplex embedded hydrogel system using anthocyanin and chondroitin sulfate co-pigmentation and then incorporation into a kappa-carrageenan hydrogel. The system can improve the stability of blueberry anthocyanins by providing effective protection, mainly due to low pH exposure during processing in the blueberry beverage industries. The chondroitin sulfate group on the nanocomposite generates a hydrophilic surface and offers good colloidal stability. Because of its high charge density, chondroitin sulfate can preserve the stacking structure of anthocyanins as a stabilizer (Xie et al., 2020).

Snack bars enriched with chondroitin sulfate are made by melting confectioner's peanut butter ingredients mixed with high fructose corn syrup and molasses. The liquid mixture was then supplemented with salt, chondroitin sulfate, glucosamine sulfate and hyaluronic acid. The sugar, whey protein, rice flour and soy protein are then added until homogeneous. The mixture is extruded at room temperature, cut into individual bars and then coated with melted chocolate confectionery before cooling. The final bar weighs around 70 grams and has roughly 300 calories (Stone, 2010).

10.6.4 Potential Application for Personal Care and Cosmetics

Chondroitin sulfate is commonly used in skin conditioning agents or hair treatments because of its capacity to moisturize, hea, and soothe skin, as well as its anti-inflammatory characteristics and ability to rebuild the intercellular matrix. Chondroitin sulfate is a substance found in skin care and cosmetics. Chondroitin sulfate is generated from muco-polysaccharides and has the ability to produce collagen and elastin, which can be used directly or incorporated into skin care products externally (Newseed, 2015). Treatment with chondroitin sulfate promotes the proliferation of keratinocytes and fibroblasts, as well as the migration and synthesis of fibroblast extracellular matrix components. Chondroitin sulfate also promotes the production of type I procollagen via activating extracellular signal-regulated kinase pathways. The chondroitin sulfate therapy improves skin wound healing and regeneration using a full-thickness skin wound model and an aging skin model. As a result, chondroitin sulfate can be used in therapeutic settings to improve skin aging (Min et al., 2020).

Chondroitin sulfate is used in hair care products like shampoo and conditioner to keep hair moist and silky and prevent it from drying out. Chondroitin sulfate is commonly used in rejuvenating serums, sun lotions, lip balms, pain relief creams and a variety of other products. In addition to skin and hair treatments, different grades of sodium chondroitin sulfate are used in beauty and personal care products, with 20%, 40%, and 80% being the most common (Newseed, 2015).

10.7 CONCLUSIONS

Marine chondroitin sulfate has good prospects to be used for various purposes because it is safe for consumption and there are no objections to its use and consumption for cultural, religious and harmful disease reasons. Marine biota that have been explored for the extraction and isolation of chondroitin sulfate by various methods are still limited, and many of them need upscaling and commercialization studies. The health-related efficacy of chondroitin sulfate has been widely revealed, not only for osteoarthritis which has been widely known to the public, so it is likely to encourage an increase in the need for chondroitin sulfate. Chondroitin sulfate has the potential to be applied for the production of food supplements, nutraceuticals and personal care products by taking advantage of its biocompatible, anti-inflammatory, biodegradable, non-immunogenic and non-toxic properties, in addition to the health benefits that have been demonstrated.

REFERENCES

Abdallaha, M.M., N. Fernándeza, A.A. Matiasa, and M.R. Bronze. 2020. Hyaluronic acid and Chondroitin sulfate from marine and terrestrial sources: Extraction and purification methods. *Carbohydrate Polymers* 243 (2020) 116441:1–11. https://doi.org/10.1016/j.carbpol.2020.116441.

Adebowale, A.O., Z. Liang, and N.D. Eddington. 2000. Nutraceuticals, a call for quality control of delivery systems. *Journal of Nutraceuticals Functional and Medical Foods* 2 (2): 15–30.

Aghamohammadi, D., N. Dolatkhah, F. Bakhtiari, F. Eslamian, and M. Hashemian. 2020. Nutraceutical supplements in management of pain and disability in osteoarthritis: A systematic review and meta-analysis of randomized clinical trials. *Scientifc Reports*. (2020) 10:20892 https://doi.org/10.1038/s41598-020-78075-x.

Agiba, A.M. 2017. Nutracetical formulation containing glucosamine and chondroitin sulphate in the treatment of osteoarthritis: Emphasis on clinical efficacy and formulation challenges. *International Journal of Current Pharmaceutical Review and Research* 9 (2): 1–7.

Agiba, A.M., M. Nasr, S. Abdel-Hamid, A.B. Eldin, and A.S. Geneidi. 2018. Enhancing the Intestinal Permeation of the Chondroprotective Nutraceuticals Glucosamine Sulphate and Chondroitin Sulphate Using Conventional and Modified Liposomes. *Current Drug Delivery* 15 (6): 907–916.

Archiando, D. 2020. Chondroitin Sulfate. https://lifepack.id/chondroitin-sulfate/ (accessed September 7, 2021) (*in Bahasa Indonesia*).

Barrow, C.J. 2010. Marine nutraceuticals: Glucosamine and omega-3 fatty acids New trends for established ingredients. *AgroFOOD Industry Hi-Tech* 21 (2): 46–41.

BfR. 2019. Risk assessment of chondroitin sulfate in food supplements. BfR opinion no 040/2018 of 7 December 2018. *BfR-Stellungnahmen* (2018) H. 040 DOI 10.17590/20190521–141041.

Blanco, M., J. Fraguas, C.G. Sotelo, R.I. Pérez-Martín, and J.A. Vázquez. 2015. Production of chondroitin sulphate from head, skeleton and fins of *Scyliorhinus canicula by*-products by combination of enzymatic, chemical precipitation and ultrafiltration methodologies. *Marine Drugs* 13: 3287–3308.

Bottegoni, C., R.A.A. Muzzarelli, F. Giovannini, A. Busilacchi, and A. Gigante. 2014. Oral chondroprotection with nutraceuticals made of chondroitin sulphate plus glucosamine sulphate in osteoarthritis. *Carbohydrate Polymers* 109: 126–138.

Chen, S., C. Xue, L. Yin, Q. Tanga, G. Yu, and W. Chai. 2011. Comparison of structures and anticoagulant activities of fucosylated chondroitin sulfates from different sea cucumbers. *Carbohydrate Polymers* 83: 688–696.

Clegg, D.O., D.J. Reda, and C.L. Harris et al. 2006. Glucosamine, chondroitin sulfate, and the two in combination for painful knee osteoarthritis. *The New England Journal of Medicine* 354 (8): 795–808.

Cohen, M., R. Wolfe, T. Mai, and D. Lewis. 2003. A randomized, double blind, placebo controlled trial of a topical cream containing glucosamine sulfate, chondroitin sulfate, and camphor for osteoarthritis of the knee. *The Journal of Rheumatology* 30: 523–528.

Cunhaa, A.L., L.G. Oliveiraa, L.F. Maia, L.F.C. Oliveira, Y.M. Michelacci, J. Adriano, and J.A.K. Aguiar. 2015. Pharmaceutical grade chondroitin sulfate: Structural analysis and identification of contaminants in different commercial preparations. *Carbohydrate Polymers* 134: 300–308. http://dx.doi.org/10.1016/j.carbpol.2015.08.006

Dierker, T., C. Shao, T. Haitina, J. Zaia, A. Hinas, and L. Kjellén. 2016. Nematodes join the family of chondroitin sulfate-synthesizing organisms: Identification of an active chondroitin sulfotransferase in Caenorhabditis elegans. *Scientific Reports* 6: 34662. doi: 10.1038/srep34662.

Fernández-Martín, S., A. González-Cantalapiedra, F. Muñoz, M. García-González, M. Permuy, and M. López-Peña. 2021. Glucosamine and chondroitin sulfate: Is there any scientific evidence for their effectiveness as disease-modifying drugs in knee osteoarthritis preclinical studies?—A systematic review from 2000 to 2021. *Animals* 11 (6), June: 1608. doi: 10.3390/ani11061608.

Garnjanagoonchorn, W., L. Wongekalak, and A. Engkagul. 2007. Determination of chondroitin sulfate from different sources of cartilage. *Chemical Engineering and Processing* 46: 465–471.

Goto, M., S. Ito, Y. Kato, S. Yamazaki, K. Yamamito, and Y. Katagata. 2011. Anti-aging effects of extracts prepared from salmon nasal cartilage in hairless mice. *Molecular Medicine Reports* 4: 779–784.

Grand View Research. 2021. *Chondroitin Sulfate Market Size, Share & Trends Analysis Report by Source (Bovine, Swine, Poultry), by Application (Nutraceuticals, Pharmaceuticals, Animal Feed, Personal Care & Cosmetics), and Segment Forecasts, 2021–2028.* www.grandviewresearch.com/industry-analysis/chondroitin-sulfate-market (accessed August 12, 2021).

Han, L.K., M. Sumiyoshi, T. Takeda, et al. 2000. Inhibitory effects of chondroitin sulfate prepared from salmon nasal cartilage on fat storage in mice fed a high-fat diet. *International Journal of Obesity* 24: 1131–1138.

Hanindika, D., M.A. Alamsjah, and N.E. Sugijanto. 2014. The isolation development of chondroitin sulphate from cuttlebone (*Sepia phraonis*), ray (*Raja* sp.), and shark (*Carcharinus falciformes*) cartilages. *Jurnal Ilmiah Perikanan dan Kelautan* 6 (2): 129–132 (*in Bahasa Indonesia with English Abstract*).

He, G., Y. Yin, X. Yan, and Q. Yu. 2014. Optimisation extraction of chondroitin sulfate from fish bone by high intensity pulsed electric fields. *Food Chemistry* 164: 205–210.

Henrotin, Y., M. Mathy, C. Sanchez, and C. Lambert. 2010. Chondroitin sulfate in the treatment of osteoarthritis: From in vitro studies to clinical recommendations. *Therapeutic Advances in Musculoskeletal Disease* 2(6): 335–348.

Hu, S.W., Y.Y. Tian, Y.G. Chang, Z.J. Li, C.H. Xue, and Y.M. Wang. 2014. Fucosylated chondroitin sulfate from sea cucumber improves glucose metabolism and activates insulin signaling in the liver of insulin-resistant mice. *Journal of Medicinal Food* 17 (7): 749–757.

Hungerford, D.S., and L.C. Jones. 2003. Glucosamine and chondroitin sulfate are effective in the management of osteoarthritis. *The Journal of Arthroplasty* 18 (3—Suppl. 1): 5–9.

Huskisson, E.C. 2008. Glucosamine and chondroitin for osteoarthritis. *The Journal of International Medical Research* 36 (6): 1–19.

Ibáñez-Sanz, G., A.D. Villanueva, L.V. Marqués, E. Gracia, N. Aragonés, R.O. Requena, J. Llorca, J. Vidán, P. Amiano, P. Nos, G.F. Tardón, R. Rada, M.D. Chirlaque, E. Guinó, V.D. Batista, G.C. Vinyals, B.P. Gómez, B.M. Pozo, T.D. Sotos, J. Etxeberria, A. Molinuevo, B.Á. Cuenllas, M. Kogevinas, M. Pollán, and V. Moreno. 2018. Possible role of chondroitin sulphate and glucosamine for primary prevention of colorectal cancer. Results from the MCC-Spain study. *Scientific Reports* 8: 2040. doi: 10.1038/s41598-018-20349-6

Joseph, N. 2021. Chondroitin. *Hello Sehat* 12/1/2021. https://hellosehat.com/herbal-alternatif/herbal/chondroitin-adalah/ (accessed August 12, 2021) (*in Bahasa Indonesia*).

Katta, J., Z. Jin, E. Ingham, and J. Fisher. 2009. Chondroitin sulphate: An effective joint lubricant? *Osteoarthritis and Cartilage* 17: 1001–1008.

Kobayashia, T., I. Kakizakia, H. Nozakab, and T. Nakamura. 2017. Chondroitin sulfate proteoglycans from salmon nasal cartilage inhibit angiogenesis. *Biochemistry and Biophysics Reports* 9: 72–78.

Konovalova, I., V. Novikov, Y. Kuchina, and N. Dolgopiatova. 2019. Technology and properties of chondroitin sulfate from marine hydrobionts. *International applied research conference "Biological Resources Development and Environmental Management", KnE Life Sciences*, 305–314. DOI: 10.18502/kls.v5i1.6075.

Krichen, F., H. Bougatef, F. Capitani, et al. 2018. Purification and structural elucidation of chondroitin sulfate/dermatan sulfate from Atlantic bluefin tuna (*Thunnus thynnus*) skins and their anticoagulant and ACE inhibitory activities. *RSC Advances* 8: 37965–37975.

Lamari, F.N., and N.K. Karamanos. 2006. Structure of chondroitin sulfate. *Advances in Pharmacology* 53: 33–48.

Lee, K.B., J.S. Kim, S.T. Kwak, W. Sire, J.H. Kwak, and Y.S. Kim. 1998. Isolation and identification of chondroitin sulfates from the mud snail. *Archives of Pharmacal Research* 21 (5): 555–558.

Lestari, R., M.Z. Wanda, and T. Wikanta. 2000. Acute toxicity of chondroitin sulfate extracted from shark cartilage. *Jurnal Penelitian Perikanan lndonesia* 6 (1): 65–71. (*in Bahasa Indonesia with English Abstract*).

Li, A., and Q. Xu. 2004. Joint-protecting beverages and/or foods and their preparation. US Patent No. US 2004/0198695A1.

Li, J., E.M. Sparkenbaugh, G. Su, et al. 2020. Enzymatic synthesis of chondroitin sulfate E to attenuate bacteria lipopolysaccharide-induced organ damage. *ACS Central Science* 6: 1199–1207.

Lomonte, A.B.V., E. Gimenez, A.C. da Silva, et al. 2021. Treatment of knee osteoarthritis with a new formulation of a fixed-dose combination of glucosamine sulfate and bovine chondroitin: A multicenter, randomized, single-blind, non-inferiority clinical trial. *Advances in Rheumatology* 61 (7). https://doi.org/10.1186/s42358-021-00165-9.

Maccari, F., F. Ferrarini, and N. Volpi. 2010. Structural characterization of chondroitin sulfate from sturgeon bone. *Carbohydrate Research* 345: 1575–1580.

Maia, L.F., T.A. Gonzagaa, R.G. Carvalhob, et al. 2016. Monitoring of sulfated polysaccharide content in marine sponges by Raman spectroscopy. *Vibrational Spectroscopy* 87: 149–156.

Medgadget. 2021. *Chondroitin Sulfate Market Changing Dynamics of Competition with Forecast to 2030*. www.medgadget.com/2021/02/chondroitin-sulfate-market-changing-dynamics-of-competition-with-forecast-to-2030.html (accessed August 12, 2021).

Min, D., S. Park, H. Kim, et al. 2020. Potential anti-aging effect of chondroitin sulfate through skin regeneration. *International Journal of Cosmetic Science* 42 (5): 520–527.

Murado, M.A., J. Fraguas, M.I. Montemayor, J.A. Vázquez, and P. González. 2010. Preparation of highly purified chondroitin sulphate from skate (Raja clavata) cartilage by-products. Process optimization including a new procedure of alkaline hydroalcoholic hydrolysis. *Biochemical Engineering Journal* 49 (1): 126–132.

Myron, P., S. Siddiquee, and S.A. Azad. 2014. Fucosylated chondroitin sulfate diversity in sea cucumbers: A review. *Carbohydrate Polymers* 112: 173–178.

Nader, H.B., T.M.P.C. Ferreira, J.F. Paiva, et al. 1984. Isolation and structural studies of heparan sulfates and chondroitin sulfates from three species of molluscs. *The Journal of Biological Chemistry* 259 (3): 1431–1435.

Nanjing University of Science and Technology. 2008. Method for improving chondroitin sulfate quality by protein isoelectric point. Chinese Patent No. CN101348814A.

Narvy, S.J., and C.T. Vangsness. 2010. Critical appraisal of the role of glucosamine and chondroitin in the management of osteoarthritis of the knee. *Nutrition and Dietary Supplements* 2010 (2): 13–25.

Newseed. 2015. *Applications and Uses of Chondroitin Sulfate*. www.foodsweeteners.com/applications-and-uses-of-chondroitin-sulfate/ (accessed September 10, 2021).

Pacini, S., and M. Ruggiero. 2017. Description of a novel probiotic concept: Implications for the modulation of the immune system. *American Journal of Immunology* 13 (2): 107–113.

Panagos, C.G., D.S. Thomson, C. Moss, et al. 2014. Fucosylated chondroitin sulfates from the body wall of the sea cucumber *Holothuria forskali*. *J. Biological Chemistry* 289 (41): 28284–28298.

Peter, M.G. 2012. *Chondroitinsulfate*, RÖMPP [Online]. Verlag KG, Stuttgart.] https://roempp.thieme.de/lexicon/RD-03-01650.

Rubio-Terrés, C., M.B. Pineda, M. Herrero, C. Nieto, and D. Rubio-Rodríguez. 2020. Analysis of the health and budgetary impact of chondroitin sulfate prescription in the treatment of knee osteoarthritis compared to NSAIDs and COXIBs. *Clinico Economics and Outcomes Research* (12): 505–514.

Sawitzke, A.D., H. Shi, and M. Finco, et al. 2008. The effect of glucosamine and/or chondroitin sulfate on the progression of knee osteoarthritis: A GAIT report. *Arthritis Rheum* 58 (10): 3183–3191.

Schenck, R.C. 2000. Nutraceuticals: Glucosamine and Chondroitin Sulfate. *Relias Media*, June 26, 2000. www.reliasmedia.com/articles/46440-nutraceuticals-glucosamine-and-chondroitin-sulfate (accessed September 14, 2021).

Seiler, R., H.Z. Oo, D. Tortora, et al. 2017. An oncofetal glycosaminoglycan modification provides therapeutic access to cisplatin-resistant bladder cancer. *European Urology* 72 (1): 142–150.

Siagian, C. 2014. Effects of glucosamine-chondroitine sulfate, glucosamine-chondroitine sulfate-methylsulfonylmethane and placebo in patients with knee osteoarthritis kellgren lawrence grade I-II: A double blind randomized controlled study. Master Thesis. Faculty of Medicine—University of Indonesia. Jakarta (*in Bahasa Indonesia with English Abstract*).

Šimánek, V., V. Křen, J. Ulrichová, and J. Gallo. 2005. The efficacy of glucosamine and chondroitin sulfate in the treatment of osteoarthritis: Are these saccharides drugs or nutraceuticals? *Biomedical Papers* 149 (1): 51–56.

Singh, J.A., S. Noorbaloochi, R. MacDonald, and L.J. Maxwell. 2015. Chondroitin for osteoarthritis. *Cochrane Database of Systematic Reviews* (1). Art. No.: CD005614. DOI: 10.1002/14651858.CD005614.pub2.

Sopyan, I., G.T. Moeljadi, S.R. Utami, and M. Abdassah. 2017. Chondroitin in transdermal patch and its main physical properties. *IJPST* 1 (1): 23–28. DOI: 10.15416/ijpst.v1i1.10425.

Stone, K.R. 2010. Cartilage enhancing food supplements and methods of preparing the same. US Patent No. US 7,851,458 B2.

Suiraoka, I.P. 2012. *9 Degenerative Diseases and Preventive Perspective*. Denpasar: Politeknik Kesehatan (*in Bahasa Indonesia*).

Sulityowati, W., A.T. Indhira, A. Arbai, and E. Yatmasari. 2015. Glucosamine and chondroitin sulphate content of shark cartilage (*Prionace glauca*) and its potential as anti-aging supplements. *International Journal of ChemTech Research* 8 (10): 163–168.

Syed, Q.A., A. Ishaq, U.U. Rahman, S. Aslam, and R. Shukat. 2017. Pulsed electric field technology in food preservation: A review. *Journal of Nutritional Health & Food Engineering* 6 (5): 168–172.

Takeda, T., M. Majima, and H. Okuda. 1998. Effects of chondroitin sulfate from salmon nasal cartilage on intestinal absorption of glucose. *Journal of Japan Society of Nutrition and Food Science* 51: 213–217 (*in Japanese with English Abstract*).

Tamura, J., K. Arima, A. Imazu, N. Tsutsumishita, H. Fujita, M.Yamane, and Y. Matsumi. 2009. Sulfation patterns and the amounts of chondroitin sulfate in the diamond squid, *Thysanoteuthis rhombus*. *Biosci. Biotechnol. Biochem* 73 (6): 1387–1391.

Tatara, Y., I. Kakizaki, S. Suto, H. Ishioka, M. Negishi, and M. Endo. 2015. Chondroitin sulfate cluster of epiphycan from salmon nasal cartilage defines binding specificity to collagens. *Glycobiology* 25 (5): 557–569.

Theoharides, T.C., P. Patra, W. Boucher, et al. 2000. Chondroitin sulphate inhibits connective tissue mast cells. *British Journal of Pharmacology* 131: 1039–1049.

Toffoletto, O., A. Tavares, D.E. Casarini, B.M. Redublo, and A.B. Ribeiro. 2005. Pharmacokinetic profile of glucosamine and chondroitin sulfate association in healthy male individuals. *ACTA Ortop Bras* 13 (5): 235–237.

Transparency Market Research. 2017. *Chondroitin Sulfate Market*. www.transparencymarketresearch.com/chondroitin-sulfate-market.html (accessed September 15, 2021).

Ustyuzhanina, N.E., M.I. Bilan, A.S. Dmitrenok, et al. 2020. Fucosylated chondroitin sulfates from the sea cucumbers *Paracaudina chilensis* and *Holothuria hilla*: Structures and anticoagulant activity. *Marine Drugs* 18, 540: 1–10. doi:10.3390/md18110540.

Utami, P., S.J.R. Kalangi, and T.F. Pasiak. 2012. The role of glucosamine in osteoarthritis. *Jurnal Biomedik* 4 (3): S29–34 (*in Bahasa Indonesia with English Abstract*).

van Blitterswijk, W.J., J.C.M. van de Nes, and P.I.J.M. Wuisman. 2003. Glucosamine and chondroitin sulfate supplementation to treat symptomatic disc degeneration: Biochemical rationale and case report. *BMC Complementary and Alternative Medicine* 3. www.biomedcentral.com/1472-6882/3/2.

Václavíková, E., and Kvasnička, F. 2015. Quality control of chondroitin sulphate used in dietary supplements. *Czech Journal of Food Sciences* 33 (2): 165–173.

Vázquez, J.A., J. Fraguas, R. Novoa-Carvallal, et al. 2018. Isolation and chemical characterization of chondroitin sulfate from cartilage by-products of blackmouth catshark (*Galeus melastomus*). *Marine Drugs* 16, 344: 1–15. doi:10.3390/md16100344.

Vázquez, J.A., I. Rodríguez-Amado, M.I. Montemayor, J. Fraguas, M.P. González, and M.A. Murado. 2013. Chondroitin sulfate, hyaluronic acid and chitin/chitosan production using marine waste sources: Characteristics, applications and eco-friendly processes: A review. *Marine Drugs* 11: 747–774.

Volpi, N., and F. Maccari. 2005. Glycosaminoglycan composition of the large freshwater mollusc bivalve *Anodonta anodonta*. *Biomacromolecules* 6: 3174–3180.

Volpi, N. 2009. Quality of different chondroitin sulfate preparations in relation to their therapeutic activity. *JPP* 61: 1271–1280.

Volpi, N. 2019. Chondroitin sulfate safety and quality. *Molecules* 24, 1447: 1–13. DOI: 10.3390/molecules24081447.

Vrana, N.E., N. Builles, V. Justin, et al. 2008. Development of a reconstructed cornea from collagen—chondroitin sulfate foams and human cell cultures *IOVS* 49 (12): 5325–5331.

Wang, Y., L.H. Ye, S.Z. Ye, S.J. Liu, Z.Y. Liu, and C.Y. Wu. 2017. Lipid-lowering and anti-oxidation effects of chondroitin sulfate prepared from squid cartilage in hypercholesterolemia mice. *International Journal of Clinical and Experimental Medicine* 10 (2): 2230–2240.

Wikanta, T., R. Lestari, and. M.Z. Wanda 2000. Shark cartilage: The effect of chondroitin sulfate intake on the stomach mucouse protection of white rats. *Jurna! Penelitian Perikanan Indonesia* 6 (1): 28–35 (*in Bahasa Indonesia with English Abstract*).

Wikanta, T., R. Perwita, P. Sarnianto, and M. Murdinah. 2002. Isolation and characterization of glycosaminoglycan compound from stingray (*Trygon sephen*) organ. *Jumal Penelitian Perikanan lndonesia* I (6): 11–20 (*in Bahasa Indonesia with English Abstract*).

Xie, C., Q. Wang, R. Ying, Y. Wang, Z. Wang, and M. Huang. 2020. Binding a chondroitin sulfate-based nanocomplex with kappa-carrageenan to enhance the stability of anthocyanins. *Food Hydrocolloids* 100: 105448. https://doi.org/10.1016/j.foodhyd.2019.105448 R.

Xuan, M., L. Pengju, W. Ning, L. Jing, and W. Ailing. 2020 Extraction method of shark chondroitin sulfate. China—Patent Application Number of 202010452711.8.

Ye, J., B. Huang, and P. Gong. 2021. Nerve growth factor-chondroitin sulfate/ hydroxyapatite-coating composite implant induces early osseointegration and nerve regeneration of peri-implant tissues in Beagle dogs. *Journal of Orthopaedic Surgery and Research* 16: 51. https://doi.org/10.1186/s13018-020-02177-5.

Yin, Y.G., Y.Z. Han, and Y. Han. 2006. Pulsed electric field extraction of polysaccharide from *Rana temporaria chensinensis* David. *International Journal of Pharmaceutics* 312: 33–36.

You, H. 2021. *Chondroitin Testing for Sports Nutrition and Joint Health Supplements*. www.eurofinsus.com/food-testing/resources/chondroitin-testing-for-sports-nutrition-and-joint-health-supplements/. (accessed September 11, 2021).

Zaki, A. 2013. *Knee Osteoarthritis Pocket Book*. Bandung: Celtics Press. (*in Bahasa Indonesia*).

Zhu, X., L. Sang, D. Wu, J. Rong, and L. Jiang. 2018. Effectiveness and safety of glucosamine and chondroitin for the treatment of osteoarthritis: A meta-analysis of randomized controlled trials. *Journal of Orthopaedic Surgery and Research* 13: 170. https://doi.org/10.1186/s13018-018-0871-5.

11 Polysaccharides from Marine Micro- and Macro-Organisms

A Novel Natural Product Approach

Kannan Kamala, Pitchiah Sivaperumal, and Gopal Dharani

CONTENTS

11.1 INTRODUCTION

One-third of the earth is covered by oceans, and it is a hot spot for natural product derivatives with great biodiversity. Also, marine natural products are chemically diverse with promising bioactivity. Every year clinical trials of marine natural products keep rising and several products are accepted as therapeutic agents (Wang et al., 2018). Out of various bioactive compounds from the marine environment, polysaccharides are available in large amounts,

DOI: 10.1201/9781003303916-11

having multifunctional bioactivity (Gantt et al., 2011). Polysaccharides are a complex organic substance which is ubiquitous in sources like microbes, algae, and animals. Generally, it consists of monomers connected by glycosidic linkages with different types of substituents i.e. acetates, amino acids, ethers, lactates, phosphates, pyruvate, and sulfates. Sporadically the diverse extra chemical side chains in polymers contribute to their various functions. The regular repeating units of monomers include monopolysaccharide, heteropolysaccharide, homopolysaccharides, etc. The size of polysaccharides is 50 Da and up (Casillo et al., 2018). In all living organisms, carbon is the most important element followed by nitrogen, oxygen, and sulfur. Sulfur is present in proteoglycan, polysaccharides, and glycolipids (Bochenek et al., 2013). The concentrations of sulfate ions in the marine environment range between 25 and 28 mM, which is a better value than the fresh water and terrain soil (Helbert, 2017). The sulfated polysaccharides are abundantly available in the ocean; sulfated polysaccharides (heparan sulfate/chondroitin sulfate) are available in the animal cell, seagrass, and seaweeds (Olsen et al., 2016). Extensive polysaccharides are produced by the cytoplasm of the living organisms through enzymatic processes. In addition, polysaccharide synthesis is instigated by cell organelles, cytoplasmic membrane, cytoplasm, etc. (Madigan et al., 1999). Intracellular polysaccharide glycogen, starch, and cyanopycin production take place in the cytoplasm of the cell. Hence the polysaccharide was very limited and determined through cell mass. In addition, the extracellular biopolymer dextran, alginate, chitosan, and xanthan produced outside of the cell can be differentiated without cell lyses in the bulky production of polysaccharides (Alexander et al., 2001). Furthermore, polysaccharide stores energy, conserves and expresses the genetic information of the cell, and communicates between environment and cell. Different organisms will produce different types of polysaccharides depending on their input unit. The genetic manipulation of polymers from marine organisms carries a massive therapeutic and biotechnological application like tissue culture and pharmacological drug development (Johansson et al., 2012). Biodegradable polymers are flexible as nanofibers, clays, and cellulose whiskers; therefore, they are used in electrospinning (Schiffman and Schauer, 2008). Moreover, marine organisms are still remaining unexplored in various aspects, and most of the marine organism is consisted of substances which exhibit potential novel bioactive-oriented products. A rising number of bioactive compounds lead to improving the interest in extensive preclinical studies in cancer biology (Dembitsky et al., 2005; Silva et al., 2012). Hence, the present chapter covers the source of polysaccharides and their biomedical and other technological applications.

11.2 MARINE POLYSACCHARIDES AND THEIR BIOLOGICAL ACTIVITIES

Marine organisms are known to produce various biopolymeric substances which are classified into polysaccharides, amino acids, and proteins (McNaught and Wilkinson, 1997). Among the polysaccharides, the sulfated polymer glycosaminoglycans are different from the other polymer by their constitution of hexamine and hexose unit that forms proteoglycans in the organism (Hardingham and Aosang, 1992). Microbes are well known to grow everywhere on the earth, especially in the marine environment, which is the reason for the major primary production which includes polysaccharides and other organic substances (Abreu and Taga, 2016; Field et al., 1998). Extracellular polymeric substances from microbes are used in the solubilization of hydrophobic organic chemicals and cationic species (Santschi et al., 1998). These cationic polysaccharides are endorsed with a high level of uronic acid with different amino acids COO^-, SO_4 and C-O- (Bhaskar and Bhosle, 2005). Also, the hydrophobic and hydrophilic nature of the polymers depends on the amino acid, peptide, and proteins connected with polysaccharides (Gutierrez et al., 2009). In addition, marine polysaccharides are thick in nature, stable, and good emulsifiers (Kaplan, 1998). Recently, silicon with alginate in rechargeable batteries was shown to perform eight times more than the batteries with a graphite anode (Ryou et al., 2013). Furthermore, marine polysaccharides have stable clips, substitute human tissue, manufacture of the biodegradable capsules, and potential wound healing activity. Heterogeneous biodegradable polysaccharide starch carries the drug in the form of foam into the human tissue.

Marine polysaccharides are agar, carrageenan, alginate, and chitin which all have a parallel chemical structure with distinct properties. Most of the traditional heparin and fucans are sulfated polysaccharide, which has been used for anticoagulant and antithrombotic activity. Bioactivity of other polysaccharides has also been investigated from microbes and macro-organisms like green seaweed (*Chaetomorpha anteninna*), red seaweed (*Gracilaria corticate*), and brown seaweed (*Gracilaria corticate*) in Indian waters which exhibits anticoagulant, antiviral, antithrombic activity, and other bioactivities (Shanmugam and Mody, 2000; Adhikari et al., 2005; Chattopadhyay and Ray, 2005). Further bioactivity of polysaccharides is discussed in this chapter in detail.

11.3 MARINE POLYSACCHARIDES FROM MACRO-ORGANISMS

11.3.1 Ulvan

The sulfated heteropolysaccharide ulvan consists of the L-iduronic acid/D-glucuronic acid bound with sulfated L-rhamnose with D-glucose and D-xylose traces (Lahaye, 1998; Wijesekara and Karunarathna, 2017). The chemical composition of ulvan depends on the source where it was extracted and its geographical location (Qi et al., 2012). But generally, it is extracted from *Ulva pertusa* and *Ulva lactuca* (Yaich et al., 2017). Ulvan has the potential to reduce the level of total cholesterol (69%), triglycerides (46%), and total lipid (30%) in hypercholesterolemic animal models (Borai et al., 2015). Higher sulfate content ulvan has exhibited significant antioxidant and antihyperlipidemic activity than the natural ulvan (Qi et al., 2015). Ulvan can be modified by the addition of a functional group to improve the bioactivity, for example, benzoylated and acetylated ulvan has exhibited potential antioxidant activity compared to natural ulvan (Rizk, 2016; Qi et al., 2006). The chelating ability of acetylated ulvan was stronger than the benzoylated ulvan in terms of sulfate content (Qi et al., 2015, Yaich et al., 2017).

11.3.2 Fucoidan

Fucoidan derived from brown seaweed *Fucus vesiculosus* and consists of L-fructose with a linear backbone. It is a well-studied marine polysaccharide and reduces abdominal aortas and atherosclerotic plaques in the aortic arch by the modification of lipid uptake metabolism and descending thoracic aorta (Johnston et al., 2003; Wang et al., 2016; Yokota et al., 2016; Park et al., 2016). Fucoidan has a higher level of anti-inflammatory effect (Preobrazhenskaya et al., 1997). The fucoidan decreases the velocity leukocyte in capillaries up to 51% in pulmonary microvessels of rabbits (Li et al., 2016). A few animal studies revealed that fucoidan has the ability to prevent diabetes (Liang et al., 2016; Cui et al., 2014). Anticoagulant activity of fucoidan was greatly dependent on the extracted source (seaweed) position of the sulfate groups, sulfate content, and their molecular weight (Li et al., 2008; Cumashi et al., 2007; Pomin, 2014).

11.3.3 Laminarin Sulfate

Laminarin sulfate has 20–25 disaccharides composed of D-glucopyranose bonded with glycosidic linkages (Nelson and Lewis, 1974). It is predominantly found in seaweed *Laminariales* and *Fucales* species (Holdt and Kraan, 2011). Chemically sulfated laminarin inhibits the heparinase, which is associated with various diseases like thrombosis, plaque morphology in diabetic patients (Baker et al., 2012), vascular injury (Baker et al., 2009), discarding the glycocalyx in septic shock (Schmidt et al., 2012), and stenting (Baker et al., 2009) and enhance sthe removal of proteoglycans on the cell surface (Jung et al., 2016; Yang et al., 2007). In addition, it has exhibited anticoagulant activity (Shanmugam and Mody, 2000), free radical scavenging activity (Tsiapali et al., 2001; Choi et al., 2012), and reduction of lipids (Shanmugam and Mody, 2000). Hepatotoxicity in a rat model of metabolic syndrome and obesity was reduced by the dietary supplement of laminarin (Szilagyi et al., 2015; Castoldi et al., 2017; Cuevas et al., 2017).

11.3.4 Rhamnan Sulfate

Chlorophtes are the major source of rhamnan sulfate basically constituted by L-rhamnose linked with carbons (Harada and Maeda, 1998). Limited work has been reported with this compound which has been used in cardiovascular disease treatment (Li et al., 2012). Due to the low molecular weight, it has significant anticoagulant activity, and concentrated rhamnan sulfate has shown an effect on fibrin polymerization (Li et al., 2012; Shammas et al., 2017).

11.3.5 Alginate

Alginate was discovered in 1881 by Stanford, and he claimed alginate as a pharmaceutical agent in his patent. Commercially alginate was extracted from *Macrocystis pyrifera* by Kelco in 1929, but the worldwide production did not take place until 1959 (Barbosa et al., 2019). Generally, alginate was used as a solidifying agent that has biomedical applications and personal care uses (Tonnesen and Karlsen, 2002: Brownlee et al. 2005). The water retains capacity, gelling, and viscosifying properties, and the stabilizing quality will enhance the commercial quality of alginate (Gomez et al., 2009). In addition, alginate took advantage of the specific biological processes of hypolipidemic hypocholesterolemic effects (Smit, 2004). Also the alginate demands in the area of biotechnology and biomedicinal research have been examined. Chitosan performs well in cationic drug release by hydrogel formation into the stable complex. Also, chitosan supports the tight junction properties and allows the drugs to interact with the cell membrane (Farkas, 1990). The fabricated chitosan nanoparticles are used in oral consumption (Prego et al., 2005, 2006). Marine-derived chitosan improved the drug transport into the tight junction between the epithelial cells (Ranaldi et al., 2002; Yeh et al., 2011; Sonaje et al., 2011). The amine and hydroxyl group in chitosan will enhance the hydrogen bond interaction with drugs and mucus by electrostatic interaction with sulfonic and sialic acids of the mucus layer, which is proven by the mucosal drug delivery research (Senel et al., 2000; Dyer et al., 2002; Sonaje et al., 2011; Dash et al., 2011). The mucoadhesive activity of chitosan is the reason for the prolonged activity of drugs on the surface membrane of the intestine (Qi et al., 2005). In addition, chitosan has a wide range of antimicrobial and antioxidant activity (Leuba and Stossel, 1986; Park et al., 2003; Schnurch and Thiomers, 2005). The low-molecular-weight chitosan nanoparticles penetrate into the cell membrane to bind with DNA and prevent messenger RNA (mRNA) synthesis and DNA transcription (Greene, 1977; Sudarshan et al., 1992). The mucoadhesive property of chitosan is nontoxic in nature and is the reason to use it as a nasal spray, buccal drug delivery, and vaginal tablet to deliver vaccines (Farkas, 1990).

11.3.6 Carrageenan

The galactan family of sulfate polysaccharide consists of an ester sulfate group at 15–40% and replaces the unit of anhydro-galactose, D-galactose, with α and β glycosidic linkages dictated by the activity of carrageenan, and it is isolated into three forms which are iota (ι), kappa (κ), and lambda (λ) composed with a sulfate group on galactose (Necas and Bartosikova, 2013; Knutsen et al., 1994; Silva et al., 2010; Usov, 1998). The carrageenan viscosity with water forms the gel without temperature which extracted from the rhophytes from the Atlantic Ocean (Gates, 2014; Hernandez-Ledesma and Herrero, 2014). It is used for the activity of lipid reduction, a cough remedy, blood coagulation, food processing, atherosclerosis, and pharmaceutical applications (Necas and Bartosikova, 2013). It has shown atherosclerotic development in rats, also lowering the level of lipid, phospholipids, and cholesterol in rabbits (Murata, 1961, 1962; Ito and Tsuchiya, 1972). In addition, triglyceride and serum cholesterol were reduced in humans at 8 weeks (Sokolova et al., 2014). Furthermore, carrageenan has been used for its antithrombin and anticoagulation activity (Kindness et al., 1979, 1980; McMillan et al., 1979; Guven et al., 2009; Silva et al., 2010).

11.3.7 Agar

Agar from the seaweed was reported by the Chinese in 300 AD. Later on, the freeze-thaw agar extraction method was developed by the Japanese in the seventeenth century, and 1880 it was used for the industrial purpose of the formation of reversible gel (Kingsbury, 1984). The linear polysaccharide agar is derived from the *Geledium* and *Gracilaria* in large quantities and in a small quantity from *Graciliaropsis*, *Afeltia*, and *Pterocladia* (Kingsbury, 1984; Laurienzo, 2010). Alternating units of anhydro D-galactose and L-galactose units are connected with alpha and beta glycosidic bonds (Li et al., 2014). It has been used as a water-soluble dietary fiber to influence lipid metabolism in rats (Jimenez-Escrig and Sanchez-Muniz, 2000).

11.3.8 Chitosan

The linear polysaccharide chitosan is attained by the deacetylation of chitin, which was first reported in 1884. The degree of acetylation will differentiate the chitin and chitosan residues from the exoskeleton of the crustaceans and arthropods. After the deacetylation process, the final product with less than 50% is chitosan is soluble in acidic solution (Younes and Rinaudo, 2015). The presence of the acetyl group in the side chain of chitosan provides their physical property. The biomedical application of chitosan was studied, and it has proven anti-inflammatory activity and antimicrobial activity on the infected wounds (Schnurch and Dunnhaupt,2012).

11.4 MARINE BIO-POLYSACCHARIDES FROM MICROBES

The primary exopolysaccharides are documented from the marine microbes *Alteromonas*, *Bacillus*, *Enterobacter*, *Halomonas*, *Planococcus*, *Pseudoalteromonas*, and *Rhodococcus*. Polysaccharides from marine bacteria are composed of organic and inorganic molecules, amino acid, uronic acid, hexo, and pento-sugars (Manivasagan and Kim, 2014). Polysaccharides from *Saccharophagus degradans* were studied with various carbon sources including xylose and starch, and *Vibrio furnissi* isolated from the coastal area of Goa has exhibited an enormous yield in the exponential growth phase (Gonzalez Garcia et al., 2015, Bramhachari et al., 2007). Similarly, *Enterobacter cloacae* from marine sediment has produced polysaccharides that have potential emulsifying activity (Iyer et al., 2005). Marine bacterial polysaccharides are acquiring stable physical and chemical parameters by reason of application in biomedical research and food industries (Priyanka Singh and Ena Gupta, 2020). Polysaccharide from fungi has more industrial application by their novel function and structure (Seviour et al., 1992). Polysaccharide from *Fusarium oxysporum* was composed of glucose mannose and galactose at the ratio of 1.33:1:1.33 with the molecular weight of 61.2 KDa. This polysaccharide has multiple side chains with galactofuranose residues, which shown great antioxidant activity (Chen et al., 2015). Sun et al. reported the polysaccharide from *Penicillium* sp. F23 was primarily composed of different amounts of glucose and galactose and mannose exhibited antioxidant activity (Sun et al., 2009). Phoma herbarum has produced a potential amount of polysaccharides consist of D-glucan linked with less quantity of glucopyranosyl and residues of glucuronic acid. It has application in the field of macrophage receptors, immunomodulatory effect on dendritic cells, T cells, and antitumor potential (Yang et al., 2005; Chen et al., 2009, 2014).

11.4.1 Dextran

Generally, various qualities of dextrans are extracted from *Leuconostoc mesenteroides*. The marine bacterium *Acetobacter* and *Streptococcus* also produce dextran (Niven et al., 1941; Hehre and Hamilton, 1951; Qader et al., 2005). Mostly it is used in the food industry for thickening the food product; it also improves moisture preservation, inhibits the crystallization of sugar, and maintains the flavor and appearance of different foodstuffs (Qader et al., 2005; Naessens et al., 2005; Purama and Goyal, 2005).

11.4.2 Hyaluronic Acid

It is composed of the repeating units of disaccharides including D-glucuronic acid and N-acetyl D glucosamine without protein bounds (Lindahl et al., 2015). It is a multifunctional glue, having a smooth viscous property used in various medicinal fields in humans. In particular it is used in joint disease, ophthalmological surgical aid, and wound healing (Goa and Benfield, 1994; Lee et al., 2003). It is semi-flexible in nature with high molecular weight, which decreases the unique viscoelastic property of the solution in the ophthalmological field (Lapcik et al., 1998). As well, it has an enormous beneficial claims in rheumatology, tissue engineering, pulmonary pathology, and orthopedic surgery (Giji and Arumugam, 2014). In addition, it has a crucial role in the transport of macromolecules from cell to bacteria and tissue penetration and hydration by the absorbing capacity of water molecules (Garg and Hales, 2004). The scaffold preparation of hyaluronic acid with chitosan has been employed in chronic ulcer cure and wound dressing (Abdel-Mohsen et al., 2017; Abdelrahman et al., 2020).

11.5 CONCLUSION

The marine ecosystem provides an increasing variety of natural products that might be used for different pharmaceutical purposes. In particular, marine bioactive polysaccharides from micro- and macro-organisms are significant for numerous benefits in new drug discovery. Marine polysaccharides are acquiring stable physical and chemical parameters by reason of application in biomedical research and food industries. In addition, the diverse chemical nature is providing effective and low-cost drug development. Furthermore, marine polysaccharides have stable nature, substitute human tissue, manufacture biodegradable capsules, and accelerating wound healing properties.

ACKNOWLEDGMENT

The authors are grateful to Saveetha Dental College and Hospital, SIMATS, and the second author acknowledges SERB-DST for moral support through TARE scheme (File no: TAR/2019/000143).

REFERENCES

Abdel-Mohsen, A. M., J. Jancar, R. M. Abdel-Rahman, L. Vojtek, P. Hyrsl, M. Duskova, and H. Nejezchlebova. 2017. A novel in situ silver/hyaluronan bio-nanocomposite fabrics for wound and chronic ulcer dressing: In vitro and in vivo evaluations. *International Journal of Pharmaceutics* 520(1–2): 241–253.

Abdelrahman, R. M., A. M. Abdel-Mohsen, M. Zboncak, J. Frankova, P. Lepcio, L. Kobera, and J. Jancar. 2020. Hyaluronan biofilms reinforced with partially deacetylated chitin nanowhiskers: Extraction, fabrication, in-vitro and antibacterial properties of advanced nanocomposites. *Carbohydrate Polymers* 235: 115951.

Abreu, N. A. and M. E. Taga, 2016. Decoding molecular interaction in microbial communities. *FEMS Microbiol.Rev.*, 40(5): 648–663.

Adhikari, U., C. Mateu, E. B. Damonte, and B. Ray. 2005. *Proceedings of CARBO XX Carbohydrate Conference*, November 24–26. Lucknow: Lucknow University.12 pp.

Alexander, S. 2001. Thermo responsive polymer colloids for drug delivery and cancer therapy. *Macromolecule Bioscience* 11: 1722–1734.

Baker, A. B., W. J. Gibson, V. B. Kolachalama, M. Golomb, L. Indolfi, C. Spruell, et al. 2012. Heparanase regulates thrombosis in vascular injury and stent-induced flow disturbance. *Journal of the American College of Cardiology* 59: 1551–1560.

Baker, A. B., A. Groothuis, M. Jonas, D. S. Ettenson, T. Shazly, E. Zcharia, et al. 2009. Heparanase alters arterial structure, mechanics, and repair following endovascular stenting in mice. *Circulation Research* 104: 380–387.

Barbosa, A. I., A. J. Coutinho, S. A. C. Lima, and S. Reis. 2019. Marine polysaccharides in pharmaceutical applications: Fucoidan and Chitosan as key players in the drug delivery match field. *Marine Drugs* 17(654): 1–21.

Bhaskar, P. V., and N. B. Bhosle. 2005. Microbial extracellular polymeric substances in marine biogeochemical processes. *Current Science* 88 45–53.

Bochenek, M., Etherington, G. J., Koprivova, A., Mugford, S. T., Bell, T. G., Malin, G. and S. Kopriva, 2013. Transcriptomic analysis of the sulfate deficiency response in the marine microalga Emiliania huxleyi. *New Phytol.* 199, 650–662.

Borai, I. H., M. K. Ezz, M. Z. Rizk, M. El Sherbiny, A. A. Matloub, N. F. Aly, et al. 2015. Hypolipidemic and anti-atherogenic effect of sulphated polysaccharidesfrom the green alga Ulva fasciata. *International Journal of Pharmacological Science Review and Research* 31: 1–12.

Bramhachari, P. V., P. B. Kishor, R. Ramadevi, R. Kumar, B. R. Rao, and S. K. Dubey. 2007. Isolation and characterization of mucous exopolysaccharide produced by *Vibrio furnissii* strain VB0S3. *Journal of Microbiolal Biotechnol*ogy 17: 44–51.

Brownlee, I. A., A. Allen, J. P. Pearson, P. W. Dettmar, M. E. Havler, M. R. Atherton, and E. Onsoyen. 2005. Alginate as a source of dietary fiber. *Critical Review on Food Science and Nutrition* 45 497–510.

Casillo, A., R. Lanzetta, M. Parrilli and M. M. Corsaro, 2018. Exopolysaccharide from marine and marine extremeophilic bacteria: structures, properties, ecological roles and applications. *Mar. Drugs.*, 16(2): 1–34.

Castoldi, A., V. Andrade-Oliveira, C. F. Aguiar, M.T. Amano, J. Lee, M. T. Miyagi, et al. 2017. Dectin-1 activation exacerbates obesity and insulinresistance in the absence of MyD88. *Cell Repl* 19: 2272–2288.

Chattopadhyay, K., T. Ghosh, C. A. Pujol, M. J. Carlucci, E. B. Damonte and B. Ray, 2007. Ploysaccarides from Grracilaria corticata:Saulfation, chemical characterization and anti HSV activities. *Int. J. Biol. Macrmol.*, 43: 346–351.

Chen, S., R. Ding, Y. Zhou, X. Zhang, R. Zhu, and X. D. Gao, 2014. Immunomodulatoryeffects of polysaccharide from marine fungus Phomaherbarum YS4108 on T Cells and Dendritic Cells. *Mediators Inflamm.* Volume 2014, Article ID 738631, 13.

Chen, S., D. K. Yin, W. B. Yao, Y. D. Wang, Y. R. Zhang, and X. D. Gao. 2009. Macrophage receptors of polysaccharide isolated from a marine filamentous fungus *Phoma herbarum* YS4108. *Acta Pharmacol Sin* 30: 1008–1014.

Chen, Y. L., W. J. Mao, H. W. Tao, W.M. Zhu, M. X. Yan, X. Liu, and T. Guo. 2015. Preparation and characterization of a novel extracellular polysaccharide with antioxidant activity, from the mangrove-associated fungus *fusarium oxysporum. Marine Biotechnology* 17: 219–228.

Choi, J. I., H. J. Kim, J. H. Kim, and W. H. Lee. 2012. Enhanced biological activities of laminarin degraded by gamma-ray irradiation. *Journal of Food Biochemistry* 36: 465–469.

Cuevas, A. M., M. Lazo, I. Zuniga, F. Carrasco, J. J. Potter, V. Alvarez, et al. 2017. Expression ofMYD88 in adipose tissue of obese people: Is there some role in the development of metabolic syndrome? *Metabolic Syndrome and Related Disorders* 15: 80–85.

Cui, W., Y. Zheng, Q. Zhang, J. Wang, L. Wang, W. Yang, et al. 2014. Low-molecular-weight fucoidan protects endothelial function and ameliorates basal hypertension in diabetic Goto-Kakizaki rats. *Lab Investigation* 94: 382–393.

Cumashi, A., N. A. Ushakova, M. E. Preobrazhenskaya, A. DIncecco, A. Piccoli, L. Totani, et al. 2007. A comparative study of the anti-inflammatory, anticoagulant, antiangiogenic, and antiadhesive activities of nine different fucoidans from brown seaweeds. *Glycobiology* 17: 541–552.

Dash, M., F. Chiellini, R. M. Ottenbrite, and E. Chiellini. 2011. Chitosan A versatile semi-synthetic polymer in biomedical applications. *Progress in Polymer Science* 36: 981–1014.

Dembitsky, V. M., T. A. Gloriozova, and V. V. Poroikov. 2005. Novel antitumor agents: Marine sponge alkaloids, their synthetic analogs and derivatives. *Mini Reviews in Medical Chem*istry 5: 319–336.

Dyer, A. M., M. Hinchclie, P. Watts, J. Castile, I. Jabbal-Gill, R. Nankervis, A. Smith, and L. Illum. 2002. Nasal delivery of insulin using novel chitosan based formulations: A comparative study in two animal models between simple chitosan formulations and chitosan nanoparticles. *Pharmaceutical Research* 19: 998–1008.

Farkas, V. 1990. Fungal cell walls: Their structure, biosynthesis and biotechnological aspects. *Acta Biotechnology* 10: 225–238.

Field, C. B., M. J. Behrenfeld, J. T. Randerson and P. Falkowski, 1998. Primary production of the biosphere: Integrating terrestrial and oceanic components. *Science*, 281: 237–240.

Gantt, R. W., P. P. Pain and J. S. Thorson, 2011. Enzymatic methods for glycol (diversification/randomization) of drugs and small molecules. *Nat. Prod. Rep.*, 11: 1811–1853.

Garg, H. G., and C. A. Hales. 2004. *Chemistry and Biology of Hyaluronan*. Elsevier.

Gates, K. W. 2014. Bioactive compounds from marine foods: Plant and animal sources, by Blanca Hernández-Ledesma and Miguel Herrero (Editors). *Journal of Aquatic Food Production and Technology* 23: 313–317.

Giji, S., and M. Arumugam. 2014. Isolation and characterization of hyaluronic acid from marine organism. *Advanced Food and Nutritional Research* 72: 61–77.

Goa, K. L., and P. Benfield. 1994. Hyaluronic-acid—a review of its pharmacology and use as a surgical aid in ophthalmology, and its therapeutic potential in joint disease and wound-healing. *Drugs* 47: 536–566.

Gomez, C. G., M. V. P. Lambrecht, J. E. Lozano, M. Rinaudo, and M. A. Villar. 2009. Influence of the extraction-purification conditions on final properties of alginates obtained from brown algae (*Macrocystis pyrifera*). *International Journal of Biological Macromolecules* 44: 365–371.

Gonzalez Garcia, Y., A. Heredia, J. C. Meza-Contreras, F. M. E. Escalante, R. M. Camacho-Ruiz, and J. Cordova. 2015. Biosynthesis of extracellular polymeric substances by the marine bacterium *Saccharophagus degradans* under different nutritional conditions. *International Journal of Polymer Science*.

Greene, W. H. 1977. Biochemistry of antimicrobial action. *Yale Journal of Biology and Medicine* 50: 87–88.

Gutierrez, A. F. W., R. J. Zaldivar and S. G. Contreras. 2009. Effect of various levels of digestible energy and protein in the diet on the growth of gamitana (*Colossoma macropomum*) Cuvier 1818. Rev. Invest. Vet. Peru., 20 (2): 178–186.

Guven, K. C., Y. Ozsoy, and O. N. Ulutin. 2009. Anticoagulant, fibrinolytic and antiaggregant activity of carrageenans and alginic acid. *Botanica Marina* 34: 429–432.

Harada, N., and M. Maeda. 1998. Chemical structure of antithrombin-active rhamnan sulfate from Mono-stromnitidum. *Biosci Biotechnol Biochemistry* 62: 1647–1652.

Hardingham, T. E., and A. J. Aosang. 1992. Proteoglycans: Many forms and many functions. *FASEB Journal* 6(3): 861–870.

Hehre, E. J., and D. M. Hamilton. 1951. The biological synthesis of dextran from dextrin. *Journal of Biological Chemistry* 192: 161–174.

Helbert, W., 2017. Marine polysaccharide sulfatases. *Front. Mar. Sci.*, 4: 1–10.

Hernandez-Ledesma, B., and M. Herrero. 2014. *Bioactive Compounds from Marine Foods: Plant and Animal Sources*. Ed. N. J. Wiley, 1st Edition. Hoboken: Blackwell publication, 464 pp.

Holdt, S. L., and S. Kraan. 2011. Bioactive compounds in seaweed: Functional food applications and legislation. *Journal of Applied Phycology* 23: 543–597.

Ito, K., and Y. Tsuchiya. 1972. The effect of algal polysaccharides on the depressing of plasma cholesterol level in rats. In *Proceeding of the Seventh International Seaweed Symposium*. Tokio University Press, 451–454.

Iyer, A., K. Mody, and B. Jha. 2005. Characterization of an exopolysaccharide produced by a marine, *Enterobacter cloacae*. *Indian Journal of Experimental Biology* 43: 467–471.

Jimenez-Escrig, A., and F. J. Sanchez-Muniz. 2000. Dietary fibre from edible seaweeds: Chemical structure, physicochemical properties and effects on cholesterol metabolism. *Nutracutical Research* 20: 585–598.

Johansson, C., J. Bras, I. Mondragon, P. Nechita, D. Plackett, P. Simon, D. G. Svetec, S. Virtanen, M. G. Baschetti, C. Breen, and S. Aucejo. 2012. Renewable fibers and bio-based materials for packaging applications—a review of recent developments. *Bioresources* 7(2): 2506–2552.

Johnston, T. P., Y. Li, A. S. Jamal, D. J. Stechschulte, and K. N. Dileepan. 2003. Poloxamer 407-induced atherosclerosis in mice appears to be due to lipid derangements and not due to its direct effects on endothelial cells and macrophages. *Mediat Inflamm* 12: 147–155.

Jung, O., V. Trapp-Stamborski, A. Purushothaman, H. Jin, H. Wang, R. D. Sanderson, et al. 2016. Heparanase-induced shedding of syndecan-1/CD138 in myeloma and endothelial cells activates VEGFR2 and an invasive phenotype: Prevention by novel synstatins. *Oncogenesis* 5: e202.

Kaplan, D. 1998. *Biopolymers from Renewable Resources*. Ed. D. L. Keplan. Medford: Tufts University.

Kindness, G., W. F. Long, and F. B. Williamson. 1979. Enhancement of antithrombin III activity by carrageenans. *Thrombosis Research* 15: 49–60.

Kindness, G., W. F. Long, and F. B. Williamson. 1980. Anticoagulant effects of sulphated polysaccharides in normal and antithrombin III-deficient plasmas. *Brzilian Journal of Pharmacology* 69: 675–677.

Kingsbury, J. M. 1984. The biology of Seaweed—Lobban, Cs, Wynne, Mj. *Bioscience* 34: 334.

Knutsen, S. H., D. E. Myslabodski, B. Larsen, and A. I. Usov. 1994. A modified system of nomenclature for red Algal Galactans. *Botanica Marina* 37: 163–170.

Lahaye, M. 1998. NMR spectroscopic characterisation of oligosaccharides from two Ulva rigida ulvan samples (Ulvales, Chlorophyta) degraded by a lyase. *Carbohydrate Research* 314: 1–12.

Lapcik, L., L. Lapcik, S. de Smedt, J. Demeester, and P. Chabrecek. 1998. Hyaluronan: Preparation, structure, properties, and applications. *Chemical Reviews* 98(8): 2663–2684.

Laurienzo, P. 2010. Marine polysaccharides in pharmaceutical applications: An overview. *Marine Drugs* 8: 2435–2465.

Lee, S. B., Y. M. Lee, K. W. Song, and M. H. Park. 2003. Preparation and properties of polyelectrolyte complex sponges composed of hyaluronic acid and chitosan and their biological behaviors. *Journal of Applied Polymer Science* 90(4): 925–932.

Leuba, J. L., and P. Stossel. 1986. Chitosan and other polyamines: Antifungal activity and interaction with biological membranes. In *Chitin in Nature and Technology*, eds. R. Muzzarelli, C. Jeuniaux, and G. W. Gooday. Boston, MA: Springer, 215–222.

Li, B., F. Lu, X. Wei, and R. Zhao. 2008. Fucoidan: Structure and bioactivity. *Molecules* 13: 1671–1695.

Li, H., W. Mao, Y. Hou, Y. Gao, X. Qi, C. Zhao, et al. 2012. Preparation, structure and anticoagulant activity of a low molecular weight fraction produced by mild acid hydrolysis of sulfated rhamnan fromMonostroma latissimum. *Bioresource Technology* 114: 414–418.

Li, M., G. Li, L. Zhu, Y. Yin, X. Zhao, C. Xiang, et al. 2014. Isolation and characterization of an agaro-oligosaccharide (AO)-hydrolyzing bacterium from the gut microflora of Chinese individuals. *PLoS One* 9: e91106.

Li, X., J. Li, Z. Li, Y. Sang, Y. Niu, Q. Zhang, et al. 2016. Fucoidan from Undaria pinnatifida prevents vascular dysfunction through PI3K/Akt/eNOSdependent mechanisms in the l-NAME-induced hypertensive rat model. *Food Function* 7: 2398–2408.

Liang, Z., Y. Zheng, J. Wang, Q. Zhang, S. Ren, T. Liu, et al. 2016. Low molecular weight fucoidan ameliorates streptozotocin-induced hyper-responsiveness of aortic smooth muscles in type 1 diabetes rats. *Journal of Ethnopharmacology* 191: 341–349.

Lindahl, U., J. Couchman, K. Kimata, and J. D. Esko. 2015. *Proteoglycans and Sulfated Glycosaminoglycans*. Cold Spring: Essentials of Glycobiology.

Madigan, M. T., J. M. Martinko, and J. Parker. 1999. *Brock Biology of Microorganisms*, 9th Edition. Upper Saddle River, NJ: Prentice-Hall Inc.

Manivasagan, P., and S. K. Kim. 2014. Extracellular polysaccharides produced by marine bacteria. *Advances in Food Nutrional Research* 72: 79–94.

McMillan, R. M., D. E. MacIntyre, and J. L. Gordon. 1979. Stimulation of human platelets by carrageenans. *Journal of Pharmacy and Pharmacology* 31: 148–152.

McNaught, A. D., and A. Wilkinson. 1997. *Compodium of Chemical Terminology, IUPAC*. Oxford: Black Well Scientific Publications.

Murata, K. 1961. Inhibitory effect of sulfated mono- and polysaccharides on experimental hyperlipemia and atherosclerosis in rabbits. *Japanese Heart Journal* 2: 198–209.

Murata, K. 1962. The effects of sulfated polysaccharides obtained from seaweeds on experimental atherosclerosis. *Journal of Gerontology* 17: 30–36.

Naessens, M., A. Cerdobbel, W. Soetaert and E. J. Vandamme, 2005. *Leuconostoc* dextransucrase and dextran: production, properties and applications. *J. Chem. Technol. Biotechnol.*, 80: 845–860.

Necas, J., and L. Bartosikova. 2013. Carrageenan: A review. *Veterinary Medicine* 58: 187–205.

Nelson, T. E., and B. A. Lewis. 1974. Separation and characterization of the soluble and insoluble components of insoluble laminaran. *Carbohydrate Research* 33: 63–74.

Niven, C. F., K. L. Smiley, and J. M. Sherman, 1941. The production of large amount of a polysaccharide by streptococcus salivarius. *Journal Bacteriology* 41: 479–484.

Olsen, J. L., Rouzé, P., Verhelst, B., Lin, Y. C., Bayer, T., Collen, J., Dattolo, E., De Paoli, E., Dittami, S., Maumus, F., Michel, G., Kersting, A., Lauritano, C., Lohaus, R., Topel, M., Tonon, T., Vanneste, K., Amirebrahimi, M., Brakel, J., Bostrom, C., Chovatia, M., Grimwood, J., Jenkins, J. W., Jueterbock, A., Mraz, A., Stam, W. T., Tice, H., Bornberg-Bauer, E., Green, P. J., Pearson, G. A., Procaccini, G., Duarte, C. M., Schmutz, J., Reusch, T. B., Van de Peer, Y., (2016). The genome of the seagrass *Zostera marina* reveals angiosperm adaptation to the sea. *Natu.*, 530: 331–335.

Park, J., M. Yeom, and D. H. Hahm. 2016. Fucoidan improves serum lipid levels and atherosclerosis through hepatic SREBP-2-mediated regulation. *Journal of Pharmacological Science* 131: 84–92.

Park, P. J., J. Y. Je, and S. K. Kim. 2003. Free radical scavenging activity of chitooligosaccharides by electron spin resonance spectrometry. *Journal of Agriculture and Food Chemistry* 51: 4624–4627.

Pomin, V. H. 2014. Anticoagulant motifs of marine sulfated glycans. *Glycoconjugate Journal* 31: 341–344.

Prego, C., M. Fabre, D. Torres, and M. J. E. Alonso. 2006. Cacy and mechanism of action of chitosan nanocapsules for oral peptide delivery. *Pharmaceutical Research* 23: 549–556.

Prego, C., M. Garcia, D. Torres, and M. J. Alonso. 2005. Transmucosal macromolecular drug delivery. *Journal of Control Release* 101: 151–162.

Preobrazhenskaya, M. E., A. E. Berman, V. I. Mikhailov, N. A. Ushakova, A. V. Mazurov, et al. 1997. Fucoidan inhibits leukocyte recruitment in a model peritoneal inflammation in rat and blocks interaction of P-selectin with its carbohydrate ligand. *Biochemistry and Molecular Biology International* 43: 443–451.

Priyanka, S., and G. Ena. 2020. Role of marine microbial polysaccharides in sustainable environmental security; In *Environmental Challenges and Issues in Present Scenario*, eds. P. K. Singh, A. Srivastava, V. Srivastava, and R. Kumar, Springer Nature Singapore Pte Ltd., Chapter 4.16–23pp.

Purama, R. K. and A. Goyal, 2005. Dextransucrase production by *Leuconostoc mesenteroides. Indian J. Microbiol.*, 2: 89–101.

Qader, S. A., L. Iqbal, A. Aman, E. Shireen, and A. Azhar. 2005. Production of dextran by newly isolated strains of *leuconostoc mesenteroides* PCSIR-3 and PCSIR-9. *Turkish Journal of Biochemistry* 31: 21–26.

Qi, H., X. Liu, J. Zhang, Y. Duan, X. Wang, and Q. Zhang. 2012. Synthesis and antihyperlipidemic activity of acetylated derivative of Ulvan from Ulva pertusa. *International Journal of Biological Macromolecules* 50: 270–272.

Qi, H., Q. Zhang, T. Zhao, R. Hu, K. Zhang, and Z. Li. 2006. In vitro antioxidant activity of acetylated and benzoylated derivatives of polysaccharide extracted from Ulva pertusa (Chlorophyta). *Bioorg Med Chem Letters* 16: 2441–2445.

Qi, L. F., Z. R. Xu, Y. Li, X. Jiang, and X. Y. Han. 2005. In vitro e_ects of chitosan nanoparticles on proliferation of human gastric carcinoma cell line MGC803 cells. *World Journal of Gastroenterol* 11: 5136–5141.

Ranaldi, G., I. Marigliano, I. Vespignani, G. Perozzi, and Y. Sambuy. 2002. The e_ect of chitosan and other polycations on tight junction permeability in the human intestinal Caco-2 cell line (1). *Journal of Nutrional Biochemistry* 13: 157–167.

Rizk, M. Z. 2016. The anti-hypercholesterolemic effect of Ulvan Polysaccharide extracted from the green alga Ulvafasciataon Aged Hypercholesterolemic rats. *Asian Journal of Pharmacological and Clinical Research* 9: 165–176.

Ryou, M. H., S. Hong, M. Winter, H. Lee, and J. W. Choi. 2013. Improved cycle lives of $LiMnO_4$ cathodes in lithium ion batteries by an alginate biopolymer from seaweed. *Journal of Material Chemistry A* 1: 15224–15229.

Santschi, P. H., L. Guo, J. C. Means, and M. Ravichandran. 1998. Natural organic matter binding of trace metal and trace organic contaminants in estuaries. In *Biogeochemistry of Gulf of Mexico Estuaries*, eds. T. S. Bianchi, J. R. Pennock, and R. Twilley. New York: Wiley, 347–380.

Schiffman, J. D., and C. L. Schauer. 2008. A review: Electrospinning of biopolymer nanofibers and their applications. *Polymer Reviews* 48(2): 317–352.

Schmidt, E. P., Y. Yang, W. J. Janssen, A. Gandjeva, M. J. Perez, L. Barthel, et al. 2012. The pulmonary endothelial glycocalyx regulates neutrophil adhesion and lung injury during experimental sepsis. *Natural Medicine* 18: 1217–1223.

Schnurch, B. A., and S. Dunnhaupt. 2012. Chitosan-based drug delivery systems. *Europian Journal of Pharm. Biopharm* 81: 463–469.

Schnurch, B. A., and A. Thiomers, 2005. A new generation of mucoadhesive polymers. *Advanced Drug Delivery Reviews* 57: 1569–1582.

Senel, S., M. J. Kremer, S. Kas, P. W. Wertz, A. A. Hincal, and C. A. Squier. 2000. Enhancing e_ect of chitosan on peptide drug delivery across buccal mucosa. *Biomaterials* 21: 2067–2071.

Seviour, R. J., S. J. Stasinopoulos, D. P. F. Auer, and P. A. Gibbs. 1992. Production of pullulan and other exopolysaccharides by filamentous fungi. *Critical Reviews on Biotechnol*ogy 12: 279–298.

Shammas, A. N., H. Jeon-Slaughter, S. Tsai, H. Khalili, M. Ali, H. Xu, et al. 2017. Major limb outcomes following lower extremity endovascular revascularization in patients with and without diabetes mellitus. *Journal of Endovascular Theropy* 24:376–382.

Shanmugam, M., and K. H. Mody. 2000. Heparinoid-active sulphated polysaccharides from marine algae as potential blood anticoagulant agents. *Current Science in India* 79: 1672–1683.

Silva, F. R. F., C. M. P. G. Dore, C. T. Marques, M. S. Nascimento, N. M. B. Benevides, H. A. O. Rocha, et al. 2010. Anticoagulant activity, paw edema and pleurisy induced carrageenan: Action of major types of commercial carrageenans. *Carbohydrate Polymer* 79: 26–33.

Silva, T. H., A. Alves, B. M. Ferreira, M. Oliveira, L. L. Reys, R. J. F. Ferreria, R. A. Sousa, S. S. Silva, J. F. Mano, and R. L. Reis. 2012. Materials of marine origin: A review on polymers and ceramics of biomedical interest. *International Material Reviews*, 1–32.

Smit, A. J. 2004. Medicinal and pharmaceutical uses of seaweed natural products: A review. *Journal of Applied Phycology* 16(4): 245–262.

Sokolova, E. V., A. O. Byankina, A. A. Kalitnik, Y. H. Kim, L. N. Bogdanovich, T. F. Soloveva, et al. 2014. Influence of red algal sulfated polysaccharides on blood coagulation and platelets activation in vitro. *Journal of Biomedical Materials Research A* 102: 1431–1438.

Sonaje, K., K. J. Lin, M. T. Tseng, S. P. Wey, F. Y. Su, E. Y. Chuang, C. W. Hsu, C. T. Chen, and H. W. E. Sung. 2011. Ects of chitosan-nanoparticle-mediated tight junction opening on the oral absorption of endotoxins. *Biomaterials* 32: 8712–8721.

Sudarshan, N. R., D. G. Hoover, and D. Knorr. 1992. Antibacterial action of chitosan. *Food Biotechnology* 6: 257–272.

Sun, H. H., W. J. Mao, Y. Chen, S. D. Guo, H. Y. Li, X. H. Qi, Y. L. Chen, and J. Xu. 2009. Isolation, chemical characteristics and antioxidant properties of the polysaccharides from marine fungus *Penicillium sp*. F23–2. *Carbohydrate Polymer* 78: 117–124.

Szilagyi, K., M. J. Gijbels, S. Van der Velden, S. E. Heinsbroek, G. Kraal, M. P. De Winther, et al. 2015. Dectin-1 deficiency does not affect atherosclerosis development in mice. *Atherosclerosis* 239: 318–321.

Tonnesen, H. H., and J. Karlsen. 2002. Alginate in drug delivery systems. *Drug Development and Industrial Pharmacy* 28: 621–630.

Tsiapali, E., S. Whaley, J. Kalbfleisch, H. E. Ensley, I. W. Browder, and D. L. Williams.2001. Glucans exhibit weak antioxidant activity, but stimulate macrophage free radical activity. *Free Radical and Biological Medicine* 30: 393–402.

Usov, A. I. 1998. Structural analysis of red seaweed galactans of agar and carrageenan groups. *Food Hydrocolloids* 12: 301–308.

Wang, W., R. Chen, Z. Luo, W. Wang and J. Chen, 2018. Antimicrobial activity and molecular docking studies of novel anthroquinone from marine derived fungus. *Aspergillus versicolor*. Nat.Prod.Res., 32: 558–563.

Wang, X., L. Pei, L. Haibo, Q. V. Kai, W. Xian, J. Liu, et al. 2016. Fucoidan attenuates atherosclerosis in LDLR-/- mice through inhibition of inflammation and oxidative stress. *International Journal of Clinical and Experimental Pathology* 9: 6896–6904.

Wijesekara, I., and W. K. D. S. Karunarathna. 2017. *Usage of Seaweed Polysaccharidesas Nutraceuticals*. Nugegoda: Elsevier, 341–348.

Yaich, H., A. B. Amira, F. Abbes, M. Bouaziz, S. Besbes, A. Richel, et al. 2017. Effect of extraction procedures on structural, thermal and antioxidant properties of ulvan from Ulva lactuca collected in Monastir coast. *International Journal of Biological Macromolecules* 105: 1430–1439.

Yang, X. B., X. D. Gao, F. Han, B. S. Xu, Y. C. Song, and R. X. Tan. 2005. Purification, characterization and enzymatic degradation of YCP, a polysaccharide from marine filamentous fungus Phoma herbarumYS4108. *Biochimie* 87: 747–754.

Yang, Y., V. Macleod, H.Q. Miao, A. Theus, F. Zhan, J. D. Shaughnessy Jr, et al. 2007. Heparanase enhances syndecan-1 shedding: A novel mechanism for stimulation of tumor growth and metastasis. *Journal of Biological Chemistry* 282: 13326–13333.

Yeh, T. H., L. W. Hsu, M. T. Tseng, P. L. Lee, K. Sonjae, Y. C. Ho, and H. W. Sung. 2011. Mechanism and consequence of chitosan-mediated reversible epithelial tight junction opening. *Biomaterials* 32 6164–6173.

Yokota, T., K. Nomura, M. Nagashima, and N. Kamimura. 2016. Fucoidan alleviates high-fat diet-induced dyslipidemia and atherosclerosis in ApoE(shl) mice deficient in apolipoprotein E expression. *Journal of Nutrional Biochemistry* 32:46–54.

Younes, I., and M. Rinaudo. 2015. Chitin and chitosan preparation from marine sources. Structure, properties and applications. *Marine Drugs* 13: 1133–1174.

12 Fucoidan

Bioactivity Properties and Ulcer Healing Applications

Ellya Sinurat, Dina Fransiska, Nurhayati, and Hari Eko Irianto

CONTENTS

12.1 INTRODUCTION

Prof. Kylin of Uppsala University in Sweden was the first to discover fucoidan extracted from brown algae in 1913. Initially called as "fucoijin," the chemical was later renamed "fucoidan" in accordance with the worldwide sugar naming system. It is a sulfated polysaccharide that occurs in brown algae such as limu moui, bladderwrack, mosque, kombu, and wakame and accounts for 0.3–1.5% of the wet weight of brown seaweeds (Atashrazm et al., 2015; Zayed et al., 2020).

Echinoderms and other marine invertebrates have also been found to have fucoidan in various forms (Mulloy et al., 1994). Sulfated (SPs) and nonsulfated macromolecules generated mostly from micro and macroalgae are marine

DOI: 10.1201/9781003303916-12

polysaccharides. Chemically, among macroalgae differ in molecular weight, monosaccharide content, sulfate concentration and position, resulting in a wide spectrum of pharmacological effects (Zayed et al., 2020). Fucoidan is a kind of fucose-containing sulfated polysaccharide with a backbone made up of alternating (1→3) and (1→4)-linked -l-fucopyranosyl residues, as well as a backbone made up of (1→3) and (1→4)-linked-l-fucopyranosyl residues. On the other hand, sulfated galactofucans have backbones composed of (1→6), (1→2), and/or -d-galactopyranosyl units of d-mannopyranosyl with fucose or fuco-oligosaccharide branching, as well as mannose, glucuronic acid, galactose, xylose, or glucose replacements (Shen et al., 2018).

The benefits of fucoidan have been widely known, so the demand increases from year to year. The growing public awareness of health has also led to an increase in the need for fucoidan. The worldwide fucoidan market was worth USD 32 million in 2020 and is anticipated to be worth USD 41 million by the end of 2027, increasing at a compound annual growth rate (CAGR) of 3.4% between 2021 until 2027. Asia is the most populous consumer area. In 2017, 6751 Kg of fucoidan was consumed. China and Japan are Asia's leading and second-largest consumers, respectively. In 2017, China accounted for 38.54% of Asia's total fucoidan consumption. The United States is another major consumer nation on a worldwide scale. In 2017, the United States used 5248 Kg of fucoidan, accounting for 36.32% of total consumption (WBOC, 2021). Taking into account the excellent future prospects of fucoidans, this chapter focuses on discussing resources, chemical structure, characteristics, bioactivity, and ulcer healing applications of fucoidan.

12.2 FUCOIDAN

12.2.1 Sources of Fucoidan

Fucoidan is a sulfated polysaccharide containing a high percentage of L-fucose. Fucoidan has been shown to have various biological functions, including to preserve cell wall integrity, hydrate tissues, aid in cell communication, regulate osmotic pressure, and operate as an organism's defense system (Deniaud-Bouët et al., 2017). This biopolymer was previously reported as fucoidan isolated from brown seaweed (*Phaeophyceae*) 108 years ago (Asanka Sanjeewa & Jeon, 2021). According to the International Union of Pure and Applied Chemistry (IUPAC) system, the polymer is now known as fucoidan. However, it is also known as sulfated fucan, fucan, and fucosan (Etman et al., 2020). Brown algae are commonly available as a culinary product in many Asian cuisines, including Thai, Vietnamese, and Korean. It is also gaining popularity in Western countries and other parts of the world (Chang et al., 2015). Sulfated polysaccharides in animals were first reported by Vasseur et al. (1948), who said that fucans compounds were detected in sea urchins eggs. Li et al. (2018) and Yu et al. (2014) also reported that sea cucumbers contain fucoidan. The predominant polysaccharide component of sea cucumbers is sulfated fucoidan and fucosylated glycosaminoglycan as a minor component (Qin et al., 2018). Research on fucoidan in sea cucumbers, including *Phylloporus proteus* (Qin et al. 2018) and *Holothuria tubulosa* (Chang et al., 2016). Humans consume sea cucumber for nutritional needs. It is abundant in the Mediterranean Sea and the eastern Atlantic Ocean (Chang et al., 2015).

Mulloy et al. (1994) investigated the differences between the sulfated fucans found in echinoderms and those found in brown algae. Echinoderm sulfated fucans have a linear structure that contains fucose and sulfate groups. The four residues are (1 → 3)-linked a-L-fucopyranosyl units that differ only by specific sulfation patterns at the O(2) and O(4) positions. Brown algal sulfated fucans have tremendous structural heterogeneity with branching and a variety of substituents. There are more 3-linked and sulfated units at positions O(4), but the structure is irregular, as observed for fucans from echinoderms (Usov & Bilan, 2009).

Besides brown seaweeds and echinoderms, fucoidan-like compounds are also detected in seagrasses (Kannan et al., 2013). The sulfated polysaccharide isolated from the *Halodule pinifolia* contained monomers linked to fucoidan, including mannuronic acid, fucose, and a high concentration of uronic acid.

12.2.2 Chemical Structure

The structure of fucoidans is complicated and varied, making generalization difficult (Sanjeewa & Jeon, 2019). In fucoidan, some monosaccharide molecules have a sulfate group attached. The fucose backbone has branches attached to other monosaccharides such as galactose (glactofucan), rhamnose (rhamnofucan) or galactose and rhamnose (rhamno galactofucan) (Etman et al., 2020), mannose, glucose, xylose, and arabinose (Bilan et al., 2002), and uronic acid (van Weelden et al., 2019).

The two main backbones of fucoidan are (1 → 3)-linked-L-fucose residues or rotating (1 → 3)- and (1 → 4)- linked-L-fucose units with sulfate groups to the oxygens (Etman et al., 2020; Olatunji, 2020), as shown in Figure 12.1. This polymer is quite heterogeneous since sulfate ester groups can be substituted freely at the 2, 3, and/or 4 positions of fucopyranose units (Ale et al., 2011; Fernando et al., 2020). Brown algae can synthesize highly branched polysaccharides

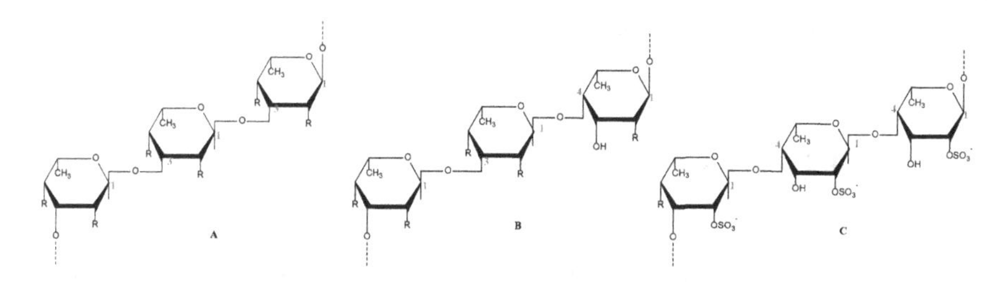

FIGURE 12.1 Structures of fucoidans with (1 → 3) linked fucose units (A), (1 → 3) and (1 → 4) linked fucose units (B), and (1 → 4) linked fucose with sulfate group attachments.

whose structure varies depending on their taxonomic status, which results in this heterogeneous structure.

The structure of fucoidan is highly variable depending on the species, harvest season, and the extraction method used (Etman et al., 2020; Olatunji, 2020). These variables affect the degree of branching, sulfation, polymer chain structure, and monosaccharide units present in a particular fucoidan. The difference in chemical structure results in variations in the bioactivity of fucoidans. Like other polymers found in algae, the fucoidan level in brown algae exhibits more complex seasonal variations in structure and bioactivity (Fletcher et al., 2017).

Methods for characterization of fucoidan include gas chromatography-mass spectrometry (GC-MS), matrix-assisted laser desorption/ionization-time of flight (MALDI-TOF) mass spectrometer, and nuclear magnetic resonance (NMR). Sea cucumber fucoidans have a simple structure and species relationship. They are branched in *Apostichopus japonicus* and linear in *Isostichopus badionotus.* Seaweed fucoidans have branched structures that vary between species (Chang et al., 2016). Different fucoidan sources have different sulfite, acetyl, and uronic acid contents (Fitton et al., 2015).

12.2.3 Characteristics of Fucoidan

Fucoidan crude extracts obtained from *Sargassum binderi* Sonder were 4.02% in terms of yield (Saepudin et al., 2018). The crude fucoidan contained 74.25% fucose, 0.28% uronate acid, 10.29% sulfate, and 5.55% protein. According to Peng et al. (2018), *Kjellmaniella crassifolia* fucoidan contains 71.68% carbohydrate and 20.04% sulfate, with 31.89% fucose and 23.54% galactose. Wei et al. (2019) also reported that brown algae had 77.4% fucose and 13.9% galactose, respectively.

Fucoidans classified their molecular weights as low-, medium-, and high-molecular-weight fucoidans. Less than 10 kDa molecular weights, 10–10,000 kDa molecular weights, and above 10,000 kDa molecular weights are defined as low-molecular-weight fucoidans. When coupled with chemotherapy, high-molecular-weight fucoidans offer superior anticancer properties, but low-molecular-weight fucoidans boost therapeutic efficacy and minimize adverse effects (Olatunji, 2020). Diverse fucoidan sources had varying monosaccharide composition, molecular weight, and sulfation degree (Ale & Meyer, 2013). The extraction technique affects yields, natural products, and structure (Ale & Meyer, 2013; Usov & Bilan, 2009).

12.3 BIOACTIVITY OF FUCOIDAN

12.3.1 Anticoagulant Activity

Anticoagulants have been widely employed in blood therapy during dialysis and surgery, as medicine of disseminated intravascular coagulation and thrombosis in many disorders, and in vitro blood tests. Fucoidan specimens obtained from *Fucus evanescens* were separated from proteins and polyphenols, diacetylated, and depolymerized by fucoidanase to investigate their biological activity. Deacetylation did not affect fucoidan's ability to inhibit thrombin and factor Xa, while purifying proteins and polyphenols had no impact (Lapikova et al., 2008).

In several tests, researchers discovered that low molecular weight fucoidan fractions had stronger anticoagulant activity and that sulfate concentration and sulfate total played a key role in the anticoagulant activity. The anticoagulant process, on the other hand, was complicated, and the low molecular weight of fucoidan fractions might affect both internal and extrinsic coagulant factors. The low molecular weight fucoidan fractions demonstrated increased anticoagulant activity in some assays, and sulfate concentration and sulfate/total ratios were critical factors in anticoagulant activity. The anticoagulant process, on the other hand, was complicated, and the low molecular weight fucoidan fractions might affect both internal and extrinsic coagulant factors (Wang et al., 2010).

Dürig et al. (1997) investigated that the low molecular weight fucoidan combines powerful anticoagulant and fibrinolytic characteristics with just moderate platelet stimulating effects, making it a good candidate for further research.

Heparin is a polysaccharide that is highly sulfated and widely used in anticoagulant therapy. It has been used to demonstrate the ability of sulfated polysaccharides to interfere with biological systems. Fucoidan has shown great anticoagulant action and huge therapeutic development potential. It can prevent the activity of coagulation factors in the coagulation pathways by interacting with antithrombin (Jung et al., 2007).

12.3.2 Antioxidant Activity

Research for new natural plant-derived antioxidants has increased. It is unclear whether plant elements are linked to a lower risk of chronic diseases, but antioxidants appear to play a key part in plant medicine's protective effect (Saeed et al., 2012).

Fucoidan extracted from *Sargassum tenerrimum* using two different methods, i.e., the first method (DFM1) and the second method (DFM2), yielded 3.68% and 1.09%, respectively. The IC_{50} value for DFM1 DPPH root recovery activity was 1.93 times higher than that of DFM2. Ion chelation activity Fe 2 was 78.3% for DFM1 and 89.4% for DFM2 at 10 mg/ml. The depolymerized pure fucoidan showed a markedly superior antioxidant capacity (Ashayerizadeh et al., 2020). The fucoidan was obtained with three sulfated polysaccharides from *L. japonica* as antioxidative (Wang et al., 2008). Their antioxidant strength was very different. The majority of fractions were higher antioxidants activity than fucoidan in vitro. The sulfate content correlated positively with the recovery capacity of the superoxide radicals. Rasio sulfate content is an indicator for hydroxyl radicals and metal ion-chelation, applied in future experiments to understand the relationship between its chemical properties and antioxidant activity. Fucoidan has two fractions heteropolysaccharides according to chemical analysis of with galactose as the predominant component. There is a positive correlation between the ability of superoxide radicals and sulfate content.

Results of studies with in vitro models showed that the sulfate/fucose content ratio was an effective indicator of the samples antioxidant activity.

At 1000 g/mL of fucoidan isolated from brown seaweed, *Sargassum polycystum* (Palanisamy et al., 2017) found overall antioxidant activity (65.3, 0.66%), the highest 2,2-diphenyl-1-picrylhydrazyl (DPPH) scavenging activity (61.2, 0.33%), and lowering ability (67.56, 0.26%).

Koh et al. (2019) *Undaria pinnatifida*, a brown seaweed species, produces fucoidan with several bioactive qualities, including antioxidant and anticancer effects. To compare crude and purified fucoidan used a fucoidan standard with a molecular weight cut-off of 300 kDa (FSTd).

Compare raw fucoidan (F crude) and purified fucoidan with a molecular weight cut-off of 300 kDa (F300). Fucoidan standard had a much higher sulfate concentration. The secondary antioxidant abilities of fucoidan samples from New Zealand are similar to butylated hydroxyanisole (BHA). However, with low molecular weight, fucoidan has a much higher capacity than BHA. A study by Liu et al. (2020) showed evidence for the effects of extraction procedures on the physicochemical properties, conformational behaviors, and antioxidant activities of *S. fusiforme* fucoidans.

12.3.3 Immunomodulator Activity

Fucoidans immunomodulatory effects have been studied in a variety of experimental models. Fucoidan compound agents have shown promise for development as a supplemental immunomodulator. Fucoidan compound agents could boost nonspecific immunity, specific immunity and regulate the secretion of cytokines like tumor necrosis factor (TNF), granulocyte-macrophage colony-stimulating factor (GM-CSF), interleukin (IL)-4, and IL-10 linked to immunity *in vivo* (Peng et al., 2019).

Two mechanisms, immunostimulation, and immunosuppression involve immune modulation. Various illnesses of the immune system are affected by this intricate mechanism. To prevent the adverse reactions of medications that produce immunosuppression, immunomodulators can be employed as immunostimulators. Low molecular weight fucoidan and high molecular weight fucoidan were analyzed for their chemical compositions and evaluated for their immunostimulatory properties *in vitro* (Yoo et al., 2019) also examined the efficacy of high molecular weight fucoidan as an immunostimulator in immunosuppressed rodents with a high *in vitro* function. Under immunosuppressive conditions, high molecular weight fucoidan is an efficient immunostimulant (Yoo et al., 2019).

At low concentrations, fucoidans (extracted from *Macrocystis pyrifera* and *Undaria pinnatifida*) slowed human neutrophil apoptosis significantly, whereas fucoidans from *Fucus vesiculosus* and *Ascophyllum nodosum* delayed apoptosis at higher doses. *M. pyrifera* fucoidan also increased natural killer (NK) cell activation and cytotoxicity against yeast artificial chromosome-1 (YAC-1) cells. *M. pyrifera* fucoidan also caused the most activation of spleen dendritic cells (DCs) and T cells and ovalbumin (OVA)-specific immunological responses compared to other fucoidans. *M. pyrifera* produces fucoidan, which has the ability to increase

NK cell activation, DC maturation, cytotoxic T lymphocytes (CTL activity), immunological responses Th1, antigen-specific antibodies, and T cell memory (Zhang et al., 2015).

12.3.4 Anti Inflammatory

Lee et al. (2012) found that fucoidan from *Ecklonia cava* was extracted with enzymes, potentially more beneficial than water extraction. The monosaccharide levels varied greatly, with fucose (53.1–77.9%) and galactose (10.1–32.8%) accounting for the majority of the sugars, along with a tiny quantity of xylose (4.0–8.2%), glucose (0.8–2.2%,) and rhamnose (2.3–4.5%). These fucoidans found one or two subfractions with average molecular weights ranging from 18 to 103 g/mol. These fucoidans dramatically reduced nitro oxide (NO) generation in Ralph and William's (RAW) 264.7 macrophage cells stimulated by lipopolysaccharide (LPS) by suppressing the expression of cyclooxygenase-2 (COX-2), inducible nitric oxide synthase (iNOS), and pro-inflammatory cytokines such IL-6, TNF-α, and IL-1ß. As a result, the current findings imply that *E. cava* fucoidan could be a beneficial treatment approach for various inflammatory illnesses.

Manikandan et al. (2020) discovered that fucoidan extracted from *Turbinaria decurrens* has anti nociceptive and anti inflammatory properties in a formalin induced pawedema mouse model. In the formalin-induced inflammatory edema state, fucoidan decreased licking time, implying an anti nociceptive effect, and lowered the extent of paw swelling. In formalin-injected mice's paw edema tissue, extracted fucoidan drastically reduced MDA while increasing superoxide dismutase (SOD), catalase (CAT), glutathione peroxidase (GPx), and glutathione-S-transferase (GST) and reducing glutathione (GSH) activities. Fucoidan's anti inflammatory activities have been attributed to its ability to alter the concentrations of enzymatic antioxidants, pro-inflammatory cytokines, and master regulator NF-B. In IC-21 macrophages, fucoidan decreased LPS induced cytotoxicity in a dose dependent way.

Using an activity test with RAW 264.7 macrophage cells produced by LPS and zebrafish, a single *Saccharina japonica* fucoidan was found to have a significant anti-inflammatory effect. Fucoidan reduced the generation of nitric oxide (NO) and cytokines such as TNF-, IL-1, and IL-6 in vitro, according to the results of the anti-inflammatory action. Down-regulation of signal pathways such as mitogen-activated protein kinase (MAPK) and nuclear factor kappa light chain enhancer of activated B cells (NF-κB) was a part of the mechanism. Fucoidan reduced cell death and nitro oxide (NO) and reactive oxygen species (ROS) generation in LPS-exposed zebrafish in an in vivo experiment (Ni et al., 2020).

12.3.5 Antiviral Activity

Sulfated polysaccharides are well-known antiviral agents with high potency. Their antiviral activity has been linked to the anionic properties of the molecules, which can inhibit virus adsorption. Antiviral tests have shown that all isolated fucoidans have had strong antiviral activity against the herpes simplex virus type 2 (HSV-2) infection, with EC_{50} values within 0.027–0.123 ug/mL; fact, the best selectability index was seen in the viscozyme extracted macromolecules (Alboofetileh et al., 2019).

Sargassum henslowianum purified polysaccharide (SHAP)-1 and SHAP-2 were isolated from *S. henslowianum* using ion exchange and gel filtration column chromatography. They are made up of fucose and galactose in a 3:1 ratio and contain 31.9 percent sulfate. By plaque reduction assay, the IC_{50} values of fucoidan against herpes simplex virus type 1 (HSV-1) were estimated to be 0.89 and 0.82 g/mL, respectively, and as low as 0.48 g/mL against HSV-2. Fucoidans' antiviral mechanism may involve at least blocking HSV-2 virion adsorption to host cells (Sun et al., 2020). Fucoidans isolated from three brown seaweed species, *Sargassum mcclurei, Sargassum polycystum*, and *Turbinara ornata*, taken from Nha Trang Bay, Vietnam, showed similar antiviral activity with a mean IC_{50} ranging from 0.33 to 0.7 g/ml and no cell toxicity. The anti–human immunodeficiency virus (HIV) action of fucoidans was found to be unrelated to sulfate content, and the proper placement of sulfate groups in the fucoidan backbones was likewise unrelated to antiviral activity. When fucoidans were pre-incubated with the virus, but not with the cells, and not after infection, they suppressed HIV-1 infection, preventing the early steps of HIV entry into target cells (Thuy et al., 2015). Fucoidans are produced in moderate amounts by the brown seaweed *Scytosiphon lomentaria*. Typical strongly sulfated galactofucans with no noticeable evidence of chemical heterogeneity, as well as fractions with larger amounts of xylose, mannose, and uronic acids are formed by cetrimide separation. In the majority of the subfractions of the succeeding extracts, fucose is the most significant monosaccharide. There are a few 2-O-acetylated fucose units present. The galactofucan fractions have a high and selective antiviral activity against HSV-1 and HSV-2, whereas the uronofucoidans were inert (Ponce et al., 2019).

12.3.6 Anti Cancer Activity

Cancer is a group of abnormally growing diseases that can invade or spread to other areas of the body (Somasundaram et al., 2016). Fucoidan, a naturally occurring component of brown algae, fights various types of cancer by targeting apoptotic key molecules (Atashrazm et al., 2015). There are a wide variety of types of cancer, but all have common cellular and molecular behavior. The majority of cancer chemotherapy medications are designed to target cancer cells' common unregulated pathways. Furthermore, the cytotoxic effects of these substances impact a wide range of healthy tissues. Fucoidan exhibits anti cancer potential in several cancer types by targeting critical apoptotic molecules. It can also protect human from the side effects of chemotherapy and radiation. As a result, the synergistic effect of fucoidan with currently available anti-cancer drugs is a hot topic (Atashrazm et al., 2015). Fucoidan's anti cancer properties have been demonstrated in vivo and in vitro in various cancers. Nonetheless, it has received little attention in clinical trials for its anti cancer properties. Fucoidan operates through multiple mechanisms, including cell cycle arrest, apoptosis, and immune system activation. Additional activities of fucoidan have been described, includes immune system-mediated inflammation, oxidative stress, and stem cell mobilization, all of which may be associated with the observed anti-cancer characteristics (Kwak, 2014).

Fucoidan cytotoxic effect revealed an increased percentage (90.4 ± 0.25%) of inhibition against the Michigan Cancer Foundation-7 (MCF-7) cell line at 150 g/mL, with an estimated IC_{50} of 50 g/ml (Palanisamy et al., 2017). Fucoidan from *Sargassum cinereum* has a cytotoxic impact on the colon cancer cell line Human colon carcinoma (HCT)-15 cell lines. The 3-(4,5-dimethylthiazol-2-yl)-2,5-diphenyltetrazolium bromide (MTT) test was used to determine the cytotoxicity effect. After 24 hours of incubation with 75 0.9037 g/mL, fucoidan extract caused approximately 50% cell death against HCT-15 (Somasundaram et al., 2016).

Thinh et al. demonstrated that when used at up to 200 g/mL for 48 hours, fucoidan produced from brown algae, thoroughly purified *S. mcclure* was less cytotoxic. However, it did stop DLD-1 colon cancer cells from forming colonies (Jin et al., 2010; Thinh et al., 2013) while it inhibited cell proliferation. These findings suggest that fucoidan apoptotic activity on cancer cells may be cell type-specific. Fucoidan examined against apoptosis in human promyeloid leukemic cells, as well as fucoidan-mediated signaling pathways. In human promyelocytic leukemic cell line–60 (HL-60), NB4, and Tamm-Horsfall Protein 1 (THP-1) cells, fucoidan triggered apoptosis, but not in K562. Fucoidan administration caused caspase-8, -9, and -3 activation, Bid cleavage, and changes in mitochondrial membrane permeability in HL-60 cells. Depletion of mitogen-activated protein kinase, as well as inhibitors of MAPK kinase 1 (MEK1) and c Jun NH2-terminal kinase (JNK), substantially prevented fucoidan induced apoptosis, procaspase cleavage, and alterations in the permeability of the mitochondrial membrane (Jin et al., 2010).

12.3.7 Anti-Ulcer Activity

Sinurat and Rosmawaty (2015) studied the action of fucoidan isolated from *Sargassum crassifolium*, which is native to Binuangeun, Banten, Indonesia. In vivo tests on fucoidan extract were performed on mice. The control (without fucoidan) and fucoidan therapy groups were observed for 16 days. Fucoidan was given at concentrations of 100, 200, 300, and 400 ppm. Aspirin was given to mice with pre-treated fucoidan 400 ppm as a stomach ulcer induction on the fourteenth day. The results of a histopathology study in the stomach tissue of mice demonstrated that 100 ppm fucoidan could prevent gastric ulcers caused by aspirin irritation at 400 ppm. The presence of fucoidan was linked to a rise in the mucus layer of the gastric mucosa.

Fucoidan's impact on aspirin-induced ulcers in rats was studied, taking biochemical and immunological characteristics into account. The state of glycogen storage in stomach tissue, as well as histological alterations, were investigated. In ulcer-induced rats, fundamental biochemical measures revealed considerable changes in aspartate (AST) transaminases and alanine (ALT) transaminases. In addition, there were minor changes in blood urea nitrogen levels and cholesterol. Glycogen storage is altered in oxyntic cells; histopathological analysis revealed neutrophil infiltration and inflammation. The immediate changes of ALT, AST, cytokines, and stomach glycogen were inhibited by fucoidan administration, which provided significant protection against ulcers. The fucoidan-pretreated group, on the other hand, had increased serum interferon (INF)-ɥ. These data imply that fucoidan's anti-ulcer properties may help to protect the stomach mucosa from inflammatory cytokine mediated oxidative damage (Choi et al., 2010).

12.4 ULCER HEALING APPLICATIONS

12.4.1 Basic Methods of Anti–Gastric Ulcer Properties

Gastric ulcers are a global health issue, and one of the leading causes of recurrence is poor healing (Escobedo-Hinojosa et al., 2018). A variety of factors can cause it. Gastric mucosa can be damaged by stress, alcohol, *Helicobacter pylori*, and nonsteroidal anti inflammatory medications. Despite the fact that aspirin has been used as a nonsteroidal anti inflammatory medicine for over a century, its side effects have been identified as one of the most common causes of stomach ulcers (Sinurat & Rosmawaty, 2015; Wu et al., 2016). According to a previous study, aspirin-induced stomach ulcers are characterized by increased gastric acid secretion and inflammatory cytokines (Hu et al., 2020; Zinkievich et al., 2010). Additionally, aspirin-induced ulcers are associated with increased reactive oxygen species (ROS), which damage stomach tissue by accelerating lipid peroxidation and inhibiting antioxidant enzymes, resulting in more severe gastrointestinal injury (Pohle et al., 2001).

In general, stomach ulcers cause inflammatory damage to the mucosa. The cytokines tumor necrosis factor (TNF), interleukin (IL)-1, IL-6, and IL-10 have a pathogenic role in either the immediate inflammatory response or the severity of gastric ulcers (Suheryani et al., 2017). The findings revealed that different doses of KF significantly lowered serum levels of TNF-, IL-1, and IL-6. In contrast, anti-inflammatory mediator IL-10 was raised in all *Kjellmaniella crassifolia* (KF)–treated groups, similar to Li et al.'s findings (Li & Ye, 2015). These studies showed that KF's stomach anti-inflammatory actions were linked to inhibiting the inflammatory response (Hu et al., 2020).

12.4.2 Ulcer Gastric Experiment in Mice

The ulcer is a frequent gastrointestinal illness that gives patients great pain, disrupts their daily routines, and creates mental anguish. It is more common among those who are always in a rush, worry a lot, and eat curries. Inflammation of the mucosa and tissue that protects the gastrointestinal system characterizes peptic ulcer disease. Peptic ulcers cause damage to the mucus membrane that typically shields the esophagus, stomach, and duodenum from gastric acid and pepsin (Maury et al., 2012).

The inflammatory breach in the skin or mucus membrane that lines the stomach or duodenum is an ulcer. The most frequent upper gastrointestinal (GI) tract condition is peptic ulcer. Ulceration occurs when the natural equilibrium is disrupted by either increased aggression or decreased mucosal resilience (Gregory et al., 2009). The acid peptic disease includes hyperacidity, gastroesophageal reflux, and stress-induced mucosal erosions. Peptic ulcers are caused by a localized loss of gastric and duodenal mucosa. Peptic ulcer disease is becoming more common throughout the population due to people's bad eating habits. People who take non-steroidal anti inflammatory drugs and smokers and alcoholics have a higher infection rate. Although western medications are efficient, they have side effects, such as diarrhea, dizziness, muscle soreness, and headache, limiting their use. Herbal medicines are effective in clinical and experimental research (Anbarasan et al., 2019; Bujanda, 2000).

The initiating factor of the infection process caused by *H. pylori* is the attachment of the causative agent to mucoid cells of the stomach epithelium. Its ability to attach to oligosaccharide components of particular phospholipids on stomach cell membranes defines it. Other possible *H. pylori* binding sites include extracellular matrix components such as laminin and fibronectin and different types of collagen. *H. pylori* have adhesins attached to the membrane, allowing the agent to get closer to the host cells (Boren, 2014). Microorganisms also require adhesion because they evade mechanical removal and create conditions for invasion, persistence, and multiplication. Almost all polysaccharides, including fucoidans, have anti-adhesive characteristics obtained from both terrestrial and marine sources. Pathogens from numerous taxonomic groups and eukaryotic cells are inhibited by these substances and their synthesized molecular fragments (Besednova et al., 2015).

Exogenous variables include excessive alcoholic use, indiscriminate use of nonsteroidal anti-inflammatory medications, stress, smoking, and *H. pylori* infection; endogenous factors include HCl, pepsin, biliary reflux, lipid peroxidation, and the generation of reactive oxygen species, whereas exogenous factors include biliary reflux, biliary reflux, lipid peroxidation, and the formation of ROS (Alqasoumi et al., 2009). Protective components include the mucus-bicarbonate barrier, surface phospholipids, prostaglandins (PGs), NO, mucosal blood flow, cell renewal, growth hormones, and antioxidant enzymes (Prabha et al., 2008). The generation of ROS, which can impair epithelial cell integrity, is increased by oxidative stress, which is prevalent in the development of gastric ulceration. The endogenous antioxidants may be overwhelmed if there is an excess of ROS metabolites produced. Furthermore, during gastric ulceration, ROS accumulates neutrophils in the mucosal tissues. Pro-inflammatory cytokines have been demonstrated to activate neutrophils and contribute significantly to ulcer damage in studies (Brito et al., 2018; Li et al., 2017).

Gastric ulcers and duodenal ulcers are the two most prevalent forms of peptic ulcers. These designations indicate the location of the ulcer. Gastric ulcers are ulcers that form in the stomach. The duodenum, which is the initial segment of the small intestine, is where duodenal ulcers occur. Gastric and duodenal ulcers may occur at the same time in an individual. The presence of acid and peptic activity in gastric juice and the deterioration of mucosal defenses are factors in peptic ulcers formation. The stomach and the first few millimeters of the duodenum are the most prevalent sites for ulcers. Acute peptic ulcers affect tissues down to the submucosa, and lesions can be single or numerous. The epithelium and muscular layers of the stomach wall are both penetrated by chronic peptic ulcers (Rambhai & Sisodia, 2018).

Fucoidan can bind to *H. pylori* and flush it out of the GI tract. It has an excellent protective effect on the GI mucosa and may reduce the risk of gastritis, gastric ulcers, and gastric cancer. Szabo et al. (1995) evaluated the anti-peptic action, the regulating activity of fibroblast primary growth factor, and the inflammatory properties of fucoidan to determine their anti-ulcerative potential. Nonsulfated polysaccharides like mannan and dextran showed anti-peptic effects. However, fucoidan and other sulfated polysaccharides, such as dextran sulfate, agar, and carrageenan, did not. At pH 7.4 and pH 4.0, all sulfated polysaccharides investigated, apart from chondroitin sulfate, inhibited the loss of basic fibroblast growth factor (bFGF) bioactivity. Overall, the results suggest that fucoidan is a safe substance with a gastric protective hazard (Shen et al., 2018).

Mice used in the bioassay with different fucoidan feeding concentrations ranging from 100 mg/kg body weight to 400 mg/kg body weight. The usage of fucoidan at a concentration of 100 ppm was able to prevent the presence of stomach ulcers, according to the results of in vivo assays. Aspirin induced gastric ulcers are thought to be harmful to the stomach. Aspirin is a medicine that contains a powerful acid that could cause stomach ulceration. Following the surgery on the mice, the pH values were observed. All of the therapies included aspirin, which has a pH of 4.3. Physically viewed the histopathology test results (Sinurat & Rosmawaty, 2015), as illustrated in Figure 12.2.

According to the hepatology data, blood clots generated against providing aspirin without fucoidan (Figure 12.2B), showing that the ulcer existed in the stomach tissue. There are no blood clots in the different performances (Figure 12.2A without aspirin and Figure 12.2C with aspirin and fucoidan). It indicates that because fucoidan does not affect the stomach tissue lining, it can prevent gastric ulcers by building a barrier in the mucosa. A thick mucus

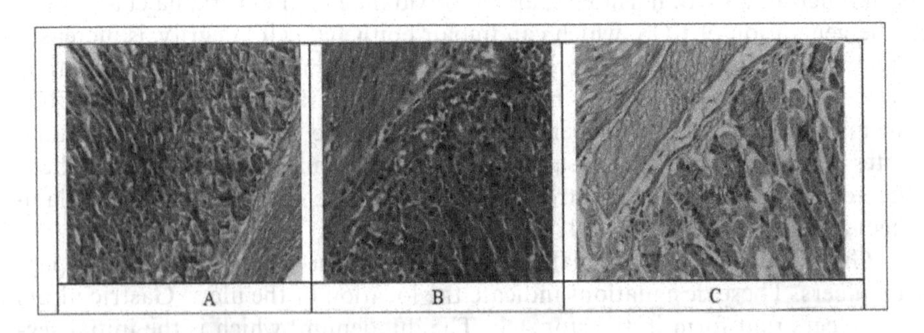

FIGURE 12.2 Test histopathology in gastric tissue of a normal ulcerated laboratory mice.

coating is particularly common on the mucous membrane, synthesized by high cylindrical epithelial cells. The gastric mucosa is a polypeptide with two purposes: it lubricates food masses to facilitate transportation within the stomach and acts as a protective coating on the lining epithelium of the stomach cavity. This protective barrier protects the stomach from being digested by its protein-lysine enzymes. It can migrate from the underlying mucosa into the surface layer according to bicarbonate secretion. The acidity of the mucous layer, or hydrogen ion concentration, is balanced in the vicinity of the epithelium and becomes acidic toward the luminal layer. While gastrointestinal mucus is excluded from the surface epithelium, little pits termed foveolae gastrique can be observed with a magnifying glass (Sinurat & Rosmawaty, 2015).

These data back the theory that fucoidan reduces aspirin induced inflammatory cytokine production and stomach mucosal injury. Because ulcer healing is such a complicated process with so many factors, this investigation is still in its early stages to identify its effect of the amount. The production of reactive metabolites has the potential to cause mucosal injury. During the healing phase, mucus production improves, protecting the ulcer crater from endogenous and exogenous aggressors such as stomach fluids and oxidants and exogenous damaging substances such as non-steroidal anti inflammatory drugs. Mucus production improves during the healing phase, protecting the ulcerative crater from endogenous aggressors, including gastric secretions and oxidants damaging exogenous agents and damaging exogenous agents, including non-steroidal anti inflammatory drugs (Chang et al., 2005; Sinurat & Rosmawaty, 2015).

12.5 FUCOIDAN'S HEALTH BENEFITS AND POTENTIAL APPLICATIONS

Functional food ingredients sourced from the sea have long been recognized as important to human health and nutrition. Designing novel functional products and health supplements from marine materials have such a lot of potential. Many marine-derived components, such as fucoidan, have been utilized as active ingredients in producing functional foods and health supplements. A diverse variety of biological processes linked to natural compounds obtained from marine sources have the ability to increase the medical benefits of such compounds not only with food products but also in the nutraceutical and pharmaceutical sectors. The chemical composition of fucoidan is extensively sulfated, and it appears to be responsible for several in vitro biological activities (Wijesinghe & Jeon, 2012). Fucoidan applications in the pharmaceutical, nutraceutical, cosmeceutical, and functional food industries have sparked considerable attention. As people seek distinct health enhancing benefits from natural products, producers and clients are more interested in these areas. In this context, studies published over the last decade have yielded reliable scientific data supporting the healing of Fucoidan. For example, a marine derived component has been used as an active ingredient in developing functional products and health supplements. The properties of fucoidan derived from brown algae have bioactive compounds, including anti-cancer, immunomodulatory, anti inflammatory, and other bioactivity. Fucoidan has already been available as a health supplement over the world to improve physical health, defending the immune system, and improving overall health, in the form of liquids, pills, and tablets (Shen et al., 2018).

12.6 CONCLUSIONS

Fucoidan is a sulfated polysaccharide that can be found in brown seaweed and sea cucumber. The bioactivities of fucoidan, such as anticoagulants, antioxidants, immunomodulator, anti inflammatory, antiviral, anticancer, and a variety of other unique bioactivities, are beneficial to preventing GI ulcers and healing gastric ulcers by fucoidan reduces aspirin induced inflammatory cytokine production and stomach mucosal injury. Mucus production improves during the healing phase. Intracellular instigators like stomach fluids and oxidants and exogenous damaging chemicals like NSAIDs are protected from the ulcer crater. Fucoidan's health activities as a natural food element obtained from the sea are directly linked to its chemical properties. Fucoidan properties depends on the type of macroalgae used, the conditions of extraction, and the chemical and physical treatment of this one of a kind product. Because of its various health benefits, fucoidan is currently commonly utilized in manufacturing functional foods.

REFERENCES

Alboofetileh, M., Rezaei, M., Tabarsa, M., Rittà, M., Donalisio, M., Mariatti, F., You, S. G., Lembo, D., & Cravotto, G. (2019). Effect of different non-conventional extraction methods on the antibacterial and antiviral activity of fucoidans extracted from Nizamuddinia zanardinii. *International Journal of Biological Macromolecules*, *124*, 131–137. https://doi.org/10.1016/j.ijbiomac.2018.11.201.

Ale, M. T., & Meyer, A. S. (2013). Fucoidans from brown seaweeds: An update on structures, extraction techniques and use of enzymes as tools for structural elucidation. *RSC Advances*, *3*(22), 8131–8141. https://doi.org/10.1039/c3ra23373a.

Ale, M. T., Mikkelsen, J. D., & Meyer, A. S. (2011). Important determinants for fucoidan bioactivity: A critical review of structure-function relations and extraction methods for fucose-containing sulfated polysaccharides from brown seaweeds. *Marine Drugs*, *9*(10), 2106–2130. https://doi.org/10.3390/md9102106.

Alqasoumi, S., Al-Sohaibani, M., Al-Howiriny, T., Al-Yahya, M., & Rafatullah, S. (2009). Rocket "Eruca sativa": A salad herb with potential gastric anti-ulcer activity. *World Journal of Gastroenterology*, *15*(16), 1958–1965. https://doi.org/10.3748/wjg.15.1958.

Anbarasan B, Anitha Therese G, Jayapriya S, Anbu N, K. K. (2019). Anti-ulcer activity of Medicinal Herbs—A Review. *International Journal of Current Research in Chemistry and Pharmaceutical Sciences*, *6*(4), 27–32. https://doi.org/10.22192/ijcrcps.

Asanka Sanjeewa, K. K., Jayawardena, T. U., Kim, H. S., Kim, S. Y., Shanura Fernando, I. P., Wang, L., Abetunga, D. T. U., Kim, W. S., Lee, D. S., & Jeon, Y. J. (2019). Fucoidan isolated from *Padina commersonii* inhibit LPS-induced inflammation in macrophages blocking TLR/NF-κB signal pathway. *Carbohydrate Polymers*, *224*(June), 115195. https://doi.org/10.1016/j.carbpol.2019.115195.

Ashayerizadeh, O., Dastar, B., & Pourashouri, P. (2020). Study of antioxidant and antibacterial activities of depolymerized fucoidans extracted from *Sargassum tenerrimum. International Journal of Biological Macromolecules, 151*, 1259–1266. https://doi.org/10.1016/J.IJBIOMAC.2019.10.172.

Atashrazm, F., Lowenthal, R. M., Woods, G. M., Holloway, A. F., & Dickinson, J. L. (2015). Fucoidan and cancer: A multifunctional molecule with anti-tumor potential. *Marine Drugs, 13*(4), 2327–2346. https://doi.org/10.3390/md13042327.

Besednova, N. N., Zaporozhets, T. S., Somova, L. M., & Kuznetsova, T. A. (2015). Review: Prospects for the use of extracts and polysaccharides from marine algae to prevent and treat the diseases caused by *Helicobacter pylori. Helicobacter, 20*(2), 89–97. https://doi.org/10.1111/hel.12177.

Bilan, M. I., Grachev, A. A., Ustuzhanina, N. E., Shashkov, A. S., Nifantiev, N. E., & Usov, A. I. (2002). Structure of a fucoidan from the brown seaweed *Fucusevanescens* C.Ag. *Carbohydrate Research, 337*, 719–730.

Boren, T. (2014). Helicobacter pylori; Multitalented adaptation of binding properties. *Anticancer Research, 34*(10).

Brito, S. A., de Almeida, C. L. F., de Santana, T. I., da Silva Oliveira, A. R., do Nascimento Figueiredo, J. C. B., Souza, I. T., de Almeida, L. L., da Silva, M. V., Borges, A. S., de Medeiros, J. W., da Costa Silva Neto, J., de Cássia Ribeiro Gonçalves, R., Kitagawa, R. R., Sant'Ana, A. E. G., Rolim, L. A., de Menezes, I. R. A., da Silva, T. G., Caldas, G. F. R., & Wanderley, A. G. (2018). Antiulcer activity and potential mechanism of action of the leaves of *Spondias mombin* L. *Oxidative Medicine and Cellular Longevity, 2018*. https://doi.org/10.1155/2018/1731459.

Bujanda, L. (2000). The effects of alcohol consumption upon the gastrointestinal tract. *American Journal of Gastroenterology, 95*(12), 3374–3382. https://doi.org/10.1016/S0002-9270(00)02140-7.

Chang, C. C., Pan, S., Lien, G. S., Liao, C. H., Chen, S. H., & Cheng, Y. S. (2005). Deformity of duodenal bulb, gastric metaplasia of duodenal regenerating mucosa and recurrence of duodenal ulcer: A correlated study. *World Journal of Gastroenterology, 11*(12), 1802–1805. https://doi.org/10.3748/wjg.v11.i12.1802.

Chang, Y., Hu, Y., Yu, L., McClements, D. J., Xu, X., Liu, G., & Xue, C. (2016). Primary structure and chain conformation of fucoidan extracted from sea cucumber *Holothuria tubulosa. Carbohydrate Polymers, 136*, 1091–1097. https://doi.org/10.1016/j.carbpol.2015.10.016.

Choi, J. il, Raghavendran, H. R. B., Sung, N. Y., Kim, J. H., Chun, B. S., Ahn, D. H., Choi, H. S., Kang, K. W., & Lee, J. W. (2010). Effect of fucoidan on aspirin-induced stomach ulceration in rats. *Chemico-Biological Interactions, 183*(1), 249–254. https://doi.org/10.1016/j.cbi.2009.09.015.

Deniaud-Bouët, E., Hardouin, K., Potin, P., Kloareg, B., & Hervé, C. (2017). A review about brown algal cell walls and fucose-containing sulfated polysaccharides: Cell wall context, biomedical properties and key research challenges. *Carbohydrate Polymers, 175*, 395–408. https://doi.org/10.1016/j.carbpol.2017.07.082.

Dürig, J., Bruhn, T., Zurborn, K. H., Gutensohn, K., Bruhn, H. D., & Béress, L. (1997). Anticoagulant fucoidan fractions from *Fucus vesiculosus* induce platelet activation in vitro. *Thrombosis Research, 85*(6), 479–491. https://doi.org/10.1016/S0049-3848(97)00037-6.

Escobedo-Hinojosa, W. I., Gomez-Chang, E., García-Martínez, K., Guerrero Alquicira, R., Cardoso-Taketa, A., & Romero, I. (2018). Gastroprotective mechanism and ulcer resolution effect of *Cyrtocarpa procera* methanolic extract on ethanol-induced gastric injury. *Evidence-Based Complementary and Alternative Medicine, 2018*. https://doi.org/10.1155/2018/2862706.

Etman, S. M., Elnaggar, Y. S. R., & Abdallah, O. Y. (2020). Fucoidan, a natural biopolymer in cancer combating: From edible algae to nanocarrier tailoring. *International Journal of Biological Macromolecules*, *147*(xxxx), 799–808. https://doi.org/10.1016/j.ijbiomac.2019.11.191.

Fernando, I. P. S., Dias, M. K. H. M., Madusanka, D. M. D., Han, E. J., Kim, M. J., Jeon, Y. J., & Ahn, G. (2020). Fucoidan refined by Sargassum confusum indicate protective effects suppressing photo-oxidative stress and skin barrier perturbation in UVB-induced human keratinocytes. *International Journal of Biological Macromolecules*, *164*, 149–161. https://doi.org/10.1016/j.ijbiomac.2020.07.136.

Fitton, J. H., Dell'Acqua, G., Gardiner, V. A., Karpiniec, S. S., Stringer, D. N., & Davis, E. (2015). Topical benefits of two fucoidan-rich extracts from marine macroalgae. *Cosmetics*, *2*(2), 66–81. https://doi.org/10.3390/cosmetics2020066.

Fletcher, H. R., Biller, P., Ross, A. B., & Adams, J. M. M. (2017). The seasonal variation of fucoidan within three species of brown macroalgae. *Algal Research*, *22*, 79–86. https://doi.org/10.1016/j.algal.2016.10.015.

Hu, Y., Ren, D., Song, Y., Wu, L., He, Y., Peng, Y., Zhou, H., Liu, S., Cong, H., Zhang, Z., & Wang, Q. (2020). Gastric protective activities of fucoidan from brown alga *Kjellmaniella crassifolia* through the NF-κB signaling pathway. *International Journal of Biological Macromolecules*, *149*, 893–900. https://doi.org/10.1016/j.ijbiomac.2020.01.186.

Jin, J. O., Song, M. G., Kim, Y. N., Park, J. I., & Kwak, J. Y. (2010). The mechanism of fucoidan-induced apoptosis in leukemic cells: Involvement of ERK1/2, JNK, glutathione, and nitric oxide. *Molecular Carcinogenesis*, *49*(8), 771–782. https://doi.org/10.1002/mc.20654.

Jung, W. K., Athukorala, Y., Lee, Y. J., Cha, S. H., Lee, C. H., Vasanthan, T., Choi, K. S., Yoo, S. H., Kim, S. K., & Jeon, Y. J. (2007). Sulfated polysaccharide purified from *Ecklonia cava* accelerates antithrombin III-mediated plasma proteinase inhibition. *Journal of Applied Phycology*, *19*(5), 425–430. https://doi.org/10.1007/s10811-006-9149-0.

Kannan, R. R. R., Arumugam, R., & Anantharaman, P. (2013). Pharmaceutical potential of a fucoidan-like sulphated polysaccharide isolated from *Halodule pinifolia*. *International Journal of Biological Macromolecules*, *62*, 30–34. https://doi.org/10.1016/j.ijbiomac.2013.08.005.

Koh, H. S. A., Lu, J., & Zhou, W. (2019). Structure characterization and antioxidant activity of fucoidan isolated from Undaria pinnatifida grown in New Zealand. *Carbohydrate Polymers*, *212*(February), 178–185. https://doi.org/10.1016/j.carbpol.2019.02.040.

Kwak, J. Y. (2014). Fucoidan as a marine anticancer agent in preclinical development. *Marine Drugs*, *12*(2), 851–870. https://doi.org/10.3390/md12020851.

Lapikova, E. S., Drozd, N. N., Tolstenkov, A. S., Makarov, V. A., Zvyagintseva, T. N., Shevchenko, N. M., Bakunina, I. U., Besednova, N. N., & Kuznetsova, T. A. (2008). Inhibition of thrombin and factor Xa by fucus evanescens fucoidan and its modified analogs. *Bulletin of Experimental Biology and Medicine*, *146*(3), 328–334.

Lee, S. H., Ko, C. I., Ahn, G., You, S., Kim, J. S., Heu, M. S., Kim, J., Jee, Y., & Jeon, Y. J. (2012). Molecular characteristics and anti-inflammatory activity of the fucoidan extracted from Ecklonia cava. *Carbohydrate Polymers*, *89*(2), 599–606. https://doi.org/10.1016/j.carbpol.2012.03.056.

Li, S., Li, J., Mao, G., Wu, T., Hu, Y., Ye, X., Tian, D., Linhardt, R. J., & Chen, S. (2018). A fucoidan from sea cucumber *Pearsonothuria graeffei* with well-repeated structure alleviates gut microbiota dysbiosis and metabolic syndromes in HFD-fed mice. *Food and Function*, *9*(10), 5371–5380. https://doi.org/10.1039/c8fo01174e.

Li, W., Wang, X., Zhi, W., Zhang, H., He, Z., Wang, Y., Liu, F., Niu, X., & Zhang, X. (2017). The gastroprotective effect of nobiletin against ethanol-induced acute gastric lesions in mice: Impact on oxidative stress and inflammation. *Immunopharmacology and Immunotoxicology*, *39*(6), 354–363. https://doi.org/10.1080/08923973.2017.1379088.

Li, X. jing, & Ye, Q. fa. (2015). Fucoidan reduces inflammatory response in a rat model of hepatic ischemia-reperfusion injury. *Canadian Journal of Physiology and Pharmacology*, *93*(11), 999–1005. https://doi.org/10.1139/cjpp-2015-0120.

Liu, J., Wu, S. Y., Chen, L., Li, Q. J., Shen, Y. Z., Jin, L., Zhang, X., Chen, P. C., Wu, M. J., Choi, J. il, & Tong, H. Bin. (2020). Different extraction methods bring about distinct physicochemical properties and antioxidant activities of *Sargassum fusiforme* fucoidans. *International Journal of Biological Macromolecules*, *155*, 1385–1392. https://doi.org/10.1016/j.ijbiomac.2019.11.113.

M Gregory, K. P., Vithalrao, G., & Franklin, V. K. (2009). Anti-ulcer (ulcer-preventive) activity of ficus arnottiana Miq. (Moraceae) leaf methanolic extract centre for the research and technology of agro-environment and biological sciences (CITAB), department of biology, university of minho, gualtar cam. *American Journal of Pharmacology and Toxicology*, *4*(3), 89–93.

Manikandan, R., Parimalanandhini, D., Mahalakshmi, K., Beulaja, M., Arumugam, M., Janarthanan, S., Palanisamy, S., You, S. G., & Prabhu, N. M. (2020). Studies on isolation, characterization of fucoidan from brown algae *Turbinaria decurrens* and evaluation of it's in vivo and in vitro anti-inflammatory activities. *International Journal of Biological Macromolecules*, *160*, 1263–1276. https://doi.org/10.1016/j.ijbiomac.2020.05.152.

Maury, P. K., Jain, S. K., & Alok, N. L. (2012). A review on antiulcer activity. *International Journal of Pharmaceutical Sciences and Research*, *3*(8), 2487.

Mulloy, B., Ribeiro, A. C., Alves, A. P., Vieira, R. P., & Mourão, P. A. S. (1994). Sulfated fucans from echinoderms have a regular tetrasaccharide repeating unit defined by specific patterns of sulfation at the 0–2 and 0–4 positions. *Journal of Biological Chemistry*, *269*(35), 22113–22123. https://doi.org/10.1016/s0021-9258(17)31763-5.

Ni, L., Wang, L., Fu, X., Duan, D., Jeon, Y. J., Xu, J., & Gao, X. (2020). In vitro and in vivo anti-inflammatory activities of a fucose-rich fucoidan isolated from *Saccharina japonica*. *International Journal of Biological Macromolecules*, *156*, 717–729. https://doi.org/10.1016/j.ijbiomac.2020.04.012.

Olatunji, O. (2020). *Aquatic biopolymers understanding their industrial significance and environmental implications*. Springer International Publishing.

Palanisamy, S., Vinosha, M., Marudhupandi, T., Rajasekar, P., & Prabhu, N. M. (2017). Isolation of fucoidan from *Sargassum polycystum* brown algae: Structural characterization, in vitro antioxidant and anticancer activity. *International Journal of Biological Macromolecules*, *102*, 405–412. https://doi.org/10.1016/j.ijbiomac.2017.03.182.

Peng, Y., Song, Y., Wang, Q., Hu, Y., He, Y., Ren, D., Wu, L., Liu, S., Cong, H., & Zhou, H. (2019). In vitro and in vivo immunomodulatory effects of fucoidan compound agents. *International Journal of Biological Macromolecules*, *127*, 48–56. https://doi.org/10.1016/j.ijbiomac.2018.12.197.

Peng, Y., Wang, Y., Wang, Q., Luo, X., He, Y., & Song, Y. (2018). Hypolipidemic effects of sulfated fucoidan from Kjellmaniella crassifolia through modulating the cholesterol and aliphatic metabolic pathways. *Journal of Functional Foods*, *51*(October), 8–15. https://doi.org/10.1016/j.jff.2018.10.013.

Pohle, T., Brzozowski, T., Becker, J. C., Van Der Voort, I. R., Markmann, A., Konturek, S. J., Moniczewski, A., Domschke, W., & Konturek, J. W. (2001). Role of reactive oxygen metabolites in aspirin-induced gastric damage in humans: Gastroprotection by vitamin C. *Alimentary Pharmacology and Therapeutics*, *15*(5), 677–687. https://doi.org/10.1046/j.1365-2036.2001.00975.x.

Ponce, N. M. A., Flores, M. L., Pujol, C. A., Becerra, M. B., Navarro, D. A., Córdoba, O., Damonte, E. B., & Stortz, C. A. (2019). Fucoidans from the phaeophyta *Scytosiphon lomentaria:* Chemical analysis and antiviral activity of the galactofucan component. *Carbohydrate Research*, *478*(March), 18–24. https://doi.org/10.1016/j.carres.2019.04.004.

Prabha, T., Dorababu, M., Goel, S., Agarwal, P. K., Singh, A., Joshi, V. K., & Goel, R. K. (2008). Effect of methanolic extract of *Pongamia pinnata* Linn seed on gastro-duodenal ulceration and mucosal offensive and defensive factors in rats. *Indian Journal of Experimental Biology*, *47*(8), 649–659.

Qin, Y., Yuan, Q., Zhang, Y., Li, J., Zhu, X., Wen, J., Liu, J., Zhao, L., Zhao, J., & Zhao, L. (2018). Enzyme-assisted extraction optimization, characterization and antioxidant activity of polysaccharides from sea cucumber *Phyllophorus proteus*. *Molecules*, *23*(3), 1–19. https://doi.org/10.3390/molecules23030590.

Rambhai, P. A., & Sisodia, S. S. (2018). Indian medicinal plants for treatment of ulcer: Systematic review. *UK Journal of Pharmaceutical Biosciences*, *6*(6), 38. https://doi.org/10.20510/ukjpb/6/i6/179237.

Saeed, N., Khan, M. R., & Shabbir, M. (2012). Antioxidant activity, total phenolic and total flavonoid contents of whole plant extracts *Torilis leptophylla* L. *BMC Complementary and Alternative Medicine*, *12*. https://doi.org/10.1186/1472-6882-12-221.

Saepudin, E., Sinurat, E., & Suryabrata, I. A. (2018). Depigmentation and characterization of fucoidan from brown seaweed *Sargassum binderi* sonder. *IOP Conference Series: Materials Science and Engineering*, *299*(1), 79–87. https://doi.org/10.1088/1757-899X/299/1/012027.

Sanjeewa, K. K. A., & Jeon, Y. J. (2021). Fucoidans as scientifically and commercially important algal polysaccharides. *Marine Drugs*, *19*(6), NA. https://doi.org/10.3390/md19060284.

Shen, P., Yin, Z., Qu, G., & Wang, C. (2018). Fucoidan and its health benefits. In *Bioactive seaweeds for food applications: Natural ingredients for healthy diets*. Elsevier Inc. https://doi.org/10.1016/B978-0-12-813312-5.00011-X.

Sinurat, E., & Rosmawaty, P. (2015). Evaluation of fucoidan bioactivity as anti gastric ulcers in mice. *Procedia Environmental Sciences*, *23*(Ictcred 2014), 407–411. https://doi.org/10.1016/j.proenv.2015.01.058.

Somasundaram, S. N., Shanmugam, S., Subramanian, B., & Jaganathan, R. (2016). Cytotoxic effect of fucoidan extracted from Sargassum cinereum on colon cancer cell line HCT-15. *International Journal of Biological Macromolecules*, *91*, 1215–1223. https://doi.org/10.1016/j.ijbiomac.2016.06.084.

Suheryani, I., Li, Y., Dai, R., Liu, X., Anwer, S., Juan, S., & Deng, Y. (2017). Gastroprotective effects of *Dregea sinensis* hemsl. (daibaijie) against Aspirin-Induced gastric ulcers in rats. *International Journal of Pharmacology*, *13*(8), 1047–1054. https://doi.org/10.3923/ijp.2017.1047.1054.

Sun, Q. L., Li, Y., Ni, L. Q., Li, Y. X., Cui, Y. S., Jiang, S. L., Xie, E. Y., Du, J., Deng, F., & Dong, C. X. (2020). Structural characterization and antiviral activity of two fucoidans from the brown algae *Sargassum henslowianum*. *Carbohydrate Polymers*, *229*(September). https://doi.org/10.1016/j.carbpol.2019.115487.

Szabo, S. J., Jacobson, N. G., Dighe, A. S., Gubler, U., & Murphy, K. M. Developmental commitment to the Th2 lineage by extinction of IL-12 signaling. *Immunity*, *2*(6): 665–675. doi: 10.1016/1074-7613(95)90011-X

Thinh, P. D., Menshova, R. V., Ermakova, S. P., Anastyuk, S. D., Ly, B. M., & Zvyagintseva, T. N. (2013). Structural characteristics and anticancer activity of fucoidan from the brown alga *Sargassum mcclurei*. *Marine Drugs*, *11*(5), 1453–1476. https://doi.org/10.3390/md11051456.

Thuy, T. T. T., Ly, B. M., Van, T. T. T., Van Quang, N., Tu, H. C., Zheng, Y., Seguin-Devaux, C., Mi, B., & Ai, U. (2015). Anti-HIV activity of fucoidans from three brown seaweed species. *Carbohydrate Polymers*, *115*, 122–128. https://doi.org/10.1016/j.carbpol.2014.08.068.

Usov, A. I., & Bilan, M. I. (2009). Fucoidans—sulfated polysaccharides of brown algae. *Russian Chemical Reviews*, *78*(8), 785–799. https://doi.org/10.1070/rc2009v078n08abeh004063.

van Weelden, G., Bobi, M., Okła, K., van Weelden, W. J., Romano, A., & Pijnenborg, J. M. A. (2019). Fucoidan structure and activity in relation to anti-cancer mechanisms. *Marine Drugs*, *17*(1). https://doi.org/10.3390/md17010032.

Vasseur, E., Setälä, K., & Gjertsen, P. (1948). Chemical studies on the jelly coat of the sea-urchin egg. In *Acta chemica scandinavica* (Vol. 2, pp. 900–913). https://doi.org/10.3891/acta.chem.scand.02-0900.

Wang, J., Zhang, Q., Zhang, Z., & Li, Z. (2008). Antioxidant activity of sulfated polysaccharide fractions extracted from *Laminaria japonica*. *International Journal of Biological Macromolecules*, *42*(2), 127–132. https://doi.org/10.1016/j.ijbiomac.2007.10.003.

Wang, J., Zhang, Q., Zhang, Z., Song, H., & Li, P. (2010). International journal of biological macromolecules potential antioxidant and anticoagulant capacity of low molecular weight fucoidan fractions extracted from *Laminaria japonica*. *46*, 6–12. https://doi.org/10.1016/j.ijbiomac.2009.10.015.

WBOC. (2021). *Global fucoidan market analysis survey 2021–2027 with top countries data industry share, size, revenue, latest trends, business boosting strategies, CAGR status, growth opportunities and forecast | with covid-19 analysis* (pp. 1–8). WBOC. Retrieved on September 23, 2021.

Wei, X., Cai, L., Liu, H., Tu, H., Xu, X., Zhou, F., & Zhang, L. (2019). Chain conformation and biological activities of hyperbranched fucoidan derived from brown algae and its desulfated derivative. *Carbohydrate Polymers*, *208*, 86–96. https://doi.org/10.1016/j.carbpol.2018.12.060.

Wijesinghe, W. A. J. P., & Jeon, Y. J. (2012). Biological activities and potential industrial applications of fucose rich sulfated polysaccharides and fucoidans isolated from brown seaweeds: A review. *Carbohydrate Polymers, 88*(1), 13–20.

Wu, Y., Hu, Y., You, P., Chi, Y. J., Zhou, J. H., Zhang, Y. Y., & Liu, Y. L. (2016). Study of clinical and genetic risk factors for aspirin-induced gastric mucosal injury. *Chinese Medical Journal*, *129*(2), 174–180. https://doi.org/10.4103/0366-6999.173480.

Yoo, H. J., You, D. J., & Lee, K. W. (2019). Characterization and immunomodulatory effects of high molecular weight fucoidan fraction from the sporophyll of *Undaria pinnatifida* in cyclophosphamide-induced immunosuppressed mice. *Marine Drugs*, *17*(8). https://doi.org/10.3390/md17080447.

Yu, L., Ge, L., Xue, C., Chang, Y., Zhang, C., Xu, X., & Wang, Y. (2014). Structural study of fucoidan from sea cucumber Acaudina molpadioides : A fucoidan containing novel tetrafucose repeating unit. *Food Chemistry*, *142*(May 2009), 197–200. https://doi.org/10.1016/j.foodchem.2013.06.079.

Zayed, A., El-Aasr, M., Ibrahim, A. R. S., & Ulber, R. (2020). Fucoidan characterization: Determination of purity and physicochemical and chemical properties. *Marine Drugs*, *18*(11), 1–31. https://doi.org/10.3390/md18110571.

Zhang, W., Oda, T., Yu, Q., & Jin, J. O. (2015). Fucoidan from *Macrocystis pyrifera* has powerful immune-modulatory effects compared to three other fucoidans. *Marine Drugs*, *13*(3), 1084–1104. https://doi.org/10.3390/md13031084.

Zinkievich, J. M., George, S., Jha, S., Nandi, J., & Levine, R. A. (2010). Gastric acid is the key modulator in the pathogenesis of non-steroidal anti-inflammatory drug-induced ulceration in rats. *Clinical and Experimental Pharmacology and Physiology*, *37*(7), 654–661. https://doi.org/10.1111/j.1440-1681.2010.05357.x.

13 Alginate and Hydrogel Applications for Wound Dressing

Dina Fransiska, Ellya Sinurat, Fera Roswita Dewi, and Hari Eko Irianto

CONTENTS

13.1 INTRODUCTION

The selection of the appropriate form of wound therapy at this time is not an easy one, and the growing need for effective wound care regularly needs to be addressed. Wound healing is influence by the severity of the lesion and suitable dressing techniques and equipment. Hydrogel is the most important of the many materials used in wound dressings because it creates a permanent wet surface in the wound and absorbs the exudates (Lazarus et al., 1994).

Wound healing is a complex and ongoing process influenced by various factors, and it requires the right environment to recover quickly. Various wound dressings, such as fiber, sponge, hydrogel, foam, hydrocolloid, and others are used for wound treatment. Hydrogels are commonly employed in the biomedical industry because they may give mechanical support and a wet environment for wounds (Zhang & Zhao, 2020).

DOI: 10.1201/9781003303916-13

As the most significant proportion of the human body, the skin protects the body from the outside. Despite their self-regeneration solid capacity, major skin flaws will not heal independently and require skin replacements to cover them. In recent years, significant progress has been achieved in skin tissue engineering to generate new skin substitutes. Because of their porous and hydrated molecular structure, hydrogels are possibilities with the most significant potential to imitate the original skin microenvironment. Hydrogels can be split into two classes based on the substance used to make them, i.e., natural and synthetic. Hydrogels can also be strengthened by adding nanoparticles, resulting in "in situ" hybrid hydrogels with improved characteristics and customized functionality. Different sensors can also be integrated into hydrogel wound dressings to offer real-time data on the wound environment (Tavakoli & Klar, 2020). Hydrogels are similar to the extracellular matrix in terms of their composition and mechanical properties. They can serve as a support material for cells during the tissue regeneration process and allow for the diffusion of nutrients, metabolites, and growth (Jirkovec et al., 2021).

Moreover, the skin performs many vital functions, including preventing water loss from the body, and it has a role in regulating body temperature. Normal human skin consists of three layers, including the epidermis, the dermis, and the hypodermis. The barrier function of the skin is provided by the epidermis, which is composed mainly of keratinocytes. They form a stratified epithelium, with basal keratinocytes at the innermost layer and the keratinized, relatively impermeable outer stratum corneum layer on the surface.

Natural polymers, such as agarose, alginates, carrageenan, hyaluronic acid, pectin, chitosan, and neutral dextran, as well as the ones of synthetic polymer, for example, polyglycolic acid (PGA), polylactic acid (PLA), polyacrylic acid, poly-e-caprolactone, polyvinylpyrrolidone (PVP), polyvinyl alcohol (PVA), polyethylene glycol (PEG), and polyurethane (PU) have proven wound healing properties in vitro and in vivo with enhanced epithelialization (Mir et al., 2018).

Alginate is a natural linear polysaccharide derived from brown algae or bacteria, consisting of repeating units of β-1,4-linked D-mannuronic acid (M) and L-guluronic acid (G) in different ratios. It is widely use in biomedical and engineering fields due to its good biocompatibility and liquid absorption capacity. Alginate-based hydrogels have been use in wound dressing, tissue engineering, and drug delivery applications for decades (Miao Zhang & Zhao, 2020). The addition of alginate to the wound dressing keeps it from sticking to fresh granulation tissue, preventing further damage during removal (Jayakumar et al., 2011). It has been used in various applications in food, biotechnology, and pharmaceutical industries, especially for the controlled delivery of drugs and other bioactive molecules (Khotimchenko et al., 2001).

Due to its favorable features, such as biocompatibility and non-toxicity, alginate is a biopolymer employed in several biomedical applications. To date, it has proven to be particularly appealing in wound healing applications. It can be adapted to materials that have wound-healing characteristics. Alginate has been utilized to develop various wound dressing materials, including hydrogels, films, wafers, foams, nanofibers, and topical formulations. Alginate wound dressings absorb excess wound fluid, maintain a physiologically moist environment, and reduce the risk of bacterial infections at the wound site. The ratio of different polymers used in combination with alginate, the kind of cross-linkers employed, the period of cross-linking, the nature of excipients utilized, the incorporation of nanoparticles, and antibacterial agents all influence the therapeutic efficacy of these wound dressings (Aderibigbe & Buyana, 2018).

Alginate hydrogels are a type of dressing placed on the skin to treat wounds that are difficult to cure, such as bedsores, venous ulcers, and diabetic wounds. In the case of smart dressing materials, alginates are frequently coupled with other polymers, displaying different capabilities than the properties of separate components, such as improved flexibility, the longer period of drug release, or greater bioavailability of the medicinal ingredient. Both of the ingredients are non-toxic and biocompatible. Only their combination, however, provides the characteristics that the perfect third-generation dressing material must possess. Although PVA has adequate resistance, its weak adhesive properties, low gas diffusion, and low fluid absorption limit its use in wound dressings. To increase PVA's therapeutic qualities, it can be coupled with alginate, collagen, or chitin derivatives that have strong absorption and provide an optimum permanent moist medium in the wound and absorb wound exudates allowing cellular activity to continue. New trends in hybrid material design have emerged recently, such as altering the hydrogel matrix with therapeutic chemicals of natural origin, primarily taken from plants (Bialik-Wąs et al., 2021).

This review focuses on the most recent advancements in the field of wound dressing based on hydrogel-alginate. Alginate hydrogels are discussed in detail, including their synthesis, manufacturing, and biological applications.

13.2 ALGINATE

Alginate is a biopolymer and polyelectrolyte classified as biocompatible, non-toxic, non-immunogenic, and biodegradable. Alginate comprises two isomeric residues, β-D-mannuronic acid (M) and α-L-guluronic acid (G). These two residues are linked together by (1 → 4) glycosidic bonds in a homogeneous or heterogeneous manner, resulting in three distinct blocks in alginate, including poly M, poly G, and poly MG (Cheng et al., 2020). The M block segment demonstrates the creation of linear and flexible consecration due to the interaction of mannuronic acid (1 → 4) guluronic acid differently. The connection (1 → 4) is formed, introducing steric barriers around the carboxyl group. Therefore, the G-block segment provides folded and stiff structural conformations responsible for the rigidity of the molecular chain (Yang et al., 2011). As a result, the G block segment provides folded and rigid structural conformations responsible for the molecular chain's rigidity. Because it is related equatorially in C-1 and C-4, Poly M has a straight conformation, similar to a somewhat flat band. The alginate structure is including the M/G ratio, uronate residue arrangement, molecular weight (MW), and acetylation rate. Species (genotype), growth circumstances or environment, extraction procedures, harvest period, and other factors influence these characteristics (Wang et al., 2019).

13.2.1 Sources of Alginate

Alginate raw materials are generally derived from brown seaweeds. The natural alginate content of brown seaweed is about 30–60% dry weight. In addition, alginate is also found in some red algae and can be produced by some bacteria, such as genus Azotobacter and bacterial exopolysaccharides that include *Pseudomonas aeruginosa*. It is made up of a series of M (M-block) and G (G-block) residues intermingled with the order MG (MG block). Although it is possible to obtain alginate from sources of algae and bacteria, the commercially available alginate has only come from brown seaweeds (Wang et al., 2019). The alginate industry produces about 30,000 metric tons per year and comprises less than 10% of the biosynthetic alginate material. Selected types of brown algae that have been extracted their alginate are shown in Table 13.1.

The partial hydrolysis of alginate is followed by fractionation to obtain alginates containing different copolymer compounds. According to the source of alginate, the predicted biosynthesis process of bacterial alginate was reported in some excellent articles. The first step in alginate production is the oxidation of carbon sources to acetyl-CoA, which enters the TCA cycle to be transformed into fructose-6-phosphate via gluconeogenesis (Figure 13.1) (Pawar & Edgar, 2012).

TABLE 13.1
Researched Sources of Alginate (Liu et al., 2019)

Sources of Alginate	Mannuronic Acid (%)	Guluronic Acid (%)	M/G Ratio	Yield (%)
Algae				
Laminaria hyperborean	30 to 56	44 to 70	0.43 to 1.28	
Macrocystis pyrifera	59 to 62	38 to 41	1.44 to 1.631	9.95±0,31
Laminaria digitata	53 to 60	40 to 47	12 to 1.500	23.95±1,23
Ascophyllum nodosum	46 to 60	40 to 54	85 to 1.500	19.22±5,68
Sargassum spp	16 to 54	46 to 84	19 to 1.181	23.18
Laminaria japonica	65 to 70—	30 to 35—	86 to 2.26—	36.89
Sargassum fluitans (alginic acid method) (Laksanawati et al., 2017)				
Sargassum hystrix (calcium method) (Diharningrum & Husni, 2018)				
Sargassum crassifolium (Sinurat & Marliani, 2017)				
Turbinaria sp. (Liyana et al., 2021)				
Turbinaria ornata (calcium method) (Laksanawati et al., 2017)				
Bacteria				
Azotobactervinelandii	44 to 92	8 to 56	0.79 to 5.66	
Pseudomona sp.	60 to 84	16 to 37	1.50 to 5.25	

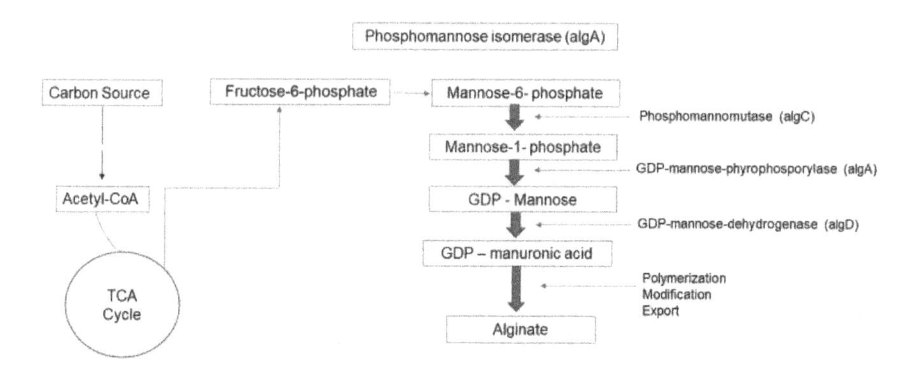

FIGURE 13.1 Bacterial alginate biosynthesis.

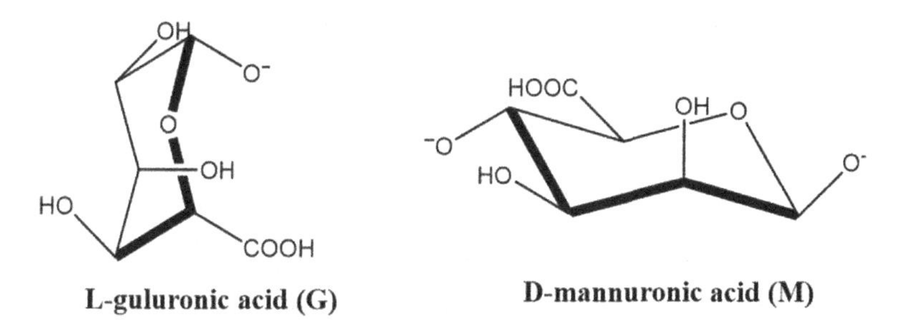

FIGURE 13.2 Chemical structure of alginate monomers: L-guluronic acid and D-mannuronic acid.

Brown seaweed is a natural alginate source, which is derived by a fractionation method that yields insoluble fractions (hydrolyzed) and insoluble fractions (unhydrolyzed/nonhydrolyzed). Non-hydrolyzed molecules have M or particularly G-rich residues, whereas hydrolyzed portions have a large number of alternating MG residues. It is, therefore, possible to hydrolyze a proposed structure consisting of M blocks, G blocks, and alternating MG blocks. Figure 13.2 depicts a characteristic structure of the alginate skeleton, with (a) indicating chain conformation and (b) indicating normal block distribution. The findings support the existence of alternative blocks M, G-block, and MG. Figure 13.3 shows a representative alginate structure from chain conformation and block distribution. The microstructure of alginates is being study using computer-based mathematical models. Only NMR 1H and 13C spectroscopy can provide a complete picture of the structure of the alginate skeleton (Pawar & Edgar, 2012).

13.2.2 Characteristics of Alginate

Alginates' physical and chemical features, such as MW, M/G ratio, M and G residue distribution, and acetylation levels, are substantially determined by

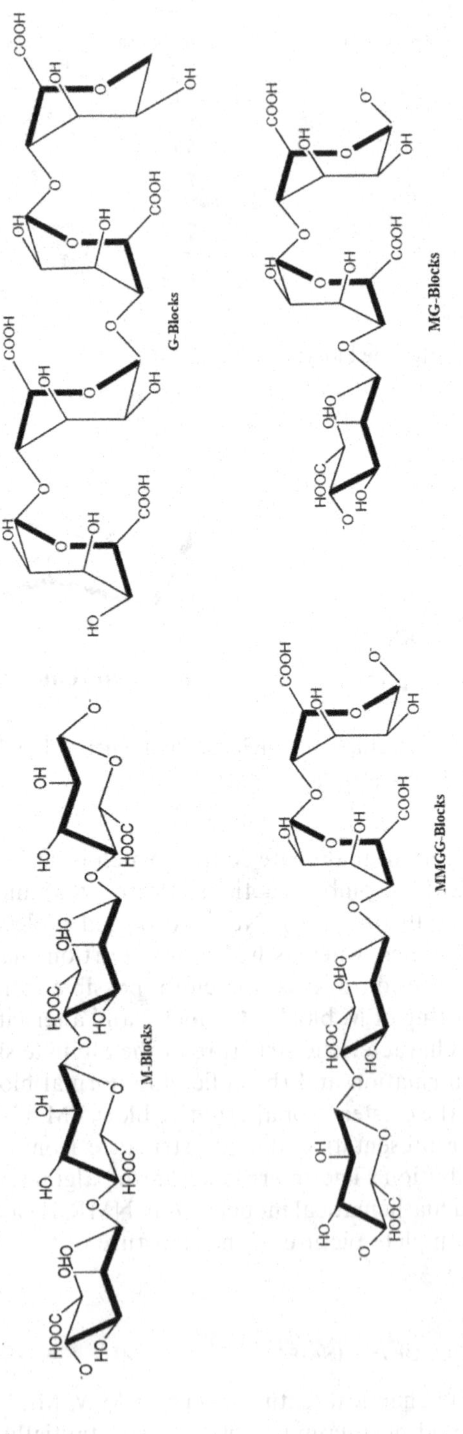

M-Blocks

G-Blocks

MMGG-Blocks

MG-Blocks

FIGURE 13.3 Representative alginate structure from chain conformation and block distribution.

their structure. This alginate's structural characteristics depend on its source and growth conditions. Alginates from *Laminaria hyperboreans*, for example, contain 30–56% M and 44–70% G, with an M/G ratio ranging from 0.43 to 1.28. *Macrocystis pyrifera* and *Laminaria digitata* alginates, on the other side, exhibit M/G ratios of 1.44 to 1.63 and 1.12 to 1.50, respectively (Table 13.1). Alginate from *Azotobacter vinelandii* acetylates at a rate of 11–30%, while alginate from *Pseudomonas* sp. acetylates at a rate of 4–57% (Pacheco-Leyva et al., 2016; Pawar & Edgar, 2012).

Viscosity, sol/gel transition qualities, and water absorption are some of the physicochemical properties of alginate (Hernández et al., 2016). Alginates with more G-G blocks have a higher solubility than those with more M-M blocks. Alginate rich in MG/GM blocks dissolves at low pH, but alginate rich in MM or GG blocks is insoluble (Liu et al., 2019). The GG content may affect the gels because of the interaction with divalent ions, such as calcium (Ca^{2+}). Higher GG alginate gels are generally more fragile and stronger than those created with higher MM content. For commercially available alginates, the MW range is between 33,000 and 400,000 Da. Longer chain alginate solutions (higher MW) have a higher viscosity than shorter chain alginate solutions at the same polymer concentration (Leirvåg, 2017). However, there was no relation between solution viscosity and MW (Hay et al., 2013). The solubility and viscosity of bacterial alginate solution are similarly affected by the presence of acetyl groups. Alginic acid is insoluble in water and organic solvents; however, alginic acid esters can produce persistent, viscous solutions (Liu et al., 2019).

A number of the physical properties of alginate contribute to its different applications. The food, cosmetics, and pharmaceutical industries employ alginate as a stabilizer, emulsifier, thickener, and gelling agent. Alginate has been utilized as a wound-healing material in sponges, hydrogels, and electrospunates, for example, to speed up skeletal muscle regeneration. Due to its adhesion, hydrophilicity, pleasant odor, ease of mixing, low cost, and long shelf life, alginate has been proven to be useful for preparing buccal, nasal, ocular, and gastrointestinal dosage forms (Varaprasad et al., 2020).

Alginate gel particles typically range in size (diameter) from >1 mm (macro), 0.2 to 1,000 mm (micro), and 0.2 mm (nano). These gel particles' chemical and mechanical properties, which typically have a high water content, can be customized depending on the type of crosslinking agent utilized. Alginate gel particles are appealing for biological applications as a natural material since they are biocompatible, non-toxic, biodegradable, and affordable (Ching et al., 2017).

Sodium alginate is used in most alginate preparations because it is more stable, has a longer shelf life, and is less quickly hydrolyzed in storage. Na-alginate is a hydrophilic and soluble matrix. It can, however, form crosslink networks with multivalent cations (such as Ca^{2+}, Ba^{2+}, and Fe^{3+}) through ionic interactions. Using polyvalent cations and crosslinking methods, water resistance, mechanical resistance, barrier characteristics, compactness, and stiffness have improved. Furthermore, the kind of cation and the methods utilized to introduce crosslinking ions have a significant impact on the process of cross-linking and the properties of Na-alginate (Zheng et al., 2016).

The physical characteristics of alginates that can be explored for the development of hydrogels for wound healing purposes are as follows.

13.2.2.1 Solubility

Three marine characteristics influence alginate solubility in water, i.e. the pH of the solvent, the ionic strength of the medium, and the presence of gelling ions in the solvent. The carboxylic acid group must be deprotonated and the pH must be above specific critical values for alginate to be soluble. The properties of the solution, such as polymer conformation, chain extension, viscosity, and hence solubility, are affected by changing the ionic strength of the medium (Pawar & Edgar, 2012; H. Zhang et al., 2021).

13.2.2.2 Cross-Linking of Ions

Hydrogels are formed when alginate chelates with cations are valent. In the presence of divalent cations, interactions between G-blocks cause them to associate and form tightly held junctions. In addition to the G-block, MG blocks play a role in the game, producing weak intersections. As a result, alginate with a high G content yields a thicker gel. There are two ways to crosslink calcium alginate. The "diffusion" method allows crosslinking ions from the outside reservoir to diffuse into an alginate solution. The second method is called "internal regulation." The ion source is placed in an alginate solution, and a controlled trigger (typically the ion source's pH or solubility) is used to prevent cross-linking ions from entering the solution. The diffusion approach creates a Ca^{2+} ion concentration gradient across the gel's thickness, whereas the internal setup creates outward uniform ion concentration. Gel perfusion sets are typically made by depositing a solution of Na-alginate in a $CaCl_2$ solution. Internal gel packages generally use insoluble calcium salts like $CaCO_3$ as a source of calcium. Slow hydrolysis lactones, such as D-glucono-d-lactone (GDL), generate pH changes that cause Ca^{2+} ions to be released internally, resulting in gel formation (Pawar & Edgar, 2012; Yamamoto et al., 2019).

13.2.2.3 The Gel of Alginic Acid

An acidic gel is created when the pH of the alginate solution is decreased below the pKa of the uronic acid in a highly regulated manner. The gel is stabilized by a network of hydrogen bonds between molecules. Two methods are commonly used to make acidic gels. Hydrolyzed lactones are slowly GDL added to the Na-alginate solution in the first approach. Proton exchange is used in the second method to convert a pre-formed Ca-alginate gel into an acidic gel (Chuang et al., 2017; Pawar & Edgar, 2012).

The gelatination process depends on Ca and the driving force. The alginate-dependent alginate glassing mechanism as an egg-box model (Cao et al., 2020). As a result, numerous researchers are beginning to delve further into the mechanism of alginate glassing, and typical egg-box models have evolved over the last 50 years. The Ca-dependent alginate gelation process is thought to be separated into three different and sequential phases (Figure 13.4, 1–3) (Cao et al., 2020): (i) in a single alginate chain, Ca^{2+} and guronic acid units

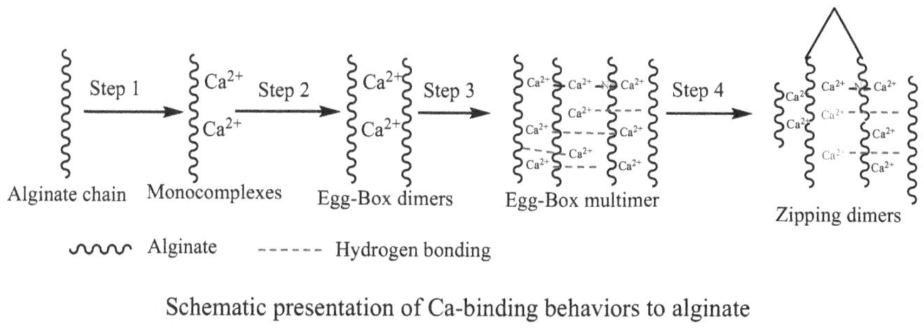

Schematic presentation of Ca-binding behaviors to alginate

FIGURE 13.4 Schematic presentation of Ca-binding behaviors to alginate.

create a monocomplex; (ii) pairing monocomplexes results in the formation of an egg box dimer; and (iii) the dimer egg box's lateral crosslink leads to the development of multimers. Egg box dimers form in a synergistic state known as "zipping". Another modified theory, as shown in Figure 13.4, is offered to interpret the Ca^{2+} and alginate further in the system with high Ca^{2+} concentrations (Wang et al., 2019). The alginate chain is bound in the fourth phase by inter-cluster separation and intracluster interaction of the egg box dimer at high Ca^{2+} concentrations. Excess Ca^{2+} is thought to break down the egg-box multimer by neutralizing the negative ions in the alginate molecule's free carboxyl group (Cao et al., 2020).

13.3 BASIC METHODS OF HYDROGEL MANUFACTURE

Various types of biopolymers have been used to make wound dressing products based on alginate. Their functional groups interact with other biopolymers easily and form a cross-linked network structure. An ideal wound dressing should have the desirable features of therapeutic options, such as offering a moist microenvironment, germ access protection, wound exudate absorption, tissue regeneration stimulation, and support to natural wound recovery processes. Alginate-based hydrogels are effective for the application of wound dressing (Liao et al., 2018). This is because alginates as bioactive compounds can improve hydrogel properties with intrinsic swelling capabilities, particularly biodegradation, optimum moisture transport via vapors, and exudate absorption. Hydrogel features a waterproof and extensively interlinked hydrophilic network of polymers (Ahmed, 2015). They were blended with natural and synthetic components. Based on hydrogels' synthetic and natural components, a three-dimensional swelling capacity network has been exhibited in various mediums (Figure 13.5) (Varaprasad et al., 2020).

In wound dressing, soft hydrogels (Figure 13.5b) can be employed to improve the healing of wounds using cellular delivery (Lee et al., 2014). Soft hydrogels, in essence, are frequently employed in chronic, necrotic, pressure, and burn wounds (Figure 13.5b). The physical (soft and steep) features of alginate hydrogels can be e-tailored through an ionic cross-connection between natural and

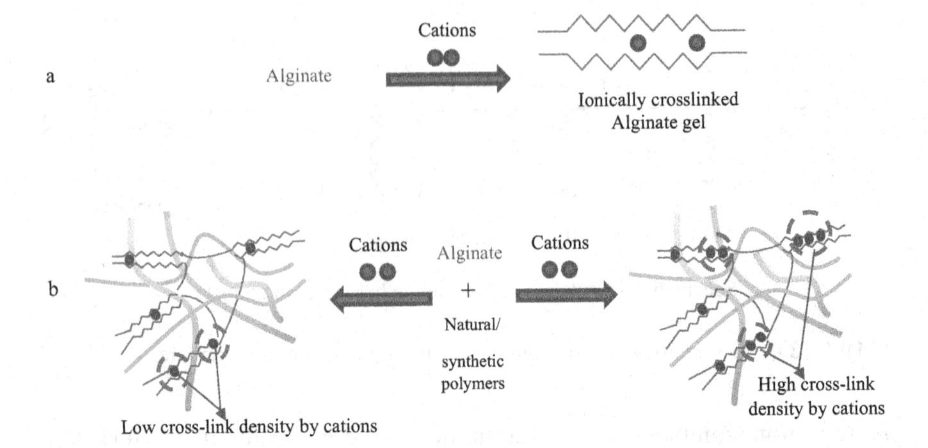

FIGURE 13.5 Ionic cross-linking of (a) an alginate gel and (b) alginate-based soft hydrogels (left) and stiff hydrogels (right) used as a hemostatic dressing material in wound dressing applications.

synthetic components (Figure 13.5b) (Galli et al., 2018). The soft (low cross-link density) and rigid (high cross-link density) characteristics of alginate hydrogels are mostly determined by the structure of hydrogels (raw materials) and how well they link together (Gharazi et al., 2018; Varaprasad et al., 2017).

13.4 THE DEVELOPMENT OF ALGINATE-BASED HYDROGEL AS A WOUND DRESSING

The hydrogel formulation using a combination between alginate and other substances (natural or/and synthetic polymers) for wound dressing has been objected to study in the last decade. Tang et al. (2019) created aqueous electrospinning honey/sodium alginate (SA)/polyvinyl alcohol (PVA) nanofibrous membranes (a 220-nm-thin emulsion in alginate microgels) for wound dressing. The particle modulus of the oil-based microgels (0–77% w/w oil total solids basis) was determined to be 150–212 Pa. The shape, chemical structure, and swelling ratio of cells were investigated. The nanofibrous membranes had smooth and homogenous three-dimensional architectures, with honey content increasing average nanofiber diameters. Fourier transform infrared spectroscopy (FTIR) spectra demonstrate that honey has been successfully integrated into nanofibers. Increasing the amount of honey in the diet lowered edema while increasing weight loss. In honey, the nano-fiber membrane's antioxidant activity was found to be concentration and time-dependent. Honey slowed the growth of gram-positive and gram-negative bacteria on nanofibers. Honey had been antibiotic activity against the MTT (3-[4,5-dimethylthiazol-2-yl]-2,5 diphenyl tetrazolium bromide) assay also confirmed the honey/SA/PVA nanofibrous membranes' good biocompatibility. Thus, honey/SA/PVA nanofibrous membranes appear to be a viable wound dressing option.

A hydrogel sheet (HS) consisting of the blended powder of alginate, chitin/chitosan, and fucoidan (ACF) has been developed as a functional wound dressing to generate a humid environment for rapid wound healing (ACF-HS; 60:20:2:4 w/w) (Murakami et al., 2010). ACF-HS gradually absorbed Dulbecco's minimal essential medium (DMEM) without maceration, and fluid absorption became consistent within 18 hours. ACF-HS should interact efficiently with and protect the wound in rats during the application, ensuring a nice moist curative environment with exudates. The wound dressing also includes qualities such as easy to apply and remove and good adherence. Skin flaws in full-density on the backs of rats were produced and mitomycin C solution (1 mg/ml in saline) was given to the wound for 10 minutes to prepare healing injuries. After the mitomycin C has been washed properly, ACF-HS was applied to curative injuries. Although ACF-HS was not stimulated to normal rat wound repair, healing-impaired wound repair was dramatically stimulated. The histological examinations showed that in the curative-impairment wounds treated by ACF-HS on day 7 significantly advanced granulation tissue and capillary formation were seen compared to those treated with calcium alginate fiber (Kaltostat Todon; Convatec Ltd., Tokyo, Japan) and the ones left untreated.

Electrospun gentamicin-loaded chitosan-alginate (CS-Alg) scaffolds were created to promote skin regeneration (Bakhsheshi-Rad et al., 2020). The porous and linked nanofibrous scaffolds had bead-free and randomly aligned continuous nanofibers. More gentamicin (Gn) added enhanced water uptake, mechanical strength, and deformation resistance, while less Gn added reduced mechanical characteristics. Antibacterial activity against Gram-negative and Gram-positive bacteria were observed in nanocomposite Gn-loaded CS-Alg scaffolds. As measured by MTT assays, cells survived better on neat CS-Alg and CS-Alg/Gn mats with lower Gn concentrations than on those with higher Gn concentrations. Burn wounds treated with CS-Alg scaffolds containing 3 wt% Gn healed faster than those treated with scaffolds containing higher or lower Gn concentrations due to increased re-epithelialization, dense collagen production, and growth of blood vessels and hair follicles. Overall, the CS-Alg/3Gn wound dressing shows significant promise in treating burn wounds.

Dermatology is focusing on developing functional and bioactive wound dressings (Nazarnezhada et al., 2020). A good wound dressing should keep the wound moist, cover the wound completely, absorb wound fluid and debris, and promote wound healing. We synthesized alginate hydrogels with varying amounts of H_2S and evaluated their biological characteristics in vitro and in vivo. The hydrogel obtained had appropriate swelling and degradation levels, as well as a medication release profile favorable to wound healing. The combination of alginate hydrogel and H_2S enhanced cell proliferation and wound healing in vitro and in vivo. The optimum concentration of H_2S was 0.5%, while larger quantities had deleterious effects on cell development and wound healing. While the specific mechanism of topical H_2S wound healing is unknown, it is critical to understand the underlying mechanism. The present investigation concluded that the alginate hydrogel produced with optimal H_2S concertation might be useful wound dressing material.

Dutra et al. (2020) successfully made transparent, soft, flexible, and mechanically resistant film wound dressings. It used 2% papain as the active agent to speed up the healing process. The films were cast using polyvinyl alcohol: calcium alginate mixes with increasing polysaccharide contents (10%, 20%, and 30% v/v). Fourier transform infrared spectroscopy (FTIR) and differential scanning calorimetry analyses were used to determine the thermal properties of the hydrogel. Tensile strength, elasticity modulus, and breakpoint elongation were measured. The effect of calcium alginate concentration on film properties such as weather resistance, swelling capacity, and mechanical properties has been studied. The stability of the papain in the films was indirectly tested using direct contact hemolysis and validated by blood agar diffusion. The XTT (tetrazolium salt XTT (2,3-bis-(2-methoxy-4-nitro-5-sulfophenyl)-2H-tetrazolium-5-carboxanilide) technique was used to assess cytotoxicity. The mixtures were miscible in the polymer concentrations tested. Increased calcium alginate content improves the weatherability and swelling properties of wound dressing films. During swelling testing, the mechanical resistance decreased without the films breaking. The films' hemolytic activity was mostly maintained throughout the study, indicating that papain was stable in the formulations. The films were non-toxic when tested on cells. The results show that PVA and calcium polymers blend can produce an interactive and bioactive wound dressing containing papain.

Amputations in diabetics commonly result in diabetic sores. Pathophysiological diabetic injuries cause decreased wound healing due to chronic inflammation and lack of tissue regeneration. Curcumin (CUR), a well-known anti-inflammatory and antioxidant, may be superior in treating diabetic wounds. However, CUR's low bioavailability and instability prevent its use. To increase the stability and solubility of CUR-CSNPs (chitosan nanoparticles), they were impregnated into a collagen scaffold for enhanced tissue regeneration applications. New nanohybrid studies were established for morphology, biodegradability, biocompatibility, in-vitro release, and in vivo wound healing studies (Karri et al., 2016). The study of nanoparticles revealed increased CUR stability and solubility. In vitro, the nano-hybrid ground improves water consumption, biocompatibility, and medicine availability. One study found that nanohybrid scaffold repaired wounds considerably faster ($p < 0.001$) than control and placebo wounds. The nanohybrid scaffold group showed complete epithelialization with thick granulated tissues, whereas the placebo scaffold group showed no compact collagen deposition and inflammatory cells. That CUR synergistic combinations (anti-inflammatory and antioxidant), chitosan (sustainable drug carrier, wound healing) and collagen (set as scaffolding wound healing) promise to tackle several pathological manifestations of diabetic injuries and enhance wound healing ability.

13.5 RECENT DEVELOPMENTS OF HYDROGELS

13.5.1 ENCAPSULATION

Whey protein encapsulation, retention, and release from alginate-based hydrogel beads were examined for the impact of pH (Zhang et al., 2016). The pH values (3, 5, and 7) used to create the hydrogel beads were selected based on the nature of the electrostatic interactions (attraction or repulsion) between the alginate and protein molecules. The injection of a whey protein/alginate mixture into a Ca^{2+} solution successfully produced protein-loaded hydrogel beads. At pH 3, there was an electrostatic attraction between the cationic protein and anionic alginate molecules, which boosted the effectiveness and retention of the beads in terms of protein encapsulation and retention. As a result, the protein molecules were better retained in the alginate beads due to their increased affinity. This study's hydrogel beads may be modified to release protein molecules when exposed to a specific pH. Increasing the pH from 3 to 7 led to protein release, which was again attributable to weakening the electrostatic interaction between protein molecules trapped in alginate hydrogel as pH rose. These findings have crucial implications for developing hydrogel beads that encapsulate, retain, and release proteins in food products and the gastrointestinal tract. It is not possible from this study to develop hydrogel beads that can slow drug release for the gastrointestinal wound.

13.5.2 Nanocomposite

Curcumin, N, O-carboxymethyl chitosan, and oxidized alginate nanocomposite hydrogel has been successfully created in situ as a novel wound dressing for cutaneous wound repair (Li et al., 2012). As a carrier, methoxy poly(ethylene glycol)-b-poly(ε-caprolactone) (MPEG-PCL) copolymer was employed with methoxy-poly-b-b-polymer (top-caprolactone) and N,O-carboxymethyl chitosan/oxidized alginate hydrogel (CCS-OA hydrogel). In vitro, the encapsulated nano curcumin was slowly released from the CCS-OA hydrogel, followed by hydrogel corrosion. Hydrogels were administered to dorsal rats to study wound healing. The usage of nano curcumin/CCS-OA hydrogel improved epidermal re-epithelialization and collagen deposition in injured tissue. On the seventh day after being injured, wound tissue content, DNA, protein, and hydroxyproline were examined. The results showed that nano curcumin and CCS-OA hydrogel can speed up wound healing. These findings suggested that the CCS-OA hydrogel could be used as a viable wound dressing.

The hydrogel dressing containing calcium alginate is effective due to its sound absorption of characteristics and tearless pain. Its application is, however, limited by its weak mechanical and non-bacterial features. A new biomimetic hydrogel dressing of 2,2,6,6-tetramethyl piperidine-1-oxyloxidized bacterial (TOBC) cellulose was intensively fitted, and is a straightforward technique of loading Zn^{2+} has been developed (Minghao Zhang et al., 2020). Due to the creation of the joint network structure, mechanical characteristics were improved by adding 20 wt% TOBC. The hydrogel possesses good antibacterial and biological qualities when the Zn^{2+} content is regulated at about 0,0001 wt%. An adequate and straightforward method for creating new biomimetic hydrogel dressings with good characteristics has been obtained. Numerous alginate-based materials for wound dressing are provided in Table 13.2.

TABLE 13.2
Various Types of Alginate-Based Wound Dressing Materials

Alginate-Based Composite Material	Application	References
Carboxymethylcellulose, alginate, gatifloxacin	Antibacterial	(Kesavan et al., 2010)
Alginate, fucoidan, chitosan	Wound healing	(Murakami et al., 2010)
Bioglass/agarose, alginate	Chronic wound healing	(Zeng et al., 2015)
Alginate, calcium gluconate crystals	Hybrid alginate hydrogel wound healing	(Liao et al., 2018)
Alginate, nanocellulose	Tissue engineering/ wound healing	(Siqueira et al., 2019)
Polyacrylamide, alginate, divalent ions (copper, zinc, strontium, calcium)	Wound healing	(Zhou et al., 2018)
Chitosan, alginate, alpha-tocopherol	Wound healing	(Ehterami et al., 2019)
Carboxymethyl chitosan, alginate	Chronic wounds	(Lv et al., 2019)
Sodium alginate, honey	Wound dressing	(Tang et al., 2019)
Nanocellulose alginate, Tempo-oxidized cellulose nanocrystals	Cell growing, wound healing	(Siqueira et al., 2019)
Alginate, chitosan, alpha-tocopherol	Wound healing	(Ehterami et al., 2019)
Alginate polyelectrolyte, carboxymethyl chitosan	Injectable hydrogel wound healing	(Lv et al., 2019)
Sodium alginate, tributylamine, tributylamine, dihexylamine	Anti-hemolytic, antibacterial wound healing	(Zare-Gachi et al., 2020)
Calcium alginate, Zn^{2+}	Antibacterial, wound dressing	(Minghao Zhang et al., 2020)
Alginate, hydrogen sulfide	Wound dressing	(Nazarnezhada et al., 2020)
Sodium alginate poloxamer 407, pluronic F-127, and polyvinyl alcohol	Wound healing	(Abbasi et al., 2020)
Sodium alginate, hydroxyethylacryl chitosan	Drug release wound healing	(Chalitangkoon et al., 2020)
Alginate, chitosan, gentamicin	Antibacterial, nanofibrous wound dressing	(Bakhsheshi-Rad et al., 2020)
Sodium alginate, carboxymethyl chitosan	Antimicrobial wound dressing	(Chen et al., 2020)

13.5.3 Nanostructured Fiber

The creation of various wound dressings relies on the production of polymeric nanofibrous meshes. Ultra-fine fibers with sizes ranging from several micrometers to a few nanometers make up these formations. Nanofibrous meshes have several intrinsic features that make them particularly promising for wound healing applications, such as large surface area and nanoporosity. The meshes can be made using a variety of processes, such as phase of separation or self-assembly. However, electrospinning is the most popular since it is a simple, cost-effective, and adaptable approach (Abrigo et al., 2014).

Numerous studies have been conducted on the use of various alginate-based formulations, including chitosan-fibrin-sodium alginate composite for wound dressings (Devi et al., 2012), alginate nanofiber-based wound dressings (Leung et al., 2014), and sodium alginate/poly(ethylene oxide) blend nanofibers (Park et al., 2010). Because chitosan and sodium alginate are present, the fibrin nature is reduced. The porosity structure of the matrix aids in the efficient absorption of wound fluids, keeping the site dry. This aids in the prevention of bacterial infection as well as faster recovery. The composite has a porous structure with pore sizes ranging from 100 to 400 mm, which is ideal for cell adhesion and skin development (Devi et al., 2012). Two cross-linking procedures, calcium solution cross-linking and glutaraldehyde double-cross-linking, were investigated as processing strategies for changing alginate nanofiber characteristics after electrospinning. Although calcium cross-linking is more particular to polysaccharides like alginate, glutaraldehyde cross-linking is a cost-effective way to increase mechanical properties in various natural polymers, including polysaccharides and proteins. An optical assessment of the impacts on nanofiber morphology revealed that nanofiber morphology could be conserved while welds between fibers were created to improve structural stability by optimizing the process. The addition of polylysine to alginate nanofibers improved fibroblast cell adhesion and proliferation (Leung et al., 2014). Because of its high viscosity and conductivity, sodium alginate cannot be electrospun alone. By combining sodium alginate with poly(ethylene oxide) (PEO) and adding lecithin as a natural surfactant, the nanofiber can be enhanced and the bead removed. The morphology of sodium alginate/PEO blend nanofibers revealed transparent fibers as the alginate level increased. During the electrospinning process, fine alginate nanofibers with smooth and homogeneous fiber were generated in SA/PEO ratios of 1/2 and 2/2 (Park et al., 2010).

Electrospinning is a simple, cost-effective, and repeatable procedure that can use both synthetic and natural polymers to solve these unique wound issues. Electrospun meshes have a large surface area, are microporous, and can be loaded with medicines or other biomolecules (Abrigo et al., 2014).

13.5.4 Conclusions

The worldwide alginate market was valued at USD 728.4 million in 2020 and is anticipated to expand at a compound annual growth rate (CAGR) of 5.0% from 2021 to 2028. The market is anticipated to expand significantly owing to rising demand for goods used in a variety of applications, including food applications as thickeners, emulsifiers, and gelling agents (Grand View Research, 2021). The increase in the need for alginate will be higher when the application of alginate for hydrogel wound dressings develops. During the projection period of 2021–2025, the hydrogel dressing market increased by $123.56 million at a CAGR of 6%. The market is being pushed by the rising frequency of chronic wounds and the increased usage of hydrogels for burn healing. Furthermore, the rising frequency of chronic wounds is expected to fuel market expansion (Research & Markets, 2021).

Efforts have been made by many scientists around the world to develop alginate-based hydrogel production technology for wound dressings and show that alginate is suitable for hydrogel production. Several formulations have been developed and show promising results. The development of new technologies, such as encapsulation, nanocomposites and nanostructured fibers, is expected to encourage better production of hydrogel-based wound dressings, especially regarding faster, safer and easier wound healing. The availability of brown seaweed supported by good farming technology is hoped to ensure the availability of raw materials for alginate production through an effective and efficient extraction process. So that the sustainable production of alginate-based hydrogel wound dressing can be guaranteed with sufficient availability of alginate as the main ingredient.

REFERENCES

Abbasi, A. R., Sohail, M., Minhas, M. U., Khaliq, T., Kousar, M., Khan, S., Hussain, Z., & Munir, A. (2020). Bioinspired sodium alginate based thermosensitive hydrogel membranes for accelerated wound healing. *International Journal of Biological Macromolecules, 155*, 751–765. https://doi.org/10.1016/j.ijbiomac.2020.03.248.

Abrigo, M., McArthur, S. L., & Kingshott, P. (2014). Electrospun nanofibers as dressings for chronic wound care: Advances, challenges, and future prospects. *Macromolecular Bioscience, 14*(6), 772–792. https://doi.org/10.1002/mabi.201300561.

Aderibigbe, B. A., & Buyana, B. (2018). Alginate in wound dressings. *Pharmaceutics, 10*(2). https://doi.org/10.3390/pharmaceutics10020042.

Ahmed, E. M. (2015). Hydrogel: Preparation, characterization, and applications: A review. *Journal of Advanced Research, 6*(2), 105–121. https://doi.org/10.1016/j.jare.2013.07.006.

Bakhsheshi-Rad, H. R., Hadisi, Z., Ismail, A. F., Aziz, M., Akbari, M., Berto, F., & Chen, X. B. (2020). In vitro and in vivo evaluation of chitosan-alginate/gentamicin wound dressing nanofibrous with high antibacterial performance. *Polymer Testing, 82*(December 2019), 106298. https://doi.org/10.1016/j.polymertesting.2019.106298.

Bialik-Wąs, K., Pluta, K., Malina, D., Barczewski, M., Malarz, K., & Mrozek-Wilczkiewicz, A. (2021). Advanced SA/PVA-based hydrogel matrices with prolonged release of Aloe vera as promising wound dressings. *Materials Science and Engineering C, 120*, 111667. https://doi.org/10.1016/j.msec.2020.111667.

Cao, L., Lu, W., Mata, A., Nishinari, K., & Fang, Y. (2020). Egg-box model-based gelation of alginate and pectin: A review. *Carbohydrate Polymers, 242*(January), 116389. https://doi.org/10.1016/j.carbpol.2020.116389.

Chalitangkoon, J., Wongkittisin, M., & Monvisade, P. (2020). Silver loaded hydroxyethylacryl chitosan/sodium alginate hydrogel films for controlled drug release wound dressings. *International Journal of Biological Macromolecules, 159*, 194–203. https://doi.org/10.1016/j.ijbiomac.2020.05.061.

Chen, K., Wang, F., Liu, S., Wu, X., Xu, L., & Zhang, D. (2020). In situ reduction of silver nanoparticles by sodium alginate to obtain silver-loaded composite wound dressing with enhanced mechanical and antimicrobial property. *International Journal of Biological Macromolecules, 148*, 501–509. https://doi.org/10.1016/j.ijbiomac.2020.01.156.

Cheng, D., Jiang, C., Xu, J., Liu, Z., & Mao, X. (2020). Characteristics and applications of alginate lyases: A review. *International Journal of Biological Macromolecules, 164*, 1304–1320. https://doi.org/10.1016/j.ijbiomac.2020.07.199.

Ching, S. H., Bansal, N., & Bhandari, B. (2017). Alginate gel particles—A review of production techniques and physical properties. *Critical Reviews in Food Science and Nutrition, 57*(6), 1133–1152. https://doi.org/10.1080/10408398.2014.965773.

Chuang, J. J., Huang, Y. Y., Lo, S. H., Hsu, T. F., Huang, W. Y., Huang, S. L., & Lin, Y. S. (2017). Effects of pH on the shape of alginate particles and its release behavior. *International Journal of Polymer Science, 2017*. https://doi.org/10.1155/2017/3902704.

Devi, M. P., Sekar, M., Chamundeswari, M., Moorthy, A., Krithiga, G., Murugan, N. S., & Sastry, T. P. (2012). A novel wound dressing material fibrin chitosan sodium alginate. *Indian Academy of Sciences, 35*(7), 1157–1163.

Diharningrum, I. M., & Husni, A. (2018). Metode ekstraksi jalur asam dan kalsium alginat berpengaruh pada mutu alginat rumput laut cokelat sargassum hystrix J. Agardh. *Jurnal Pengolahan Hasil Perikanan Indonesia, 21*(3), 532–542.

Dutra, J. A. P., Carvalho, S. G., Zampirolli, A. C. D., Daltoé, R. D., Teixeira, R. M., Careta, F. P., Cotrim, M. A. P., Oréfice, R. L., & Villanova, J. C. O. (2020). Papain wound dressings obtained from poly (vinyl alcohol)/ calcium alginate blends as new pharmaceutical dosage form : Preparation and preliminary evaluation. *European Journal of Pharmaceutics and Biopharmaceutics, 113*(December 2016), 11–23. https://doi.org/10.1016/j.ejpb.2016.12.001.

Ehterami, A., Salehi, M., Farzamfar, S., Samadian, H., Vaez, A., Ghorbani, S., Ai, J., & Sahrapeyma, H. (2019). Chitosan/alginate hydrogels containing Alpha-tocopherol for wound healing in rat model. *Journal of Drug Delivery Science and Technology, 51*(March), 204–213. https://doi.org/10.1016/j.jddst.2019.02.032.

Galli, R., Sitoci-Ficici, K. H., Uckermann, O., Later, R., Marečková, M., Koch, M., Leipnitz, E., Schackert, G., Koch, E., Gelinsky, M., Steiner, G., & Kirsch, M. (2018). Label-free multiphoton microscopy reveals relevant tissue changes induced by alginate hydrogel implantation in rat spinal cord injury. *Scientific Reports, 8*(1), 1–13. https://doi.org/10.1038/s41598-018-29140-z.

Gharazi, S., Zarket, B. C., DeMella, K. C., & Raghavan, S. R. (2018). Nature-inspired hydrogels with soft and stiff zones that exhibit a 100-fold difference in elastic modulus. *ACS Applied Materials and Interfaces, 10*(40), 34664–34673.

Grand View Research. (2021). *Alginate market size, share & trends analysis report by type (high M, high G), by product (sodium, propylene glycol), by application (pharmaceutical, industrial), by region, and segment forecasts, 2021–2028*. www.grandviewresearch.com/industry-analysis/alginate-market.

Hay, I. D., Rehman, Z. U., Moradali, M. F., Wang, Y., & Rehm, B. H. A. (2013). Microbial alginate production, modification and its applications. *Microbial Biotechnology, 6*(6), 637–650. https://doi.org/10.1111/1751-7915.12076

Hernández, A. J., Romero, A., Gonzalez-Stegmaier, R., & Dantagnan, P. (2016). The effects of supplemented diets with a phytopharmaceutical preparation from herbal and macroalgal origin on disease resistance in rainbow trout against Piscirickettsia salmonis. *Aquaculture, 454*, 109–117. https://doi.org/10.1016/j.aquaculture.2015.12.016

Jayakumar, R., Prabaharan, M., Sudheesh Kumar, P. T., Nair, S. V., & Tamura, H. (2011). Biomaterials based on chitin and chitosan in wound dressing applications. *Biotechnology Advances, 29*(3), 322–337. https://doi.org/10.1016/j.biotechadv.2011.01.005.

Jirkovec, R., Samkova, A., Kalous, T., Chaloupek, J., & Chvojka, J. (2021). *Preparation of a Hydrogel Nanofiber Wound Dressing*. 1–12.

Karri, V. V. S. R., Kuppusamy, G., Talluri, S. V., Mannemala, S. S., Kollipara, R., Wadhwani, A. D., Mulukutla, S., Raju, K. R. S., & Malayandi, R. (2016). Curcumin loaded chitosan nanoparticles impregnated into collagen-alginate scaffolds for diabetic wound healing. *International Journal of Biological Macromolecules, 93*, 1519–1529. https://doi.org/10.1016/j.ijbiomac.2016.05.038.

Kesavan, K., Nath, G., & Pandit, J. K. (2010). Sodium alginate based mucoadhesive system for gatifloxacin and its in vitro antibacterial activity. *Scientia Pharmaceutica, 78*(4), 941–957. https://doi.org/10.3797/scipharm.1004-24.

Khotimchenko, Y. S., Kovalev, V. V., Savchenko, O. V., & Ziganshina, O. A. (2001). Physical-chemical properties, physiological activity, and usage of alginates, the polysaccharides of brown algae. *Russian Journal of Marine Biology, 27*(1). https://doi.org/10.1023/A:1013851022276.

Laksanawati, R. Ustadi, & Amir Husni (2017). Pengembangan metode ekstraksi alginat Dari Rumput Laut Turbinaria Ornate. *Jphpi, 20*(2), 362–369.

Lazarus, G. S., Cooper, D. M., Knighton, D. R., Margolis, D. J., Percoraro, R. E., Rodeheaver, G., & Robson, M. C. (1994). Definitions and guidelines for assessment of wounds and evaluation of healing. *Wound Repair and Regeneration, 2*(3), 165–170. https://doi.org/10.1046/j.1524-475X.1994.20305.x.

Lee, Y., Bae, W., Lee, J. W., Suh, W., & Park, K. D. (2014). Enzyme-catalyzed in situ forming gelatin hydrogels as bioactive wound dressings: Effects of fibroblast delivery on wound healing efficacy. *Journal of Material Chemistry B, 2*, 7712–2218. https://doi.org/https://doi.org/10.1039/C4TB01111B.

Leirvåg, I. T. (2017). Strategies for stabilising calcium alginate gel beads: Studies of chitosan oligomers, alginate molecular weight and concentration (Master's thesis, NTNU).

Leung, V., Hartwell, R., Elizei, S. S., Yang, H., Ghahary, A., & Ko, F. (2014). Postelectrospinning modifications for alginate nanofiber-based wound dressings. *Journal of Biomedical Materials Research—Part B Applied Biomaterials, 102*(3), 508–515. https://doi.org/10.1002/jbm.b.33028.

Li, X., Chen, S., Zhang, B., Li, M., Diao, K., Zhang, Z., Li, J., Xu, Y., Wang, X., & Chen, H. (2012). In situ injectable nano-composite hydrogel composed of curcumin, N,O-carboxymethyl chitosan and oxidized alginate for wound healing application. *International Journal of Pharmaceutics, 437*(1–2), 110–119. https://doi.org/10.1016/j.ijpharm.2012.08.001.

Liao, J., Jia, Y., Wang, B., Shi, K., & Qian, Z. (2018). Injectable hybrid poly(ε-caprolactone)-b -poly(ethylene glycol)-b-poly(ε-caprolactone) porous microspheres/alginate hydrogel cross-linked by calcium gluconate crystals deposited in the pores of microspheres improved skin wound healing. *ACS Biomaterials Science & Engineering, 4*(3), 1029–1036. https://doi.org/10.1021/acsbiomaterials.7b00860.

Liu, J., Yang, S., Li, X., Yan, Q., Reaney, M. J. T., & Jiang, Z. (2019). Alginate oligosaccharides: Production, biological activities, and potential applications. *Comprehensive Reviews in Food Science and Food Safety, 18*(6), 1859–1881. https://doi.org/10.1111/1541-4337.12494.

Liyana, D., Kimia, P. S., Teknik, F., & Belitung, U. B. (2021). Optimasi konsentrasi *Na_2CO_3 dan suhu ekstraksi alginat dari Turbinaria sp sebagai bahan baku Polimer Elektrolit untuk DSSC Bachelor (Sarjana S-1) thesis Universitas Bangka Belitung.*

Lv, X., Liu, Y., Song, S., Tong, C., Shi, X., Zhao, Y., Zhang, J., & Hou, M. (2019). Influence of chitosan oligosaccharide on the gelling and wound healing properties of injectable hydrogels based on carboxymethyl chitosan/alginate polyelectrolyte complexes. *Carbohydrate Polymers, 205*(July 2018), 312–321. https://doi.org/10.1016/j.carbpol.2018.10.067.

Mir, M., Ali, M. N., Barakullah, A., Gulzar, A., Arshad, M., Fatima, S., & Asad, M. (2018). Synthetic polymeric biomaterials for wound healing: A review. *Progress in Biomaterials*, *7*(1), 1–21. https://doi.org/10.1007/s40204-018-0083-4.

Murakami, K., Aoki, H., Nakamura, S., Nakamura, S. ichiro, Takikawa, M., Hanzawa, M., Kishimoto, S., Hattori, H., Tanaka, Y., Kiyosawa, T., Sato, Y., & Ishihara, M. (2010). Hydrogel blends of chitin/chitosan, fucoidan and alginate as healing-impaired wound dressings. *Biomaterials*, *31*(1), 83–90. https://doi.org/10.1016/j.biomaterials.2009.09.031.

Nazarnezhada, S., Abbaszadeh-Goudarzi, G., Samadian, H., Khaksari, M., Ghatar, J. M., Khastar, H., Rezaei, N., Mousavi, S. R., Shirian, S., & Salehi, M. (2020). Alginate hydrogel containing hydrogen sulfide as the functional wound dressing material: In vitro and in vivo study. *International Journal of Biological Macromolecules*, *164*, 3323–3331. https://doi.org/10.1016/j.ijbiomac.2020.08.233.

Pacheco-Leyva, I., Pezoa, F. G., & Diaz-Barrera, A. (2016). *Leyva 2016.pdf* (p. 12). Hindawi Publishing Corporation.

Park, S. A., Park, K. E., & Kim, W. D. (2010). Preparation of sodium alginate/poly(ethylene oxide) blend nanofibers with lecithin. *Macromolecular Research*, *18*(9), 891–896. https://doi.org/10.1007/s13233-010-0909-y.

Pawar, S. N., & Edgar, K. J. (2012). Alginate derivatization: A review of chemistry, properties and applications. *Biomaterials*, *33*(11), 3279–3305. https://doi.org/10.1016/j.biomaterials.2012.01.007.

Research and Markets. (2021). *Global hydrogel dressing market 2021–2025*. www.researchandmarkets.com/reports/4895839/global-hydrogel-dressing-market-2021-2025.

Sinurat, E., & Marliani, R. (2017). Karakteristik Na-alginat dari rumput laut cokelat *Sargassum crassifolium* dengan perbedaan alat penyaring (The Characteristics of Sodium Alginate from Brown Seaweed Sargassum crassifolium with Different Filtering Tools), *Jurnal Pengolahan Hasil Perikanan Indonesia 20*, 351–361. doi: 10.17844/jphpi.v20i2.18103

Siqueira, P., Siqueira, É., de Lima, A. E., Siqueira, G., Pinzón-Garcia, A. D., Lopes, A. P., Segura, M. E. C., Isaac, A., Pereira, F. V., & Botaro, V. R. (2019). Three-dimensional stable alginate-nanocellulose gels for biomedical applications: Towards tunable mechanical properties and cell growing. *Nanomaterials*, *9*(1), 1–22. https://doi.org/10.3390/nano9010078.

Tang, Y., Lan, X., Liang, C., Zhong, Z., Xie, R., Zhou, Y., Miao, X., Wang, H., & Wang, W. (2019). Honey loaded alginate/PVA nanofibrous membrane as potential bioactive wound dressing. *Carbohydrate Polymers*, *219*(May), 113–120. https://doi.org/10.1016/j.carbpol.2019.05.004.

Tavakoli, S., & Klar, A. S. (2020). Advanced hydrogels as wound dressings. *Biomolecules*, *10*(8), 1–20. https://doi.org/10.3390/biom10081169.

Varaprasad, K., Jayaramudu, T., Kanikireddy, V., Toro, C., & Sadiku, E. R. (2020). Alginate-based composite materials for wound dressing application: A mini review. *Carbohydrate Polymers*, *236*(September 2019), 116025. https://doi.org/10.1016/j.carbpol.2020.116025.

Varaprasad, K., Raghavendra, G. M., Jayaramudu, T., Yallapu, M. M., & Sadiku, R. (2017). A mini review on hydrogels classification and recent developments in miscellaneous applications. *Materials Science and Engineering C*, *79*, 958–971. https://doi.org/10.1016/j.msec.2017.05.096.

Wang, B., Wan, Y., Zheng, Y., Lee, X., Liu, T., Yu, Z., Huang, J., Ok, Y. S., Chen, J., & Gao, B. (2019). Alginate-based composites for environmental applications: A critical review. *Critical Reviews in Environmental Science and Technology*, *49*(4), 318–356. https://doi.org/10.1080/10643389.2018.1547621.

Yamamoto, K., Yuguchi, Y., Stokke, B. T., Sikorski, P., & Bassett, D. C. (2019). Local structure of Ca2+ alginate hydrogels gelled via competitive ligand exchange and measured by small angle X-ray scattering. *Gels*, *5*(1). https://doi.org/10.3390/gels5010003.

Yang, J. S., Xie, Y. J., & He, W. (2011). Research progress on chemical modification of alginate: A review. *Carbohydrate Polymers, 84*(1), 33–39. https://doi.org/10.1016/j.carbpol.2010.11.048

Zare-Gachi, M., Daemi, H., Mohammadi, J., Baei, P., Bazgir, F., Hosseini-Salekdeh, S., & Baharvand, H. (2020). Improving anti-hemolytic, antibacterial and wound healing properties of alginate fibrous wound dressings by exchanging counter-cation for infected full-thickness skin wounds. *Materials Science and Engineering C*, *107*(May 2019), 110321. https://doi.org/10.1016/j.msec.2019.110321.

Zeng, Q., Yan, H., Haiyan, L., & Jiang, C. (2015). Design of a thermosensitive bioglass/agarose-alginate composite hydrogel for chronic wound healing. *Journal of Material Chemistry B*, *3*, 8856–8864. https://doi.org/https://doi.org/10.1039/C5TB01758K.

Zhang, H., Cheng, J., & Ao, Q. (2021). Preparation of alginate-based biomaterials and their applications in biomedicine. *Marine Drugs*, *19*(5), 1–24. https://doi.org/10.3390/md19050264.

Zhang, M., Chen, S., Zhong, L., Wang, B., Wang, H., & Hong, F. (2020). Zn2+-loaded TOBC nanofiber-reinforced biomimetic calcium alginate hydrogel for antibacterial wound dressing. *International Journal of Biological Macromolecules*, *143*, 235–242. https://doi.org/10.1016/j.ijbiomac.2019.12.046.

Zhang, M., & Zhao, X. (2020). Alginate hydrogel dressings for advanced wound management. *International Journal of Biological Macromolecules*, *162*, 1414–1428. https://doi.org/10.1016/j.ijbiomac.2020.07.311.

Zhang, Z., Zhang, R., Zou, L., & McClements, D. J. (2016). Protein encapsulation in alginate hydrogel beads: Effect of pH on microgel stability, protein retention and protein release. *Food Hydrocolloids*, *58*, 308–315. https://doi.org/10.1016/j.foodhyd.2016.03.015.

Zheng, H., Yang, J., & Han, S. (2016). The synthesis and characteristics of sodium alginate/graphene oxide composite films crosslinked with multivalent cations. *Journal of Applied Polymer Science*, *133*(27), 1–7. https://doi.org/10.1002/app.43616.

Zhou, Q., Kang, H., Bielec, M., Wu, X., Cheng, Q., Wei, W., & Dai, H. (2018). Influence of different divalent ions cross-linking sodium alginate-polyacrylamide hydrogels on antibacterial properties and wound healing. *Carbohydrate Polymers*, *197*(July 2017), 292–304. https://doi.org/10.1016/j.carbpol.2018.05.078.

14 Exopolysaccharides from Marine Microalgae

Azita Navvabi, Ahmad Homaei, Nazila Navvabi, and Philippe Michaud

CONTENTS

14.1 INTRODUCTION

Autotrophic microalgae convert solar energy, inorganic nutrients, and CO_2 into proteins, carbohydrates, lipids, nucleic acids, and other products. Heterotrophs categorizes into photoheterotroph and chemoheterotroph depending on their energy source (Perez-Garcia et al. 2011) (Figure 14.1). They absorb carbon dioxide and transform it into oxygen. One-ton microalgae are able to convert 1.83 tons of environment's carbon dioxide (Chisti 2007; Zhu 2015). The classification of microalgae based on pigment composition divides these microorganisms into nine groups. Red, brown, green and gold microalgae are the most important of these groups (Pulz and Gross 2004). Microalgae or microphytes grow in surface water where there is sunlight. Most of them by hooks are firmly attached to the ocean floor or rocks (Lee et al. 2014). Extracellular polymeric substances (EPSs) excreted by microalgae into their immediate environment are responsible for keeping cells interaction with each other (Xiao and Zheng 2016). They are able to grow alone or gathering with each other (Fuentes-Grünewald et al. 2009; Thurman and Burton 1997; Xiao and Zheng 2016). They can adapt themselves with the extreme adverse situations (Landsberg 2002; Caldwell 2009).

DOI: 10.1201/9781003303916-14

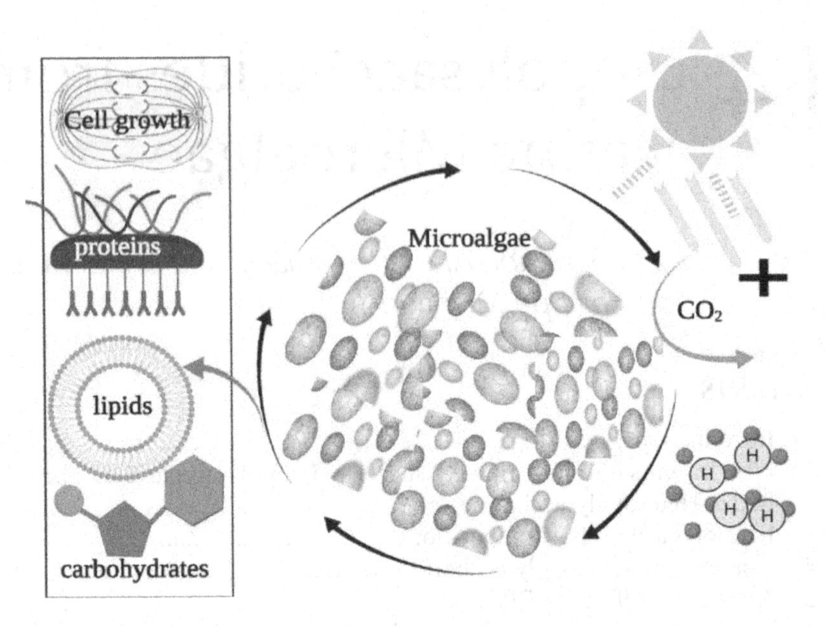

FIGURE 14.1 Conversion of solar energy, inorganic nutrients, and CO_2 by phototrophic microalgae strains into biomass.

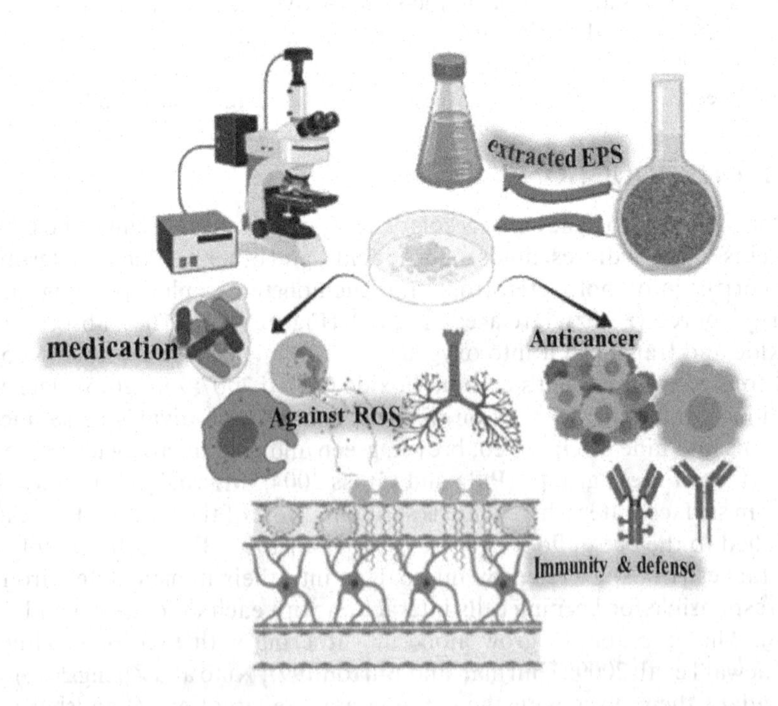

FIGURE 14.2 EPS of microalgae is screened to assess their efficiency against various cancers and infections.

Microalgae include in their molecular composition pigments, proteins, vitamins, polysaccharides, and exopolysaccharides, but up to now, there are a very few investigations carried out on the production of exopolysaccharides by microalgae due to a low amount of yield from microalgae and also to the difficulty producing them in photobioreactors (Delattre et al. 2016; Pierre et al. 2019).

Culture parameters such as temperature, irradiance, and composition of culture medium, strongly impact the EPS production by microalgae and in some cases modify their structure (composition) even if this phenomenon is often open to debate in literature (Otero and Vincenzini 2003). For harvesting them in a large scale, it is recommend using culturing and harvesting technologies to supply optimum amounts (Pulz and Gross 2004; Borowitzka 2018).

EPSs from microalgae have high molecular weights due to their constructions of various polymers (Xiao and Zheng 2016).

Most of them are linear or branched heteropolysaccharides, including up to six different monosaccharides in their structures depending on species (Bazaka et al. 2011; Delattre et al. 2016).

Monosaccharides and some non-carbohydrate structures such as pyruvate, succinate, phosphate, and acetate are part of the constituent microalgae EPSs structures (Li et al. 2001; Nicolaus et al. 2010). Analyses of the monosaccharide composition of microalgal EPS with chromatographic methods have showed their monosaccharides consist mainly of mannose, galactose, and glucose (Templeton et al. 2012).

In brief, besides these considerations, some, but not all, marine microalgae have many advantages, including easy production, high growth rate, food consumption, and medicines aspects (Figure 14.2). Microalgal extracellular polymeric substances are known as a protein moiety. Hence, EPS is carbohydrate derived by products from various strains of microalgae and algae (Raposo and Morais 2014; Zhu et al. 2015).

14.2 NEW INSIGHTS INTO FUNCTIONAL ASPECTS OF EXOPOLYSACCHARIDES

A microalgae colony consists of individual viable cells. These visible masses of microorganisms all originate from a single chain (Finkel et al. 2010; Pierre et al. 2019).

Exopolysaccharides are the metabolic products of several microorganisms. EPSs have special roles in phototrophic biofilms such as suppressor of environmental stresses, slipping mobility and forming structural integral unit of biofilms that determines their physiochemical properties (Evans 2003; Staudt et al. 2004; Rossi and De Philippis 2015).

Microalgal EPS with both hydrophobic and hydrophilic groups can be used as bioflocculants; for example, the EPS screened from *Anabaena* sp. BTA990 and *Nostoc* sp. BTA97 has shown flocculation capacity due to their composition molecules (Khattar et al. 2010; Tiwari et al. 2015).

Microalgae EPSs usually comprise 6–10 different monomers which exhibit hydrophobic properties through their peptide moieties, deoxysugars, and acetyl

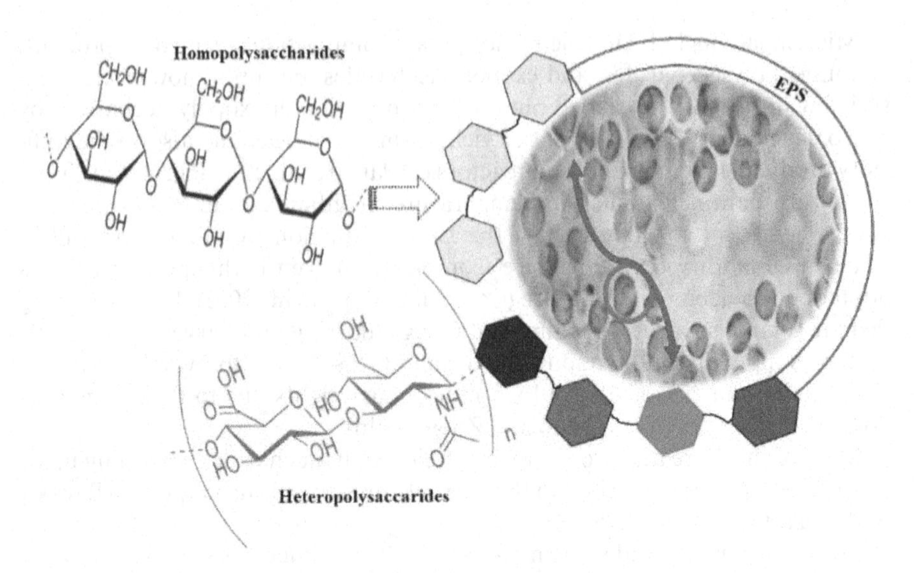

FIGURE 14.3 Types of EPS: homo- and heteropolysaccharides—classification of exopolysaccharides.

group contents (Mota et al. 2013). Most of the EPS from marine microalgae, but also from Cyanobacteria, are sulfated (Paulsen et al. 1998; Raposo et al. 2013).

Microalgae excretes EPSs in response to critical conditions, and then these released compounds enclose the cells of microalgae, although the recent reports explain that there is no consensus on this point currently (Liu et al. 2016).

Exopolysaccharides are categorized into homopolysaccharides and heteropolysaccharides (Figure 14.3).

14.2.1 Heteropolysaccharides

Various monosaccharides participate in forming heteropolysaccharides structures. Disaccharides that possess amino sugar participate in heteropolysaccharide construction (Vasudevan et al. 2013).

14.3 EXOPOLYSACCHARIDES COMPOSITION

Some EPSs possesses sulfate groups, glucuronic acid, and galacturonic acid which participate in forming a consistency and fixity of EPS (De Philippis and Vincenzini 2003; Xiao and Zheng 2016).

Organic or inorganic agents affect the chemical characteristic of EPS. The chemical composition of EPS is generally measured by using various chromatographic and mass spectrometric methods (Kumar et al. 2018).

Ester-linked acetyl and peptide moieties groups enhance emulsifying characteristic to EPS and make suitable it for commercial and industrial usages (Shepherd et al. 1995) (Table 14.1).

TABLE 14.1
The Major Constitution of Microalgal Exopolysaccharides

EPS of Strains	Compositions of EPS	References
Porphyridium	Heteroxylan composed of xylose, galactose, and glucose and other minor monosaccharides	Phlips et al. 1989
Synechococcus	glucose, galactose, and fucose	Gaignard et al. 2019
Pavlova	three or five monosaccharides, including neutral sugars and uronic acids	Gaignard et al. 2019
Tetraselmis globosa	Two monosaccharides representing 81% and 75% of the total monosaccharides, respectively	Elboutachfaiti et al. 2011
Exanthemachrysis	in glucuronic acid representing around 35% of its composition	Elboutachfaiti et al. 2011; Gaignard et al. 2019
Botryococcus braunii	Rhamnose, galactose, arabinose, fucose, xylose, Proteins	Allard and Casadevall 1990
Ankistrodesmus densus	Rhamnose, galactose, arabinose, glucose	Paulsen et al. 1998
Chlamydomonas reinhardtii	Rhamnose, galactose, arabinose, glucose, galacturonic acid, xylose, ribose	Bafana 2013
Chlorella pyrenoidosa	Rhamnose, galactose, arabinose, glucose, mannose, fucose, xylose	Maksimova et al. 2004
Dunaliella sp.	Galactose, rhamnose, xylose, mannose, and galacturonic acid with some traces of other monosaccharides glucose, arabinose, and glucuronic acid.	Mishra and Jha 2009
Heterosigma akashiwo	Rhamnose, galactose, arabinose, glucose, fucose, xylose, uronic acids	Khandeparker and Bhosle 2001

A lot of studies have explored the characteristics the effects of weather variation, intensity of light, concentrations of sulfur, phosphate, potassium, and other metal ions on morphological traits of EPS (Otero and Vincenzini 2003; Ge et al. 2014).

14.4 APPLICATION OF EXOPOLYSACCHARIDES

Microalgae possess high-value compounds such as carotenoids. Their bioactive molecules are useful for human and animal health. For example, the green unicellular flagellate *Dunaliella salina* contains high levels of natural β-carotene (Selim et al. 2018). Exopolysaccharides have anti-tumor, anti-cancer, bioflocculants, emulsifying, antioxidant, antiviral, symbiosis, and water retention properties (Selim et al. 2018) (Table 14.2 and Figure 14.4). Moreover, EPS of algae

TABLE 14.2
Biological Activities of Some EPS from Marine and Fresh Water Organisms

Exo polysaccharides Species	Phylum	Application	References
Chlorella zofingiensis	Chlorophyta	Anti-colorectal cancer, free radical scavenging	(Liu et al. 2014; Azaman 2016; Zhang, Liu et al. 2019)
Anabaena	Blue-green algae	Bioflocculants	(Prasanna et al. 2006; Tiwari et al. 2019)
Anabaena sp.	Blue-green algae	Pseudoplasticity commercial applications	(Kumar et al. 2018)
Chlorella vulgaris	Chlorophyta	Free radical scavenging, anti-colorectal cancer	(Zhang, Liu et al. 2019)
Nostoc PCC 7413	Blue-green algae	Wound healing	(Alvarez et al. 2020)
Porphyridium cruentum	Rhodophyta	Against (VSV), (ASFV), Vaccinia virus	(Abraham et al. 2014; Asgharpour et al. 2015)
Haslea ostrearia	Bacillariophyta	Suppression of tumor cell	(Bhatia 2016)
Anabaena Spiroides	Blue-green algae	Elimination of Mn(II),	(Freire-Nordi et al. 2005)
Chlorococcum	Microalgae	Cu(II), Pb(II) and Hg(II)	(Zhang, Liu et al. 2019)
Scenedesmus	Microalgae	Antitumor	(Zhang, Liu et al. 2019)
Chlorella pyrenoidosa	Microalgae	Free radicals scavenging	(Zhang, Liu et al. 2019)
FACHB-9	Microalgae	Free radicals scavenging	(Yim et al. 2004)
Gelidium cartiagenium		Antiviral activity	

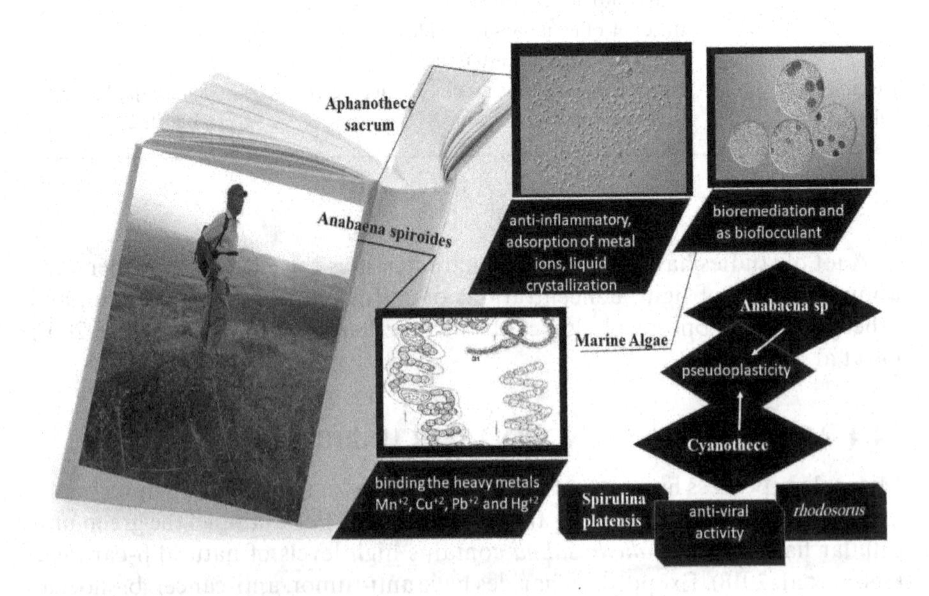

FIGURE 14.4 Application and utilizations of various EPSs of microalgae and algae strains.

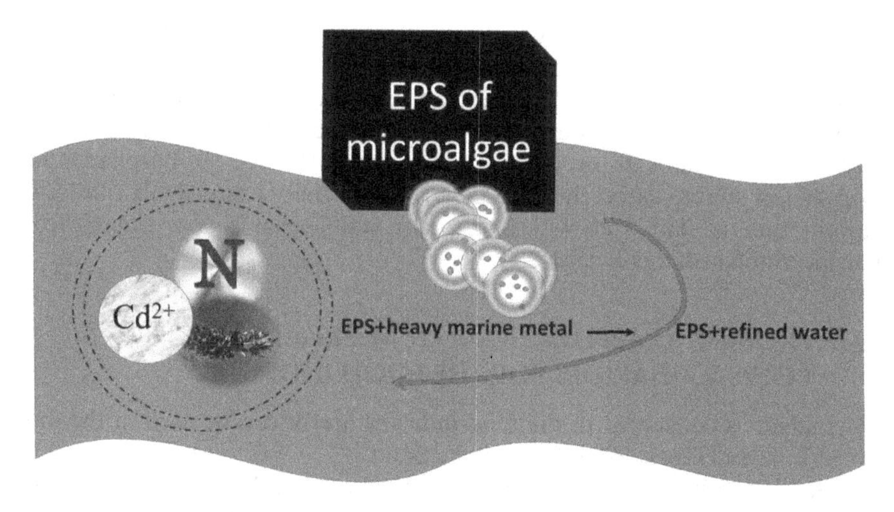

FIGURE 14.5 The role of EPSs of microalgae species in removing heavy cations

and microalgae can refine heavy toxic metal cations from seawater (Figure 14.5) (Kumar et al. 2015).

Figure 14.5 Microalgae is biosorbent of marine metal cations.

Metal cations are produced by industrial activities, pollutants from cars and shipping. They negatively influence human health leading to numerous pathologies such as cancers (Kumar et al. 2015).

EPS screened from numerous microalgae are in high demand for their novel anticancer properties (Patel et al. 2013; Raposo et al. 2013). Exopolysaccharides from microalgae *Chlorella pyrenoidosa* FACHB-9, *Scenedesmus* sp., and *Chlorococcum* sp. demonstrate antioxidant and antitumor activities, and they suppress viable colon cancer HCT116 and HCT8 cell lines (Zhang, Liu et al. 2019). *Tetraselmis suecica* has antioxidant activity and cytotoxicity activity on tumor cells (Parra-Riofrío et al. 2020).

They have defense and resistance reactions against stressed radiation such as ultraviolet (UV) (Barsanti et al. 2008; Kumar et al. 2018).

The EPS of both microalgae *Chlorella zofingiensis* and *Chlorella vulgaris* are active against free radicals and human colon cancer HCT8 cell lines (Zhang, Liu et al. 2019).

14.5 ANTICANCER EFFECTS OF EPS

Cancer occurs when abnormal cells develop and invade healthy cells in the body in an uncontrolled way. Compounds extracted from microalgae and algae demonstrated bioactive compounds (El-Hack et al. 2019).

Recently, significant improvements in treatment of colorectal cancer with natural compounds have occurred. Algal EPS appears very promising for treatment of colorectal cancer (Park et al. 2017). The mechanism of microalgae EPS anticancer activity is due their notable structural variety. For example, their molecular chemodiversity has an influence on cellular proliferation, differentiation, apoptosis, and metastasis (Dewi et al. 2018).

Microalgae categorized into the Chromista kingdom and Bacillariophyceae class were reported to produce EPS with anticancer efficacy (Gaignard et al. 2019). Microalgae can produce GA3P (glyceraldehyde 3-phosphate) and D-galactan sulfate extracellular polysaccharide that is capable of inhibiting various tumors. This EPS with anti-tumor activity has been extracted from marine dinoflagellate (Tsianta 2020).

14.6 PROMISING APPLICATIONS OF MICROALGAL POLYSACCHARIDES FOR THE FOOD INDUSTRY

Microalgae is consumed in the food industry and it can be used in the food cycle (Pulz and Gross 2004; Borowitzka 2018).

The bioactive compounds of microalgae such as EPS, vitamins, ω-3 fatty acids, peptides, proteins, pigments, and antioxidants make them suitable for potential food supplements (Pulz and Gross 2004; Borowitzka 2018).

Arthrospira platensis (*Spirulina*) contains polysaccharides and food utilization as dietary supplement and pharmaceutical food (Cicci and Bravi 2016; Sed et al. 2017).

Accumulation of EPS upon surfaces produce resulting in the formation of a biofilm that facilitate attachment and matrix formation (Pignolet et al. 2013; Pfannenstiel and Keller 2019).

14.7 ENERGY RECOVERY DERIVED FROM MARINE MICROALGAE EPS

The colonial clusters of *Botryococcus braunii* were observed in temperate, tropical oligotrophic freshwater, reservoirs, brackish ponds, and lakes around the world in all climate zones except Antarctica (Lupi et al. 1994).

The yield of conversion of light energy into total carbohydrate and EPS production was calculated by the following formula (García-Cubero et al. 2018):

$$Yx, ph(t) = \frac{Px(t)}{Iabs(t)}$$

Physicochemical characteristic of *Botryococcus braunii* oil biodiesel such as kinematic viscosity, cetane number, specific gravity, pour point, oxidation stability, and lubricity are mostly similar to diesel fuel (Soares et al. 2008). The cetane numeral is an index of quality of inflammation of a fuel. Its number is between standard diesel indicator (Soares et al. 2008).

14.8 THE CAPSULAR EPS ROLE IN THE DEFENSE SYSTEM

EPS of microalgae increase a host's immune system by increasing natural killer cell activity. *Gyrodinium impudicum KG03* sulfated EPS cause the defense role of cellular peritoneal macrophages in in vitro murine models (Bae et al. 2006; Kumar et al. 2018).

In stress situations which could be biotic or abiotic, these microorganisms secrete large amounts of EPSs, which protect them from difficult droughts and nutrient-stress conditions (Angelis et al. 2012). For example, trichomes of the *Nostoc* strain are placed in a thick layer of EPS, including a protein part in its structure and which protects the cells from the external environment (Borowitzka 2018). Moreover *Nostoc carneum* EPS presents antioxidant activity. Its mechanism is via the absorbance of O-H and N-H stretching, and C-H stretching (Hussein et al. 2015); hence antioxidants have a multi-function in human protection from different oxidative damages (Lin and Beal 2003).

14.9 ANTIMICROBIAL AND ANTIVIRAL ACTIVITIES

Research has shown that microalgal EPS helps in the inhibition of virus, pathogenic bacteria, and protozoans (Reichelt and Borowitzka 1984; Garcia et al. 2014; López et al. 2015). EPS of *Purpureum, cruentum, rhodosorus,* and *aerugineum* species have shown antiviral properties (Xiao and Zheng 2016).

The reports indicate that released polysaccharides of *Nostoc insulare* serves as antimicrobial agents (Potts 1994).

Some green microalgae, such as the *Chlorella autotrophica*, *Tetraselmis tetrathele*, *Tetraselmis suecica*, *Chlorella stigmatophora*, *Dunaliella tertiolecta,* and *Dunaliella salina,* are capable of inhibiting viral hemorrhagic septicemia virus (VHSV) and African swine fever virus (ASFV) (Fabregas et al. 1999). In addition, several microalgae from different taxonomic groups such as the diatom *Navicula directa* and *Naviculan* with antiviral activity characteristic are able to synthetize sulfated exopolysaccharides (Lee et al. 2006).

EPS of *Gyrodinium impudicum* have the ability to suppress encephalomyocarditis and mumps virus. Cytopathogenic effect of p-KG03 strain reported is due to its EPS. The factor of concentrations determine the killing of viruses (Yim et al. 2004).

Raposo et al. (2014) studied the antibacterial bioflocculants of *Dunaliella salina* EPS.

Sulfated EPS of the red microalgae *Porphyridium aerugineum* contain phycobiliproteins and phycocyanin and are active against *Escherichia coli* and *Staphylococcus aureus* (Najdenski et al. 2013; Raposo and Morais 2014).

14.10 CONCLUSION

The organisms presented in marine ecosystems have valuable properties due to their composition. Microalgal EPS are natural agents that are promising compounds. Purified exopolysaccharides from various microalgae species have different functions such as antioxidant, antitumor activities, anti-cancer, bioflocculants, antiviral, and antibacterial activity. The studies have shown EPSs obtained from microalgae *Chlorella zofingiensis* and *Chlorella vulgaris* suppress human colon cancer cell lines HCT8 in vitro.

REFERENCES

Abraham, R. E., et al. (2014). "Suitability of magnetic nanoparticle immobilised cellulases in enhancing enzymatic saccharification of pretreated hemp biomass." *Biotechnology for Biofuels* **7**(1): 90.

Allard, B. and E. Casadevall (1990). "Carbohydrate composition and characterization of sugars from the green microalga Botryococcus braunii." *Phytochemistry* **29**(6): 1875–1878.

Alvarez, X., et al. (2020). "Biochemical characterization of Nostoc sp. Exopolysaccharides and evaluation of potential use in wound healing." *Carbohydrate Polymers*: 117303.

Angelis, S. D., et al. (2012). "Co-culture of microalgae, cyanobacteria, and macromycetes for exopolysaccharides production: Process preliminary optimization and partial characterization." *Applied Biochemistry and Biotechnology* **167**(5): 1092–1106.

Asgharpour, M., et al. (2015). "Eicosapentaenoic acid from Porphyridium cruentum: Increasing growth and productivity of microalgae for pharmaceutical products." *Energies* **8**(9): 10487–10503.

Azaman, S. N. A. (2016). *Morphological, biochemical and transcriptomic characterisation of Chlorella sorokiniana and Chlorella zofingiensis during normal and stress conditions*. University of Sheffield.

Bae, S.-Y., et al. (2006). "Activation of murine peritoneal macrophages by sulfated exopolysaccharide from marine microalga Gyrodinium impudicum (strain KG03): Involvement of the NF-κB and JNK pathway." *International Immunopharmacology* **6**(3): 473–484.

Bafana, A. (2013). "Characterization and optimization of production of exopolysaccharide from Chlamydomonas reinhardtii." *Carbohydrate Polymers* **95**(2): 746–752.

Barsanti, L., et al. (2008). *Oddities and curiosities in the algal world. Algal toxins: Nature, occurrence, effect and detection*. Springer: 353–391.

Bazaka, K., et al. (2011). "Bacterial extracellular polysaccharides." In *Bacterial adhesion*. Springer: 213–226.

Bhatia, S. (2016). *Natural polymer drug delivery systems: Nanoparticles, plants, and algae*. Natural Polymer Drug Delivery Systems: Nanoparticles, Plants, and Algae.

Borowitzka, M. A. (2018a). "Biology of microalgae." In *Microalgae in health and disease prevention*. Elsevier: 23–72.

Borowitzka, M. A. (2018b). "Microalgae in medicine and human health: A historical perspective." In *Microalgae in health and disease prevention*. Elsevier: 195–210.

Chisti, Y. (2007). "Biodiesel from microalgae." *Biotechnology Advances* **25**(3): 294–306.

Cicci, A. and M. Bravi (2016). "Fatty acid composition and technological quality of the lipids produced by the microalga Scenedesmus dimorphus 1237 as a function of culturing conditions." *Chemical Engineering Transactions* **49**: 181–186.

De Philippis, R. and M. Vincenzini (2003). "Outermost polysaccharidic investments of cyanobacteria: Nature, significance and possible applications." *Recent Research Developments in Microbiology* **7**: 13–22.

Delattre, C., et al. (2016). "Production, extraction and characterization of microalgal and cyanobacterial exopolysaccharides." *Biotechnology Advances* **34**(7): 1159–1179.

Dewi, I. C., et al. (2018). "Anticancer, antiviral, antibacterial, and antifungal properties in microalgae." In *Microalgae in health and disease prevention*. Elsevier: 235–261.

El-Hack, M. E. A., et al. (2019). "Microalgae in modern cancer therapy: Current knowledge." *Biomedicine & Pharmacotherapy* **111**: 42–50.

Elboutachfaiti, R., et al. (2011). "Polyglucuronic acids: Structures, functions and degrading enzymes." *Carbohydrate Polymers* **84**(1): 1–13.

Evans, L. V. (2003). *Biofilms: Recent advances in their study and control*. CRC Press.

Fabregas, J., et al. (1999). "In vitro inhibition of the replication of haemorrhagic septicaemia virus (VHSV) and African swine fever virus (ASFV) by extracts from marine microalgae." *Antiviral Research* **44**(1): 67–73.

Finkel, Z. V., et al. (2010). "Phytoplankton in a changing world: Cell size and elemental stoichiometry." *Journal of Plankton Research* **32**(1): 119–137.

Freire-Nordi, C. S., et al. (2005). "The metal binding capacity of Anabaena spiroides extracellular polysaccharide: An EPR study." *Process Biochemistry* **40**(6): 2215–2224.

Gaignard, C., et al. (2019). "Screening of marine microalgae: Investigation of new exopolysaccharide producers." *Algal Research* **44**: 101711.

Garcia, C. P., et al. (2014). "Detection of silver nanoparticles inside marine diatom Thalassiosira pseudonana by electron microscopy and focused ion beam." *PLoS One* **9**(5): e96078.

García-Cubero, R., et al. (2018). "Production of exopolysaccharide by Botryococcus braunii CCALA 778 under laboratory simulated Mediterranean climate conditions." *Algal Research* **29**: 330–336.

Ge, H., et al. (2014). "Effects of light intensity on components and topographical structures of extracellular polysaccharides from the cyanobacteria Nostoc sp." *Journal of Microbiology* **52**(2): 179–183.

Hussein, M. H., et al. (2015). "Characterization and antioxidant activity of exopolysaccharide secreted by Nostoc carneum." *International Journal of Pharmaceutics* **11**(5): 432–439.

Khandeparker, R. D. and N. B. Bhosle (2001). "Extracellular polymeric substances of the marine fouling diatom Amphora rostrata Wm. Sm." *Biofouling* **17**(2): 117–127.

Khattar, J., et al. (2010). "Isolation and characterization of exopolysaccharides produced by the cyanobacterium Limnothrix redekei PUPCCC 116." *Applied Biochemistry and Biotechnology* **162**(5): 1327–1338.

Kumar, D., et al. (2018). "Exopolysaccharides from cyanobacteria and microalgae and their commercial application." *Current Science* **115**(2): 234–241.

Kumar, K. S., et al. (2015). "Microalgae—A promising tool for heavy metal remediation." *Ecotoxicology and Environmental Safety* **113**: 329–352.

Lee, J.-B., et al. (2006). "Antiviral sulfated polysaccharide from Navicula directa, a diatom collected from deep-sea water in Toyama Bay." *Biological and Pharmaceutical Bulletin* **29**(10): 2135–2139.

Li, P., et al. (2001). "Cyanobacterial exopolysaccharides: Their nature and potential biotechnological applications." *Biotechnology and Genetic Engineering Reviews* **18**(1): 375–404.

Lin, M. T. and M. F. Beal (2003). "The oxidative damage theory of aging." *Clinical Neuroscience Research* **2**(5–6): 305–315.

Liu, J., et al. (2014). "Chlorella zofingiensis as an alternative microalgal producer of astaxanthin: Biology and industrial potential." *Marine Drugs* **12**(6): 3487–3515.

Liu, L., et al. (2016). "Extracellular metabolites from industrial microalgae and their biotechnological potential." *Marine Drugs* **14**(10): 191.

López, A., et al. (2015). "Phenolic profile of Dunaliella tertiolecta growing under high levels of copper and iron." *Environmental Science and Pollution Research* **22**(19): 14820–14828.

Lupi, F., et al. (1994). "Influence of nitrogen source and photoperiod on exopolysaccharide synthesis by the microalga Botryococcus braunii UC 58." *Enzyme and Microbial Technology* **16**(7): 546–550.

Maksimova, I., et al. (2004). "Extracellular carbohydrates and polysaccharides of the alga Chlorella pyrenoidosa Chick S-39." *Biology Bulletin of the Russian Academy of Sciences* **31**(2): 175–181.

Mishra, A. and B. Jha (2009). "Isolation and characterization of extracellular polymeric substances from micro-algae Dunaliella salina under salt stress." *Bioresource Technology* **100**(13): 3382–3386.

Mota, R., et al. (2013). "Production and characterization of extracellular carbohydrate polymer from Cyanothece sp. CCY 0110." *Carbohydrate Polymers* **92**(2): 1408–1415.

Najdenski, H. M., et al. (2013). "Antibacterial and antifungal activities of selected microalgae and cyanobacteria." *International Journal of Food Science & Technology* **48**(7): 1533–1540.

Nicolaus, B., et al. (2010). "Exopolysaccharides from extremophiles: From fundamentals to biotechnology." *Environmental Technology* **31**(10): 1145–1158.

Otero, A. and M. Vincenzini (2003). "Extracellular polysaccharide synthesis by Nostoc strains as affected by N source and light intensity." *Journal of Biotechnology* **102**(2): 143–152.

Park, G.-T., et al. (2017). "Potential anti-proliferative and immunomodulatory effects of marine microalgal exopolysaccharide on various human cancer cells and lymphocytes in vitro." *Marine Biotechnology* **19**(2): 136–146.

Parra-Riofrío, G., et al. (2020). "Antioxidant and cytotoxic effects on tumor cells of exopolysaccharides from Tetraselmis suecica (Kylin) butcher grown under autotrophic and heterotrophic conditions." *Marine Drugs* **18**(11): 534.

Patel, A. K., et al. (2013). "Separation and fractionation of exopolysaccharides from Porphyridium cruentum." *Bioresource Technology* **145**: 345–350.

Paulsen, B. S., et al. (1998). "Extracellular polysaccharides from Ankistrodesmus densus (Chlorophyceae)." *Journal of Phycology* **34**(4): 638–641.

Perez-Garcia, O., et al. (2011). "Heterotrophic cultures of microalgae: Metabolism and potential products." *Water Research* **45**(1): 11–36.

Pfannenstiel, B. T. and N. P. Keller (2019). "On top of biosynthetic gene clusters: How epigenetic machinery influences secondary metabolism in fungi." *Biotechnology Advances*.

Phlips, E. J., et al. (1989). "Growth, photosynthesis, nitrogen fixation and carbohydrate production by a unicellular cyanobacterium, Synechococcus sp. (Cyanophyta)." *Journal of Applied Phycology* **1**(2): 137–145.

Pierre, G., et al. (2019). "What is in store for EPS microalgae in the next decade?" *Molecules* **24**(23): 4296.

Pignolet, O., et al. (2013). "Highly valuable microalgae: Biochemical and topological aspects." *Journal of Industrial Microbiology & Biotechnology* **40**(8): 781–796.

Potts, M. (1994). "Desiccation tolerance of prokaryotes." *Microbiology and Molecular Biology Reviews* **58**(4): 755–805.

Prasanna, R., et al. (2006). "Morphological, physiochemical and molecular characterization of Anabaena strains." *Microbiological Research* **161**(3): 187–202.

Pulz, O. and W. Gross (2004). "Valuable products from biotechnology of microalgae." *Applied Microbiology and Biotechnology* **65**(6): 635–648.

Raposo, M. and A. M. Morais (2014). "Bioactivity and applications of polysaccharides from marine microalgae."

Raposo, M. F. D. J., et al. (2013). "Bioactivity and applications of sulphated polysaccharides from marine microalgae." *Marine Drugs* **11**(1): 233–252.

Reichelt, J. L. and M. A. Borowitzka (1984). *Antimicrobial activity from marine algae: Results of a large-scale screening programme*. Eleventh International Seaweed Symposium. Springer.

Rossi, F. and R. De Philippis (2015). "Role of cyanobacterial exopolysaccharides in phototrophic biofilms and in complex microbial mats." *Life* **5**(2): 1218–1238.

Sed, G., et al. (2017). "Extraction and purification of exopolysaccharides from exhausted Arthrospira platensis (Spirulina) culture systems." *Chemical Engineering Transactions* **57**: 211–216.

Selim, M. S., et al. (2018). "Production and characterisation of exopolysaccharide from Streptomyces carpaticus isolated from marine sediments in Egypt and its effect on breast and colon cell lines." *Journal of Genetic Engineering and Biotechnology* **16**(1): 23–28.

Shepherd, R., et al. (1995). "Novel bioemulsifiers from microorganisms for use in foods." *Journal of Biotechnology* **40**(3): 207–217.

Soares, I. P., et al. (2008). "Multivariate calibration by variable selection for blends of raw soybean oil/biodiesel from different sources using Fourier transform infrared spectroscopy (FTIR) spectra data." *Energy & Fuels* **22**(3): 2079–2083.

Staudt, C., et al. (2004). "Volumetric measurements of bacterial cells and extracellular polymeric substance glycoconjugates in biofilms." *Biotechnology and Bioengineering* **88**(5): 585–592.

Tiwari, O. N., et al. (2015). "Characterization and optimization of bioflocculant exopolysaccharide production by cyanobacteria Nostoc sp. BTA97 and Anabaena sp. BTA990 in culture conditions." *Applied Biochemistry and Biotechnology* **176**(7): 1950–1963.

Tiwari, O. N., et al. (2019). "Purification, characterization and biotechnological potential of new exopolysaccharide polymers produced by cyanobacterium Anabaena sp. CCC 745." *Polymer* **178**: 121695.

Tsianta, A. (2020). "Pharmaceutical Applications of Eukaryotic Microalgae."

Xiao, R. and Y. Zheng (2016). "Overview of microalgal extracellular polymeric substances (EPS) and their applications." *Biotechnology Advances* **34**(7): 1225–1244.

Yim, J. H., et al. (2004). "Antiviral effects of sulfated exopolysaccharide from the marine microalga Gyrodinium impudicum strain KG03." *Marine Biotechnology* **6**(1): 17–25.

Zhang, J., et al. (2019a). "Characterization of exopolysaccharides produced by microalgae with antitumor activity on human colon cancer cells." *International Journal of Biological Macromolecules* **128**: 761–767.

Zhang, J., et al. (2019b). "Production and characterization of exopolysaccharides from Chlorella zofingiensis and Chlorella vulgaris with anti-colorectal cancer activity." *International Journal of Biological Macromolecules* **134**: 976–983.

Zhu, L. (2015). "Microalgal culture strategies for biofuel production: A review." *Biofuels, Bioproducts and Biorefining* **9**(6): 801–814.

Zhu, S., et al. (2015). "Characterization of lipid and fatty acids composition of Chlorella zofingiensis in response to nitrogen starvation." *Journal of Bioscience and Bioengineering* **120**(2): 205–209.

15 Marine Algal Secondary Metabolites Are a Potential Pharmaceutical Resource for Human Society Developments

Somasundaram Ambiga, Raja Suja Pandian, Lazarus Vijune Lawrence, Arjun Pandian, Ramu Arun Kumar, and Bakrudeen Ali Ahmed Abdul

CONTENTS

DOI: 10.1201/9781003303916-15

15.1 INTRODUCTION

The marine environment's biodiversity was a unique supply of chemical compounds that have potential to develop in industry as drugs, beauty products, food supplements, genetic probes, enzymes, active pharmaceutical ingredients and agricultural products. The oceans are likely the most important natural resource on the planet, providing sustenance primarily in the form of fish and shellfish. The sea world is a great environmental resource for so many functionally active chemical compounds due to its incredible biodiversity. Marine species exist in complex ecosystems and are subjected to cruel conditions, resulting in the production of a large range of specific and powerful active ingredients which cannot be seen anywhere else. Marine microorganisms have become a valuable source of novel therapeutic drugs due to their enormous biochemical and genetic diversity. New physiologically active compounds are abundant in marine species. Several bioactive chemicals from marine invertebrates are really generated by various microbial symbionts, which is surprising. Competition between microorganisms for marine habitat space and food is a key driver of the development of such valuable antibiotics and various helpful medicinal products. Surprisingly, microorganisms linked with marine invertebrates have been identified as promising drug development possibilities.

Most of the bioactive substances, such as proteins, peptides, and amino acids, are proteinaceous. In addition, several marine organisms, having excellent source in protein, are great initial materials for producing bioactive peptides derived from protein. These chemicals have a very different chemical structure from those found in terrestrial and microbial systems. Natural and non-natural amino acids can be found in cyclic and linear peptides, as well as depsipetides i.e., peptides in which one or more amide linkages are replaced by ester bonds. These compounds can be produced by the organism or taken from marine microbes with which it has a symbiotic relationship. Proteins that catalyze biological reactions are known as enzymes. As a result, enzymes are the essential catalytic keystone of metabolism that play a critical part in not just general health and also numerous manufacturing activities, following a millennia-old pattern. As a result, enzymes are the subject of extensive, multidisciplinary research around the world, involving not just biologists, and also process chemical engineers, designers and researchers from other domains.

The most of phyla as well as more than 90% of all living classification of organisms may be identified in the marine ecosystem due to its tremendous biodiversity. A marine enzyme could be a one-of-a-kind protein component that has never been detected in a terrestrial species, or it could be a well-known enzyme with novel features from a terrestrial origin. Furthermore, because the seas cover well over 70% of the surface of the planet, they contain a vast amount of natural products and novel biologically active molecules. Marine sources provide the most peptides and natural small molecules because marine sources account for half of all world biodiversity. Antimicrobial peptides are produced by the majority of marine species.

Bacteria and fungus are abundant in the marine environment. The number of bioactive compounds obtained from marine bacteria and fungi has massively grown in recent decades. To live in the harsh sea environment, marine bacteria develop a diverse range of secondary metabolites For examples, high stress concentration, water stress, high light intensities and cooler temperatures, all result in the emergence of a diverse range of intriguing and structurally complicated molecules. Thermophilic and archaea bacteria, for example, produce thermostable enzymes of various types. The biomedical compounds which are extracted from oceans and seas has become a large field of research in modern technologies (Jahromi and Barzkar, 2018).

Many highly constituents of bioactive compounds have been isolated from marine sources such as polysaccharides, alkaloids, peptides and polyphenols. These components have high potential in bioactive applications, and also they are predominant compounds for anticancer. Apart from bacteria, fungus, and actinomycetes, numerous other marine species including crabs, fish, snakes, prawns, algae and plants have been examined in order to tap into the aquatic world's arsenal. Scientists are particularly interested in properties such as cold adaptivity, strong salt tolerance, barophilicity, hyperthermostability and ease of large-scale growth.

15.2 BIOACTIVE COMPOUNDS FROM MARINE

15.3 ORGANISMS

15.3.1 POLYSACCHARIDES

Polysaccharides are described as a storehouse of energy and structural compounds of all living things, including marine and higher plants. In algae, polysaccharides are the most important macromolecules accounting for more than 80% of its weight. These resistant polysaccharides are known as dietary fibers, which are not digested in the body, but by the action of enzymes in the gut micro-organisms, can ferment to varying degrees (Nunraksa *et al.*, 2019). Edible seaweed contains a significantly high amount of fiber in which the polysaccharide level is higher than that of wheat bran (Garcia-Vaquero *et al.*, 2017). Polysaccharides in algae are different from polysaccharides in terrestrial plants; they include unusual polyuronides, some of which are methylated, acetylated, pyruvylated or sulfated. Among the various polysaccharides, the sulfated forms are the one most studied for their biological values (Ganesan *et al.*, 2018).

Sulfated polysaccharides (SPS) have a wider application in the pharmaceutical, nutritional and cosmetic fields and exhibit antioxidant, anticancer, anti-inflammatory, antidiabetic, anticoagulant, immunomodulatory and anti-HIV activities (Nagarajan and Mathaiyan, 2015). These activities are mainly attributed to interactions between polysaccharides and the gut microbiota to demonstrate the functional and medicinal properties of sulfated polysaccharides (Seedevi *et al.*, 2017). For example, modulation of cytokines, inhibition of apoptosis, and inhibition of protein tyrosine phosphatase are common examples of SPS and the gut microbiota.

SPS is a more potent nitric oxide collector (Viera *et al.*, 2018), and the mechanism of action of this property stems from the structural characteristics of the sulfate compound and the type of sugar attached, molecular weight and glycosidic bonds. In addition, this antioxidant property is important for preventing the formation of free radicals in cells, thereby inhibiting oxidative damage of cell membranes.

Polysaccharides compounds are also the potent group in it. The innate immune system is activated by the action of polysaccharides cytotoxic effect of mechanism. A compound sulfated L-fucose (a sulfated polysaccharides) also named Fucoidan isolated from *Ascophyllum nodosum* has been used for the treatment of colon cancer. Heparin, a polysaccharide isolated from *Dictyoperis delicatula* (a seaweed source), also helps in the treatment of colon cancer. other polysaccharides compounds such as chondroitin-4-sulfate extracted from a sea cucumber *Cucumaria frondosa* and chondroitin-6-sulfate extracted from a sea cucumber *Cucumaria frondosa* have the potential anticancer effect (Sithranga Boopathy and Kathiresan, 2010).

15.4 ALKALOIDS

The derived components from marine origin of alkaloids has been divided into indoles, halogenated indoles, phenylethylamines and also other alkaloids as four groups. A compound Brugine isolated from *Bruguieria sexangula* (a plant source) has been used in the treatment of sarcoma. In addition to that marine guanidine alkaloids, a group of bioactive compounds are found in some marine sponges. The marine sponges belonging to the genera *Clathria, Batzella, Ptilocaulis* and some other genera of marine sponges have the compounds which was put forward be chemotaxonomic markers. Especially. From the family *Crambeidae*, the marine sponges of the *Monanchora* genus, guanidine alkaloids were extracted. These extracted compounds were designated by distinctive chemical structures and also they are biologically active (Gross *et al.*, 2006).

15.5 PEPTIDES

In marine flora, various and different types of peptides have been isolated. It is reported that over 2500 new peptides have been identified. They are possessed with anti-proliferative activity. For several human cell lines such as breast, lung, bladder and pancreatic cell lines, the peptides that have purified is exhibited cytotoxic effects against abovementioned human cell lines. A compound, Sansalvamide A is isolated from the marine fungi which inhibits the protein complex formation. This compound binds to the HSP90, an N-middle domain.

As it is the mandatory for the tumour growth promotion, this compound allosteically inhibits the protein complex formation. Sansalvamide A is being used as anti-cancer therapeutic lead. This activites are exhibited against different carcinomas such as breast, pancreatic and melanoma. Other compound, Apratoxin has also been isolated from the marine bacteria *Lyngbya boulloni* has been used for the treatment of cervical cancer. Lyngbyabellin B is the compound also isolated from marine algae *Lyngbya majuscule* has been used for the treatment of Burkitt lymphoma cancer. Short while ago, two new cyclodepsipetides compound have been isolated from the marine fungus *Scopulariopsis brevicaulis* namely, Scopularide A and Scopularide B. They both have been used in the treatment of colon and pancreatic cancer cell lines. They inhibit the colon and pancreatic cancer cell line growth (Marquez *et al.*, 2002).

15.6 PROTEIN AND AMINO ACIDS

Proteins represent 5% to 47% of the dry weight of algae. Red algae have the highest protein content, while green has less and brown has the least (Černá, 2011). Of the total amino acids (aa) in algae, between 42% and 48% are essential aa (Wong and Cheung, 2000). However, the high polyphenol content of algae may reduce the protein digestibility of algae, giving a slightly lower score on the aa scale for correcting protein digestibility (Wong and Cheung, 2001). Nevertheless, algae is still a viable substitute for animal protein if the diet includes other popular vegan foods, such as soy or fungal protein.

15.7 LIPIDS

The total lipid content of algae varied from 0.60% to 4.14% (El Maghraby and Fakhry, 2015; Rodrigues *et al.,* 2015). Most algal lipids are polyunsaturated, including fatty acids (n3, or omega 3) such as docosahexaenoic and eicosapentaenoic acid. The most common monounsaturated fatty acids (n6 or omega 6) from algae are linoleic and arachidonic acids (Belattmania *et al.,* 2018). The main saturated fatty acids include palmitic and myristic acids. From a dietary point of view, n6 and n3 fatty acids are essential; however, consuming them in a disproportionate proportion can lead to chronic inflammatory diseases such as obesity, rheumatoid arthritis, nonalcoholic fatty liver disease and cardiovascular disease (Patterson *et al.,* 2012). A ratio between 2.5:1 and 4:1 (n6:n3) is often recommended to prevent chronic diseases associated with excessive consumption of n6 monounsaturated fats (Simopoulos, 2016). The n6:n3 ratio of fatty acids in algae is at this low (Biancarosa *et al.,* 2018), making them an excellent source of dietary fat (Dawczynski *et al.,* 2007).

15.8 VITAMINS

Vitamins are the essential characteristics of the essential characteristics that cause many metabolic itineraries and serve as a precursor of an adsorbent enzyme. The algae are of an edible nature and constitute the type of primordial carotenoids that have a high biological value. Vitamin A (carotenoid) found in seaweed is relatively higher than carrots. Beta carotene and lycopene derivatives are carotene derivatives, canthaxanthin, zeaxanthin, vi-axanthin, siphonaxanthine and astaxanthin are derivatives of xanthopopia (Othman *et al.*, 2018). Lutein and zeaxanthin derived from algae have been reported to protect against macular degeneration. These food colorants are extremely valuable nutritional components exhibiting metabolic activity unique to human health (Murata and Nakazoe, 2001). They have potent antioxidant activities and are closely associated with anticancer, antihypercholesterolemic and neurodegenerative diseases (Teas *et al.*, 2013). The effective inducer of apoptosis in HL60 cells was green algae siphonaxanthin when compared to fucoxanthin. Essential minerals such as calcium, iron, iodine, magnesium, phosphorus, potassium, zinc, copper, manganese, selenium and fluoride are present in algae (Misurcova, 2011; Qin, 2018).

15.9 PHYTOCHEMICALS

Algae have a wide range of secondary metabolites and are continually attracting the scientific community's interest in their potential bioactivity, compared with terrestrial plants. Phlorotannin, gallic acid, quercetin, phloroglucinol, carotenoids and its derivatives are the most studied phytochemicals in algae. Polyphenol compounds are commonly reported in all genera of algae, but their presence is most likely in brown and red algae (Abirami and Kowsalya, 2017). Phlorotannins which are assembled by polymerization of phloroglucinol units (1, 3, 5 trihydroxybenzene monomer units) to form polyphloroglucinols are the most important phenolic compounds derived from marine plants (Rajauria *et al.*, 2016). Brown algae have a high amount of phloroglucinol and are used as a beneficial nutrient for their health. These phenolic compounds are reported to exhibit potential biological activities such as antioxidant, anti-proliferative, anti-inflammatory, anti-diabetic, anti-HIV, anti-Alzheimer activity in vitro (Rajauria et al., 2013). Phlorotannin is believed to be effective against scavenging free radicals and metal chelators, with this characteristic inhibiting lipid peroxidation. The antioxidant property of phlorotannin was more effective than that of tocopherol in an in vitro linoleic model system, which showed high activity against the assay of DPPH, superoxide and peroxy radicals (Li *et al.*, 2009).

15.10 POLYPHENOLS

Polyphenols are classified into many types. They are phenolic acids, flavonoids, epigallate, epicatechin, anthocyanidins, gallic acids, tannins and catechin. Polyphenols have much potential to reduce the mitotic index. Also, the levels of cellular proteins were decrease which needed for the cancer cell proliferation and colony formation. For instance, due to the cytotoxic activities of compound, Scutellarein 4′-methylether exhibits effect of anticancer in vivo and in vitro. This compound is isolated from the marine algae *Osmundea pinnatifida*, and it is used in the treatment of choriocarcinoma cancer (Spada *et al.*, 2008).

Furthermore, a polyphenolic compound Phlorofucofuroecol A is isolated from the marine brown seaweeds and used in the treatment of cancer. Also, another polyphenolic compound Phlorofucofuroecol is isolated from the marine brown seaweeds and used in the treatment of colon cancer. Adding to it, the phenols have been inhibiting the human platelet aggregation and they also exhibited antiviral and anti-inflammatory activity. For example, an edible seaweed, *Palmaria palmata*, is rich in polyphenols have the high potential antioxidant and anticancer properties *(Shan et al., 2006)*.

15.11 ENZYMES FROM MARINE SPECIES

15.11.1 Protease

Oceans are the seed bank for hundreds of thousands of unique micro-organisms living in a rare "biosphere," and they become active anytime environmental and seasonal changes occur. Moreover, they are an unstoppable genetic reservoir for a range of enzymes and activities that can be used in pharmaceutical as well as other sectors. Antibiotics are chemotherapeutic chemicals produced by microorganisms with the ability to inhibit or kill other organisms in high concentrations. Bacteria in the marine environment produce a variety of bioactive metabolites. Proteases are biocatalysts found in all living things. Proteases, also called as proteinases or proteolytic enzymes, are widespread in nature and have degradative capabilities that catalyze the entire digestion of proteins via cleaving the polypeptide chains. They are required for cell growth and differentiation.

Protease sales account for almost 60% of all global industrial enzyme revenues. Proteases are frequently employed in today's society. Proteolytic enzymes protect the body from cancer, making them effective anticancer medicines. When a low quantity of protease enzyme is administered to a tumor, it has a specific anticancer effect. This effect was greater than mitomycin C, which was given at a lethal dose to the protease. Using a bacterial protease enzyme, it is able to totally kill tumor cells. In comparison to most anticancer drugs, the protease enzyme has an unique mechanism of function as an antineoplastic agent. These enzymes, however, do not interact with metabolism of nucleic acid, unlike other anticancer drugs. Protease enzyme, which is synthesized by marine species, has a broad range of pharmaceutical applications, including digestive medicines, new antiinflammatory agents and anticancer compounds. Since the 18th century, proteolytic enzymes, also known as peptidases, have been studied intensively. Because they are potential therapeutic molecules, recent breakthroughs in biotechnology have substantially increased research in the area. Seventy-five percent of industrial enzymes are hydrolytic in nature. Protease enzymes are one of three large types of commercial enzymes that account for over 60% of worldwide enzyme revenue. Protease enzymes have been used effectively in numerous sectors, such as biotechnological, including detergents, silver retrieval, textile processing, medical and pharmaceutical applications, silk, chemical, waste treatment for commercial purposes, pulp and paper industries, fish processing, feeds and food manufacturing industries. Figure 15.1 shows marine proteases and its biological significance.

Proteases enzymes, commonly known as biological catalysts, are responsible for a wide range of biochemical processes. They've been used in a variety of fields, especially therapeutics. The properties of molecules produced from the marine differ from those of their terrestrial counterparts. Marine microbes (epibionts and endosymbionts), which are abundant in unique environments, produce a plethora of medically and industrially essential molecules. These microbes secrete enzymes with specific characteristics like pH, metal, heat and cryo-tolerance and so on. Proteases are enzymes that break down lengthy chains of proteins into smaller fragments. Endopeptidases and exopeptidases are the two large families of proteases depending on their method of action. Exopeptidases degrade terminal amino acid positions attached to polypeptide chains, while endopeptidases catalyze the breakdown of peptide bonds in the middle portion of polypeptide chains. A further way of classifying proteases is by their optimum pH, which might be neutral, acidic, or alkaline. In terms of the active centers involved, enzymes can be classed as cysteine proteases, metalloproteases, serine proteases and aspartyl proteases.

Proteases are employed in the detergent and leather industries, as well as in pharmaceuticals, including digestive and anti-inflammatory therapies. The alkaline protease is tested like a cleansing additive after being obtained from a symbiotic bacteria present in the gland of Deshayes of a sea shipworm. Maximum production of alkaline protease (623.1 U mg-1 protein) was isolated from sea saltern (China Yellow Sea). *Bacillus mojavensis* A21, a bacterium

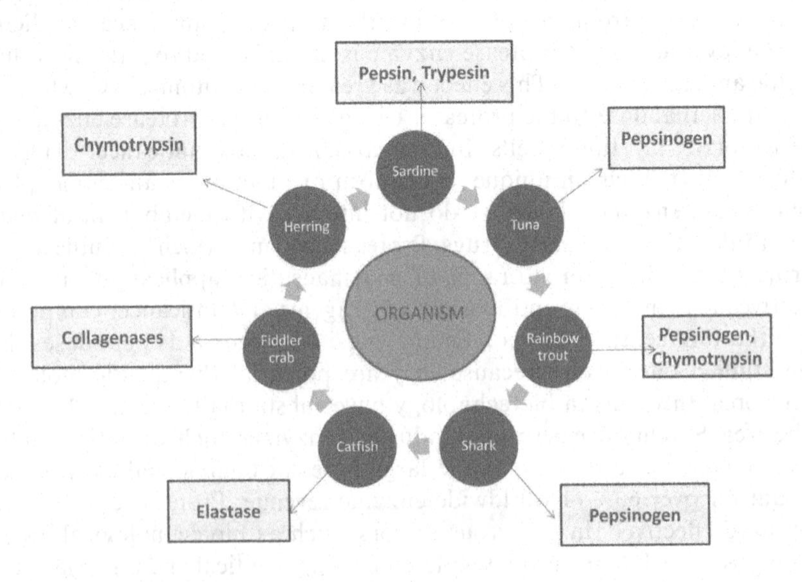

FIGURE 15.1 Animals have a variety of proteases.

that generates alkaline proteases, was isolated from seawater in 2009, and also found BM1 and BM2 detergent-stable alkaline serine-proteases. With a presence of non-surfactants, both BM1 and BM2 proteases were quite stable. Furthermore, they both showed excellent stability and activity with a number of common solid and liquid detergents.

15.12 HYDROLASES

Almelysin, a new metalloproteinase with significant efficiency in low temperatures, is also other proteinase isolated from the culture filtrate of *Alteromonas* sp. The metalloprotease secreted by *Alteromonas* sp. is essential in the strain's chitin degradation pathway. *Aeromonas salmonicida* subsp. has been found as a protamine-reducing marine bacterium obtained from marine soil. Extremophile hydrolases have benefits over chemical biocatalysts. These catalysts are non-polluting, environmentally acceptable, extremely specific, and occur in mild reaction circumstances. Such hydrolases may activate in the form of organic liquids, which is crucial for the production of single-isomer chiral medicines. These hydrolases have been used in a variety of ways. L-asparaginase is a hydrolase which produces L-aspartic and ammonia from L-asparagine. L-glutaminase activities is also present in this enzyme. Antileukemia/antilymphoma drugs made from microbial L-asparaginase preparations for biomedical applications presently account for one-third of global demand. L-asparaginases have been widely utilized in children particularly its act as chemotherapy for acute lymphoblastic leukemia, which is considerably greater than various therapeutic enzymes. L-asparaginase has been treated as an anti-tumor therapy in non-lymphoma, bovine lymphoma sarcoma, chronic lymphocytic leukemia Hodgkin's pancreatic carcinoma, lymphosarcoma, lymphosarcoma, reticulum sarcoma, acute myelomonocytic leukemia, melanoma sarcoma and acute myelocytic leukemia.

Several marine microorganisms have been found to produce L-asparaginase. *Bacillus* sp. and *Pseudomonas* sp. were the most common bacteria isolated from marine sources (Figure 15.2). In the considered habitats, the majority of *Bacillus spp.* were associated with the aerial distribution of the dormant spores, typical of the *Bacillus* genus, which has proved to be highly durable. These L-asparaginases are mostly extracellular, which in nature is not very abundant. However, this arrangement is beneficial because it allows for considerable concentration of an enzyme in the fermentation medium that is generally devoid of endotoxins, which helps downstream processing.

Amylases are enzymes that help to convert complex carbohydrates like starch into simple sugars. They are divided into three groups: alpha-amylase, beta-amylase, and gamma-amylase. γ-amylase, is most effective in acidic conditions. Recently, researchers have discovered extracellular amylase-producing terrestrial bacteria like *Saccharomycopsis, Arxula adeninivorans, Candida japonica, Saccharomycopsis, Lipomyces, Filobasidium capsuligenum,* and *Schwanniomyces.*

With the improvement of marine research and technology, scientists have identified an increasing number of marine microbes capable of producing

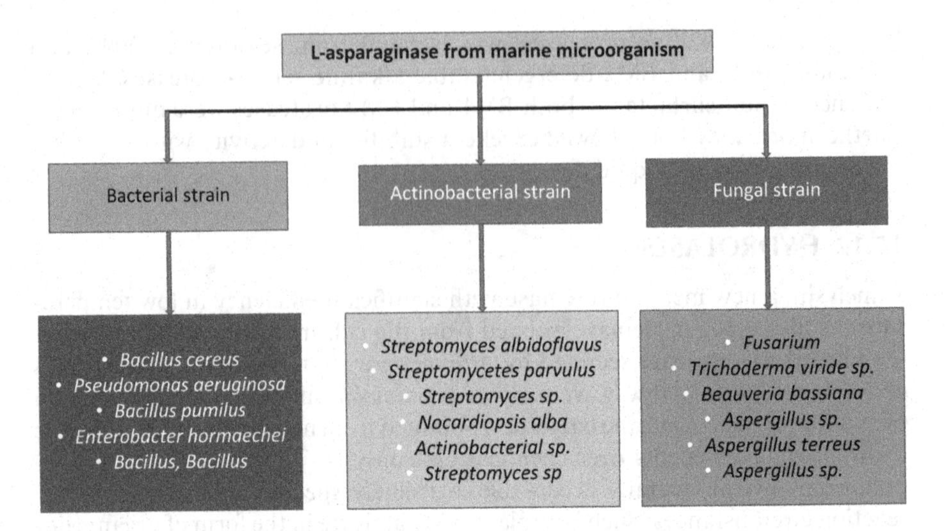

FIGURE 15.2 Production of L-asparaginase from marine microorganism.

amylase. The extracellular amylase-producing yeast strain from marine like *Aureobasidium pullulans* N13d was obtained from Pacific Ocean deep sea sediments. The medium contains 2% sugar, 0.35% peptone, and 0.15% malt extract to develop a new amylase from marine *Streptomyces sp.* D1. Furthermore, a new amylase from *Mucor* sp., which is connected with the *Spirastrella sp.*, marine sponge, has an optimal pH of 5.0 and a temperature of 60°C.

Fucoidan is made up of xylose, fucose, arabinose, galactose, uronic acid and mannose that is a complex sulfated polysaccharide with a molecular weight of less than 1000 Da. The fucoidan oligosaccharide, which acts as an activator for human epidermal keratinocytes. Fucoidanase from marine Vibrio, enzyme that can catalyze the hydrolysis of substrates into small molecules, and produces three types of enzymes. *Pseudomonas atlantica* and *Pseudoalteromonas carrageenovora*, both marine bacteria, were grown in media for three days with fucoidan as the only source of carbon, with substrate consumption of 31.5% and 29.9%, respectively.

Bacillus circulans was found in Tokyo Bay sea mud. *Bacillus circulans* cannot grow in normal growth media, but it can grow and produce new glucanase once the medium is suitably diluted. This glucanase can break down the glucan's -1,3 and-1,6 bonds. A novel glucanase has been discovered from the marine *Bacillus*, with maximum activity at 37°C, and this feature is excellent for dental and various health care. There are 117 marine bacteria that may produce β-mannanase, including *Klebsiella, Pseudomonas, Vibrio, Bacillus, Alcaligenes, Moraxella, Aeromonas* and *Enterobacter*, and others.

15.13 OXIDOREDUCTASES

Marine fungus like *Caldariomyces fumago* synthesizes chloroperoxidase is distinctive among peroxidases in that it has a cysteinic thiolate as a fifth axial ligand of the heme rather than an imidazole ligand. This enzyme is extremely versatile: it catalyzes not only peroxidase reactions, but also catalase and monooxygenase reactions, and it's almost unusual, in that in the presence of H_2O_2 and halide ions, it can catalyze halogenation reactions. A few oxidoreductases are now available for use in the textile, food, and other fields, and many are being actively developed for future commercialization. Industrial-technical, specialty chemical synthesis, environmental, food, pharmaceutical, and personal healthcare are some of the industries where oxidoreductase is used. Oxidoreductase-based biocatalysts fit well with the creation of highly effective, economic, and ecologically friendly industries because they are specialized, energy-saving, and biodegradable.

15.14 LIPASES

Lipases catalyze the conversion of oils and fats into free fatty acids, diacylglycerols, monoglycerols, and glycerol. It is also useful in various processes, including esterification, transesterification, and aminolysis. Lipases have recently received a lot of attention, as indicated by the growing amount of knowledge about them in current research. Several microbial lipases are also commercially accessible, with the majority of them being used in detergents, food flavoring, paper manufacturing industry, cosmetic industry, and organic synthesis, as well as various industrial uses.

In Europe, enzyme detergents now account for 90% of the market, whereas in Japan, they account for about 80%. While they work under mild circumstances, they are extremely strong in organic solvents. Lipases are useful biocatalysts with high substrate specificity. Pelagic fishes have been the principal target of fisheries as a result of the development of marine resources: such species are inventive and potential. Moreover, because these species have a high fat content, humans must deal with particular challenges in terms of fish processing, preservation, and marketing. Traditional fish degreasing procedures include alkaline processing, extraction and expression, but utilizing lipases has tremendous advantages over these approaches. As a result, lipase use in the field of fish processing is causing increasing attention. In 1935, microbial lipase was first discovered in *Penicillium oxalicum* and *Aspergillus flavus* (Table 15.1).

15.15 CHITINASE AND CHITOSANASE

Crustaceans are common arthropods used in biotechnological and molecular research. Crabs, shrimp, lobsters, krill, and barnacles are all found in this category. β -N-acetylhexosaminidase and chitinase are two forms of chitinolytic enzymes found in the liver, pancreas, and crustacean integument. Chitinolytic mechanisms in the integument have been shown to play an important role in molting and ingestion of chitin-containing foods in the hepatopancreas. While marine zooplankton are intended to shed on a regular basis, there is a considerable amount of abandoned chitin, which can be a major carbon source and energy source for chitin-degrading microbes' development and reproduction. Chitin synthesis in the entire marine biocycle is estimated at approximately 2.3 million metric tons per year. Researchers

TABLE 15.1
Production of Lipase by Marine Organisms

Marine Microorganism	Species	Application
Bacterium	*Bacilus* sp. S23	Biodiesel production and digest the pollutants
	Aeromonas sp.	Hydrolyse long length esters
	Psychrobacter sp.	
Microalgal sp.	*Tetraselmis*	Transesterification reactions towards biodiesel production biodiesel
	Nannochloropsis	
Fungi	*Aspergillus awamori, Rhizopus, Mucor, Geotrichum, Scopulariopsis, Trichoderma, Endosmosis, Penicillium, Cryptococcus victoriae*	Potential for use in industries for the production of extracellular lipase, Improved paper manufacturing, In food industries used for production of esters, In paper manufacturing lipase breaks down TG in to remove pitch from wood
Cichlid fish	Nile tilapia	Flavor and synthesis of specialty lipids such as omega-3 fatty acids
Marine invertebrate	*Hexaplex trunculus*	Absorption of nutrients, food storage and excretion, production of esters

have discovered a broad variety of microorganisms capable of producing chitinase, such as *Arthrobacter, Clostridium Penicillium, Serratia, Rhizopus, Sporocytophaga, Pseudomonas, Bacillus, Enterobacter, Klebsiella, Flavobacterium, Streptomyces, Aspergillus, Chromobacterium, Myxobacter, Vibrio fluvialis, Vibrio alginolyticus, Aeromonas hydrophila, Vibrio mimicus, Listonella anguillarum and Vibrio parahaemolyticus.* Direct application of chitinolytic enzymes (e.g., antifungals) and hydrolysis of chitin into chitooligosaccharides are the two main applications of chitinolytic enzymes. In this case, the chitooligosaccharides' chemical structure has a significant impact on their function.

Chitin and chitooligosaccharides had hypolipidemic effects, lowering LDL cholesterol and TG concentration in blood, and angiotensin I–converting enzyme (ACE) inhibitor had antihypertensive efficacy. Furthermore, chitin promotes wound healing by boosting macrophage formation and cytokine release. Chitinase is an antifungal agent that can be taken in association with antifungal medicines to treat fungal infections. Enzymes are increasingly being used in various aspects of the food industry, and chitinases offer a wide variety of applications in this area. Because of their antibacterial properties, chitin compounds prevent the proliferation of spoilage bacteria in foods, inhibiting food contamination. In agriculture, chitinases are fascinating enzymes. The chitinous recycling of the waste into biofertilizers is the most notable aspect. Pollution of agricultural soil with fungal infections might render it unprofitable at times. Using large-scale biopesticides to manage fungal infections and pests is one option, but it comes at a high cost. Biological control is an effective technique to mitigate such issues. Chitinases are used to treat a variety of illnesses, including pathogenic fungi, viral diseases, and insect pests, as antifungal, insecticidal, or antiparasitic medicines.

15.16 THERAPEUTIC PROPERTIES

15.16.1 Effect on Glucose Metabolism

Many new drugs have been developed to treat diabetes, including oral hypoglycemic drugs as well as insulin mimics. Bioactive compounds in algae have been shown to be safe and effective against type 2 diabetes, which reverses carbohydrate metabolizing enzymes (Abirami and Kowsalya, 2013). alkaloids, flavonoids, carotenoids, polyphenols and phlorotannins present in algae have been shown to have hypoglycemic effects. Fucoxanthin significantly promoted insulin sensitivity and reduced blood sugar in diabetic rats. algae activate beta cell production and limits the secretion of glucagon by alpha cells, unlike blood sugar (Kim *et al.*, 2015).

15.17 EFFECT ON CELL PROLIFERATION

Cell proliferation is the process in which the number of cells increases due to cell division and cell growth, which often occurs in tumors or cancers. Evidence suggests that algae may act as an antiproliferative by inducing maturation of dendritic cells, combining with other cytokines, and modulating the human immune system (Lowenthal and Fitton, 2015). Macrophages are activated by membrane receptors specifically TLR4, CD14, CR3 and SR, leading to the production of cytokines such as IL-12 and IFN, which enhance the activation of NK cells, which in turn stimulates the activation of T cells (Kellogg *et al.*, 2015). In addition, secondary metabolites of algae, such as root bark tannins, flavonoids, catechins, carotenoids, quercetin, and myricetin, have been shown to have relative anticancer activity. Epidemiological studies also show that, compared to other parts of the world, eating seaweed can reduce the incidence of ovarian cancer, breast cancer, and endometrial cancer in the Japanese population (Murata and Nakazoe, 2001).

15.18 EFFECT ON ADIPOSE TISSUE

Obesity is defined as the unwanted accumulation of body fat and white adipose tissue (WAT), which inhibits the secretion of cytokines in adipose tissue and leads to a variety of other disorders such as diabetes, high cholesterol and stroke (Namvar *et al.*, 2012). Thermogenesis plays an important role in the regulation of the mechanisms of obesity. Similarly, administration of algae to rats reduced plasma leptin and epididymal adipose tissue levels (Grasa-López *et al.*, 2016). In addition, algae significantly reduced adipocyte size, fasting blood sugar and insulin levels in obese rats (Gammone and D'Orazio, 2015). Algae with fucoxanthin inhibited fat absorption and serum triglyceride levels in an in vivo model and have also been shown to have anti-obesity effects in mice (Kang *et al.*, 2012).

15.19 EFFECT ON LIPID METABOLISM

Epidemiological studies confirm the strong link between non-starch polysaccharides (fiber) and lipid metabolism, and a diet rich in polysaccharides improves colon health (Qi *et al.*, 2012). Long-term use of algae (1% in the diet) uses a receptor pathway activated by an increase in peroxisomes pursued by oxidation and gluconeogenesis (Austin *et al.*, 2018). ulvan, carrageenan, alginate, fucoidan, and fucoxanthin are secondary algal metabolites reported with this property (Chater *et al.*, 2016). Ulvan has the ability to bind to lipid molecules, reducing total and low-density lipoprotein (LDL) cholesterol and increasing serum high-density lipoprotein (HDL) cholesterol (Qi *et al.*, 2012). Therefore, the molecular weight (MW) of algal polysaccharides also plays a role in regulating the ratio of lipoproteins in lipids. Carrageenan has the ability to alter the structure and organoleptic quality of lipids, thus minimizing the total absorption of fat in food (Ganesan *et al.*, 2019). The main reason for this characteristic is the presence of non-protein amino acids similar to taurine in red algae (i.e. 1.01.3 g of taurine/100 g) (Yang *et al.*, 2017). Although taurine is a nonessential amino acid, the ingestion of sulfated polysaccharides increases the excretion of bile acids in the feces and reduces blood cholesterol (Matanjun *et al.*, 2010). In addition, the inclusion of algal meal in laboratory animals increased fecal fat and decreased fat digestibility in animal models. Nori and Wakame algae have been shown to enhance fermentation (i.e. short chain fatty acids) in the colon, inhibiting lipid emulsification (Kellogg *et al.*, 2015). Chylomicron metabolism has been shown to be high in the algal diet due to enhanced arylesterase activity involved in lipoprotein metabolism and inhibition of lipopero oxidation in LDL (Ruqqia *et al.*, 2015).

This tendency is common in hypercholesterolemia where hepatic peroxidation and HDL uptake are increased by the cholesterol reverse transport pathway. The global report shows that algae trigger an important response in lipid metabolism including (1) altering the emulsification of fats in bile acids, (2) disrupting micelle formation, (3) making altered lipase enzymes, (4) binding to the cholesterol site and (5) improved colonic bacterial fermentation. Besides polysaccharides, other bioactive substances such as polyphenols, fucoxanthin and polyunsaturated fatty acids from different algae species also affect lipid metabolism through a different mode of action.

15.20 ANTI-DIABETIC ACTIVITY

People globally realize that marine foods contribute to human health promotion. A diet rich in marine items is expected to result in a lower incidence of diabetes, cancer, and obesity. Many studies have found that bioactive chemicals from marine species, such as Fucoxanthin, *Astaxanthin*, marine collagen peptides, dieckol, and krill oil (Table 15.2) have a beneficial effect on metabolic dysfunction (diabetes and obesity).

TABLE 15.2
Applications of Marine-Derived Bioactive Compounds

Sl. No	Compounds	Source	Applications
1	**Fucoxanthin** 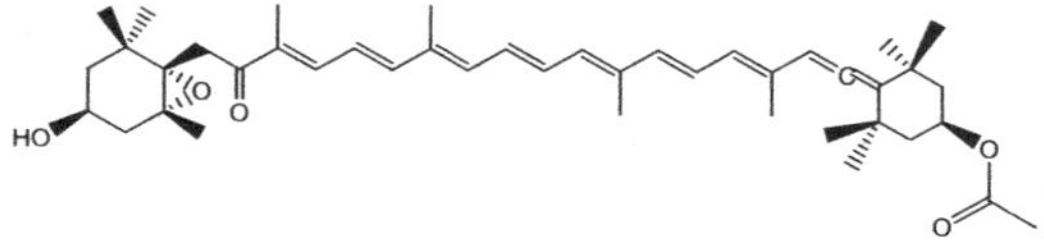	It is mainly isolated from the marine brown seaweeds like as brown seaweeds (*Phaeophyceae*) and diatoms (*Bacillariophyta*).	antidiabetic activity, antioxidant activity, anti-cancer, anti-diabetic and anti-photoaging properties
2	**Astaxanthin** 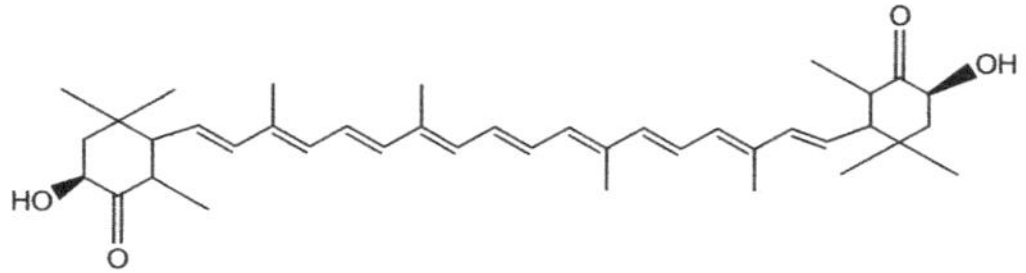	It is found in the marine algae *Hematococcus pluvialis*, *Chlorella zofingiensis*, and *Chlorococcum sp.*, and the red yeast *Phaffia rhodozyma*.	powerful antioxidant, anti-tumor, anti-diabetic, anti-inflammatory, and cardioprotective properties
3	**Marine Collagen Peptides** 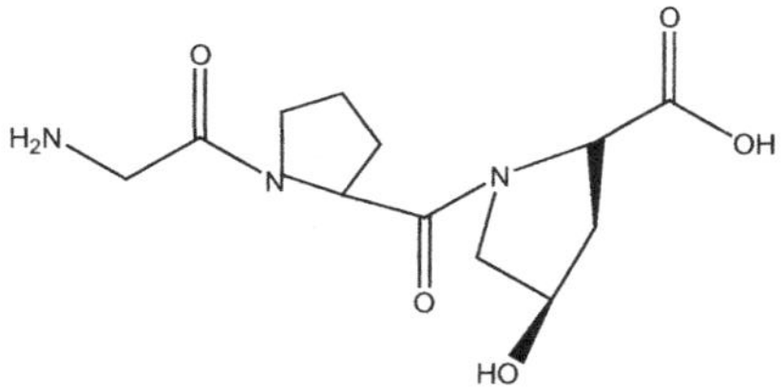	They are compounds of low-molecular-weight peptides derived from the skin of deep-sea fish like chum salmon (*Oncorhynchus keta*)	Antidiabetic activity, antioxidative activity, anti-hypertension, anti-skin aging, anti-ulcer and could improve lipid metabolism

(*Continued*)

TABLE 15.2 *(Continued)*
Applications of Marine-Derived Bioactive Compounds

Sl. No	Compounds	Source	Applications
4	**Polyphenol** 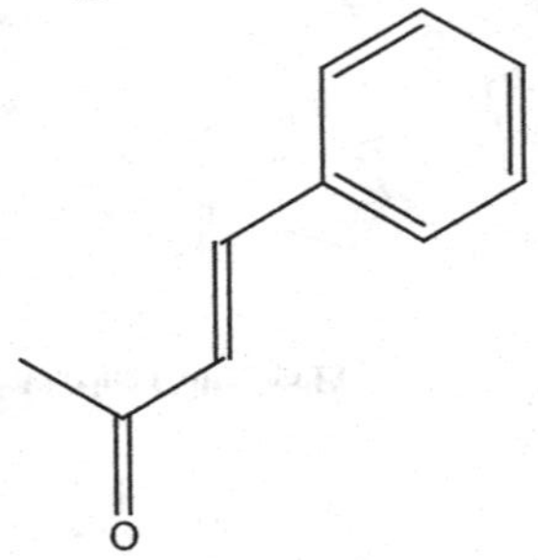 O Phenolic acids CH_2OH CH_2OH Flavonoids Lignans Stillbenes Isoflavones	**Dieckol**, a kind of phlorotannin, is a marine algal polyphenolic compounds that have been isolated from brown seaweed *Ecklonia* cava.	Antidiabetic activity, antioxidative activity

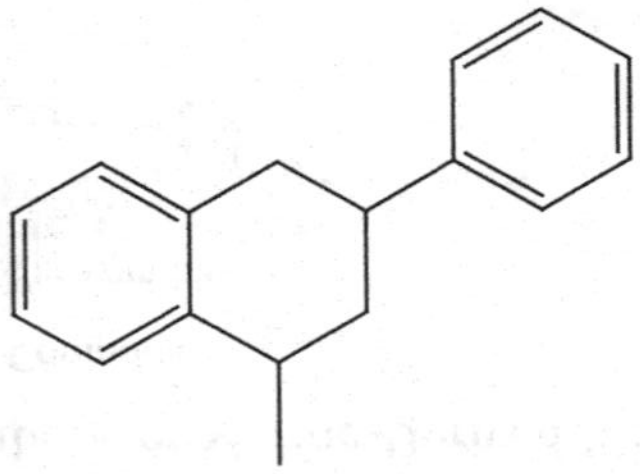

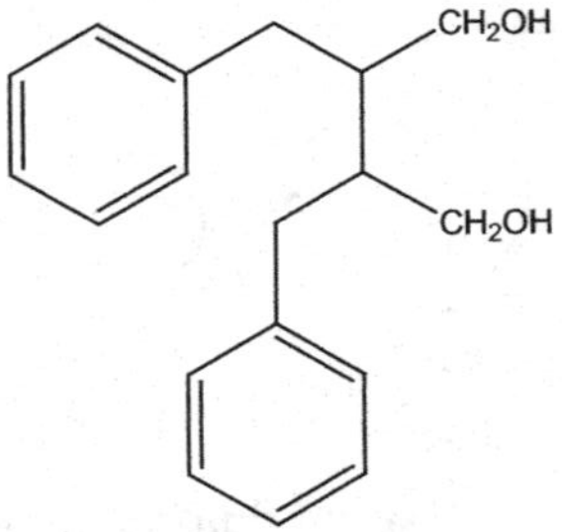

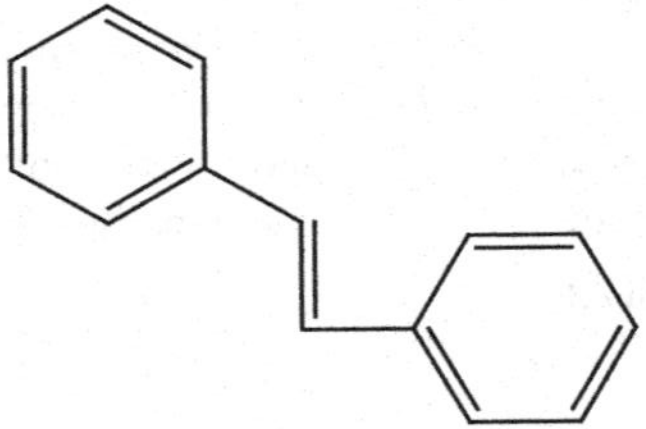

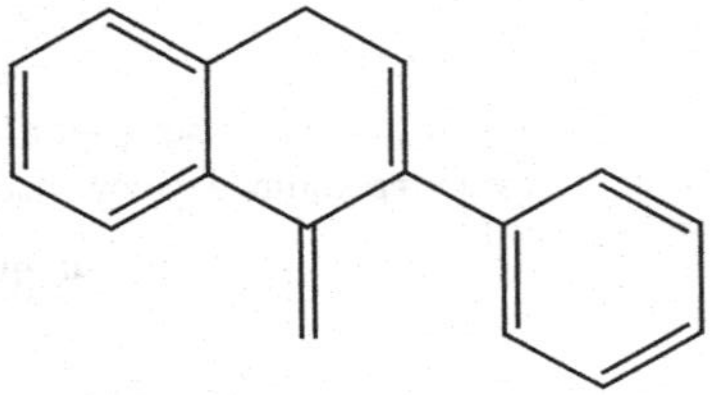

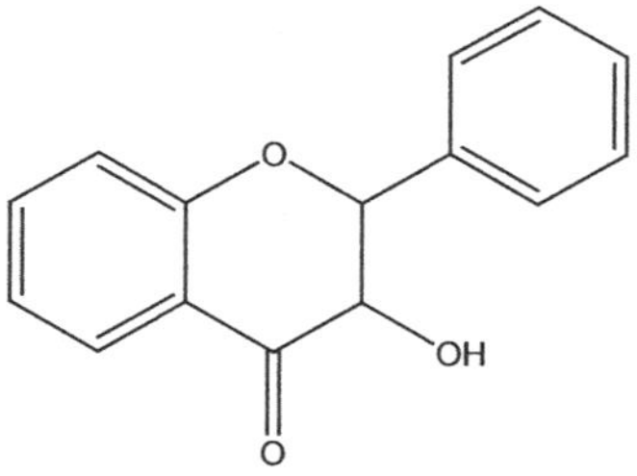

	Flavonols Flavones Flavonones Anthocyanidines (structures with O, OH labels)		
5	**Krill Oil** 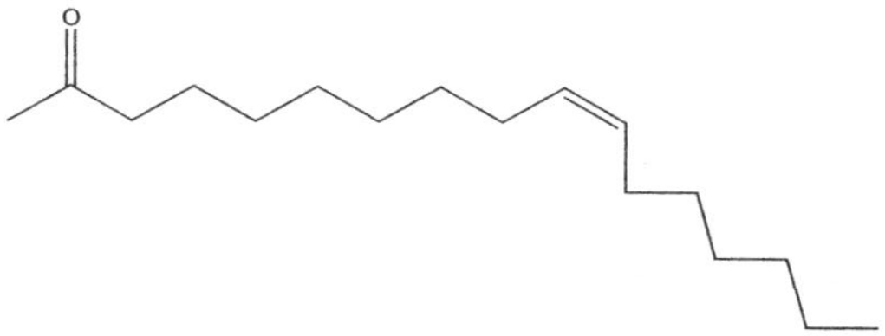	It is extracted from Antarctic krill, *Euphausia superba*	anti-inflammation, antioxidation, antihyperlipidemic effect, antihyperglycemic effect, modulation of the endocannabinoid system and cardioprotective effect.

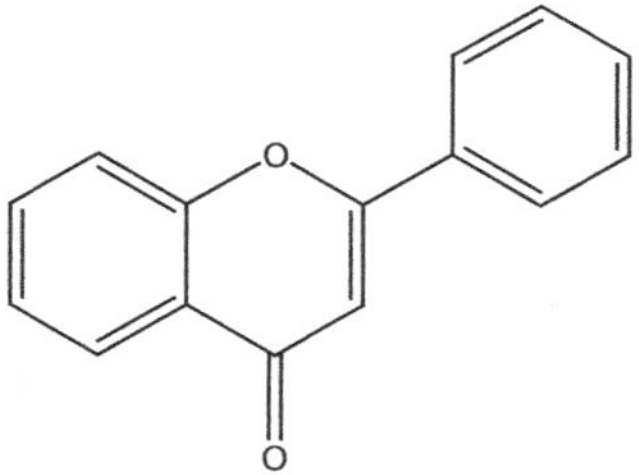

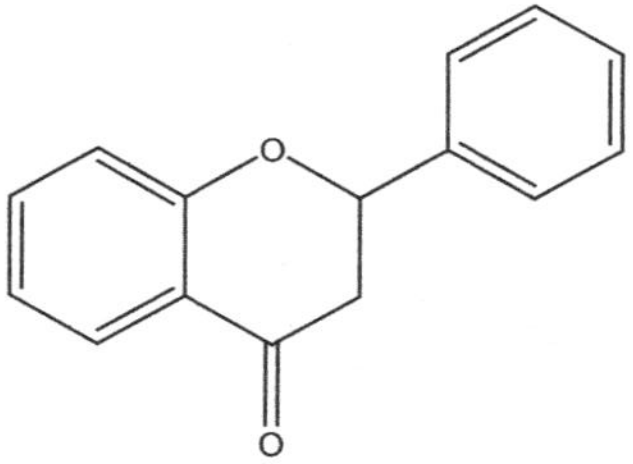

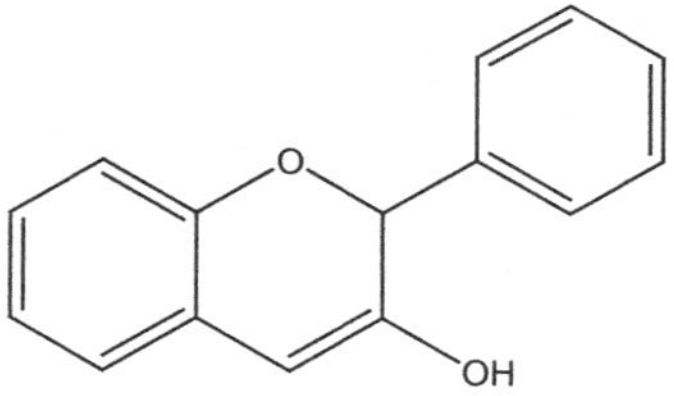

Diabetes is one of the disorders of metabolic disorders. Many people have been affecting from this disorder. Due to the obvious growing percentage of diabetic patients and the limited number of anti-diabetic medications, the investigation for novel molecules, specifically from marine sources, has drawn great interest from the research community. Microorganisms such as cyanobacteria and actinomycetes, and marine fungi have been investigated for the anti-diabetic bioactivities. Glucosidase in bacteria enzyme is involved in the degradation of polysaccharides as well as the processing of glycoproteins and glycolipids, making it a promising target for diabetes and obesity treatments (Pandey *et al.*, 2013). Bacteria linked with the marine sponge *Aka coralliphaga* produced a huge number of glucosidase inhibitors. Marine actinomycetes have produced a number of enzyme inhibitors and other useful substances (e.g., *Streptomyces sp.*).

Marine fungus also have been examined for anti-diabetic activities. Brown algae are also high in bioactive chemicals that have been shown to have considerable health benefits. In contrast to fucoxanthin, fucosterol, sulfated polysaccharides, and phlorotannins have been detected in the majority of the brown algae species, with the highest potential applications as an anti-diabetic agent (Figure 15.3). Anti-diabetic activity in microalgae has also been investigated. Microalgae are photosynthetic eukaryotes that make up a significant portion of phytoplankton in both marine and freshwater environments. Astaxanthin from *Chlorella zofingiensis* had better antioxidant, anti-glycoxidative and anti-glycative properties, implying that this microalga could be an useful food supplement and a viable diabetes patient preventive agent (Sun *et al.*, 2011). In contrast to *Chlorella spp.*, the terpene dysidine (Figure 15.4) from the sponge *Dysidea sp.*, is currently in preclinical studies for diabetes therapy.

15.20.1 Metabolites—Beneficial Activities from Marine Algae

In aquatic environments, the primary producers of oxygen are marine algae. They serve all other organisms in the marine food chain. There are two major groups where marine algae could be divided. They are macroalgae (seaweeds) and microalgae. They both performs a prolific sources of the bioactive

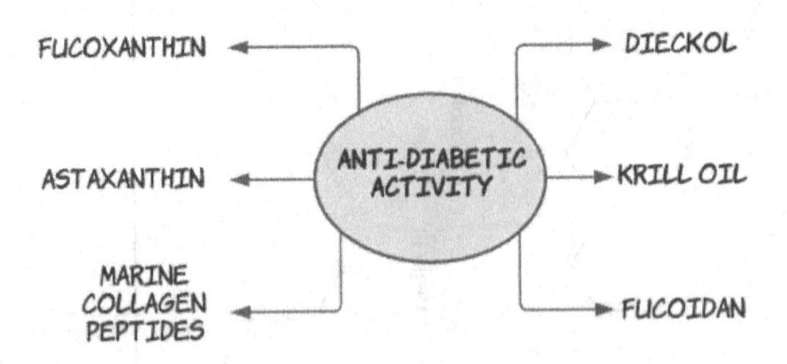

FIGURE 15.3 Bioactive compounds for anti-diabetic activity.

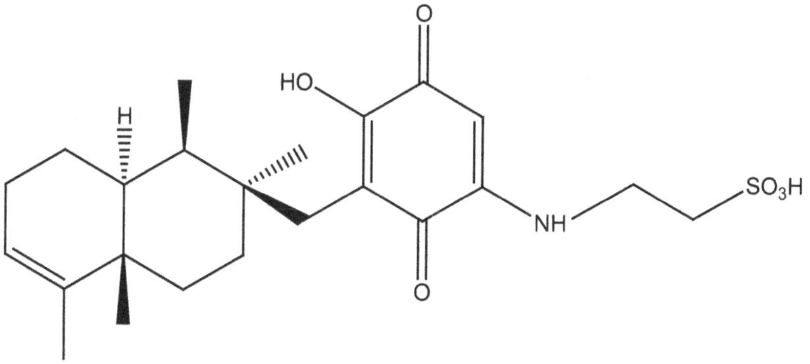

FIGURE 15.4 Compound dysidine isolated from the sponge *Dysidea* sp.

substances. So that many studies and researches have been progressing for examining their medicinal aspects. In recent times, many new metabolites from the marine sources have been reported.

15.21 MARINE MICROALGAE: BLUE-GREEN ALGAE (CYANOBACTERIA)

Marine microalgae basically constitute the phytoplankton. They are broadly classified into three main groups. They are Cyanobacteria, *Bacillariophyta* and *Dinophyceae*. Many bioactive compounds have been shown a high antiviral and anti-HIV activities. A short while ago, a novel natural anti-AIDS drugs have been found. It has derived from *Lyngbya lagerhaimanii* and *Phormidium tenue*. Another compound, calcium spirulan, has shown high antiviral activity which has isolated from *Spirulina platensis*. Antifouling compounds with antibiotic activity is produced by the strains of some species of cyanobacteria. For instance, compounds such as ulithiacyclamide and patellamides A and C are generally known for antimalarial and antitumor activity. Recently, dolastatin 13 and lyngbyastatins 5–7 were the novel two bioactive compounds which were isolated from *Lyngbya spp.* The novel bioactive compounds of cyanobacteria are strong in antifungal, anti-inflammatory, antitumor and antibiotic activities, which helps in the synthesis of molecules for the pharmaceutical applications.

15.22 MARINE MACROALGAE

Marine macroalgae, otherwise known as seaweeds, are mostly found in the tropical waters and intertidal regions. These macroalgae are multicellular organisms. They differ in their morphological types and sizes. They are classified based on their photosynthetic pigments into red, green, and brown algae. Recently, over 3,200 products have been isolated from marine macroalgae. The compounds that have derived from the marine macroalgae, have been useful in the field of medicine such as antioxidant, anticoagulant, antitumor, antifouling,

TABLE 15.3
Beneficial Applications of Marine Macroalgae and Microalgae

Sl. No	Bioactive Compounds/ Metabolites	Sources	Beneficial Applications
1	Noscomin	*Nostic commune*	Antibiotic activity
2	Dolastatin 10	*Symploca sp.*	Antitumor activity
3	Bis-bromoindoles	*Rivularia* sp.	Powerful anti-inflammatory activity
4	Malyngamide F acetate	*L. majuscule*	Anti-inflammatory activity
5	Malyngamide D and malyngamide D acetate	*L. majuscule*	Antibiotic activity
6	Carrageenan	Genera *Chondrus, Iridea*	Gastric ulcers and duodenal ulcers
7	Fucoidan	*Gracilaria corticata*	Activity against breast and colorectal cancer
8	Phenolic compounds	*Rhodomela confervoids, Polysiphonia urceolata*	Antidiabetic activity
9	Galactan sulfate	*Agardhiella tenera*	Antiviral activity (against HIV-1 and HIV-2)
10	Mycosporine-like amino acids	*Laurencia, Halymenia*	Antioxidant activity
11	Monoterpene, halomon	*Portieria hornemannii*	Cytotoxic action

antibacterial, antifungal, etc., also it is stated that the red seaweeds are used in the treatments of diarrhea and gastritis (Table 15.3). Not only that, green and brown seaweeds have been used in the various treatments of diseases such as rheumatic diseases, skin diseases, gastric ulcers, goiter, etc.

15.23 ANTICANCER DRUGS FROM MARINE FLORA

Cancer is one of the dreadful diseases in humans. In the treatments of cancer, for some instances, many drugs are causing side effects. Natural compounds from the medicinal plants have been gained results in the treatment of cancer. it is reported that the drugs selected and approved for the treatment of cancer is almost 60% of its natural origin. Even though terrestrial plants possess natural products for cancer treatment, marine floras also possess natural bioactive compounds and the marine ecosystem would become a more invaluable source in the future for novel compounds. Marine flora includes flowering plants, microflora, microalgae, and macroalgae. It is a fact that our globe has almost 71% of biodiversity in the water. This enormous resource of marine flora and algae alone hand out a great opportunity for the novel drug discovery. Many findings suggest that marine flora natural compounds are useful and are effective in treating various diseases.

15.23.1 Marine Bacteria

Marine bacteria produces many secondary metabolites which have been useful in the various sectors of pharmaceutical industries. Its pharmaceutical products shows high anti-inflammatory activities such as topsentins, manoalide and pseudopterosins and also anticancer activities such as bryostatins, discodermolide, and sarcodictyin. It also shows high antibiotic activities such as marinone. Probiotic bacteria like *Bifidobacteria a*and *Lactobacilli* produce the anticancer substances. That the compound from marine *Halomonas* sp. is cytotoxic hydroxyphenylpyrrole dicarboxylic acids, i.e., 3-(4-hydroxyphenyl)-4-phenylpyrrole-2,5-dicarboxylic acid (HPPD-1), 3,4-di-(4-hydroxy-phenyl) pyrrole-2,5-dicarboxylic acid (HPPD-2) and the indole derivatives 3-(hydroxyacetyl)-indole, indole-3-carboxylic acid, indole-3-carboxaldehyde, and indole-3-acetic acid (Erba *et al.*, 1999).

15.24 CONCLUSION AND FUTURE PROSPECTS

The secondary metabolites such as fucoxanthin, phlorotannin, fucoidan, laminarin, carrageenan, alginate, and agar are the potential source of functional compounds which are derived from algae. These compounds are widely used in food applications due to their different properties to improve the quality of food. Although algae have been used as a functional ingredient in commercial applications as a stabilizer, emulsifier, thickener, texture modifier and phytochemical enriched with vitamins and phytochemical fibers, considerable effort is required to establish their role and application in them. healthy foods for direct consumption.

Several in vitro studies have demonstrated the effectiveness of food products fortified with bioactive compounds of algae; however, the most difficult factor in the food industry is development. New products can attract consumers when unfamiliar products approach them. In addition, it is equally important to raise public awareness and focus on algae to develop a more readily available and complete alternative to daily food with medicinal effects.

REFERENCES

Abirami, R.G., S. Kowsalya, Antidiabetic activity of Ulva fasciata and its impact on carbohydrate metabolism enzymes in alloxan induced diabetic rats, *International Journal of Research in Pharmacology and Phytochemistry* 3(3) (2013): 136–141.

Abirami, R.G., S. Kowsalya, Quantification and correlation study on derived phenols and antioxidant activity of seaweeds from gulf of mannar, *Journal of Herbs, Spices & Medicinal Plants* 23(1) (2017): 9–17.

Austin, C., D. Stewart, J.W. Allwood, G.J. McDougall, Extracts from the edible seaweed, Ascophyllum nodosum, inhibit lipase activity *in vitro*: Contributions of phenolic and polysaccharide components, *Food & Function* 9(1) (2018): 502–510.

Belattmania, Z., A. Engelen, H. Pereira, E. Serrao, L. Custódio, J. Varela, R. Zrid, A. Reani, B. Sabour, Fatty acid composition and nutraceutical perspectives of brown seaweeds from the Atlantic coast of Morocco, *International Food Research Journal* 25: 1520–1527.

Biancarosa, I., I. Belghit, C.G. Bruckner, N.S. Liland, R. Waagbø, H. Amlund, S. Heesch, E.J. Lock, Chemical characterization of 21 species of marine macroalgae common in Norwegian waters: Benefits of and limitations to their potential use in food and feed, *Journal of the Science of Food and Agriculture* 98: 2035–2042.

Černá M, Chapter 24—Seaweed proteins and amino acids as nutraceuticals, in: *Advances in Food and Nutrition Research* (Ed. by S.-K. Kim), pp. 297–312. Academic Press, San Diego, CA, 2011.

Chater, P.I., M. Wilcox, P. Cherry, A. Herford, S. Mustar, H. Wheater, I. Brownlee, C. Seal, J. Pearson, Inhibitory activity of extracts of Hebridean brown seaweeds on lipase activity, *Journal of Applied Phycology* 28(2) (2016): 1303–1313.

Dawczynski, C., R. Schubert, G. Jahreis, Amino acids, fatty acids, and dietary fibre in edible seaweed products, *Food Chemistry* 103 (2007): 891–899.

El Maghraby, D.M., E.M. Fakhry, Lipid content and fatty acid composition of Mediterranean macroalgae as dynamic factors for biodiesel production, *Oceanologia* 57 (2015): 86–92.

Erba, E., D. Bergamaschi, S. Ronzoni, M. Faretta, S. Taverna, M. Bonfanti, C.V. Catapano, G. Faircloth, J. Jimeno, M. D'incalci, Mode of action of thiocoraline, a natural marine compound with anti-tumor activity, *Britsh Journal of Cancer* 80 (1999): 971. doi: 10.1038/sj.bjc.6690451.

Gammone, M.A., N. D'Orazio, Anti-obesity activity of the marine carotenoid fucoxanthin, *Marine Drugs* 13(4) (2015): 2196–2214.

Ganesan, A.R, M. Shanmugam, R. Bhat, Producing novel edible films from semi refined carrageenan (SRC) and ulvan polysaccharides for potential food applications, *International Journal of Biological Macromolecules* 112 (2018): 1164–1170.

Ganesan, A.R., M. Shanmugam, R. Bhat, Quality enhancement of chicken sausage by semi- refined carrageenan, *Journal of Food Processing and Preservation* (2019): e13988.

Garcia-Vaquero, M., G. Rajauria, J. O'Doherty, T. Sweeney, Polysaccharides from macroalgae: Recent advances, innovative technologies and challenges in extraction and purification, *Food Research International* 99 (2017): 1011–1020.

Grasa-López, A., Á. Miliar-García, L. Quevedo-Corona, N. Paniagua-Castro, G. EscalonaCardoso, E. Reyes-Maldonado, M.-E. Jaramillo-Flores, Undaria pinnatifida and fucoxanthin ameliorate lipogenesis and markers of both inflammation and cardiovascular dysfunction in an animal model of diet-induced obesity, *Marine Drugs* 14(8) (2016): 148.

Gross, H., D.E. Goeger, P. Hills, S.L. Mooberry, D.L. Ballantine, T.F. Murray, F.A. Valeriote, W.H. Gerwick, Lophocladines, bioactive alkaloids from the red alga *Lophocladia* sp, *Journal of Natural Products* 69 (2006): 640–644. doi: 10.1021/np050519e.

Jahromi, S.T., N. Barzkar, Future direction in marine bacterial agarases for industrial applications. *Applied Microbiology and Biotechnology* 102 (2008): 6847–6863. doi: 10.1007/s00253-018-9156.

Kang, S.I., H.-S. Shin, H.-M. Kim, S.-A. Yoon, S.-W. Kang, J.-H. Kim, H.-C. Ko, S.-J. Kim, Petalonia binghamiae extract and its constituent fucoxanthin ameliorate high-fat diet-induced obesity by activating AMP-activated protein kinase, *Journal of Agricultural and Food Chemistry* 60(13) (2012): 3389–3395.

Kellogg, J., D. Esposito, M.H. Grace, S. Komarnytsky, M.A. Lila, Alaskan seaweeds lower inflammation in RAW 264.7 macrophages and decrease lipid accumulation in 3T3-L1 adipocytes, *Journal of Functional Foods* 15 (2015): 396–407.

Kim, K.T., L.E. Rioux, S.L. Turgeon, Alpha-amylase and alpha-glucosidase inhibition is differentially modulated by fucoidan obtained from Fucus vesiculosus and Ascophyllum nodosum, *Phytochemistry* 98 (2014): 27–33.

Li, Y., Z.-J. Qian, B. Ryu, S.-H. Lee, M.-M. Kim, S.-K. Kim, Chemical components and its antioxidant properties in vitro: An edible marine brown alga, Ecklonia cava, *Bioorganic & Medicinal Chemistry* 17(5) (2009): 1963–1973.

Lowenthal, R.M., J.H. Fitton, Are seaweed-derived fucoidans possible future anti-cancer agents? *Journal of Applied Phycology* 27(5) (2015): 2075–2077.

Marquez, B.L., K.S. Watts, A. Yokochi, M.A. Roberts, P. Verdier-Pinard, J.I. Jimenez, E. Hamel, P.J. Scheuer, W.H. Gerwick, Structure and absolute stereochemistry of hectochlorin, a potent stimulator of actin assembly. *Journal of Natural Products* 65 (2002): 866–871. doi: 10.1021/np0106283.

Matanjun, P., S. Mohamed, K. Muhammad, N.M. Mustapha, Comparison of cardiovascular protective effects of tropical seaweeds, kappaphycus alvarezii, Caulerpa lentillifera, and Sargassum polycystum, on high-cholesterol/high-fat diet in rats, *Journal of Medicinal Food* 13(4) (2010): 792–800.

Misurcova L, Chemical composition of seaweeds, in: *Handbook of Marine Macroalgae: Biotechnology and Applied Phycology* (Ed. by S.-K. Kim), pp. 171–192. John Wiley & Sons, Hoboken, NJ, 2011.

Murata, M., J. Nakazoe, Production and Use of Marine AIgae in Japan, Japan Agricultural Research Quarterly: *JARQ* 35(4) (2001) 281–290.

Nagarajan, S., M. Mathaiyan, Emerging novel anti HIV biomolecules from marine Algae: An overview, *Journal of Applied Pharmaceutical Science* 5 (2015): 153–158.

Namvar, F., S. Mohamed, S.G. Fard, J. Behravan, N.M. Mustapha, N.B.M. Alitheen, F. Othman, Polyphenol-rich seaweed (*Eucheuma cottonii*) extract suppresses breast tumour via hormone modulation and apoptosis induction, *Food Chemistry* 130(2) (2012): 376–382.

Nunraksa, N., S. Rattanasansri, J. Praiboon, A. Chirapart, Proximate composition and the production of fermentable sugars, levulinic acid, and HMF from *Gracilaria fisheri* and Gracilaria tenuistipitata cultivated in earthen ponds, *Journal of Applied Phycology* 31(1) (2019): 683–690.

Othman, R., N. Amin, M. Sani, N. Fadzillah, Carotenoid and chlorophyll profiles in five species of Malaysian seaweed as potential halal active pharmaceutical ingredient (API), International, *Journal on Advanced Science, Engineering and Information Technology* 8(4–2) (2018): 1610–1616.

Pandey, S., A. Sree, S.S. Dash, D.P. Sethi, L. Chowdhury, Diversity of marine bacteria producing beta-glucosidase inhibitors. *Microbial Cell Factories* 12 (2013). doi: 10.1186/1475-2859-12-35.

Patterson, E., R. Wall, G. Fitzgerald, R. Ross, C. Stanton, Health implications of high dietary omega-6 polyunsaturated fatty acids, *Journal of Nutrition and Metabolism* (2012): 16.

Qi, H., L. Huang, X. Liu, D. Liu, Q. Zhang, S. Liu, Antihyperlipidemic activity of high sulfate content derivative of polysaccharide extracted from Ulva pertusa (Chlorophyta). *Carbohydrate Polymers* 87(2) (2012): 1637–1640.

Qin, Y, Applications of bioactive seaweed substances in functional food products, in: *Bioactive Seaweeds for Food Applications: Natural Ingredients for Healthy Diets* (Ed. by Y. Qin), pp. 111–134. Academic Press, San Diego, CA, 2018.

Rajauria, G., B. Foley, N. Abu-Ghannam, Identification and characterization of phenolic antioxidant compounds from brown Irish seaweed Himanthalia elongata using LC-DAD–ESI-MS/MS, *Innovative Food Science & Emerging Technologies* 37, Part B (2016): 261–268.

Rodrigues, D., A.C. Freitas, L. Pereira, T.A. Rocha-Santos, M.W. Vasconcelos, M. Roriz, L.M. Rodríguez-Alcalá, A.M. Gomes, A.C. Duarte, Chemical composition of red, brown and green macroalgae from Buarcos Bay in central west coast of Portugal, *Food Chemistry* 183 (2015): 197–207.

Ruqqia, K., V. Sultana, J. Ara, S. Ehteshamul-Haque, M. Athar, Hypolipidaemic potential of seaweeds in normal, triton-induced and high-fat diet-induced hyperlipidaemic rats, *Journal of Applied Phycology* 27(1) (2015): 571–579.

Seedevi, P., M. Moovendhan, S. Viramani, A. Shanmugam, Bioactive potential and structural chracterization of sulfated polysaccharide from seaweed (*Gracilaria corticata*), *Carbohydrate Polymers* 155 (2017): 516–524.

Shan, X.U., L. Li, Z. Li-Qun, L. Zhuo, Q. Li-Li, C. Qi, X. Chang-Fen, Reversal effect of 4′-methylether-scutellarein on multidrug resistance of human choriocarcinoma JAR/VP16 cell line. *Progress in Biochemistry and Biophysics* 33 (2006): 1061–1073.

Simopoulos, A.P., An increase in the omega-6/omega-3 fatty acid ratio increases the risk for obesity. *Nutrients* 8 (2006): 128.

Sithranga Boopathy, N., K. Kathiresan, Anticancer drugs from marine flora: An overview, *Journal of Clinical Oncology* 18 (2010). doi: 10.1155/2010/214186.

Spada, P.D.S., G.G.N. de Souza, G.V. Bortolini, J.A.P. Henriques, M. Salvador, Antioxidant, mutagenic, and antimutagenic activity of frozen fruits. *Journal of Medicinal Food* 11 (2008): 144–151. doi: 10.1089/jmf.2007.598.

Sun, Z., J. Liu, X. Zeng, J. Huangfu, Y. Jiang, M. Wang, F. Chen, Astaxanthin is responsible for antiglycoxidative properties of microalga *Chlorella zofingiensis*. *Food Chem* 126 (2011): 1629–1635. doi: 10.1016/j.foodchem.2010.12.043.

Teas, J., S. Vena, D.L. Cone, M. Irhimeh, The consumption of seaweed as a protective factor in the etiology of breast cancer: Proof of principle, *Journal of Applied Phycology* 25(3) (2013): 771–779.

Viera, A., P. Gálvez, M. Roca, Bioaccessibility of marine carotenoids, *Marine Drugs* 16(10) (2018): 397.

Wong, K.H., P.C. Cheung, Nutritional evaluation of some subtropical red and green seaweeds: Part I—proximate composition, amino acid profiles and some physico-chemical properties, *Food Chemistry* 71 (2000): 475–482.

Wong, K.H., P.C.K. Cheung, Nutritional evaluation of some subtropical red and green seaweeds Part II. *In vitro* protein digestibility and amino acid profiles of protein concentrates, *Food Chemistry* 72 (2001): 11–17.

Yang, T.H., H.-T. Yao, M.-T. Chiang, Red algae (*Gelidium amansii*) hot-water extract ameliorates lipid metabolism in hamsters fed a high-fat diet, *Journal of Food and Drug Analysis* 25(4) (2017): 931–938.

16 Nanoparticles of Marine Origin and Their Potential Applications

Fatemeh Sedaghat, Morteza Yousefzadi, and Reza Sheikhakbari-Mehr

CONTENTS

16.1 INTRODUCTION

Nanotechnology is one of the 21st century's most emergent fields of science and technology. It is one of the most important and fast-growing research areas using biosynthetic and environmentally friendly technology for nanoparticles of synthesis (NPs). It deals with nanoparticles of 1–100 nm size in one dimension and has entered nearly every utility field and opened up new horizons of possible nanoparticle applications. NPs are very small in size (<100 nm), and have a very

DOI: 10.1201/9781003303916-16

high ratio of surface to volume. Due to their fascinating properties like physical, mechanical, optical, magnetic electronics, and sensing, NPs have drawn particular attention from the scientific community and differ considerably in many respects from their bulk counterparts. NPs are used extensively in various disciplines including food, cosmetics, medicine, engineering, agriculture, and energy. Nanoparticles show completely new or enhanced properties, and their specific uses are controlled by their size, distribution, and morphology [Singh, 2020].

In fact, by the end of 2020, the global economy is expected to have around $3.1 trillion from nanotechnology contribution. Although in nature, the existence of the nano-form is not new for mankind. They are evident in glacial ice cores 10,000 years old, historical sediments, volcanic dust, and many other incidences as the result of chemical weathering [Saxena, 2016; Kahru and Ivask, 2012; Iqbal et al., 2020].

Reliable and environmentally friendly synthesis of metallic nanoparticles is a significant nanotechnology goal. Nanoparticles are synthesized by different chemical and physical methods. Several manufacturing techniques are in use that employ atomic, molecular, chemical, and particulate processing in a vacuum or liquid medium, but most of these techniques are expensive, as well as inefficient in materials and energy use. Therefore, second-generation nanotechnology is focused on clean technologies that minimize possible environmental and human health risks associated with manufacture and fabrication; there is an ever-growing demand for the development of clean, non-toxic, and environmentally benign synthesis procedures [Daniel and Astruc, 2004; Manivasagan and Kim, 2015].

There has been a convergence in recent years between biological-based technologies, green chemistry, and nanotechnology. This convergence aims to create new materials and production processes which reduce or eliminate the use of dangerous substances (Fawcett et al., 2017). Due to their unique particle size and shape dependence and their physical, chemical, and biological properties, biologically synthesized nanoparticles are of great interest in the field of biology and medicine. Microorganisms including bacteria, fungi, and algae have been suggested as possible environmentally friendly nano-factories for metal nanoparticle synthesis [Daniel and Astruc, 2004; Manivasagan and Kim, 2015].

Marine bio-nanotechnology represents an exciting and emerging research area. The marine ecosystem is biologically complex and has great potential for nanoscience and nanotechnologies. Marine organisms produce a remarkable 1–100 nm nanoparticles which form nano fabric structures such as seashells, pearls, and fish bones. Diatoms and sponges are constructed with the nanostructured cover of silica and coral reefs with calcium, arranged in remarkable architectures. Dolphins and whales have rough skin surface due to the presence of a nano ridge. These ridges enclose a pore size of 0.2 μm^2 which is below the size of marine fouling organisms and hence, there is no attachment of bio-foulers. Despite great potential, the progress in research in marine bio-nanotechnology is not adequate. Most of the studies on the biosynthesis of nanoparticles have been restricted to terrestrial organisms [Asmathunisha and Kathiresan, 2013].

The biologically diverse marine environment covers approximately 70 percent of the surface of the earth and is largely unexplored. The researchers have focused on the synthesis of nanoparticles from marine sources in recent years, and as such, they are both biocompatible and biodegradable which includes seashells, pearls, and fish bones, and the particles ranged from 1 to 100 nm size. Recent studies have shown that several marine plants can perform as bio-factories for the production of nanoparticles [Fawcett et al., 2017].

The main purpose of the chapter is to provide detailed advances in the biosynthesis and application of environmentally friendly nanomaterials from marine organisms. Therefore, we first summarize the applications of nanoparticles from marine organisms. In the second part, the biosynthesis of nanoparticles by marine microorganisms is reviewed.

16.2 GREEN SYNTHESIS OF NANOPARTICLES

The nanotechnology field is one of the most exciting research fields of modern material science. There is currently an increasing need to develop environmentally friendly and sustainable methods for the synthesis of nanomaterials that do not use toxic chemicals in synthesis protocols to prevent adverse effects in medical applications [Manivasagan and Kim, 2015].

Different synthetic methods have been employed for the preparation of NPs with diverse morphology and size. Although these methods have resulted in superior NPs, still a key understanding of improved manufacturing process is required which could be exploited at the industrial and commercial level to have better built, long-lasting, cleaner, safer, and smarter products like home appliances, communication technology, medicines, transportation, agriculture, and industries. Therefore, the main focus is to design NPs using environmentally benign approaches. These provide solutions to growing challenges related to environmental issues [Sharma et al., 2015].

The manufacture of nanoparticles can be broadly defined into two approaches, as shown in Figure 16.1. The first is the top-down approach and involves a material substantially reduced in size through physical or chemical processes. The resulting particle size, shape, and surface structure depend heavily on the technique employed during size reduction. The second method, bottom-up, builds nanoparticles by assembling atoms, molecules, and smaller particles or monomers. Unfortunately, many of the chemical and physical processes used in both approaches suffer from several disadvantages, such as low conversion rates of materials, are technically complex, require high energy requirements, and are relatively expensive. Also, many of these processes employ harmful chemicals such as reducing agents, organic solvents, and stabilizing non-biodegradable agents [Fawcett et al., 2017; Birnbaum and Pique, 2011].

Accordingly, research in recent years has focused on manufacturing nanomaterial via nanotechnology-based processes that promote the principles of green chemistry and reduce or eliminate the use of hazardous chemicals. Thus, ecofriendly green nanotechnology-based processes for the manufacture of nanoparticles have attracted considerable interest worldwide. To emphasize this alternative approach, recent research has focused on using biological entities to synthesize a wide variety of nanoparticles. Biosynthesis via unicellular and multicellular biological entities such as actinomycetes, bacteria, fungus, marine algae, plants, viruses, and yeast offer alternative eco-friendly approaches for producing nanoparticles. Each of these biological entities, to varying degrees, can perform as natural bio-factories for producing particular nanoparticles. Each of the biological entities has active molecules and compounds that can act as reducing agents and stabilizing agents to synthesize nanoparticles with diverse sizes, shapes, compositions, and physicochemical properties [Fawcett et al., 2017; Shah et al., 2015].

The biological approach includes different types of microorganisms that have been used to synthesize different metallic NPs, which has advantages over

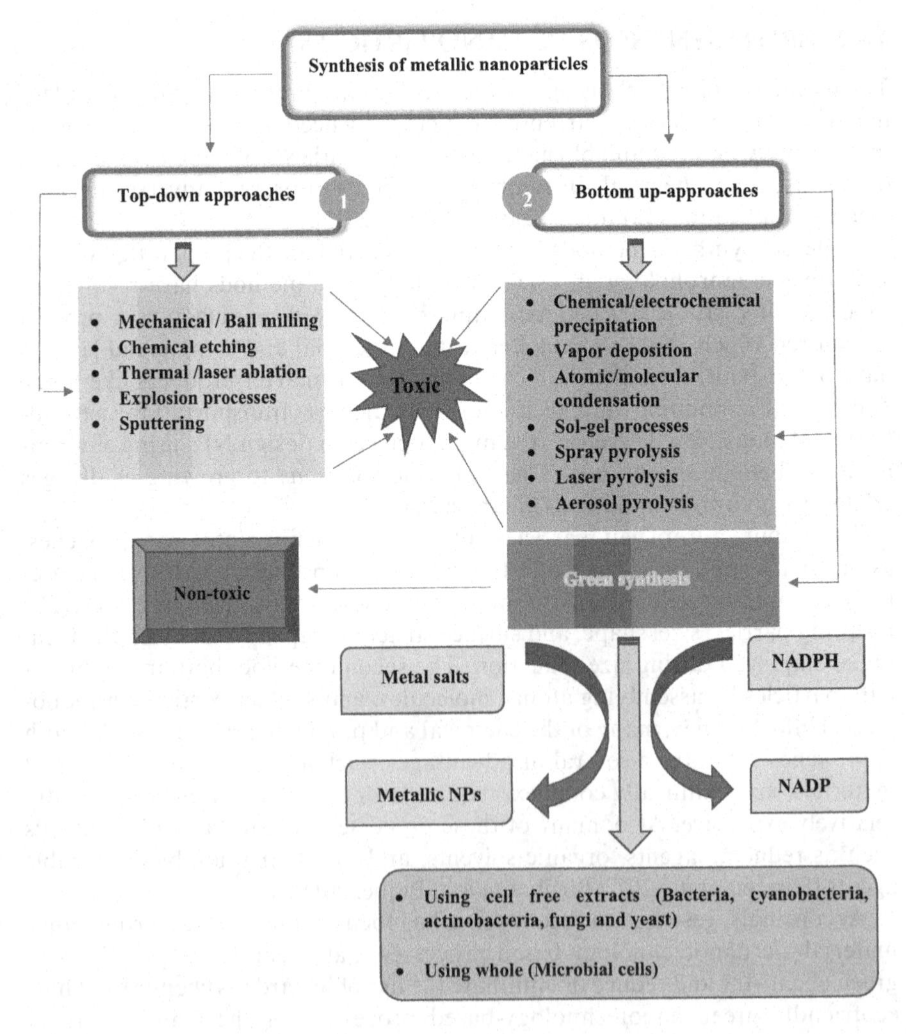

FIGURE 16.1 Synthesis of metallic nanoparticles.

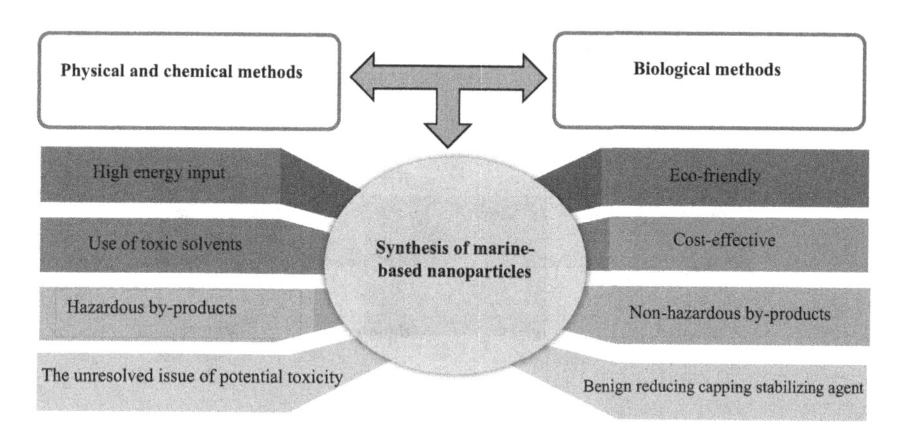

FIGURE 16.2 The disadvantages of physical and chemical methods in comparison with the advantages of biological methods.

other chemical methods as this is greener, energy-saving, and cost-effective, as seen in Figure 16.2. The coating of biological molecules on the surface of NPs makes them biocompatible in comparison to the NPs prepared by chemical methods. The biocompatibility of bio-inspired NPs offers very interesting applications in biomedicine and related fields. At present, there is no detailed understanding of the mechanisms behind the formation of nanoparticles by an extract from marine organisms. However, studies have revealed that many of the marine biomolecules can act as biocatalysts to assist in the reduction of precursor metal salts to nucleate metal and metal oxide nanoparticles, while other larger amphiphilic biomolecules act as surfactants, which direct and control nanoparticle growth [Fawcett et al., 2017; Sharma et al., 2015].

As shown in Figure 16.3, due to the presence of different bioactive molecules in cellular extracts of various organisms like plant, fungi, bacteria, and algae, these extracts are used in the synthesis of biogenic nanoparticles. These bioactive molecules can function as reducing, capping, and/or stabilizing agents during the synthesis [Gautam et al., 2019].

Biomolecules present in marine extracts also can influence particle size, morphology, composition, and physicochemical properties of the synthesized nanoparticles. Ultimately, it is the effective control of these properties that define the quality of the nanoparticles. However, a considerable amount of research is needed to identify and determine the role of specific biomolecules involved in the formation process and the influence of individual biomolecules in dictating nanoparticle growth mechanisms [Fawcett et al., 2017; Mittal et al., 2013].

It is believed that the growth of nanoparticles in solution begins with metal ions being converted from their mono or divalent oxidation states to zero-valent states. This is followed by metal ion reduction and subsequent nucleation. Following initial nucleation, a kinetically controlled process takes place in which smaller neighboring particles attach to low energy faces of the forming crystal to create larger thermodynamically stable nanoparticles. As growth

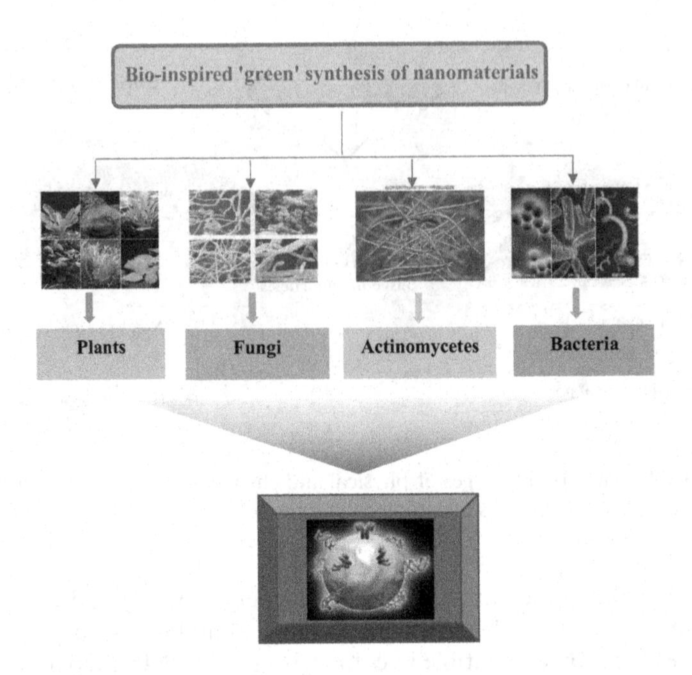

FIGURE 16.3 Schematic representation of the biogenic synthesis of nanoparticles.

progresses, biomolecules contained within marine extract act as natural surfactants (capping agents) on specific facets of the forming crystal. The adsorption of the capping agents and their subsequent interactions on the crystal facets reduce interfacial energy and lower surface tension. The modified surface's properties of the crystal facet tend to influence the orientation and assembly of subsequent growth. Thus, growth occurs in preferential planes and explains the morphologies seen in biosynthesized nanoparticles. Typical morphologies produced via biosynthesis include cubes, hexagons, pentagons, rods, spheres, triangles, and wires [Fawcett et al., 2017; Chiu et al., 2013].

The biogenic synthesis of nanoparticles is relatively straightforward at room temperature and begins by mixing a metal salt solution with an aqueous solution containing either marine organism's extract. Reduction starts immediately, and a color change in the reaction mixture indicates the formation of nanoparticles [Fawcett et al., 2017; Nagarajan and Kuppusamy, 2013]. Over time, the small neighboring nucleonic particles in the reaction mixture start agglomerating to form larger and more thermodynamically stable nanoparticles. The aggregation and self-assembly of the nucleonic particles are assisted by biomolecules present in the marine organism's extract. During self-assembly, the most energetically favorable and stable particle shapes are formed. Typical shapes formed include cubes, hexagons, pentagons, rods, spheres, triangles, and wires [Fawcett et al., 2017].

Controlling experimental parameters such as (1) extract concentration (2) metal salt concentration (3) reaction time (4) reaction solution pH and

(5) temperature will determine the quality, size, and shape of the biosynthesized nanoparticles [Shah et al., 2015]. Since these parameters have a significant effect on the quality and properties of the synthesized nanoparticles [Fawcett et al., 2017; Mittal et al., 2013].

16.3 APPLICATIONS OF NANOPARTICLES FROM MARINE ORGANISMS

Nanotechnology leads to the development of revolutionary products and technologies for a wide range of applications in the field of medical and healthcare, consumer products such as cosmetics, agriculture, information technology, energy production and storage, aerospace engineering, transportation, vehicles and infrastructures, material science, food, water and environment, instruments, and security. Nanoparticles have unique physical, chemical, and biological properties compared to larger-sized particles of the same substance. Owing to the small dimension, the surface area of the nanoparticles is maximized, leading to the maximum reactivity to the weight ratio. These properties provide a greater surface area to react on targets, superior chemical and biological reactivity, efficiently uptake into individual cells, and even cell organelles. In recent years the number of products containing nanoparticles and their potential applications in various fields continues to grow exponentially. The products containing nanoparticles are already being used or tested in various consumer products such as sunscreens, composites, electronic devices, chemical catalysts, and medical devices. In biology, nanoparticles can be used in disease diagnostic kits, biological imaging, antimicrobial agents, drug delivery systems, biomarkers, cell labeling, and nano drugs for the treatment of diseases [Singh, 2020].

Metallic nanoparticles offer a promise for the development of novel nanomedicines. Nanomedicines should be able to overcome the limitations of human diseases at the nanoscale level at which biomolecules are acting. Metallic nanoparticles are usually applied to the antibacterial drug carriers. Nano-based formulations of various antimicrobial drugs have been shown to improve either pharmacokinetics or antibacterial efficiency by achieving sustained release directly at the infection site. Drug encapsulation and delivery via nanoparticles may also help prevent adverse effects. The development of nanodevices using marine biomaterials and their use in the wide array of various applications on living organisms have recently attracted the attention of biologists toward bio-nanotechnology [Manivasagan et al., 2016].

16.4 ANTIMICROBIAL ACTIVITY

Various in vitro antimicrobial studies have shown that the MNPs check the growth of various microbial species. For such antimicrobial properties, the size and material used for the synthesis of MNPs are very crucial. During the past few years, the pathogenic microbes are developing resistance against various relatively higher doses of antibiotics, so there is a considerable threat to the health of our society. Various microbes have developed antibiotic-resistant strains against narrow as well as broad-spectrum antibiotics such as penicillin, sulfonamide, vancomycin, and methicillin. So, to overcome the risk of antibiotics resistance in microbes, there is an urgent need to find alternative ways to kill such drug-resistant bacteria. The nanoparticles can be seen as a hope for the killing of such pathogenic bacteria. Due to their novel and advanced features, metal nanoparticles appear to be a multifaceted material as compared to their larger particles from which they are derived. The important features exhibited by MNPs are greater surface-area-to-volume ratio, antimicrobial activity, thermal conductivity, nonlinear-particles performance, and chemical steadiness. The greater surface area to volume ratio of MNPs is crucial for their catalytic properties [Kamal et al., 2010]. Due to these properties, MNPs are also useful in the drug-delivery system, antisense, tissue engineering, medical diagnostics, gene therapy applications, etc. The antibacterial properties of silver nanoparticles (AgNPs) are also due to the large surface area of AgNPs, which tend to come in contact with a large number of microbial cells in contrast to their large size particles [Pandey et al., 2020].

The antimicrobial study of the bionanomaterials was performed because of the increase in novel strains of microorganisms that are resistant to most potent antibiotics resulting in novel research into the well-known activity of gold and gold-based compounds, such as gold nanoparticles (AuNPs). The effect was dependent on size and dose, and more distinct against pathogenic bacteria and fungi [Manivasagan et al., 2016].

AgNPs exhibit potential antimicrobial properties against infectious microbes such as *Escherichia coli*, *Bacillus subtilis*, *Vibrio cholerae*, *Pseudomonas aeruginosa*, and *Staphylococcus aureus*. The application of nanomaterials as new antimicrobials provides novel modes of action on different cellular targets in comparison with existing antibiotics. Multiple drug resistance to traditional antibiotics has created a great requirement for the development of new antimicrobial agents. Bacteria are classified as gram-negative or gram-positive. The peptidoglycan is the key component of the bacterial cell wall. Gram-negative bacteria have only a thin peptidoglycan layer (~2–3 nm) between their two membranes, while Gram-positive bacteria lack the outer membrane (substituted by a thick peptidoglycan layer). Smaller sized NPs disrupt the function of the membrane (such as permeability or respiration) by attaching to its surface and subsequently, penetrating the cell and cause further damage by interacting with the DNA. The antimicrobial properties of Ag encourage its use in biomedical applications, animal husbandry, food packaging, water purification, cosmetics, clothing, and numerous household products. Now, Ag is the engineered nanomaterial most commonly used in consumer products. Clothing, respirators, household water filters, contraceptives, antibacterial sprays, cosmetics, detergent, dietary supplements, cutting boards, socks, shoes, cell phones laptop keyboards, and toys are among the retail products that purportedly exploit the antimicrobial properties of Ag nanomaterials. Several researchers investigated the antimicrobial efficacy against different bacterial and fungal pathogens [Ramkumara et al., 2016; Franci et al., 2015; Prabhu and Poulose, 2012].

16.5 ANTICANCER ACTIVITY

With the advent of nanotechnology, different types of nano-sized materials have been developed for the wider applications from material science to biological science in general and nanomedicine in particular. NPs with the size ranges between 1 and 100 nm are mainly explored for the diagnosis and treatment of human cancers which led to the new discipline of nano-oncology. Recently, biologically synthesized metal NPs were investigated against a different type of cancer cell lines under in vitro conditions and showed potent activity against respective cancer cells [Ramkumara et al., 2016; Yezhelyev et al., 2006].

Nanoparticles have a greater surface area per weight than larger particles, and this property makes them more reactive to certain other molecules, and they are used or being evaluated for use in many fields. Quantum dots are the crystalline nanoparticles used to identify the location of cancer cells in the body. Gold nanoparticles allow heat from infrared lasers to detect cancer tumors. The iron oxide nanoparticles are used in better diagnosis of tumors by magnetic resonance imagining (MRI) scans. Once the nanoparticles are attached to the tumor, their magnetic property enhances the images of the scan. In addition to diagnosis, the nanoparticles are used in drug delivery and removal of new tumors. Magnetic nanoparticles that attach to cancer cells in the bloodstream may allow the cancer cells to be removed before they establish new tumors. The nanoparticles coated with proteins can be attached to damaged portions of arteries. This can allow the delivery of drugs directly to the damaged portion of arteries to fight against cardiovascular disease. Intravenous injection of gold nanoparticles (~2 nm in diameter) can enhance radiotherapy (x-rays) and results in the eradication of subcutaneous mammary tumors in mice and it is proved to increase survival of the test animal to 86 percent as against 20 percent with x-rays alone. Gold nanoparticles are non-toxic to mice and are cleared from the body through the kidney [Manivasagan and Kim, 2015; Asmathunisha and Kathiresan, 2013)].

Chitosan is an important component of the exoskeleton of marine crustaceans and fungi. The chitosan nanoparticles exhibit potential antibacterial activity. They do have optical sensing properties and can also easily bind with other molecules. This has the potential to develop nanotechnology devices in detecting avian flu, breast cancer, and also toxic substances in the air, water, and soil. The chitosan nanomaterial incorporated with the drug can be prepared in microcapsules. These will open with specific biochemical activity related to the tumor. Since the chitosan nanoparticles can be easily detected as the fluorescence when exposed near infra-red light. Using the technique, even tiny breast cancer of 1–2 mm can be detected [Manivasagan and Kim, 2015].

16.6 DISEASE DIAGNOSTIC APPLICATIONS

Disease diagnostics is a crucial step for curing disease. Conventionally diagnostic methods depend on the appearance of symptoms after illness for most disorders, which delay the treatment period. Therefore, it is the primary objective to detect disease early for better treatment. Nanotechnology currently plays an important role in the development of disease diagnosis available, resulting in much higher sensitivity and better efficiency and economy. Several nanomaterials such as quantum dots (semiconductor nanoparticles), gold nanoparticles, and iron oxide are being investigated to construct nanosensors designed for the diagnosis of diseases. Several research studies have shown that nanoparticle-based techniques are essential in detecting and diagnosing cancerous cells and virus-infected cells such as HIV and anthrax virus [Singh, 2020].

16.7 DRUG DELIVERY

The drug delivery system is important for achieving the desired therapeutic effect of medicines and the development of new drugs. The drug delivery system maintains the concentration and stability of drugs over time, which is helpful to sustain the release of drugs at the target site [Singh, 2020].

16.8 ANTIBIOFILM ACTIVITY

Several studies revealed that "nano-functionalized materials" inhibit bacterial adhesion and biofilm formation on the surfaces by coating techniques, impregnation, or embedding nano-materials. Inorganic nano-sized materials such as Ag, CuO, and ZnO are a potential source of antimicrobial properties, which inhibit cell adhesion and destabilize the biofilm matrix. Nanostructured metal ions displayed higher efficiency because of their large surface area, which is inversely proportional to their particle size. Earlier reports demonstrated that ZnO, CuO, and AgNPs coated on surfaces of interest inhibit the biofilm formation. The cellular mechanisms underlying microbial biofilm formations have not yet been completely understood. Targeting novel receptors involved in biofilm formation is the best strategy to control problems caused by the biofilm in marine environments. There is a significant need to find out the eco-friendly antibiofilm products from the natural sources of both terrestrial and marine environments [Ramkumara et al., 2016; Lellouche et al., 2009].

16.9 ANTIFOULING PROPERTIES

Numerous anti-biofouling measures such as mechanical, chemical, and biological methods are in practice but their effects on the anti-biofouling are not remarkable. On contrary, the commercially available antifouling paints are highly toxic to the unintentional aquatic organisms. Application of antifouling compounds from natural sources is considered one of the best replacement options for the antifouling processes [Ramkumara et al., 2016].

It has been reported that the nanoparticle coating composed of titanium dioxide (TiO_2) and silver nanoparticles prevents the biofouling of ship hulls. The biofouling of ship hulls is a very serious problem in fuel consumption, biocorrosion, and ultimately heavy economic loss to the shipping industry throughout the world [Manivasagan and Kim, 2015].

A wide variety of marine natural products from seagrasses, seaweeds, mangroves, coral reefs, sponges, and their associated organisms proved to be an excellent source of bioactive compounds and a wide range of secondary metabolites, many of which exhibited a broad spectrum of antifouling activity against marine biofilm-forming microbes, algal spore adherence, mussel (phenoloxidase activity) and barnacles to the artificial substratum. Compared to the earlier anti-foulers, nanomaterials such as Ag, TiO2, ZnO, MnO, MgO, Ag/Pd, Ag/TiO_2, Ag/ZnO, TiO2/ZnO, Ti/Mn, Cu/Mn, etc., are efficient in inhibiting the bacterial adhesion, biofilm formation, and macrofouler attachment due to their effective property and large specific surface area, which is inversely proportional to their particle size and shape. Ag, CuO, ZnO, and magnesium fluoride NPs have good antimicrobial properties and also reduced the cell adhesion and destabilized the biofilm matrix. Previous reports demonstrated that ZnO, CuO, and AgNPs coated on solid surfaces inhibit the biofilm formation, and particularly AgNPs act as potential antimicrobial agents against different marine biofilm-forming microorganisms [Ramkumara et al., 2016; Prakash et al., 2015].

16.10 REMOVAL OF TOXIC CHEMICALS

Nanotechnology plays a significant role in the elimination of toxic chemicals found in soil. Several environmental protection agencies were worked on contaminated sites to test the efficiency of nanoparticles, especially nano zerovalent iron for removal or degradation of environmental contaminants. Stevenson et al. [2017] evaluated the effect of sulfurized nano zerovalent iron (FeSSI) on common freshwater alga (*Chlamydomonas reinhardtii*) against cadmium toxicity. They found that FeSSI absorbed cadmium from the water body and lighten the toxicity of cadmium on *C. reinhardtii* for more than a month. Iron and other nanoparticles from iron rust can be used to remove contaminants such as pesticides, heavy metals, and radionuclides from the soil [Singh, 2020].

16.11 ORGANIC POLLUTANT REMOVAL

Water pollution with organic pollutants has significantly increased as a result of the expansion of urban, agricultural, industrial, and manmade activities. Organic pollutants are comprised of broad groups of chemical pollutants mainly made up of carbon and hydrogen with smaller amounts of other atoms such as halogens, nitrogen, sulfur, and phosphorous. Xenobiotics like pesticides, insecticides, synthetic dyes, pharmaceuticals, aromatic hydrocarbons, halogenated hydrocarbons, and phenols are typical organic pollutants [Gautam et al., 2019].

16.12 AGRICULTURE

Nanotechnology has been used in agriculture at all levels from germination to storage. New innovative techniques such as precise farming; enhancing the ability of plants to absorb nutrients; disease detection; and control, effective processing, storage, and packaging enhance the quality of agriculture, which is essential in the current scenario to fulfill the food demand of the world. Nanoparticles are being developed as biofertilizer in agriculture to overcome the limitations of conventional farming. Nanoparticles have the potential to enhance the fertility of the soil by providing nutrients to plants at an optimum level. Nanostructured materials having nutrients essential for plant growth and development in aqueous solution and hydrogels are being studied for use in growing plants or crops [Singh, 2020].

16.13 REMOVAL OF HEAVY METALS

Heavy metals are well known for their unfavorable impact on the environment and human health. Heavy metals like cadmium (Cd), lead (Pb), arsenic (Ar), chromium (Cr), mercury (Hg), and nickel (Ni) are highly poisonous and toxic even at low concentration due to their tendency for bioaccumulation. The continuous increase in the usage of heavy metal ions over the last few decades has resulted in an enlarged flux of metallic substances in the aquatic and terrestrial environment. Acid mine drainage, industrial and domestic effluents, agricultural run-off are the major contributors to heavy metal contamination of water [Gautam et al., 2019].

16.14 INSECTICIDES AND PESTICIDES

Crop loss to the tune of 30 percent in plants is due to insect pests infesting several crop plants. The use of chemical insecticides and pesticides in crop protection disturbs soil health, water bodies, and finally, it affects human health. The potential application and benefits of nanotechnology are enormous. Nanotechnology in agriculture plays an important role in the slow-release effects which includes pest control with increased shelf-life to various applications in the agricultural fields. More numbers of nanoparticles have been developed using marine organisms like plants, animals, microbes, etc., for a variety of applications. But very few findings were reported for the insect pest management. It needs more attention for crop protection, to meet the satisfactory level of production, and to increase our economic status of the country. The agricultural application of nanotechnology can suggest the development of efficient and potential implications for overcoming the management of pests in crops. Nanoparticles can be used in the formulations of pesticides, insecticides, insect repellents, pheromones, and fertilizers [Asmathunisha and Kathiresan, 2013; Singh et al., 2015].

16.15 FOOD INDUSTRY

The most significant and challenging aspect of the food industry is to supply safe, contamination-free, and enhanced nutritional value of food to the consumers. The nanoparticle is used in various fields of the food industry, including the processing of food, packaging of food, enhance the nutritional value of food and promotes food safety, detection of foodborne pathogens, and shelf-life extension of food and food products [Singh, 2020].

16.16 BIOSYNTHESIS OF NANOPARTICLES BY MARINE MICROORGANISMS

The bacterial cells perform various bio-processes like biomineralization, bioleaching, and bioaccumulation to solubilize the metal ions by altering their oxidation state through reduction and/or oxidation. The ability of biotransformation of metals has fueled the interest of researchers to utilize the bacterial cells as potential nano-factories for the fabrication of NPs. Bacterial cells can synthesize metallic nanoparticles by employing their defensive and protective strategy against soluble metal ions. The relative ease of manipulation of bacterial cells makes them suitable for intracellular or extracellular biosynthesis of nanoparticles. The bacterial synthesis is slower as compared to the chemical or physical methods. However, the period of formation of nanoparticles may be reduced by coupling it with physical energy sources like microwave and ultrasound [Gautam et al., 2019].

Microorganisms are known to synthesize inorganic nanoparticles such as gold, silver, calcium, silicon, iron, gypsum, and lead, in nature either inside or outside cells. At present, microbial methods in the synthesis of nanomaterials of varying compositions are extremely limited and confined to metals, some metal sulfide, and very low oxides. All these are restricted to the microorganisms of terrestrial origin [Asmathunisha and Kathiresan, 2013].

Marine microorganisms, such as bacteria, cyanobacteria, actinobacteria, yeast, and fungi are tiny organisms that live in marine ecosystems (Figure 16.4).

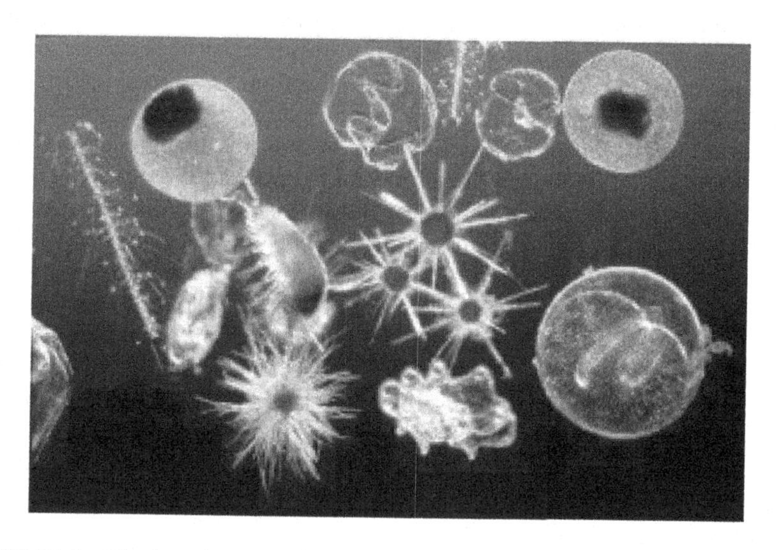

FIGURE 16.4 Marine microorganisms.

These microorganisms are the best sources of metabolite producers and are very important from an industrial point of view. Marine microorganisms are prokaryotic and eukaryotic cells that live in the ocean and account for > 98% of ocean biomass. Marine microorganisms are ubiquitous in the marine environment as well as extreme environments (e.g., hypersaline) and thrive at a wide range of acidity, alkalinity, temperatures, and salinity [Manivasagan et al., 2016].

Marine microbes have the potential ability to synthesis nanoparticle for the reason that the marine microbes exist in the sea bottom, over millions of years in the past for reducing the vast number of inorganic elements deep in the sea. It is important to study the marine microbes for the biosynthesis of nanoparticles and to elucidate biochemical pathways that lead to metal ion reduction by the different classes of microbes to develop nanoparticles. The biosynthesis of nanoparticles with the use of microorganisms depends on culture conditions and hence standardizing these conditions for the high synthesis of nanoparticles is necessary. Many marine microorganisms are known to produce nanostructured mineral crystals and metallic nanoparticles with properties similar to chemically synthesized materials, while they have strict control over the size, shape, and composition of particles [Asmathunisha and Kathiresan, 2013].

Synthesis of nanoparticle using microbes offers better size control through compartmentalization in the periplasmic space and vesicles. The rate of intracellular particle formation and therefore, the size of the nanoparticles could, to an extent, be manipulated by controlling parameters such as pH, temperature, substrate concentration, and time of exposure to the substrate. Marine microbes play several important roles in the synthesis of nano-based drugs for human life improvement. Additionally, nanoparticles synthesized by microorganisms tend to be stabilized by peptides such as phytochelatins, thus preventing aggregation. These short peptides are synthesized in response to heavy metal stress and have been implicated as a universal mechanism to sequester metal ions in bacteria and fungi [Asmathunisha and Kathiresan, 2013; Singh et al., 2015].

In recent years, using marine microorganisms for the green synthesis of nanoparticles is still frequently emerging to understand the molecular mechanisms of biosynthesized nanoparticles. Biosynthesis of metallic nanoparticles using bacteria and fungi has attracted a special interest compared with the biosynthesis of metallic nanoparticles using actinobacteria and yeast because a well-developed technology is available for biosynthesis using bacteria and fungi than actinobacteria and yeast. Therefore, using marine microorganisms to synthesize nanoparticles is safer because they are not pathogenic to humans. However, marine microorganisms remain relatively unexplored for metallic nanoparticle synthesis from a human perspective [Manivasagan et al., 2016; Zhang et al., 2011; Hulkoti and Taranath, 2014]. A selection of biosynthesized nanoparticles by marine microorganisms and their biological activity is shown in Table 16.1.

TABLE 16.1
A Selection of Biosynthesized Nanoparticles via Marine Microorganisms and Their Biological Activity

Organisms	Species	Nanoparticle	Biological Activity	Ref.
Cyanobacteria	*Spirulina platensis*	Silver Gold Bimetallic	-	[Asmathunisha and Kathiresan, 2013] [Pandey et al., 2020] [Singh et al., 2015]
	Nostoc ellipsosporum	Gold	-	[Saxena, 2016] [Pandey et al., 2020]
	Oscillatoria willei	Silver	-	[Asmathunisha and Kathiresan, 2013] [Pandey et al., 2020] [Singh et al., 2015]
	Microcoleus cthonoplastes	Gold	-	[Saxena, 2016]
	Turbinaria onoides	Gold	Antimicrobial	[Singh et al., 2015]
	Lyngbya majuscule	Gold	-	[Pandey et al., 2020] [Manivasagan et al., 2016]
	Phormidium valderianum	Gold	-	[Saxena, 2016]
	Ph. tenue	Cadmium	-	[Asmathunisha and Kathiresan, 2013]
	Plectonema boryanum UTEX 485	Platinum Palladium Silver	-	[Pandey et al., 2020]
	Valderianum sp.	Silver	-	[Pandey et al., 2020]
	Gloeocapsa sp.	Silver	-	[Pandey et al., 2020]
	Phormidium sp.	Silver	-	[Pandey et al., 2020]
	Lyngbya sp.	Silver	-	[Pandey et al., 2020]
	Limnothrix sp. 37–2–1	Silver	-	[Pandey et al., 2020]
	Anabaena sp. 66–2	Silver	-	[Pandey et al., 2020]
	Synechocystis sp.*48–3*	Silver	-	[Pandey et al., 2020]
	Spirulina subsalsa	Gold	-	[Pandey et al., 2020]
Bacteria	*E. coli*	Silver	Antimicrobial	[Asmathunisha and Kathiresan, 2013] [Singh et al., 2015]
	V. alginolyticus	Silver	-	[Manivasagan et al., 2016]
	Marinobacter pelagius	Gold	-	[Manivasagan et al., 2016]
	Klebsiella pneumoniae	Gold	-	[Manivasagan et al., 2016]

(Continued)

TABLE 16.1 ***(Continued)***
A Selection of Biosynthesized Nanoparticles via Marine Microorganisms and Their Biological Activity

Organisms	Species	Nanoparticle	Biological Activity	Ref.
	Saccharophagus degradans	Manganese	-	[Manivasagan et al., 2016]
	B. subtilis	Silver	Antifungal	[Singh et al., 2015]
	Shewanella algae	Silver	-	[Manivasagan et al., 2016]
	P. fluorescens	Silver	Antimicrobial	[Singh et al., 2015]
	P. aeruginosa	Silver	Antifungal Antimicrobial	[Asmathunisha and Kathiresan, 2013] [Singh et al., 2015]
	P. aeruginosa	Cadmium	Bioremediation	[Manivasagan et al., 2016]
Yeast	*Pichia capsulata*	Silver	-	[Asmathunisha and Kathiresan, 2013] [Singh et al., 2015]
	Candida sp. VITDKGB	Silver	Antimicrobial	[Manivasagan et al., 2016]
	Saccharomyces cerevisiae	Manganese	-	[Manivasagan et al., 2016]
	Yarrowia lipolytica	Gold	-	[Manivasagan et al., 2016]
	Rhodosporidium diobovatum	Lead	-	[Asmathunisha and Kathiresan, 2013] [Singh et al., 2015]
Fungi	*Penicillium fellutanum*	Silver	-	[Asmathunisha and Kathiresan, 2013] [Singh et al., 2015]
	Rhizopus oryzae	Gold	-	[Manivasagan et al., 2016]
	Thraustochytrium sp.	Silver	-	[Asmathunisha and Kathiresan, 2013]
	Aspergillus niger	Silver	Antimicrobial	[Asmathunisha and Kathiresan, 2013] [Singh et al., 2015]
	A. flavus	Silver	-	[Vijayan et al., 2018]
Actinomycetes	*Streptomyces hygroscopicus*	Gold	-	[Singh et al., 2015]
	Streptomyces sp.	Gold	Antimalarial	[Singh et al., 2015]
	Nocardiopsis sp.	Gold	Antimicrobial Antioxidant cytotoxic	[Manivasagan et al., 2016]
	Thermoactinomyces sp.	Silver	Antibacterial	[Manivasagan et al., 2016]

16.17 BIOSYNTHESIS OF NANOPARTICLES BY MANGROVE AND SALT MARSH PLANTS

Plants are an important, safe, and easily available source for nanoparticles synthesis with broad variability of metabolites that may aid in reduction. A number of plants are being currently investigated for nanoparticle synthesis for their efficacy and so much research has been done with plants concerning phytochemicals. The main phytochemicals responsible for their activity have been identified as terpenoids, flavones, ketones, aldehydes, amides, and carboxylic acids. In this regard, plants and plant part extracts-based biosynthesis is cost-effective and eco-friendly. The environmental conditions of the marine ecosystem and the characterization of marine plants are extremely different from the terrestrial ecosystem. Marine environmental conditions are extremely diverse from terrestrial and are an excellent source of various types of bioactive compounds. Therefore, the marine plants might produce different types of bioactive compounds including polyphenols, flavonoids, alkaloids, and tannins. The marine plants have antibacterial, anti-plasmodial, antiviral, antioxidant, anticancer activities and also proved to have a high content of secondary metabolites [Asmathunisha and Kathiresan, 2013; Singh et al., 2015, Ravikumar et al., 2011].

Particularly, the biosynthesis of nanoparticles from mangroves and mangrove associates are very limited [Asmathunisha and Kathiresan, 2013; Singh et al., 2015, Ravikumar et al., 2011]. Costal plants, especially mangroves and mangrove associates, are a good source of nanoparticles (Figure 16.5). Nevertheless, studies emphasizing the synthesis of NPs using marine plants such as seagrass, seaweeds, mangroves, and plants of coastal origin are scanty. An efficient and eco-friendly one-pot green synthesis of AgNPs using mangrove leaf buds has been reported [Ramkumara et al., 2016, Palaniappan et al., 2015, Kumar et al., 2013]. Table 16.2 shows a selection of biosynthesized nanoparticles via mangrove, seagrass, and salt marshes and their biological activity.

FIGURE 16.5 Mangrove and salt marsh plants.

TABLE 16.2
A Selection of Biosynthesized Nanoparticles via Mangrove, Seagrass, and Salt Marshes and Their Biological Activity

Organisms	Species	Nanoparticle	Biological Activity	Ref.
Mangroves	*Xylocarpus mekongensis*	Silver	Antimicrobial	[Asmathunisha and Kathiresan, 2013]
	Rhizophora mucronata	Silver	Larvicidal Antimicrobial	[Asmathunisha and Kathiresan, 2013] [Singh et al., 2015]
	Avicennia marina	Silver	Antimicrobial	[Singh et al., 2015]
	Rhizophora apiculata	Silver	Antibacterial	[Singh et al., 2015]
Salt marshes	*Sesuvium portulacastrum*	Silver	Antimicrobial	[Asmathunisha and Kathiresan, 2013]
	Suaeda monoica	Silver	Anticancer	[Singh et al., 2015]
Seagrass	*Cymodocea serrulata*	Silver	Anticancer	[Vijayan et al., 2018]
	C. serrulata	Silver	Antioxidant Cytotoxicity	[Vijayan et al., 2018]
	Syringodium isoetifolium	Silver	Antimicrobial Hemolytic	[Vijayan et al., 2018]

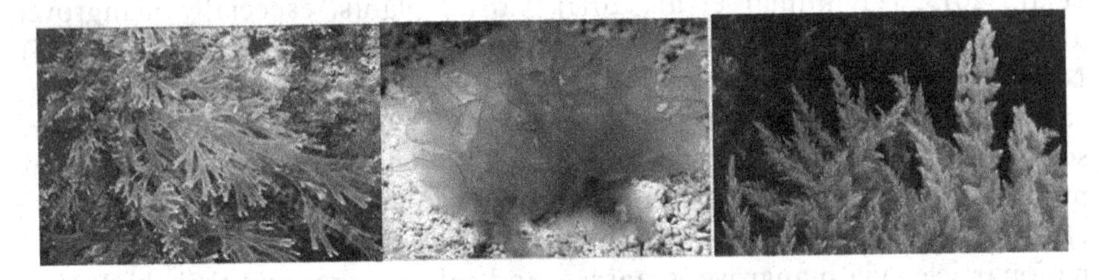

FIGURE 16.6 Marine algae (brown, green, and red algae).

16.18 BIOSYNTHESIS OF NANOPARTICLES BY MARINE ALGAE

Marine algae are very diverse and exist almost everywhere on earth (Figure 16.6). Algae are widely used in food, medicine, and manufacturing industries. These photoautotrophic organisms are rich in bioactive molecules (such as polysaccharides (alginate, laminarin, fucoidan), polyphenols, carotenoids, fiber, protein, vitamins, and minerals) having carboxyl, hydroxyl, and amine functional groups which serve as reducing as well as a capping agent to form metal NPs. This leads to the origin of a new promising field of research called "phyco-nanotechnology" [Asmathunisha and Kathiresan, 2013; Gautam et al., 2019; Singh et al., 2015].

Algae extracts are continuously explored for the production of metallic nanoparticles because it eliminates the elaborate process of maintaining cell cultures, and nanoparticles are produced extracellularly. Algae are also being used as a "bio-factory" for the synthesis of metallic nanoparticles. Among different genres of reductants, seaweeds have distinct advantages because of their high metal uptake capacity, low cost, and macroscopic structure [Manivasagan and Kim, 2015; Davis et al., 2003].

The use of natural renewable marine resources like seaweed-derived polysaccharides has been exploited for many years by the food industry. The cell walls present in seaweeds are composed of polysaccharides. These polysaccharides mainly consist of small sugar units linked via glycosidic bonds, which have hydrophilic surface groups such as carboxyl, hydroxyl, and sulfate groups. Typical polysaccharides found in seaweeds include agar, alginate, carrageenan, fucoidan, and laminarin. Other biomolecules found in the cell walls include proteins and enzymes. The presence of these bioactive materials has attracted considerable interest in recent years and has resulted in the creation of food products, renewable bioenergy, and biomedical applications [Fawcett et al., 2017; Smit, 2004; Venkatesan et al., 2016]. In Table 16.3 a selection of biosynthesized nanoparticles by marine algae and their biological activity is presented.

TABLE 16.3
A Selection of Biosynthesized Nanoparticles via Marine Algae and Their Biological Activity

Organisms	Species	Nanoparticle	Biological Activity	Ref.
Diatoms	*Navicula atomus*	Gold	-	[Asmathunisha and Kathiresan, 2013]
	Eolimna minima	gold	-	[Saxena, 2016]
	Diadesmis gallica	Gold	-	[Asmathunisha and Kathiresan, 2013]
Seaweed	*Sargassum ilicifolium*	Silver	Cytotoxicity Nano-biofertilizer Antibacterial	[Pudake et al., 2019] [Singh et al., 2015]
	S. muticum	Zinc	Anticancer Anti-inflammatory Antioxidant Antidiabetic Anti-allergic	[Vijayan et al., 2018] [Pudake et al., 2019]
	S. wightii	Gold Silver	Antibacterial	[Asmathunisha and Kathiresan, 2013] [Manivasagan et al., 2016]
	S. myriocystum	Zinc	Antimicrobial	[Vijayan et al., 2018]
	T. conoides	Silver Gold	Antibiofilm	[Singh et al., 2015]
	S. cinereum	Silver	Antibacterial	[Asmathunisha and Kathiresan, 2013] [Singh et al., 2015]
	Caulerpa scalpelliformis	Silver	Mosquitocidal	[Vijayan et al., 2018]

(Continued)

TABLE 16.3 ***(Continued)***
A Selection of Biosynthesized Nanoparticles via Marine Algae and Their Biological Activity

Organisms	Species	Nanoparticle	Biological Activity	Ref.
	Gelidiella acerosa	Silver	Antifungal	[Asmathunisha and Kathiresan, 2013] [Singh et al., 2015]
	Fucus vesiculosus	Gold	Biosorption	[Manivasagan and Kim, 2015] [Asmathunisha and Kathiresan, 2013]
	Cladosiphon okamuranus	Gold	-	[Asmathunisha and Kathiresan, 2013]
	Chaetomorpha linum	Silver	Antibacterial	[Dharma-Wardana et al., 2015]
	Padina gymnospora	Silver	Antibacterial	[Singh et al., 2015]
	Colpomenia sinuosa	Silver	Anti-diabetic	[Singh et al., 2015]
	Ulva fasciata	Silver	Antibacterial	[Asmathunisha and Kathiresan, 2013] [Singh et al., 2015] [Pudake et al., 2019]
	U. lactuca	Silver Zinc	Antibacterial Mosquito larvicide	[Singh et al., 2015] [Pudake et al., 2019]
	Stoechospermum marginatum	Gold	Antibacterial	[Singh et al., 2015]
	Bryopsis plumosa	Palladium/ silver	Antibiofilm Anticancer	[Vijayan et al., 2018]
	Centroceras clavulatum	Silver	Larvicidal	[Vijayan et al., 2018]
	Hypnea musciformis	Silver Gold	Pesticidal Nano-fungicides	[Vijayan et al., 2018] [Pudake et al., 2019]
	Gracilaria corticata	Gold	Antimicrobial Antioxidant	[Singh et al., 2015]
	P. boergessenii	Silver	Nanofungicides	[Pudake et al., 2019]
	Amphiroa anceps	Silver	Nano-biofertilizer	[Pudake et al., 2019]
	Ch. antennina	Silver	Seed germination Nano-biofertilizer Antibacterial	[Pudake et al., 2019]
	P. boryanum UTEX 485	Gold	Antibacterial	[Pudake et al., 2019]
	Klepsormidium flaccidum	Gold	Antibacterial	[Pudake et al., 2019]
	U. intestinalis	Gold	Antibacterial	[Pudake et al., 2019]
	Galaxaura elongata	Gold	Antibacterial	[Pudake et al., 2019]
Microalgae	*Spirrulina platensis*	Silver	Antibacterial	[Pudake et al., 2019]
	Chlorella vulgaris	Silver	Antibacterial	[Pudake et al., 2019]

16.19 BIOSYNTHESIS OF NANOPARTICLES BY MARINE ANIMALS

Marine animal-mediated NPs synthesis is very sparse due to the wide availability of renewable sources such as marine microbes and planktons. The presence of nanostructures on shark skin gave a new opening for the advancements in marine nanotechnology in synthesis and designing of used nanomaterial for biomedical applications [Ramkumara et al., 2016; Dean and Bhushan, 2010].

Fish oil has nutritional value and the presence of a permissible limit of silver nanoparticles in the oil might enhance its efficacy—an idea that may open many avenues in the field of nanobiotechnology. The use of cod liver fish oil has been shown to produce silver nanoparticles as a reducing agent as well as a surfactant. Presence of carboxylate ions and amine groups in the fish oil triggers in situ generations of organically capped silver nanoparticles. The marine sponge, *Acanthella elongate*, is shown to produce gold nanoparticles, and this process is attributed to water-soluble organics present in the sponge extract [Asmathunisha and Kathiresan, 2013].

Dolphins and whales have rough skin surface due to the presence of a nano ridge. These ridges enclose a pore size of 0.2 μm^2 which is below the size of marine fouling organisms and hence there is no attachment of biofuels. Nanoscaled structures found on shark skin are in a "brick-and-mortar" arrangement, like a micro-architecture on nacre (mother of pearl), and paved a way for the latest advances in the production of synthetic designed materials, in particular, to be used in biomedical applications [Asmathunisha and Kathiresan, 2013; Singh et al., 2015]. A selection of biosynthesized nanoparticles via marine animals and their biological activity is shown in Table 16.4.

16.20 FUTURE PERSPECTIVES

Nanoparticles have attracted considerable interest in recent years and accordingly have been extensively reported on in the literature. The unique size and shape-dependent physicochemical surface properties make nanoparticles more interactive and reactive to certain chemical species compared to their

TABLE 16.4
A Selection of Biosynthesized Nanoparticles via Marine Animals and Their Biological Activity

Organisms	Species	Nanoparticle	Biological Activity	Ref.
Sponges	*Acanthella elongata*	Gold	-	[Asmathunisha and Kathiresan, 2013]
Fish	Cod liver oil	Silver	-	[Asmathunisha and Kathiresan, 2013]
Oyster	*Saccostrea cucullata*	Silver	Antimicrobial	[Singh et al., 2015]

bulk scale counterparts. The novel properties have been extensively investigated and evaluated for a wide range of applications in several fields [Fawcett et al., 2017].

The marine ecosystem has captured major attention in recent years, as they contain valuable resources that are yet to be explored much for the beneficial aspects of human life. Synthesis of nanoparticles with the help of marine resources accomplishes the need for safe, stable, and environment-friendly particles since it involves a diverse marine ecosystem that is freely available and this biological synthesizing method does not involve harmful solvents and reduced downstream processing steps which shrink the cost for their synthesis. An important challenge in nanoparticle synthesizing technology is to tailor the properties of nanoparticles by controlling their size and shape. Using marine organisms and their bioactive substances, the biosynthesis of nanoparticles extracellularly would be constructive if it is produced in a controlled manner to their size and shape. Nanoparticles of desired size and shape have been obtained successfully using living organisms—from simple unicellular organisms to highly complex eukaryotes [Singh et al., 2015].

The field of nanobiotechnology is still in its infancy, and more research needs to be focused on the mechanistic of nanoparticle formation from the marine resources, which may lead to fine-tuning the process, ultimately leading to the synthesis of nanoparticles with strict control over the size and shape parameters. Therefore, it needs collaborative research of various disciplines to develop simple and cost-effective techniques to improve the quality of life [Singh et al., 2015].

REFERENCES

Asmathunisha N, Kathiresan K (2013) A review on the biosynthesis of nanoparticles by marine organisms. *Colloids Surf B Biointerfaces* 103: 283–287.

Birnbaum AJ, Pique A (2011) Laser-induced extra-planar propulsion for three-dimensional microfabrication. *Appl Phys Lett* 98: 134101–134106.

Chiu CY, Ruan L, Huang Y (2013) Bimolecular specificity-controlled nanomaterial synthesis. *Chem Soc Rev* 42: 2512–2527.

Daniel MC, Astruc D (2004) Gold nanoparticles: Assembly, supramolecular chemistry, quantum size-related properties, and applications toward biology, catalysis, and nanotechnology. *Chem. Rev* 104: 293–346.

Davis TA, Volesky B, Mucci A (2003) A review of the biochemistry of heavy metal biosorption by brown algae. *Water Res* 37: 4311–4330.

Dean B, Bhushan B (2010) Shark-skin surfaces for the fluid-drag reduction in turbulent flow: A review. *Math Phys. Eng. Sci* 368: 4775–4806.

Dharma-Wardana MWC, Amarasiri SL, Dharmawardene N, Panabokke CR (2015) Chronic kidney disease of unknown aetiology and ground-water ionicity: Study based on Sri Lanka. *Environ. Geochem. Health* 37: 221–231.

Fawcett D, Verduin JJ, Shah M, Sharma SB, Poinern GEJ (2017) Review of Current Research into the Biogenic Synthesis of Metal and Metal Oxide Nanoparticles via Marine Algae and Seagrasses. *J Nanosci.* https://doi.org/10.1155/2017/8013850.

Franci G, Falanga A, Galdiero S, Palomba L, Rai M, Morelli G, Galdiero M (2015) Silver nanoparticles as potential antibacterial agents. *Molecules* 20: 8856–8874.

Gautam PK, Singh A, Misra K, Kumar Sahoo A, Samanta SK (2019) Synthesis and applications of biogenic nanomaterials in drinking and wastewater treatment. *J Environ Manage* 231: 734–748.

Hulkoti NI, Taranath T (2014) Biosynthesis of nanoparticles using microbes—a review. *Colloids Surf B Biointerfaces* 121:474–483.

Iqbal J, Abbasi BA, Ahmad R, Shahbaz A, Zahra SA, Kanwa S, Munir A, Rabbani A, Mahmood T (2020) Biogenic synthesis of green and cost-effective iron nanoparticles and evaluation of their potential biomedical properties. *J. Mol. Struct* 1199 (2020): 126979.

Kahru A, Ivask A (2012) Mapping the dawn of nanoecotoxicological research. *Acc Chem Res* 46 (3): 823–833.

Kamal A, Bharathi EV, Ramaiah MJ, Dastagiri D, Reddy JS, Viswanath A, . . . Sastry GN (2010) Quinazolinone linked pyrrolo [2, 1-c] [1, 4] benzodiazepine (PBD) conjugates: Design, synthesis and biological evaluation as potential anticancer agents. *Bioorg. Med. Chem* 18: 526–542.

Kumar P, Govindaraju M, Senthamilselvi S, Premkumar K (2013) Photocatalytic degradation of methyl orange dye using silver (Ag) nanoparticles synthesized from *Ulva lactuca*. *Colloids Surf B Biointerfaces* 103: 658–661.

Lellouche J, Kahana E, Elias S, Gedanken A, Banin E (2009) Antibiofilm activity of nanosized magnesium fluoride. *Biomaterials* 30: 5969–5978.

Manivasagan P, Kim S (2015) *Biosynthesis of Nanoparticles Using Marine Algae: A Review. Marine Algae Extracts: Processes, Products, and Applications*. Edited by Se-Kwon Kim and Katarzyna Chojnacka. Wiley-VCH Verlag GmbH & Co.

Manivasagan P, Nam SY, Oh J (2016) Marine microorganisms as potential biofactories for synthesis of metallic nanoparticles. *Crit. Rev. Microbiol* 42:1007–1019.

Mittal AK, Chisti Y, Banerjee UC (2013) Synthesis of metallic nanoparticles using plant extracts. *Biotechnol Adv* 31: 346–356.

Nagarajan S, Kuppusamy KA (2013) Extracellular synthesis of zinc oxide nanoparticle using seaweeds of Gulf of Mannar, India. *J Nanobiotechnology* 11: 39–45.

Palaniappan P, Sathishkumar G, Sankar R (2015) Fabrication of nano-silver particles using *Cymodocea serrulata* and its cytotoxicity effect against human lung cancer A549 cells line. *Spectrochim. Acta A* 138: 885–890.

Pandey SN, Verm I, Kumar M (2020) Cyanobacteria: The potential source of biofertilizer and synthesizer of metallic nanoparticles. *Advances in Cyanobacterial Biology*. https://doi.org/10.1016/B978-0-12-819311-2.00023-1.

Prabhu S, Poulose K (2012) Silver nanoparticles: Mechanism of antimicrobial action, synthesis, medical applications, and toxicity effects. *Int. Nano Lett* 2: 1–10.

Prakash S, Ramasubburayan R, Iyapparaj P, Arthi APR, Ahila NK, Ramkumar VS, Immanue G, Palavesam A (2015) Environmentally benign antifouling potentials of triterpene-glycosides from *Streptomyces fradiae*: A mangroveisolate. *RSC Adv* 5: 29524–29534.

Pudake RN, Chauhan N, Kole CH (2019) *Nanoscience for Sustainable Agriculture*. Springer Nature Switzerland.

Ramkumara VS, Prakashb S, Ramasubburayanc R, Pugazhendhid A, Gopalakrishnane K, Rajendran RB (2016) Seaweeds: A resource for marine bionanotechnology. *Enzyme Microb Technol* 95: 45–57.

Ravikumar S, Inbaneson SJ, Suganthi P, Venkatesan M, Ramu A (2011) Mangrove plant source as a lead compound for the development of new antiplasmodial drugs from the South East coast of India. *Parasitol. Res* 108: 1405–1410.

Saxena P (2016) Phyco-nanotechnology: New horizons of gold nano-factories. *Proc Natl Acad Sci*. DOI: 10.1007/s40011-016-0813-0.

Shah M, Fawcett D, Sharma S, Tripathy SK, Poinern GEJ (2015) Green synthesis of metallic nanoparticles via biological entities. *Materials* 8: 7278–7308.

Sharma D, Kanchi S, Bisetty K (2015) Biogenic synthesis of nanoparticles: A review, Arab. *J. Chem.* http://dx.doi.org/10.1016/j.arabjc.2015.11.002.

Singh CR, Kathiresan K, Anandhan S (2015) A review on marine based nanoparticles and their potential applications. *Afr J Biotechnol* 14: 1525–1532.

Singh D (2020) Cyanobacteria as a source of nanoparticle: Application and future Projections. *Advances in Cyanobacterial Biology*. https://doi.org/10.1016/B978-0-12-819311-2.00021-8.

Smit AJ (2004) Medicinal and pharmaceutical uses of seaweed natural products: A review. *J Appl Phycol* 16: 245–262.

Stevenson LM, Adeleye AS, Su Y, Zhang Y, Keller AA, Nisbet RM (2017) Remediation of cadmium toxicity by sulfidized nano-iron: The importance of organic material. *ACS Nano* 10: 10558–10567.

Venkatesan J, Anil S, Kim SK, Shim MS (2016) Seaweed polysaccharide-based nanoparticles: Preparation and applications for drug delivery. *Polymers* 8: 1–25.

Vijayan R, Joseph S, Mathew B (2018) Green synthesis of silver nanoparticles using *Nervalia zeylanica* leaf extract and evaluation of their antioxidant, catalytic, and antimicrobial potentials. *Particulate Science and Technology*. DOI: 10.1080/02726351.2018.1450312.

Yezhelyev MV, Gao X, Xing Y, Al-Hajj A, Nie S, O'Regan RM (2006) Emerging use of nanoparticles in diagnosis and treatment of breast cancer. *Lancet Oncol* 7: 657–667.

Zhang X, Yan S, Tyagi R, Surampalli R (2011) Synthesis of nanoparticles by microorganisms and their application in enhancing microbiological reaction rates. *Chemosphere* 82: 489–94.

Index

R

S

T

U

V

W

Printed in the United States
by Baker & Taylor Publisher Services